Advances in Analysis of Food and Beverages

Advances in Analysis of Food and Beverages

Topic Editors

Anna M. Kaczmarek
Constantinos G. Tsiafoulis

MDPI

Basel • Beijing • Wuhan • Barcelona • Belgrade • Novi Sad • Cluj • Manchester

Topic Editors
Anna M. Kaczmarek
Department of Food Quality and Safety Management
Poznan University of Life Sciences
Poznan
Poland

Constantinos G. Tsiafoulis
Departament of Chemistry
University of Ioannina
Ioannina
Greece

Editorial Office
MDPI AG
Grosspeteranlage 5
4052 Basel, Switzerland

This is a reprint of the Topic, published open access by the journals *Beverages* (ISSN 2306-5710), *Foods* (ISSN 2304-8158), *Molecules* (ISSN 1420-3049), *Nutrients* (ISSN 2072-6643) and *Separations* (ISSN 2297-8739), freely accessible at: https://www.mdpi.com/topics/QT1KS93SDC.

For citation purposes, cite each article independently as indicated on the article page online and as indicated below:

Lastname, A.A.; Lastname, B.B. Article Title. *Journal Name* **Year**, *Volume Number*, Page Range.

ISBN 978-3-7258-7869-7 (Hbk)
ISBN 978-3-7258-7870-3 (PDF)
https://doi.org/10.3390/books978-3-7258-7870-3

Contents

About the Editors

Anna M. Kaczmarek

Anna M. Kaczmarek, Department of Food Quality and Safety Management, Poznań University of Life Sciences, Poznan, Poland. Kaczmarek's research focuses on food quality and authenticity, analytical approaches in food science, chemometric data analysis, and consumer-related aspects of novel foods. She is particularly interested in the application of advanced analytical and statistical methods to the evaluation of food composition, quality, and safety. Her recent work has addressed both instrumental and data-driven approaches in food research, with emphasis on methodological rigor and practical relevance for food analysis and quality management.

Constantinos G. Tsiafoulis

Constantinos G. Tsiafoulis, NMR Centre, Network of Research Supporting Laboratories, Laboratory of Analytical Chemistry, Department of Chemistry, University of Ioannina, Ioannina, Greece. He also serves as a SEP member (Collaborative Teaching Personnel) at the Hellenic Open University. Tsiafoulis' research focuses on the development of methods and their application in the analysis of food matrices, studying food content, authenticity, quality, and safety. He is particularly interested in Nuclear Magnetic Resonance spectroscopic methods in *-omics* approaches and multivariate data analysis. His recent work focused on method development, application, and analysis on food matrices, among others, as well as on extra virgin olive oil—its phenolic compounds and classification through secoiridoids—and on milk authentication through NMR-based metabolomics of the lipid fraction.

Preface

The analysis of food and beverages remains one of the most dynamic and rapidly developing areas of contemporary food research. As analytical technologies continue to advance, researchers are gaining increasingly powerful tools to investigate the composition, quality, authenticity, and functionality of food systems with greater precision and depth than ever before. At the same time, the growing complexity of global food production and supply chains has created a clear need for robust and application-oriented analytical strategies that can respond to both scientific and practical challenges.

This Reprint brings together a selection of contributions published within the Topic "Advances in Analysis of Food and Beverages" and reflects the diversity of current research in this field. The included articles address a wide range of analytical questions and food-related matrices, while collectively illustrating the central role of modern instrumental methods in advancing food and beverage science. Chromatographic and spectroscopic approaches, metabolomic investigations, volatile compound profiling, and chemometric tools are all represented in this volume, demonstrating how analytical research continues to expand not only in technical sophistication but also in relevance to quality control, authenticity assessment, process monitoring, and product innovation.

One of the strengths of this Reprint is the breadth of perspectives it offers. Although the individual studies focus on different products, compounds, and methodological approaches, they share a common scientific aim: to generate reliable knowledge through well-designed analytical investigation and to translate that knowledge into meaningful insights for research and practice. In this sense, the volume highlights both the methodological richness of the field and its strong connection to real-world needs in food analysis.

We hope that this Reprint will serve as a useful resource for researchers, analysts, and professionals working in analytical chemistry, food science, and related disciplines. By bringing these contributions together in a single volume, we also hope to underline the continuing importance of analytical excellence in supporting the quality, integrity, and scientific understanding of food and beverages.

Anna M. Kaczmarek and Constantinos G. Tsiafoulis
Topic Editors

beverages

MDPI

Article

Varietal Authentication of Brunello di Montalcino Wine Using a Minimal Panel of DNA Markers

Maxwell K. Kibor [1], Monica Scali [2] and Rita Vignani [2,*]

[1] Department of Zoology and Entomology, University of Pretoria, Pretoria 0002, South Africa; maxyekibor@gmail.com

[2] Department of Life Sciences, University of Siena, 53100 Siena, Italy; monica.scali@unisi.it

* Correspondence: rita.vignani@unisi.it

Abstract: Wine DNA fingerprinting (WDF), retrieved from the amplification of a wider panel of Simple Sequence Repeat (SSR) marker mappings in the *Vitis vinifera* L. genome, was used to assess the monovarietal nature of Brunello di Montalcino wine. The reliability of the varietal assessment was carried out by estimating the PI values associated with resolutive unrooted dendrograms depicting the correct varietal nature of different wines. As few as five SSR DNA markers associated with a PI value of one over a million or less, PI $\leq 10^{-6}$, can identify the purity of Sangiovese against Merlot, Pinot Noir, Cabernet Sauvignon, Primitivo (Zinfandel), and genetic variants of the Sangiovese as plant references. WDF was used on other monovarietal wines obtained from Cabernet Sauvignon, Merlot, Chardonnay, and Pinot Noir to test the feasibility of the method. In blended wines, the test was able to trace the main varietal component in a three-variety blend, keeping the varietal fingerprint detectable when the main variety was at least 75% (v/v). The data confirm how local genetic variants of Sangiovese can be tracked in commercial wines, becoming, at wine makers' demand, part of an evidence synthesis of geographical origin.

Keywords: DNA markers; wine DNA fingerprinting; simple sequence repeats; Sangiovese; geographical origin

1. Introduction

The quality, safety, and authenticity of food and beverages are currently an essential added cultural value in the agrifood industry globally. This is specifically linked to human health in contemporary society, which is increasingly aware of food consumption [1]. Additionally, due to the globalization of dietary markets and the increase in movements of various food products from one country to another, consumers are becoming more concerned with the geographical origin of their food and beverages [2]. Several countries have adopted quality, safety, and authenticity standards to safeguard their agrifood industries. Italy is currently the leading country in food and beverage authentication globally, followed by Spain, the United States of America, the United Kingdom, Germany, France, China, Portugal, Switzerland, and Greece [3]. There is an increasing interest in research to develop innovative analytical tools to control fraud and adulteration of food and beverages, such as wine [4].

Wine is an added-value beverage made using the fermented grape juice from grapevines (*Vitis vinifera* L.). It is made up of a complex mixture of compounds. The complex compounds contribute to the quality of wines because they are responsible for color, taste, and aroma [5,6]. The quality, composition, value, and functional properties of wine are influenced by the variety, geographical origin, and farming systems [7].

Consumers demand more informative labels on product origin, production process, and plant-derived components. Since the 1960s, wine quality has been granted by DOC and DOCG labels, from the moment that the geographical origin and varietal components were major targets of fraudulent practices in the wine-making industry. In addition to fraudulent practices, the wine and grape sector faces challenges of climate change.

Conservation of natural variability in the grapevine germplasm is essential for adaptation to the environment, and it is equally important for developing food and beverage authentication strategies. Food and wine authenticity is a crucial key factor that is rooted in the natural variability in vineyards because it provides important elements concerning the geographical origin of wine grape varieties. The availability of analytical tools for wine authentication protects consumers against fraud and preserves wine identity in the interest of producers. The European Union put in place regulations and laws, including Council Regulation NO. 1234/2007, Regulation EU NO. 1151/2012, NO. 664/2014, NO. 668/2014, which aimed to govern the entire European food sector [8]. Despite these laws and regulations, the wine sector continues to face major challenges due to competition for market demand and a trend of producing wines exhibiting good and high-quality flavors.

International wines, such as Brunello di Montalcino, are an example of wines that are expected to face fraud and adulteration in the local and international market [9]. Classical wine fraud and adulteration include change of labels, dilution with water, addition of alcohol, color, and flavor substances, blending with lesser quality wines, and misrepresentation of variety and geographical origin [10]. Brunello di Montalcino wine is an Italian wine produced from the Sangiovese grapevine variety in Tuscany, Italy [11]. The wine has fostered the economy of the Tuscany region and Italy as a whole. The wine fetches a high market price and is among the wines that have received recognition as quality and prestigious wines both at the national and international levels [12].

Due to the importance of quality and authenticity requirements, several biotechnology methodologies for the determination of quantitative and qualitative compounds in wine have been established to confirm wine products. They include mineral analysis (spectrometry), organic profiling (chromatography), isotopic ratios, and DNA analysis [13]. Other analytical methods include capillary electrophoresis and sensory analysis [14]. DNA analysis is the most powerful methodology for determining the risk of adulteration of wines. In comparison to other types of markers, like proteins, which are affected by environmental factors, DNA is thought to be more consistent, even given external variations. Additionally, DNA markers are stable and numerous in number and, therefore, can form a wide basis to provide reliable information [15]. The idea and history of wine biological origin authentication through DNA testing dates back to the DNA fingerprinting and identification of grapevine cultivars, with emphasis on the selection process, which, from alien accessions of wild grapes, brought the differentiation of actual *V. vinifera* varieties [16]. WDF and other multi-analytical approaches have proved to be very promising tools for wine varietal authentication [13,15,17–22]. WDF, a method for detecting genotype rarity in grapevine varieties, uses nuclear genomic DNA amplification to reconstruct individual genotypes. The validation of WDF relies on the statistical validation of allelic detection from wine that is able to match the existing grapevine datasets [23]. Genotyping-based methodologies in wine and forensic medicine applications are closely related, as they depict DNA profiles of contributors in difficult conditions [17]. DNA from wine is often degraded and present in small quantities, requiring the definition of low template DNA (LT-DNA). The high value and reliability of molecular testing in forensics, even in extreme conditions, reinforce the idea of using DNA as a varietal authentication marker for wine quality. SSR-based genotyping is self-regulated, but a predictive key for correctly addressing wine DNA analysis could save time and money [24]. DNA fingerprinting technology is widely accepted for grapevines by the International Organization of Vine and Wine (OIV). The technique is PCR-based and utilizes SSR markers and CE detection systems for the characterization of grapevine germplasm diversity [25]. The utilization of SSR markers has been effective in identification, percentage assignment, and genetic mapping of *Vitis vinifera* genomes [26]. SSRs are universally distributed throughout eukaryotic genomes, and there are many studies on agronomically important plant species using SSRs [27]. DNA fingerprinting has been applied successfully for wine traceability. The term WDF was first introduced by Vignani and Scali (2016) [28], and several prior papers used DNA-based methods for wine varietal authentication [29]. According to the WDF definition, the SSR-based approach, in conjunction with statistical elaboration, is used for wine varietal authentication. WDF provides information on the probability of the presence of a particular variety of grapes in a wine bottle.

The aim of this study was to propose a basis for the development of a molecular single-test kit for Brunello di Montalcino varietal authentication and for developing wine databanks by using bioinformatics. These can be achieved through integrated wet lab procedures with statistical validation of genetic tests. In addition, the molecular genetic variability set of data provided from the wet lab molecular analysis of wines, which includes experimental (IN7, CB17 Sangiovese wines) and commercial monovarietal wines, enables the tracking of wine origin. Evidence of geographical origin is among the information that wine makers demand and can be fulfilled by tracking grapevine genetic variants in different wines. Small labs may consider gaining the ability to carry out wine quality assessment through the integrated molecular biology and bioinformatics approach to determine the varietal composition of wines with increasing complexity. This reduces operational costs by reducing unnecessary analysis. Furthermore, a low level of expertise and technical requirements are needed. This work builds upon the data and findings from a previously published work [17]. In comparison to the previous study, this study demonstrates the efficiency and accuracy of using a minimal panel of 5 SSR markers compared to the 10 SSR markers used in the authentication of Brunello di Montalcino wine.

2. Materials and Methods

2.1. Wine and Grapevine References

The wine list and relative description are reported in Table 1. The wines included experimental and commercial wines. The experimental Sangiovese wines (CB17 and IN7), obtained by the small-scale fermentation of grapevine berries given at harvest (2014) by Case Basse di Gianfranco Soldera (Montalcino, Italy), were kindly provided by the University of Florence. The commercial Brunello and Rosso di Montalcino wines were kindly provided by local wine makers (Tenuta Le Potazzine, Tenuta Caprili, and Azienda Agricola Palazzo, Montalcino, Siena, Italy). A Brunello di Montalcino blended with 5% Pinot Noir was included as an experimental wine.

To test the method on wines other than Brunello di Montalcino, we used the following monovarietal control experimental wines: 100% Sauvignon Blanc, 100% Merlot, 100% Cabernet Sauvignon, and 100% Chardonnay.

To test the limit of detection of the main variety in a plurivarietal wine, we used several blended wines (Wine Institute, San Francisco, CA, USA). In detail, the blended wines, containing up to three varieties each, were the following: 75% Cabernet Sauvignon + Tempranillo, 76% Merlot + Rubired + Rubi Cabernet, 76% Chardonnay + Viognier, Symphony, and 76% Sauvignon Blanc + French Colombard + Chardonnay.

Plant references were obtained from the National Italian Grapevine Register held at CREA (Conegliano Veneto, TV, Italy). Plant references and relative genotypes are conserved and updated as ex situ data DNA collection by the SERge-GEnomics S.r.L. spin-off company

(Siena, Tuscany, Italy). The grapevines used as references were Sangiovese, Pinot Noir, Cabernet Sauvignon, Merlot, CB17-Sangiovese genetic variant, IN7-Sangiovese genetic variant, Sauvignon Blanc, Chardonnay, Tempranillo, Rubired, Rubi Cabernet, Viognier, French Colombard, and Symphony. SSR genotypes were either obtained in our laboratory or retrieved from the open data available in databanks.

Table 1. List of wines.

Wine Name	Vintage	Year of Analysis	Origin	Type (E = Experimental) (C = Commercial)
1. IN7	2014	2019	University of Florence	E
2. CB17	2014	2019	University of Florence	E
3. 95% Sangiovese + 5% Pinot Noir	2014	2019	Siena, Italy	E
4. Brunello di Montalcino	2014	2019	Montalcino, Siena, Italy	C
5. Rosso di Montalcino	2014	2019	Montalcino, Siena, Italy	C
6. 100% Sauvignon Blanc	2018	2019	Siena, Italy	E
7. 100% Merlot	2018	2019	Siena, Italy	E
8. 100% Cabernet Sauvignon	2018	2019	Siena, Italy	E
9. 100% Chardonnay	2018	2019	Siena, Italy	E
10. 75% Cabernet Sauvignon + Tempranillo	2016	2019	USA	C
11. 76% Merlot + Rubired + Rubi Cabernet	2016	2019	USA	C
12. 76% Chardonnay + Viognier, Symphony	2016	2019	USA	C
13. 76% Sauvignon Blanc + French Colombard + Chardonnay	2016	2019	USA	C

2.2. DNA Extraction from Grapevine Leaves

DNA from grapevine leaves was extracted using a DNAeasy Plant mini kit (Qiagen, Milano, Italy), following the manufacturer's instructions. Homogenization was performed by grinding 100 mg of fresh leaves in a 1.5 mL Eppendorf tube with a plastic pestle at room temperature. The DNA was then diluted in 100 µL elution buffer (10 mM Tris-HCl pH 8.5).

2.3. DNA Purification from Wines

DNA was purified from the wines following the protocol by Bigliazzi et collaborators [30], starting from 400 mL of wine. Quantification of *Vitis vinifera* DNA was performed following a validated protocol [31].

2.4. SSR-Genotyping and Allele Sizing

Wine DNA fingerprinting (WDF) was obtained by genotyping the *V. vinifera* genomic DNA from the wine, as described in [17].

2.5. DNA Amplification

The DNA extracted from the wines and plants was genotyped by using SSR markers VVS2 [32], VVMD7, VVMD21, VVMD24, VVMD25, VVMD27, VVMD31, VVMD32, VVMD34, and VVMD36 [33], and Vrzag21, Vrzag47, Vrzag64, Vrzag79, and Vrzag83 [34]. Polymerase chain reaction was carried out in double replicas in a PCR plate of 96 wells at a total volume of 12.5 µL, containing 2.5 µL DNA, 0.25 mM of dNTPs, 0.25 µM of each primer (one of them fluorescently labeled), 1× Green GoTaq reaction buffer containing 1.5 mM $MgCl_2$, and 0.1 U GoTaq Flexi DNA polymerase (Promega, Milano, Italy). The PCR conditions included denaturation steps for 5 min at 95 °C, 39 cycles of 30 s at 95 °C, 30 s at 48–56 °C, depending on the specific marker, 1 min at 72 °C, and an extension stage at 72 °C for 10 min. Grape DNA can be genotyped using the same protocol.

2.6. Capillary Electrophoresis

The polymorphism of the SSRs was examined by allele sizing on a capillary electrophoresis. First, 1 μL of PCR product and 10 μL loading mix, containing highly deionized formamide and 0.2 μL internal standard (GS-400 Rox Standard, Thermofisher, Monza, Italy), were denatured at 95 °C for 5 min and kept on ice for 2 min before loading into an AB3130 genetic analyzer (Applied Biosystems, Waltham, MA, USA) machine. Electrophoresis was performed with the following parameter settings: separation temperature of 60 °C, injection voltage of 1.6 kV, injection time of 8 s, run voltage of 15 kV, and run time. At least two technical replicates were performed for each sample of DNA extracted.

2.7. Genotyping

The allele size was obtained using Gene Marker Version 2.2.0 software (SoftGenetics, State College, PA, USA). The PI was calculated to estimate the discrimination power of each panel of markers in discriminating varieties. The PI values were calculated by use of Identity software, Identity 4.0, Centre for Applied Genetics, University of Agricultural Sciences, Vienna [34]. Each sample was processed in triplicate.

2.8. Consistency of WDF After Data Merging

WDF analysis was performed according to [30]. Each experiment was repeated at least three times, and the data from each experiment were gathered in one major dataset. An estimate of the discrimination power of the test in resolving varieties was determined by calculating the PI using the minimal number of markers able to depict the varietal nature and propaedeutically to produce graphic outputs.

3. Results

3.1. Estimating Informativeness of WDF Testing: Setting a Minimal Panel of Markers

The extractable DNA out of a five-year aged wine is estimated by use of RT-PCR in the range of 30–40 ng/mL, and it is usually used for downstream genotyping applications with standard panels consisting of 10–16 SSR loci in order to reconstruct source vine identity [30]. The discrimination power of the SSR panel of markers was calculated with Identity software by setting the WDF marker panel consisting of seven SSR markers using Sangiovese, Sangiovese-IN7, Sangiovese-CB17, Merlot, Pinot Noir, Cabernet Sauvignon, and Zinfandel as grapevine references. The estimated PI value for the correct genetic profiling of monovarietal wines using the seven SSR markers was calculated to be in the order of about one over ten billion (PI = 1.04×10^{-10}) [17].

To reduce the number of tests, the panel was decreased to five SSRs. The WDF performed on both the experimental wines (IN7 and CB17) and the Brunello di Montalcino, Rosso di Montalcino, and SG-C wines using the five-SSR panel suitable for the Sangiovese-based monovarietal is referred to as the minimal panel, and it is associated with a discriminatory power in the order of 1 over a million or less (PI $\leq 10^{-6}$) (Table 2).

To simulate a two-variety blend or an experimental mixing of Pinot Noir in a Sangiovese almost pure wine, a mock Brunello di Montalcino wine, added with 5% Pinot Noir, was tested by WDF. The WDF carried out using five SSR markers allowed for the detection of Pinot Noir-specific alleles at four SSR markers out of five (Table 3).

The data demonstrate a significant correspondence between the genotype that can be obtained from wine DNA and the standard genotype profile of the respective vines, confirming the reliability of the varietal nature of wine prediction, even using the minimal panel of markers (Table 2). The monovarietal nature of the Brunello and Rosso di Montalcino wines is confirmed by the consistency in observing, in all markers, a systematic lack of multi-allelic profiles, indicative of multiple biological sources. More in detail, this panel seems to effectively differentiate monovarietal Sangiovese wines from other experimental monovarietal wines, such as CB17, IN7, and SG-C, all deriving from varietal variants of Sangiovese in historical vineyards in Montalcino.

Table 2. The data shown represents the first molecular assay using a minimal panel of markers, potentially adjustable to a one-tube PCR test that is designed for varietal authentication of high-quality Brunello di Montalcino and Rosso di Montalcino wines. The IN7, CB17, and SG-C wines show the allelic variants of the respective Sangiovese vineyards linked to the presence of a few historical forms of local plant strains. Alleles matching between wines and plant references are in bold.

	VVS2	VVMD25	VVMD27	VrZag21	VrZag83
Grapevine type					
Sangiovese	135–135	241–241	179–185	202–204	191–195
Wines					
IN7	–	**238**–241	179–185	204–204	191–**201**
CB17	135–**146**	241–**255**	185–**189**	–	191–**201**
SG-C	135–**145**	241–**254**	–	204–**206**	191–195
Rosso di Montalcino		241–241	179–185	202–204	191–195
Brunello di Montalcino	135–135	241–241	179–185		191–195

Table 3. Brunello di Montalcino mock wine added with 5% Pinot Noir. Alleles matching Pinot Noir are in bold.

	100% Sangiovese	100% Pinot Noir	95% Sangiovese + 5%Pinot Noir
VVMD25	241–241	241–249	241–**249**
VrZag21	202–204	201–207	**201**–204–**207**
VrZag83	191–195	190/202 or 189–201	191–**201**
VVS2	135–135	140–154	135–**140–154**
VVMD27	179–185	185–189	

3.2. Varietal Assessment in the Control Monovarietal and Blended Wines

The detection of minor varietal components in blended wines by use of SSR markers is linked to a low risk of losing genetic information (low allele drop risk). As already shown, to address the reliability of the genetic test for the Brunello di Montalcino wine, the system was tested on the experimental wine in which 5% of Pinot Noir was added. Six informative alleles matching Pinot Noir and not Sangiovese were detected (alleles in bold, Table 3). To assess the detection limits of minor varietal components in the other wine types, four additional wines (wines 10, 11, 12, and 13) were used. The composition of the varietal blends was only partially known, reflecting the real situation of most commercial wines. The scope of the characterization of the four blended wines was specifically to determine if minor quantities of several varieties added to the blend would impair the identification of the major variety. Table 4 shows the allele profiles that were detected by Random Match Probability on the blended wines.

Table 4. Allele profiles detected in the blended wines. The alleles in bold are those that do not match the main varietal component genotype. The wines include the initial CS = Cabernet Sauvignon, M = Merlot, C = Chardonnay, and SB = Sauvignon Blanc. The number of SSR full loci or single alleles (in brackets) detected by Random Match Probability matching the major variety are reported in bold at the bottom of each column.

	Wine 10			Wine 11			Wine 12			Wine 13		
	CS + Tem-pranillo	Control wine 100% CS	CS Plant	**M** + Rubired, Ruby Cabernet, etc.	Control wine 100% M	M Plant	**C** + Viognier, Symphony, etc.	Control wine 100% C	C Plant	**SB** + French Colombard + Chardonnay, etc.	Control wine 100% SB	SB Plant
SSR markers												
VVMD27				**194**		185 189				175	175–189	175–189
VVMD25	240–249	240	240–249	240	240–249	240–249	240–255	240–255	240–255	240–248	240–248	240–248
VVMD21			249–258	243–249		243–249	249	249	249	249	243–249	243–249
VVMD24	208–216	208–216	208–216	208–212–**214**		208–212	208–216	208–216	208–216	216–**208**	216	216
VVMD32			240	240		240		240	240–271		240–256	240–256
VVMD31			206–210	212	212	212–216			214–216	216	210–216	210–216
VVMD36			252–262			252			252–274		262	262
VVMD34			239–247	239		239			239		238–247	238–247
VVMD7			238			238						238–256
VVS2	141		141–153	139	139	139–153	139	139	139–145	136	136	136–153
ZAG47	152–166		152–166		166	166	166	158–166	158–166	153–166	153–166	153–166
ZAG79			245			245			243			245
ZAG21	201		201–207	201–**220**		201			201–207	205–207	205–207	205–207
ZAG83	201		201	195		195–201		189	189–201		187–201	187–191–201
ZAG64			139–160			139–160		240–255	160–164			
	4 loci+2(1/2)			**5 loci+3(1/2)**			**3 loci+3(1/2)**			**4 loci+4(1/2)**		

Both in the blended and control wines, the data show a tendency of allele drop risk in the amplified SSR loci, which appears likely to be marker-dependent. Depending on the type of blend, the allele drop might be critical if the "lost" allele tracks the presence of the main component. For instance, the Cabernet Sauvignon wines tend to lose alleles 249, 153, and 207; the Merlot wines tend to lose alleles 249, 216,153, and 201; the Chardonnay wines tend to lose alleles 145, 158, 201, and 271; and the Sauvignon Blanc wines tend to lose alleles 189, 243, 210, and 153. Allele drop risk does not seem to increase in relation to the complexity of the wine blends, in agreement with the hypothesis of a marker-type-dependent effect rather than a wine-composition-related issue. Allele drop risk opens issues concerning allele call and constitutes an implicit invitation to adopt rigorous methodological guidelines, which include the use of technical replicas and a minimal threshold for DNA quality and quantity, to keep a high confidence level of the WDF test. The above components may help to assess the feasibility of undertaking WDF and support molecular evidence with specific statistical elaboration. In addition, this supports the main effort in WDF as a varietal diagnostic tool, which focuses on further improvement in the grapevine DNA extraction method from wine. This is achieved by scaling down the initial volume of wine and simplifying procedures. Notably, this would ease the production of technical replicas, which seem crucial for the validation of observed alleles in WDF profiles.

3.3. Allele Sizing Reflects Vineyard Diversity

Some local Sangiovese grapevine strains, despite the overall phenotypic asset recalling the grapevine cultivar, exhibit minor allelic differences, as shown in Table 5. Historical vineyards (>40 years) in Montalcino reveal natural diversity that may be used as an indication of the geographical origin of the wines. As a matter of fact, some alleles informative of vineyard origin may be tracked in the wines (Table 2).

Table 5. Allelic variants of Sangiovese_IN7 and CB17 potentially indicative of microgeographic origin.

	VVS2	VVMD27	VVMD25	VVMD21	VVMD24	VVMD36	Vrzag21	Vrzag83
Sangiovese	135	179–185	241	243–249	206–212	201–203	202–204	191–195
Sangiovese_CB17	135–**146**	185–**189**	241–**255**	249	206–212	203–**205**	202–**205**	191–**201**
Sangiovese_IN7	n.d.	179–185	**238**–241	**231**–243	212–**216**	**252–262**	204	191–**201**

n.d. stands for: "Not Detected".

4. Discussion

Previous work by a research team [17] demonstrated that WDF can effectively describe the varietal nature of wines after the calculation of different distance matrices by Neighbor-Joining (NJ), followed by Principal Component Analysis (PCA), in a wine-dedicated bioinformatics elaboration. The WDF of Sangiovese-based wines and experimental wines remains informative for assessing the varietal origin, even in the case of using a minimal panel of informative SSR markers. The data show how the WDF for the Sangiovese experimental and commercial wines is consistent with their varietal nature, even after merging data from multiple experiments, if the target of the test is to identify the Sangiovese profile against specific varieties, namely, Merlot, Pinot Noir, and Cabernet Sauvignon. Additionally, the PI value generated can serve as an essential tool to validate the methodology. As shown in Supplementary Table S1, the PI value associated with the use of five SSR markers ($PI \leq 1.0 \times 10^{-6}$), with respect to an extended panel of seven SSRs, appears to be stringent enough to guarantee the correct varietal assignment.

The Brunello di Montalcino wine was analyzed seven years from its plurivarietal production, and the Rosso di Montalcino five years from production. Both DOCG wines did not lose the capability of the molecular detection of the Sangiovese variety even after wine aging. To resolve a blended varietal wine containing 75% of the main variety, the PI is sensibly decreased due to the need to increase the number of SSR markers to be used for carrying out the WDF test. Namely, a PI value in the order of 10^{-7} was observed for the blended whites taken from the US market, and the order of 10^{-13} for some red wines commercially available and also produced in the US [17]. Point-of-care technology demand along the wine-making chain focuses on rapid, low-cost, and reliable analytical methodologies [35]. In a mid-sized private laboratory setting, monitoring nonconformance is essential in determining areas where further improvements can be made. DNA-based technologies are rapidly moving towards the use of consolidated biotechnology applications, which allow quick and low-cost approaches for wine varietal characterization, even in small and basic molecular biology laboratories.

WDF is highly dependent on the DNA extraction procedure from wine. The state of DNA in wine has been a subject of debate. Some authors emphasize the need to consider the development of specific molecular markers for wine and the sensitivity of detection methods since wine contains a mixture of phenolic compounds, such as acids and phenols, that interfere with DNA extraction and the downstream performance of DNA [36]. Also, fermentation introduces DNA into wine from multiple biological sources, while WDF is interested in grape DNA only. Optimal protocols for DNA extraction from wine are necessary, as it is the bottleneck of DNA analysis. As a matter of fact, it was proven that DNA extracted with optimized protocols from several wine types could be genotyped to reconstruct the varietal composition of monovarietal wines [30]. After this achievement, the varietal authentication was also used on blended wines, where the system estimates the probability of the presence of specific grapevines by bioinformatics [17]. More recently, research on wine bio-molecular integrated compositional analysis was published [15,37,38]. This work took a different multidisciplinary approach to wine authentication by demonstrating that the bioinformatics tools that validate DNA evidence in wine are common to other more "classical" compositional analyses, such as chemical profiling, and open a new perspective on wine ontology and anti-counterfeiting measures. Genotype profiles of the original grapevine varieties of Sangiovese monovarietal wines, both experimental and commercial, were identified via SSR PCR-based amplification. In addition, the data form the basis for the definition of predictive key features associated with the analysis.

Given that the authentication of wines is of major interest to the wine industry, consumers, and governmental policymakers responsible for food and wine conformity checks, we believe that the findings presented here could stimulate the high-throughput processability of the analysis focusing on wine varietal authentication. This work expands on the prior research on the varietal authentication of wines [39]. Since most certifications promote technological advancements, the proposed methodology could influence the wine certification process by contributing to the upgrading of major global building certification schemes. Additionally, the technique could boost regulatory compliance through the development of a more robust wine traceability system that addresses geographical provenance and wine variety authentication [40]. With the growing demand for wine consumption globally, the methodology could also influence consumer trust. Consumers are able to access information on wine labels linked to geographical origin [41]. For most wine samples, a significant correspondence between the genotype obtained from wine DNA and the standard plant profile of the respective vines was calculated, through PI, to be lower than one over a million (PI $\leq 10^{-6}$). The data shown here represent the first molecular assay performed on Brunello di Montalcino and Rosso di Montalcino using a minimal set of informative SSRs in WDF experimental plots to reconstruct the varietal identity of the source vine with sufficient statistical reliability.

5. Conclusions and Perspectives

The sequencing of many grape genomes, the availability of techniques to quantify nucleic acids in wine, and the innovations in bioinformatics and chemoinformatics provide the grounds to explore a new direction in wine quality and product safety assessment. Despite the fact that the DNA fingerprinting approach is reliable for wine varietal authentication [28], the efficiency of the approach should still be tested on other wine types, and the validation of allele observation is essential for organizing genetic profiling databases for wine. There is also a need to integrate multi-omics analytical approaches with other quality evaluation parameters regarding wines; information on the history, cultural, and chemical features beyond the product is also required to complete a circumstantial picture of a high-quality wine. There is also a need to simplify the results for a wider use of the authentication technique and for its acceptance by the general public. Molecular integrated approaches have played a role in predicting the composition of wines [24]. The application of the statistical validation of genetic data has ensured high-quality control standards, given that a number of analyses, such as chemical, physical, cultural, historical, and statistical analyses, provide information on the characteristics of wine products that will be part of an innovative ontology databank for wine.

In the future, to create cutting-edge block-chain databanks, molecular, chemical, and metabolomics profiling should be combined to produce extensive wine ontology databanks provided with innovative and informative wine labels. To avoid needless analysis, even in volunteer certification, the test of varietal authentication on various wines must be proven to be reliable. In this regard, this study shows that high-commercial-value monovarietal Brunello di Montalcino may be subjected to a reliable WDF test.

Supplementary Materials: The following supporting information can be downloaded at: https://www.mdpi.com/article/10.3390/beverages11030081/s1; Table S1: Among the 15 SSR markers, 7 (bold, first column) were efficiently used for WDF in Sangiovese wines in previous work [17]. A limited panel of five SSR markers (bold, third column from left) were successfully used to obtain confirmation of the Sangiovese monovarietal nature of the wines.

Author Contributions: Conceptualization, M.K.K., R.V. and M.S.; methodology and formal analysis, M.K.K., R.V. and M.S.; data curation, M.K.K., R.V. and M.S.; writing—original draft preparation, M.K.K.; writing—review and editing, M.K.K., R.V. and M.S.; supervision, R.V. and M.S. All authors have read and agreed to the published version of the manuscript.

Funding: This work was supported by a grant from REGIONE TOSCANA ("Winefinger" Project, Grant PRAF A.1.2.e), Wine Integrated Molecular Analysis WIMA working group: University of Siena; Small wineries (Montalcino, Verona) to adopt the method as a volunteer method for wine quality certification.

Institutional Review Board Statement: Not applicable.

Informed Consent Statement: Not applicable.

Data Availability Statement: The data that support the findings of this study will be openly available in a data repository upon acceptance of this manuscript.

Acknowledgments: Thanks to wine producer Case Basse di Gianfranco Soldera (Montalcino, Italy), who provided experimental Sangiovese wines (CB17 and IN7) for assessments; local wine makers (Tenuta Le Potazzine, Tenuta Caprili, and Azienda Agricola Palazzo, Montalcino, Siena, Italy), who provided commercial Brunello and Rosso di Montalcino wines; and SERge-GEnomics S.r.L. spin-off company (Siena, Tuscany, Italy), who collaborated in this project.

Conflicts of Interest: The authors declare no conflicts of interest.

References

1. Salgueiro, L.; Martins, A.P.; Correia, H. Raw materials: The importance of quality and safety. A review. *Flavour Fragr. J.* **2010**, *25*, 253–271. [CrossRef]
2. Nützenadel, A.; Trentmann, F. *Food and Globalization: Consumption, Markets and Politics in the Modern World*; Berg: New York, NY, USA, 2008.
3. Danezis, G.P.; Tsagkaris, A.S.; Camin, F.; Brusic, V.; Georgiou, C.A. Food authentication: Techniques, trends & emerging approaches. *Trends Anal. Chem.* **2016**, *85*, 123–132. [CrossRef]
4. Ulberth, F. Tools to combat food fraud—A gap analysis. *Food Chem.* **2020**, *330*, 127044. [CrossRef] [PubMed]
5. Ali, K.; Maltese, F.; Choi, Y.H.; Verpoorte, R. Metabolic constituents of grapevine and grape-derived products. *Phytochem. Rev.* **2010**, *9*, 357–378. [CrossRef] [PubMed]
6. Pickering, G.J.; Hayes, J.E. Influence of biological, experiential and psychological factors in wine preference segmentation. *Aust. J. Grape Wine Res.* **2017**, *23*, 154–161. [CrossRef]
7. Granato, D.; Magalhães Carrapeiro, M.D.; Fogliano, V.; van Ruth, S.M. Effects of geographical origin, varietal and farming system on the chemical composition and functional properties of purple grape juices: A review. *Trends Food Sci. Technol.* **2016**, *52*, 31–48. [CrossRef]
8. Michaelsen, F.; Hill, J.; Buckingham, S.; Rzepecka, J.; Chever, T.; Kane, F.; Lepeule, L.; Romieu, V.; Zappalaglio, A. *Study on Control and Enforcement Rules for Geographical Indication (GI) Protection for Non-Agricultural Products in the EU*; Publication Office of the European Union: Luxembourg, 2021.
9. Guirao Pérez, G.; Cano Fernández, V.J. Reseña de libro wine in the old world: New risks and opportunities. *Econ. Agrar. Recur. Nat.* **2016**, *4*, 166–170. [CrossRef]
10. Holmberg, L. Wine fraud. *Int. J. Wine Res.* **2010**, *2*, 105–113. [CrossRef]
11. Voullamoz, J.F.; Monaco, A.; Costantini, L.; Stefanini, M.; Scienza, A.; Grando, M.S. The parentage of 'Sangiovese', the most important italian wine grape. *Vitis* **2007**, *46*, 19–22.
12. Baker, C. The Wines of Italy. In *From Ground to Glass*; Austin Macauley Publishers: London, UK, 2023.
13. Popîrdă, A.; Luchian, C.E.; Cotea, V.V.; Colibaba, L.C.; Scutarașu, E.C.; Toader, A.M. A review of representative methods used in wine authentication. *Agriculture* **2021**, *11*, 225. [CrossRef]
14. Mac, H.X.; Pham, T.T.; Ha, N.T.T.; Nguyen, L.L.P.; Baranyai, L.; Friedrich, L. Current techniques for fruit juice and wine adulterant detection and authentication. *Beverages* **2023**, *9*, 84. [CrossRef]
15. Villano, C.; Lisanti, M.T.; Gambuti, A.; Vecchio, R.; Moio, L.; Frusciante, L.; Aversano, R.; Carputo, D. Wine varietal authentication based on phenolics, volatiles and DNA markers: State of the art, perspectives and drawbacks. *Food Control* **2017**, *80*, 1–10. [CrossRef]
16. Schnitzler-Lenoble, A.; Arnold, C.; Guibal, F.; Walter, J. Histoire de la vigne sauvage, vitis vinifera ssp. sylvestris, en camargue/wild grapevine Vitis vinifera ssp. sylvestris in Camargue, southern France. *Ecol. Mediterr.* **2018**, *44*, 53–66. [CrossRef]
17. Vignani, R.; Liò, P.; Scali, M. How to integrate wet lab and bioinformatics procedures for wine DNA admixture analysis and compositional profiling: Case studies and perspectives. *PLoS ONE* **2019**, *14*, e0211962. [CrossRef]
18. Siret, R.; Boursiquot, J.; Merle, M.; Cabanis, J.; This, P. Toward the authentication of varietal wines by the analysis of grape (*Vitis vinifera* L.) residual DNA in must and wine using microsatellite markers. *J. Agric. Food Chem.* **2000**, *48*, 5035–5040. [CrossRef] [PubMed]
19. İşçi, B.; Yildirim, H.K.; Altindişli, A. A review of the authentication of wine origin by molecular markers. *J. Inst. Brew.* **2009**, *115*, 259–264. [CrossRef]
20. Ranaweera, R.K.; Gilmore, A.M.; Capone, D.L.; Bastian, S.E.; Jeffery, D.W. Spectrofluorometric analysis combined with machine learning for geographical and varietal authentication, and prediction of phenolic compound concentrations in red wine. *Food Chem.* **2021**, *361*, 130149. [CrossRef]
21. Barrias, S.; Fernandes, J.R.; Martins-Lopes, P. Newly developed QCM-DNA biosensor for SNP detection in small DNA fragments: A wine authenticity case study. *Food Control* **2025**, *169*, 111036. [CrossRef]
22. Barrias, S.; Ibáñez, J.; Martins-Lopes, P. High resolution melting analysis of microsatellite markers applied to grapevine varietal fingerprinting throughout the wine production chain. *Food Control* **2024**, *160*, 110368. [CrossRef]
23. Vignani, R.; Scali, M.; Liò, P. Molecular Markers and Genomics for Food and Beverages Characterization. In *Handbook of DNA Profiling*; Springer: Berlin/Heidelberg, Germany, 2021; pp. 1–21.
24. Palade, M.; Popa, M.-E. Wine traceability and authenticity—A literature review. *Sci. Bulletin. Ser. F Biotechnol.* **2014**, *XVIII*, 226–233.
25. Testolin, R.; Messina, R.; Cipriani, G.; De Mori, G. SSR-based DNA fingerprinting of fruit crops. *Crop Sci.* **2023**, *63*, 390–459. [CrossRef]

26. Ikten, H.; Sari, D.; Sabir, A.; Hasan Meydan, H.; Mutlu, N. Estimating genetic diversity among selected wild grapevine genotypes from Southern Turkey by simple sequence repeat (SSR) and inter-Primer Binding Site(iPBS). *Markers. Genet. Resour. Crop Evol.* **2025**, *72*, 2361–2377. [CrossRef]
27. Phumichai, C.; Phumichai, T.; Wongkaew, A. Novel chloroplast microsatellite (cpSSR) markers for genetic diversity assessment of cultivated and wild hevea rubber. *Plant Mol. Biol. Rep.* **2015**, *33*, 1486–1498. [CrossRef]
28. Vignani, R.; Scali, M. Il Wine DNA fingerprinting (WDF): Un'analisi molecolare per accertare la composizione varietale di un vino. In *Il Wine DNA Fingerprinting (WDF): Un'analisi Molecolare per Accertare la Composizione Varietale di un Vino*; CoNaVi: North York, ON, Canada, 2016.
29. Catalano, V.; Moreno-Sanz, P.; Lorenzi, S.; Grando, M.S. Experimental review of DNA-based methods for wine traceability and development of a single-nucleotide polymorphism (SNP) genotyping assay for quantitative varietal authentication. *J. Agric. Food Chem.* **2016**, *64*, 6969–6984. [CrossRef]
30. Bigliazzi, J.; Scali, M.; Paolucci, E.; Cresti, M.; Vignani, R. DNA extracted with optimized protocols can be genotyped to reconstruct the varietal composition of monovarietal wines. *Am. J. Enol. Vitic.* **2012**, *63*, 568–573. [CrossRef]
31. Scali, M.; Spinsanti, G.; Vignani, R. Validation of a simplified small-scale DNA extraction protocol from wine by quantitative real-time PCR. *3 Biotech* **2024**, *14*, 145. [CrossRef]
32. Thomas, M.R.; Scott, N.S. Microsatellite repeats in grapevine reveal DNA polymorphisms when analysed as sequence-tagged sites (STSs). *Theor. Appl. Genet.* **1993**, *86*, 985–990. [CrossRef] [PubMed]
33. Bowers, J.E.; Dangl, G.S.; Vignani, R.; Meredith, C.P. Isolation and characterization of new polymorphic simple sequence repeat loci in grape (*Vitis vinifera* L.). *Genome* **1996**, *39*, 628–633. [CrossRef]
34. Wagner, H.; Sefc, K. *IDENTITY*, Version 1.0 [Computer Program]; Centre for Applied Genetics, University of Agricultural Sciences: Vienna, Austria, 1999.
35. Perumal, J.; Wang, Y.; Attia, A.B.E.; Dinish, U.S.; Olivo, M. Towards a point-of-care SERS sensor for biomedical and agri-food analysis applications: A review of recent advancements. *Nanoscale* **2021**, *13*, 553–580. [CrossRef]
36. Pereira, L.; Gomes, S.; Barrias, S.; Gomes, E.P.; Baleiras-Couto, M.; Fernandes, J.R.; Martins-Lopes, P. From the field to the bottle—An integrated strategy for wine authenticity. *Beverages* **2018**, *4*, 71. [CrossRef]
37. Fritz, J.; Döring, J.; Athmann, M.; Meissner, G.; Kauer, R.; Schultz, H.R. Wine quality under integrated, organic and biodynamic management using image-forming methods and sensory analysis. *Chem. Biol. Technol. Agric.* **2021**, *8*, 62. [CrossRef]
38. Crook, A.A.; Zamora-Olivares, D.; Bhinderwala, F.; Woods, J.; Winkler, M.; Rivera, S.; Shannon, C.E.; Wagner, H.R.; Zhuang, D.L.; Lynch, J.E.; et al. Combination of two analytical techniques improves wine classification by Vineyard, Region, and vintage. *Food Chem.* **2021**, *30*, 354–12953. [CrossRef] [PubMed]
39. Siret, R.; Gigaud, O.; Rosec, J.P.; This, P. Analysis of grape Vitis vinifera L. DNA in must mixtures and experimental mixed wines using microsatellite markers. *J. Agric. Food Chem.* **2002**, *50*, 3822–3827. [CrossRef] [PubMed]
40. Sun, X.; Zhang, F.; Gutiérrez-Gamboa, G.; Ge, Q.; Xu, P.; Zhang, Q.; Fang, Y.; Ma, T. Real wine or not? Protecting wine with traceability and authenticity for consumers: Chemical and technical basis, technique applications, challenge, and perspectives. *Crit. Rev. Food Sci. Nutr.* **2022**, *62*, 6783–6808. [CrossRef]
41. Thomas, A.; Pickering, G. The importance of wine label information. *Int. J. Wine Mark.* **2003**, *15*, 58–74. [CrossRef]

beverages

MDPI

Article

Optimization of Beverage Formulation with Technique for Order of Preference by Similarity to Ideal Solution

Taraneh Javanbakht

Department of Mechanical Engineering, École de Technologie Supérieure, 1100 Notre-Dame St W, Montreal, QC H3C 1K3, Canada; taraneh.javanbakht@etsmtl.ca

Abstract: This paper presents a new optimization approach for beverage formulation using the Technique for Order of Preference by Similarity to Ideal Solution (TOPSIS). The comparison and ranking of candidates would be the essential steps for the evaluation of beverage production, which in turn requires the quality control of beverage ingredients. The quality of the beverage depends on the type and amount of the ingredients used for their preparation. This emphasizes the importance of the optimization of beverage formulation, which is addressed in this paper using the TOPSIS method. The results of this investigation show that ingredients with more impact on human health (having bigger weights) could affect the rank of drinks. The second, fourth, third and first candidates ranked first, second, third and fourth before and after the changes of criterion types and weight values. Moreover, the change of the criterion type accompanied with the increase in its concentration in the beverage had a significant impact on the candidates' ranks. The similarity coefficient of the fourth candidate for which this ingredient concentration changed showed a decrease from 0.625 to 0.500. The optimization of beverage formulation with the modified TOPSIS algorithm showed the efficiency of the automated decision-making process with the obtainment of the same ranks in comparison with those of the analyses without the consideration of the candidates' colors as they were used in beverages and could not determine their quality. The results in this research paper could be used for the quality improvement of drinks in the beverage industry.

Keywords: beverage ingredients; quality control; TOPSIS; optimization; automated decision making

1. Introduction

Beverage production is an important part of the food industry, implying certain steps such as handling, storage, packaging and transport [1,2]. These steps aim to help recycling, control and reduce the rate of deterioration, provide the required protection of ingredients and make the product available for consumers, respectively [1,3].

The global beverage market includes hot drinks, milk drinks, soft drinks and alcoholic drinks [4,5]. Soft drinks often contain water, sugar, fruit juice, sweeteners, carbon dioxide, acids, flavors, colors, preservatives, emulsifiers, antioxidants and vitamins [5,6]. Hot drinks include coffee, chocolate and tea. Milk drinks have subcategories of milk or milk-included drinks. Bottled water, carbonated drinks, dilutable drinks, fruit juices, still drinks, sports drinks and other noncarbonated products with less fruit juice are categorized as soft drinks. Beer, wine, spirits, cider, sake and flavored alcoholic beverages are alcoholic drinks [5,6]. Milk and soft drinks are the most important and the most consumed among the different types of beverages.

https://doi.org/10.3390/beverages11020038

Changes in lifestyle and diet, especially the frequent consumption of soft drinks and alcoholic drinks, can lead to obesity, metabolic syndrome, dental, pulmonary, or cardiovascular diseases [7–9]. An association between the consumption of soft drinks and gastroesophageal reflux has been reported. Moreover, carbonated beverages can influence stomach function. This can lead to both mechanical and chemical effects. It is worth noting that the mechanical effect depends on the pressure exerted by gas, which is used in the beverages, and by the fluid volume on the gastric wall. There are some factors that increase insulin and glucose levels in the blood, which cause obesity and diabetes. Therefore, the consumption of sugar-sweetened soft drinks can correspond to pancreatic cancer risk [7]. In fact, elevated insulin concentrations in the blood can activate insulin-like growth factor receptors, which may lead to cancer [7]. Another investigation revealed that soft drink consumption could be associated with a higher prevalence of multiple metabolic risk factors in middle-aged adults. In fact, these individuals are at a higher risk of developing metabolic syndrome [8]. An important factor that should be considered concerning metabolic diseases, such as obesity, metabolic syndrome, diabetes, hypertension, dyslipidemias, stroke, and cardiovascular disease mortality, is the body mass index (BMI), which is the ratio between body weight, expressed in kilograms, and the square of a person's height, expressed in meters. BMI indicates health issues such as overweight or obese conditions, corresponding to BMI > 25 and BMI > 30, respectively. Changes in lifestyle and diet, especially moderation in the consumption of transformed food and sweetened beverages, can help keep a BMI below 25 and maintain good health [9].

Sweeteners used in beverage formulations can help reduce positive energy balance and blood glucose [10]. However, conflicting data gathered from the short- and long-term consumption of natural and artificial materials, for example, aspartame, sucralose, and stevia derivatives, still exist, preventing a better conclusion on their impact on human health [11]. Preclinical and clinical studies have revealed the negative effects of sweeteners on the intestinal microbiota, glucose metabolism, and appetite control [12].

Several research works have been conducted on the prediction and optimization of beverage formulation. A constrained mixture design and multiple responses have been used for the systematic study of functional beverage formulation, such as the behavior and effects of varying ratios and proportions of beverage ingredients such as fruit juice, yogurt, and functional ingredients such as plant protein, fiber, or other functional compounds [13]. In another study, an Ultra-Accelerated Shelf-Life Test method was used for predicting the shelf life of beverages. The tests were performed based on vitamin degradation accelerated with increased temperature and oxygen diffusion [14]. A research work was performed to achieve a formula design based on pre-determined parameters by choosing the criteria that were optimized. A sensory test helped determine the flavor concentration which would be added to the beverage powder. The best formula was chosen according to the highest desirability value, which was the value closest to 1.0. The optimum formula was verified with a repeated test [15]. The optimization of fermentation conditions was also previously performed by investigating endogenous walnut lipase for the manufacture of a fermented, whey-based beverage containing conjugated linoleic acid [16]. These studies have been performed for the prediction and optimization of the mentioned aspects of beverage formulations. However, none of them have provided information based on the prediction and optimization of the soft drinks' formulation according to the consideration of their ingredient characteristics.

In this paper, the analysis and optimization of soft drinks were performed with the Technique for Order of Preference by Similarity to Ideal Solution (TOPSIS), which is a decision-making method that helps classify and optimize candidates [17–20]. This technique has been used previously for the selection of beverage cafes [21], the analysis of storeowner and supplier perspective for the prediction of sustainable packaging [22], the prediction of the best match between manufacturing technologies and food and beverage specifications [23], and the selection of optimal materials for the different components of cashew juice extractors [24]. TOPSIS has some considerable advantages in comparison with other decision-making methods. The advantages of TOPSIS are as follows: (1) All the positive and negative criteria as well as alternatives are considered in this method. So, it provides a holistic decision-making approach. (2) The implementation of TOPSIS is easy. This characteristic makes researchers prefer it to other methods for performing a decision-making process. (3) A wide range of decision-making problems across academic laboratories and industries can use this versatile approach. (4) The ranking of candidates in the output of TOPSIS is represented clearly. This enables decision-makers to make informed decisions because the best alternative based on the given criteria is chosen, and the method as an objective approach for decision-making involves a systematic and balanced analysis of alternatives, which is based on the given criteria. (5) TOPSIS has gained popularity because of its simplicity as it is considered a prevalent decision-making technique for solving multi-response optimization problems [25,26].

To the author's knowledge, the prediction and optimization of beverage formulation with TOPSIS have not been performed yet. The current work is aimed at predicting and optimizing the ranks of soft drinks according to their ingredients. For this purpose, different ingredient concentrations of soft drinks are analyzed. The results of this investigation can help with a better formulation of soft drinks with applications in the beverage industry.

2. Materials and Methods

2.1. TOPSIS Method

The second updated version of the TOPSIS algorithm in Python available on the GitHub website (https://github.com/Glitchfix/TOPSIS-Python/blob/master/topsis.py (accessed on 10 January 2025) was used for the analysis of the candidates' ranking and its optimization [27,28]. In this method, the criteria were the beverages' ingredients. The profit criterion was a characteristic for which a high value translated to a better rank and a low value translated to a worse rank for the candidate. The cost criterion was a characteristic for which a low value translated to a better rank and a high value translated to a worse rank for the candidate. The candidates' ranking was the output of this method based on the candidates' closeness to ideal solutions [27,28].

Four candidates were analyzed with the unmodified TOPSIS algorithm. The first, second, third, and fourth candidates, C_1, C_2, C_3, and C_4, were considered in the second, third, fourth, and fifth rows in the evaluation matrix.

The entry in the analysis was the average of the triangular fuzzy data corresponding to the concentrations of the beverage ingredients, which were the analysis criteria (water, sugar, fruit juice, sweeteners, carbon dioxide, acids, flavors, colors, preservatives, emulsifiers, antioxidants, and vitamins, indicated as W, SU, FJ, SW, CD, A, F, C, P, E, AO, and V, in the tables of results, respectively) [5]. The data normalization was performed by dividing the average concentrations by 100 to give fuzzy values. As described previously, the TOPSIS method includes the steps below [10,12,27,29]:

1. Data normalization

The first step of TOPSIS includes the normalization of data, which is the entry of the evaluation matrix, according to the formulas below:

$$A = \begin{bmatrix} g_{11} & g_{12} & \cdots & g_{1n} \\ g_{21} & g_{22} & \cdots & g_{2n} \\ \vdots & \cdots & \ddots & \vdots \\ g_{ml} & g_{ml} & \cdots & g_{mn} \end{bmatrix} \quad (1)$$

$$r_{mn} = \frac{g_{mn}}{\left(\sum_{m=1}^{n} g_{mn}^2\right)^{1/2}} \quad (2)$$

2. Multiplication of normalized data by their weight values

In the second step, the normalized data are multiplied by their weights as presented below:

$$\tilde{V} = [v_{ij}]_{m \times n} \quad (3)$$

3. Determination of positive and negative ideal solutions

The third step includes determination of the positive and negative ideal solutions, which are used in the following step. To carry this out, we use the formulas below:

$$V^* = \max\left\{v_1^*, v_2^*, \ldots, v_j^*\right\}; j = 1, 2, \ldots, n \quad (4)$$

$$V^- = \min\left\{v_1^*, v_2^*, \ldots, v_j^*\right\}; j = 1, 2, \ldots, n \quad (5)$$

4. Determination of distances from positive and negative ideal solutions

The distances from positive and negative ideal solutions are calculated for each candidate in the fourth step. This is performed using the formula below:

$$d_i^* = \left[\sum_{i=1}^{n} \left(v_{ij} - v_j^*\right)^2\right]^{1/2} \quad (6)$$

$$d_i^- = \left[\sum_{i=1}^{n} \left(v_{ij} - v_j^*\right)^2\right]^{1/2} \quad (7)$$

5. Determination of candidates' similarity coefficients

The candidates' similarity coefficients are determined in the last step according to the formula below:

$$C_i^* = \frac{d_i^-}{d_i^* + d_i^-} \quad (8)$$

In the formula above, d_i^* and d_i^- represent the distances from the best and worst ideals, respectively.

6. Determination of candidates' ranks

The candidates were ranked according to the comparison of their similarity coefficients. The candidate with the biggest similarity coefficient was ranked in the first position and the following ones had smaller coefficients. The last candidate had the smallest coefficient.

2.2. *Modified TOPSIS*

The modified TOPSIS, having the code line below, allowed non-consideration of the color of the soft drinks for their analysis as this characteristic does not determine their quality. Four candidates were analyzed with the TOPSIS algorithm. The first, second, third, and fourth candidates, C_1, C_2, C_3, and C_4, were considered in the first, second, third, and fourth rows in the evaluation matrix. In another investigation, eight candidates, C_1, C_2, C_3, C_4, C_5, C_6, C_7, and C_8, were analyzed. The data concerning the colors of drinks were added into the last column at the right side in the evaluation matrix of the TOPSIS code.

weight_matrix[self.column_size − 1] = 0

This modification in the algorithm led to automated decision making for the candidates' ranking [30].

3. Results and Discussion

Different series of analyses were performed with the TOPSIS algorithm on the soft drinks considering their ingredients as criteria.

3.1. Analysis of Candidates

The emphasis in the analysis was on the impact of the weights and criteria on the candidates' ranks.

3.1.1. Analysis of Weight Values and Criterion Types for Ranks of Soft Drinks

In the first series of analysis, the weights of criteria were analyzed for soft drinks. Table 1 shows the evaluation matrix in which the average values of the triangular fuzzy data corresponding to the concentrations of the beverage ingredients are presented. The first row, including the abbreviations of the soft drink ingredients, corresponds to the criteria. The second, third, fourth, and fifth rows in the evaluation matrix and normalization matrix correspond to the first, second, third, and fourth candidate, respectively.

Table 1. The evaluation matrix of entry data.

W	SU	FJ	SW	CD	A	F	C	P	E	AO	V
0.83	0.10	0.05	0.01	0.00	0.0003	0.002	0.001	0.001	0.001	0.001	0.0087
0.85	0.08	0.05	0.01	0.003	0.0003	0.002	0.001	0.001	0.001	0.001	0.0057
0.81	0.12	0.05	0.01	0.003	0.0003	0.002	0.001	0.001	0.001	0.001	0.0057
0.85	0.08	0.05	0.01	0.003	0.0005	0.002	0.001	0.001	0.001	0.001	0.0055

The weight values of water, sugar, fruit juice, sweeteners, flavors, preservatives, antioxidants, and vitamins were set to 0.1, whereas those of carbon dioxide, acids, colors and emulsifiers were set to 0.05, respectively. The higher and lower values of the weights were due to the degrees of importance considered for the ingredients. The sum of all the weight values were set to 1.0.

Water, fruit juice, carbon dioxide, flavors, antioxidants and vitamins were considered as profit criteria, whereas sugar, sweeteners, acids, colors, preservatives and emulsifiers were considered as cost criteria, respectively.

The normalized matrix is presented in Table 2.

Table 2. The normalized matrix.

W	SU	FJ	SW	CD	A
0.496908	0.51847585	0.5	0.5	0.0	0.41602515
0.50888169	0.41478068	0.5	0.5	0.57735027	0.41602515
0.48493432	0.62217102	0.5	0.5	0.57735027	0.41602515
0.50888169	0.41478068	0.5	0.5	0.57735027	0.69337525

Table 2. *Cont.*

F	C	P	E	AO	V
0.5	0.5	0.5	0.5	0.5	0.66546131
0.5	0.5	0.5	0.5	0.5	0.43599189
0.5	0.5	0.5	0.5	0.5	0.43599189
0.5	0.5	0.5	0.5	0.5	0.42069393

The weighted normalized matrix is presented in Table 3.

Table 3. The weighted normalized matrix.

W	SU	FJ	SW	CD	A
0.0496908	0.05184758	0.05	0.05	0.0	0.02080126
0.05088817	0.04147807	0.05	0.05	0.02886751	0.02080126
0.04849343	0.0622171	0.05	0.05	0.02886751	0.02080126
0.05088817	0.04147807	0.05	0.05	0.02886751	0.03466876
F	**C**	**P**	**E**	**AO**	**V**
0.05	0.025	0.05	0.025	0.05	0.06654613
0.05	0.025	0.05	0.025	0.05	0.04359919
0.05	0.025	0.05	0.025	0.05	0.04359919
0.05	0.025	0.05	0.025	0.05	0.04206939

The worst and the best alternatives for the candidates are presented in the second and third rows of Table 4, respectively.

Table 4. The worst and the best alternatives.

W	SU	FJ	SW	CD	A
0.04849343	0.0622171	0.05	0.05	0.0	0.03466876
0.05088817	0.04147807	0.05	0.05	0.02886751	0.02080126
F	**C**	**P**	**E**	**AO**	**V**
0.05	0.025	0.05	0.025	0.05	0.04206939
0.05	0.025	0.05	0.025	0.05	0.06654613

Table 5 shows the values of the distances between the target alternative and the worst condition (d_i^-) and the distances between the target alternative and the best condition (d_i^*).

Table 5. The distances between the target alternative and the worst and the best conditions. d_i^* and d_i^- represent the distances from the best and worst ideals, respectively.

Candidates	d_i^-	d_i^*
C_1	0.03000632	0.03069681
C_2	0.03825995	0.02294694
C_3	0.03206215	0.03102264
C_4	0.03562549	0.02813216

Table 6 shows the values of the similarity coefficient (CC_i) and the ranks of candidates.

Table 6. The similarity coefficient and the ranks of candidates.

Candidates	CC_i	Rank
C_1	0.49431256	2
C_2	0.62509218	4
C_3	0.50823894	3
C_4	0.55876418	1

As shown in Table 6, the second, fourth, third and first candidates have the first, second, third and fourth ranks, respectively.

Figure 1 shows the distances between the target alternative and the worst and best conditions and the similarity coefficients of the candidates.

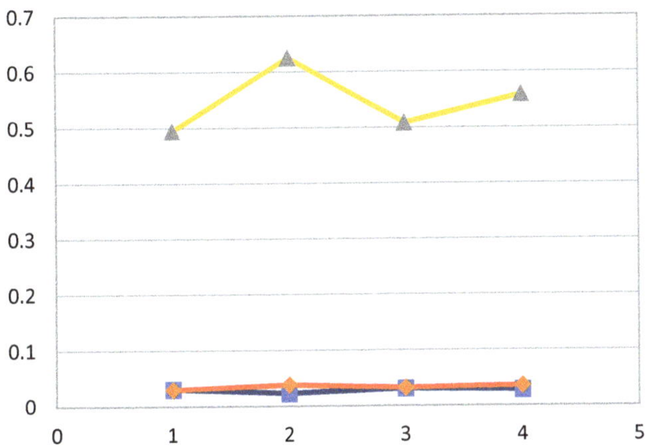

Figure 1. The distances between the target alternative and the worst and the best conditions and the similarity coefficients of the candidates. The spots from left to right correspond to the first, second, third and fourth candidates, respectively. The blue, red and yellow lines represent the distance from the best ideal, the distance from the worst ideal and the closeness coefficients of the candidates, respectively.

In the second series of analysis, the same average values of the triangular data were considered as the entry in the evaluation matrix (Table 7).

Table 7. The evaluation matrix of entry data.

W	SU	FJ	SW	CD	A	F	C	P	E	AO	V
0.83	0.10	0.05	0.01	0.00	0.0003	0.002	0.001	0.001	0.001	0.001	0.0087
0.85	0.08	0.05	0.01	0.003	0.0003	0.002	0.001	0.001	0.001	0.001	0.0057
0.81	0.12	0.05	0.01	0.003	0.0003	0.002	0.001	0.001	0.001	0.001	0.0057
0.85	0.08	0.05	0.01	0.003	0.0005	0.002	0.001	0.001	0.001	0.001	0.0055

The weight values of water, sugar, fruit juice, flavors, preservatives, emulsifiers, antioxidants and vitamins were set to 0.1, whereas those of sweeteners, carbon dioxide, acids and colors were set to 0.05, respectively. The two conditions to choose the weight values were those of the previous analysis.

Water, fruit juice, carbon dioxide, flavors, emulsifiers, antioxidants and vitamins were considered as profit criteria, whereas sugar, sweeteners, acids, colors and preservatives were considered as cost criteria, respectively.

Table 8 shows the normalized matrix.

Table 8. The normalized matrix.

W	SU	FJ	SW	CD	A
0.496908	0.51847585	0.5	0.5	0.0	0.41602515
0.50888169	0.41478068	0.5	0.5	0.57735027	0.41602515
0.48493432	0.62217102	0.5	0.5	0.57735027	0.41602515
0.50888169	0.41478068	0.5	0.5	0.57735027	0.69337525
F	**C**	**P**	**E**	**AO**	**V**
0.5	0.5	0.5	0.5	0.5	0.66546131
0.5	0.5	0.5	0.5	0.5	0.43599189
0.5	0.5	0.5	0.5	0.5	0.43599189
0.5	0.5	0.5	0.5	0.5	0.42069393

The weighted normalized matrix is presented in Table 9.

Table 9. The weighted normalized matrix.

W	SU	FJ	SW	CD	A
0.0496908	0.05184758	0.05	0.025	0.0	0.02080126
0.05088817	0.04147807	0.05	0.025	0.02886751	0.02080126
0.04849343	0.0622171	0.05	0.025	0.02886751	0.02080126
0.05088817	0.04147807	0.05	0.025	0.02886751	0.03466876
F	**C**	**P**	**E**	**AO**	**V**
0.05	0.025	0.05	0.05	0.05	0.06654613
0.05	0.025	0.05	0.05	0.05	0.04359919
0.05	0.025	0.05	0.05	0.05	0.04359919
0.05	0.025	0.05	0.05	0.05	0.04206939

The worst and the best alternatives for the candidates are presented in the second and third rows of Table 10, respectively.

Table 10. The worst and the best alternatives.

W	SU	FJ	SW	CD	A
0.04849343	0.0622171	0.05	0.025	0.0	0.03466876
0.05088817	0.04147807	0.05	0.025	0.02886751	0.02080126
F	**C**	**P**	**E**	**AO**	**V**
0.05	0.025	0.05	0.05	0.05	0.04206939
0.05	0.025	0.05	0.05	0.05	0.06654613

Table 11 shows the values of the distances between the target alternative and the worst condition (d_i^-) and the distances between the target alternative and the best condition (d_i^*).

Table 11. The distances between the target alternative and the worst and best conditions. d_i^* and d_i^- represent the distances from the best and worst ideals, respectively.

Candidates	d_i^-	d_i^*
C_1	0.03000632	0.03069681
C_2	0.03825995	0.02294694
C_3	0.03206215	0.03102264
C_4	0.03562549	0.02813216

Table 12 shows the values of the similarity coefficient (CC_i) and the ranks of candidates.

Table 12. The similarity coefficient and the ranks of candidates.

Candidates	CC_i	Rank
C_1	0.49431256	2
C_2	0.62509218	4
C_3	0.50823894	3
C_4	0.55876418	1

As shown in Table 12, the same ranks for candidates were obtained as those of the previous analysis.

Figure 2 shows the distances between the target alternative and the worst and the best conditions and the similarity coefficients of the candidates.

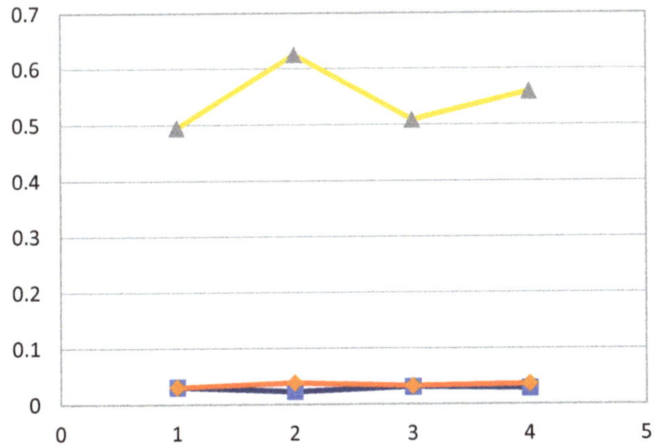

Figure 2. The distances between the target alternative and the worst and the best conditions and the similarity coefficients of the candidates. The spots from left to right correspond to the first, second, third and fourth candidates, respectively. The blue, red and yellow lines represent the distance from the best ideal, the distance from the worst ideal and the closeness coefficients of the candidates, respectively.

The comparison of the candidates' ranks obtained in the first and second series of analyses showed that the modifications of the weights and the criteria did not have an impact on the TOPSIS output.

3.1.2. Impact of Data Modification on Ranks of Soft Drinks

In the third series of analysis, the concentration of the preservatives of the fourth candidate in the evaluation matrix was 0.0015 in place of 0.001. The concentrations of the other drink ingredients were not changed. The weight values and criterion types were the same as those of the first series of analysis.

The weight values of the soft drink ingredients and the criterion types were chosen to be the same as those of the first series of analysis.

Table 13 shows the normalized matrix for this series of analysis.

Table 13. The normalized matrix.

W	SU	FJ	SW	CD	A
0.496908	0.51847585	0.5	0.5	0.0	0.41602515
0.50888169	0.41478068	0.5	0.5	0.57735027	0.41602515
0.48493432	0.62217102	0.5	0.5	0.57735027	0.41602515
0.50888169	0.41478068	0.5	0.5	0.57735027	0.69337525
F	**C**	**P**	**E**	**AO**	**V**
0.5	0.5	0.43643578	0.5	0.5	0.66546131
0.5	0.5	0.43643578	0.5	0.5	0.43599189
0.5	0.5	0.43643578	0.5	0.5	0.43599189
0.5	0.5	0.65465367	0.5	0.5	0.42069393

The weighted normalized matrix is presented in Table 14.

Table 14. The weighted normalized matrix.

W	SU	FJ	SW	CD	A
0.0496908	0.05184758	0.05	0.05	0.0	0.02080126
0.05088817	0.04147807	0.05	0.05	0.02886751	0.02080126
0.04849343	0.0622171	0.05	0.05	0.02886751	0.02080126
0.05088817	0.04147807	0.05	0.05	0.02886751	0.03466876
F	**C**	**P**	**E**	**AO**	**V**
0.05	0.025	0.04364358	0.025	0.05	0.06654613
0.05	0.025	0.04364358	0.025	0.05	0.04359919
0.05	0.025	0.04364358	0.025	0.05	0.04359919
0.05	0.025	0.06546537	0.025	0.05	0.04206939

The worst and best alternatives for the candidates are presented in the second and third rows of Table 15, respectively.

Table 15. The worst and best alternatives.

W	SU	FJ	SW	CD	A
0.04849343	0.0622171	0.05	0.05	0.0	0.03466876
0.05088817	0.04147807	0.05	0.05	0.02886751	0.02080126
F	**C**	**P**	**E**	**AO**	**V**
0.05	0.025	0.06546537	0.025	0.05	0.04206939
0.05	0.025	0.04364358	0.025	0.05	0.06654613

Table 16 shows the values of the distances between the target alternative and the worst condition (d_i^-) and the distances between the target alternative and the best condition (d_i^*).

Table 16. The distances between the target alternative and the worst and best conditions. d_i^* and d_i^- represent the distances from the best and worst ideals, respectively.

Candidates	d_i^-	d_i^*
C_1	0.03710215	0.03069681
C_2	0.04404559	0.02294694
C_3	0.03878365	0.03102264
C_4	0.03562549	0.0356035

Table 17 shows the values of the similarity coefficient (CC_i) and the ranks of candidates.

Table 17. The similarity coefficient and the ranks of candidates.

Candidates	CC_i	Rank
C_1	0.54723777	2
C_2	0.65747016	3
C_3	0.55558961	1
C_4	0.5001544	4

As shown in Table 17, the second, third, first, and fourth candidates have the first, second, third, and fourth ranks, respectively.

Figure 3 shows the distances between the target alternative and the worst and best conditions and the similarity coefficients of the candidates.

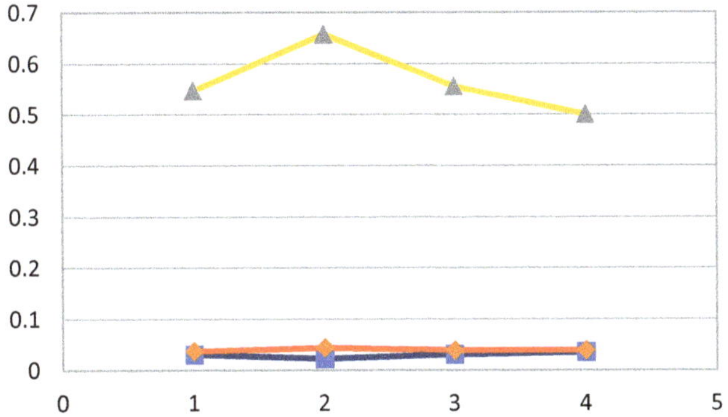

Figure 3. The distances between the target alternative and the worst and best conditions and the similarity coefficients of the candidates. The spots from left to right correspond to the first, second, third and fourth candidates, respectively. The blue, red and yellow lines represent the distance from the best ideal, the distance from the worst ideal and the closeness coefficients of the candidates, respectively.

The comparison of the candidates' ranks obtained in the first, second and third series of analyses showed that the modifications of the concentration of the preservatives of the fourth candidate had a significant impact on the TOPSIS output. This candidate was ranked in the second position in the first and second series of analyses, whereas it was ranked in the last position in the third series of analysis. The change of rank was due to the consideration of the preservatives as a cost criterion in the analysis. Moreover, the increase in their concentration for the fourth candidate had a negative impact on its rank.

3.2. Case Optimization

In this section, two series of case analyses for several drinks having different colors are presented.

The analysis of four drinks with different colors—orange, pink, red and violet—with an automated decision-making process is presented. The modified algorithm, as described in the Materials and Methods, was used for the optimization of beverage formulation. The drink's color is an acceptable characteristic of the manufactured products. Therefore, it has no impact on the distinction of drinkable and undrinkable drinks. The fuzzy values of 0.3, 0.5, 0.7 and 0.9 were used for the colors of orange, pink, red and violet, respectively. The weight values and criterion types were the same as those of the third series of analysis presented in Section 3.1.2. The explained modification in the TOPSIS method allowed the non-consideration of the drinks' colors for the candidates' ranks.

Table 18 shows the values of the distances between the target alternative and the worst condition (d_i^-) and the distances between the target alternative and the best condition (d_i^*).

Table 18. The distances between the target alternative and the worst and best conditions. d_i^* and d_i^- represent the distances from the best and worst ideals, respectively.

Candidates	d_i^-	d_i^*
C_1	0.03710215	0.03069681
C_2	0.04404559	0.02294694
C_3	0.03878365	0.03102264
C_4	0.03562549	0.0356035

Table 19 shows the values of the similarity coefficient (CC_i) and the ranks of candidates.

Table 19. The similarity coefficient and the ranks of candidates.

Candidates	CC_i	Rank
C_1	0.54723777	2
C_2	0.65747016	3
C_3	0.55558961	1
C_4	0.5001544	4

As shown in Table 19, the second, third, first, and fourth candidates have the first, second, third, and fourth ranks, respectively.

Figure 4 shows the distances between the target alternative and the worst and best conditions and the similarity coefficients of the candidates.

The comparison of the results in Tables 17 and 19 and Figures 3 and 4 reveal that the same candidates' ranks were obtained with the modified TOPSIS method due to the non-consideration of the fuzzy values for the candidates' colors. This showed the efficiency of the automated decision-making process with the TOPSIS modification.

In another investigation, eight candidates, C_1, C_2, C_3, C_4, C_5, C_6, C_7, and C_8, which were the drinks with different colors (light orange, dark orange, light pink, dark pink, light red, dark red, light violet, and dark violet), were chosen for analysis. The fuzzy values of 0.3, 0.5, 0.7, and 0.9 were used for the colors of orange, pink, red, and violet, respectively. The weight values and criterion types were the same as those of the three series of analyses presented in Section 3.1. The modified TOPSIS was used for the candidates' ranking. Table 20 shows the comparative rankings.

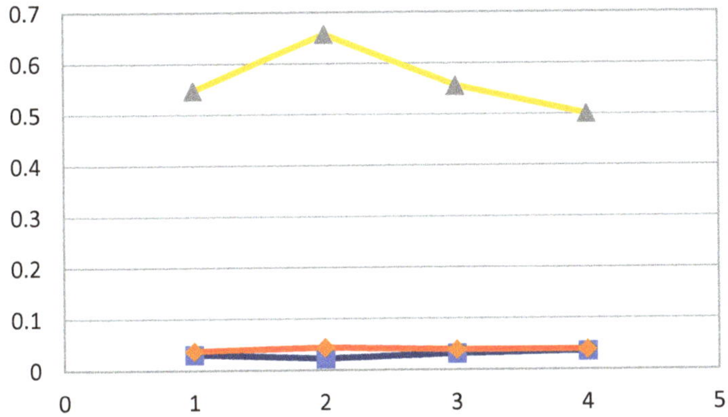

Figure 4. The distances between the target alternative and the worst and best conditions and the similarity coefficients of the candidates. The spots from left to right correspond to the first, second, third and fourth candidates, respectively. The blue, red and yellow lines represent the distance from the best ideal, the distance from the worst ideal and the closeness coefficients of the candidates, respectively.

Table 20. The comparative rankings of eight candidates.

Candidates	First Ranking	Second Ranking	Third Ranking
C_1	3	3	3
C_2	4	4	4
C_3	7	7	5
C_4	8	8	6
C_5	5	5	1
C_6	6	6	2
C_7	1	1	7
C_8	2	2	8

The modified TOPSIS made the non-consideration of the last column in the evaluation matrix, which corresponded to the candidates' colors, and the same rankings as those without this column in the matrix were obtained. This showed the efficiency of the modification in the algorithm for the automated decision-making process.

A recent study showed the optimization of a combination of polyacrylamide flocculants for clarifying raw molasses during rum fermentation using a mixture model. Temperature, oxygenation, and nutrients had an impact on critical fermentation responses [31]. The fermentation process can be applied for the manufacture of a variety of beverages, in which the release and development of lactic acid, carbon dioxide, alcohol, and flavoring compounds are required [32]. The formulations of these beverages can be predicted and optimized with TOPSIS.

Some manufacturers have used the ingredients in foods and beverages producing hydrogen peroxide [33–35]. The physicochemical properties of these materials have been investigated previously [36]. It has been revealed that the production of hydrogen peroxide in tea beverages can be dominated by catechins, with multiple factors acting synergistically, which in turn can lead to aroma deterioration and affect the quality of these beverages [37]. As an oxidizer, the consumption of hydrogen peroxide can lead to the production of free radicals, damaging proteins, DNA, and cell membrane lipids. This chemical, with a concentration-dependent toxicity, can be harmful upon dermal, ophthalmic exposure, ingestion, and wound [38]. Therefore, the presence of hydrogen peroxide in beverages can be considered as a cost criterion in a further investigation with TOPSIS.

The acidity of beverages is an accurate indicator of their erosive potential, which is determined according to their ingredients. As baseline pH values only give a measure of the initial hydrogen ion concentration, they provide no indication as to the presence of undissociated acid. Therefore, the total titratable acidity is considered a more accurate measure of the total acid content of a drink for predicting its erosive potential [39]. More beverages have an acidic pH due to their ingredients [40]. It has been revealed that acidic beverages can alter the roughness and color stability of polymerized acrylic resin and artificial teeth [41], which have diverse biomedical applications [42]. Another study has shown that acidulated drinks overwhelm the neutralizing effects of alkaline-stimulated saliva, and their frequent consumption can pervade all oral organs (tongue, teeth, gums, cheeks, palate, etc.) and oral stagnation niches [43]. The erosive effects through the decreased surface hardness and weight loss of human tooth specimens have been revealed with the analysis of acidic drinks [44]. The optimization of the acidity of beverages considering the ingredients used for their formulations is an important issue for further investigations.

The results obtained in the current work could have diverse applications in sciences and engineering as well as industrial applications. It is worth noting that the optimization of beverage formulations requires a comparison of the studied soft drinks with new formulations. Further investigations are required to determine the best formulation for other types of drinks, similarly to the analyses presented in this investigation.

The optimization of the applied model in this study as well as similar models used for the analysis of fermentation output could be performed with TOPSIS. This can help optimize the fermentation conditions for the manufacture of alcoholic drinks.

4. Conclusions

This work focused on the prediction and optimization of soft drink formulation with TOPSIS. The candidates analyzed in this investigation had different ingredient concentrations. The results of the study revealed that the modifications of the weight values and criterion types did not affect the candidates' ranks when the same concentrations were considered for their ingredients. The modification of the small weight values considered for the candidates did not have an impact on the TOPTIS output. Moreover, the change of the criterion type did not affect the candidates' ranking when the modification accompanied the modification on the low weight values. On the other hand, when a modification was performed on the concentration of the preservatives of a candidate without the change of the weight values and criterion types, the ranks changed. So, the output of TOPSIS was different from the previous series of analyses. Therefore, the choice of appropriate ingredient concentrations of soft drinks would be crucial for further investigations of these beverages. The results obtained in this work could be used in the beverage industry for the improvement of product quality.

Funding: This research received no external funding.

Data Availability Statement: No new data was created apart from those presented in this paper.

Conflicts of Interest: The author declares no conflicts of interest.

Abbreviations

The following abbreviations are used in this manuscript:

TOPSIS	Technique for Order of Preference by Similarity to Ideal
BMI	Body mass index
W	Water
SU	Sugar
FJ	Fruit juice
SW	Sweeteners
CD	Carbon dioxide
A	Acids
F	Flavors
C	Colors
P	Preservatives
E	Emulsifiers
AO	Antioxidants
V	Vitamins

References

1. Kregiel, D. Health safety of soft drinks: Contents, containers, and microorganisms. *BioMed Res. Int.* **2015**, *2015*, 128697. [CrossRef] [PubMed]
2. Ramos, M.; Valdés, A.; Mellinas, A.C.; Garrigós, M.C. New trends in beverage packaging systems: A review. *Beverages* **2015**, *1*, 248–272. [CrossRef]
3. Twede, D.; Harte, B. Logistical packaging for food marketing systems. In *Food and Beverage Packaging Technology*, 2nd ed.; Coles, R., Kiwan, M., Eds.; Wiley/Blackwell: Chichester, UK, 2011; pp. 85–105.
4. Silva, M.M.N.; Albuquerque, T.L.; Pereira, K.S.; Coelho, M.A.Z. Food additives used in non-alcoholic water-based beverages—A review. *J. Nutr. Health Food Eng.* **2019**, *9*, 109–121. [CrossRef]
5. Abdellatif, A.S.A. The beverages. *Acta Sci. Nutr. Health* **2018**, *2*, 19–27.
6. Rampersaud, G.C.; Bailey, L.B.; Kauwell, G.P.A. National survey beverage consumption data for children and adolescents indicate the need to encourage a shift toward more nutritive beverage. *J. Am. Diet. Assoc.* **2003**, *103*, 97–100. [CrossRef]
7. Cuomo, R.; Andreozzi, P.; Zito, F.P. Alcoholic beverages and carbonated soft drinks: Consumption and gastrointestinal cancer risks. *Cancer Treat. Res.* **2014**, *159*, 97–120.
8. Dhingra, R.; Sullivan, L.; Jacques, P.F.; Wang, T.J.; Fox, C.S.; Meigs, J.B.; D'Agostino, R.B.; Gaziano, J.M.; Vasan, R.S. Soft drink consumption and risk of developing cardiometabolic risk factors and the metabolic syndrome in middle-aged adults in the community. *Circulation* **2007**, *116*, 480–488. [CrossRef]
9. Schiano, C.; Grimaldi, V.; Scognamiglio, M.; Costa, D.; Soricelli, C.; Nicoletti, G.F.; Napoli, C. Soft drinks and sweeteners intake: Possible contribution to the development of metabolic syndrome and cardiovascular diseases. Beneficial or detrimental action of alternative sweeteners? *Food Res. Inter.* **2021**, *142*, 110220. [CrossRef]
10. Pepin, A.; Stanhope, K.L.; Imbeault, P. Are fruit juices healthier than sugar-sweetened beverages? A review. *Nutrients* **2019**, *11*, 1006. [CrossRef]
11. Lobach, A.R.; Roberts, A.; Rowland, I.R. Assessing the in vivo data on low/no-calorie sweeteners and the gut microbiota. *Food Chem. Toxicol.* **2019**, *124*, 385–399. [CrossRef]
12. Daly, K.; Darby, A.C.; Shirazi-Beechey, S.P. Low calorie sweeteners and gut microbiota. *Physiol. Behav.* **2016**, *164*, 494–500. [CrossRef] [PubMed]
13. Wanyonyi, M.; Kiluyi, J.W.; Okango, A.A. Optimization of a functional beverage formulation using a constrained mixture design with multiple responses. *Asian J. Prob. Stat.* **2025**, *27*, 1–16. [CrossRef]
14. Muniandy, A.; Beniyathiar, P.; Ozadali, F.; Mishra, D.K. Development of predictive model for the novel ultra-accelerated shelf-life test (UASLT) for shelf-life of packaged beverage. *LWT Food Sci. Technol.* **2024**, *210*, 116686. [CrossRef]
15. Astawan, M.; Mahdi, S.A.; Budi, A.S.; Wresdiyati, T.; Febrinda, A.E. Optimization of beverage powder formula based on the defatted tempe flour. *Food Sci. Technol.* **2023**, *11*, 33–43. [CrossRef]
16. Moslemi, M.; Moayedi, A.; Khomeiri, A.; Maghsoudlou, Y. Development of a whey-based beverage with enhanced levels of conjugated linoleic acid (CLA) as facilitated by endogenous walnut lipase. *Food Chem.* **2023**, *17*, 100547. [CrossRef]

17. Rahim, R.; Supiyandi, S.; Siahaan, A.P.U.; Listyorini, T.; Utomo, A.P.; Triyanto, W.A.; Irawan, Y.; Aisyah, S.; Khairani, M.; Sundari, S.; et al. TOPSIS method application for decision support system in internal control for selecting best employees. *J. Phys. Conf. Ser.* **2018**, *1028*, 012052. [CrossRef]
18. Javanbakht, T. Automated decision-making for water analysis. *J. Eng. Sci.* **2022**, *9*, H19–H24. [CrossRef]
19. George, J.; Singh, A.; Bhaisare, A.K. Implementation of TOPSIS technique for supplier selection. *Int. Res. J. Eng. Technol.* **2018**, *5*, 2582–2585.
20. Fahmi, M.H.I.; Amin, I.H.A. Best products with TOPSIS method and sales forecasting with weighted moving average. *J. App. Inform. Comput.* **2020**, *4*, 116–123. [CrossRef]
21. Nordin, W.N.H.W.; Zuhaimi, N.M.; Muhamad, R. The selection of beverage cafes using fuzzy TOPSIS. *J. Math. Comput. Sci.* **2021**, *7*, 129–140.
22. Draculan, K.A.E.; Santoso, B.S.; Sinawang, B.H.; Perez, W.D.D.P.; Redi, A.A.N.P.; Ho, H.-C.; Ayu, K.G.; German, J.D. Comparative analysis on beverage store owners towards sustainable packaging using multicriteria decision techniques. *IOP Conf. Ser. Earth Environ. Sci.* **2023**, *1324*, 012079. [CrossRef]
23. Bertolini, M.; Esposito, G.; Ramagnoli, G. A TOPSIS-based approach for the best match between manufacturing technologies and product specifications. *Exp. Sys. App.* **2020**, *159*, 113610. [CrossRef]
24. Emovon, I.; Aibuedefe, W.O. Fuzzy TOPSIS application in materials analysis for economic production of cashew juice extractor. *Fuzzy Inform. Eng.* **2022**, *12*, 1–18. [CrossRef]
25. Sriburum, A.; Wichapa, N.; Khanthirat, W. A novel TOPSIS linear programming model based on the Taguchi method for solving the multi-response optimization problems: A case study of a Fish scale scraping machine. *Eng. Sci.* **2024**, *29*, 1088. [CrossRef]
26. Ginting, G.; Fadlina, F.; Aan, M.; Siahaan, A.P.U.; Rahim, R. Technical approach of TOPSIS for decision making. *Int. J. Rec. Tren. Eng. Res.* **2017**, *3*, 58–64. [CrossRef]
27. Behzadian, M.; Otaghsara, K.; Yazdani, M.; Ignatius, J. A state-of the-art survey of TOPSIS applications. *Exp. Sys. App.* **2012**, *39*, 13051–13069. [CrossRef]
28. Javanbakht, T.; Chakravorty, S. Optimization of Machine Learning Algorithms for Proteomic Analysis Using TOPSIS. *J. Eng. Sci.* **2022**, *9*, E7–E11. [CrossRef]
29. Devnath, A.; Islam, M.S.; Islam, R.E. An integrated QFD-TOPSIS method for prioritization of major lean tools: A case study. *Int. J. Res. Ind. Eng.* **2020**, *9*, 65–76. [CrossRef]
30. Javanbakht, T. Modélisation et Traitement Informatique de L'inconsistance des Croyances Épistémiques. Ph.D. Thesis, University of Quebec in Montreal, Montreal, QC, Canada, 2022.
31. Mangwanda, T.W.; Johnson, J.B.; Batley, R.J.; Jackson, S.; McKeown, T.; Naiker, M. Comprehensive multi-scale optimisation of rum fermentation. *Beverages* **2025**, *11*, 17. [CrossRef]
32. Newbold, D.; Koppel, K. Carbonated dairy beverages: Challenges and opportunities. *Beverages* **2018**, *4*, 660. [CrossRef]
33. Sieg, H.; Schaar, C.; Fouquet, N.; Böhmert, L.; Thünemann, A.F.; Braeuning, A. Particle iron oxide food colorants (E 172) during artificial digestion and their uptake and impact on intestinal cells. *Toxicol. Vitr.* **2024**, *96*, 105772. [CrossRef] [PubMed]
34. Bopitiya, D.; Christensen, D.; Martin, M.; Zhang, J.; Bennett, L.E. Production of hydrogen peroxide in formulated beverages is associated with the presence of ascorbic acid combined with selected redox-active functional ingredients. *Food Chem.* **2021**, *338*, 127947. [CrossRef]
35. Rak, M.; Mendys, D.; Płatek, A.; Sitarz, O.; Stefaniuk, I.; Bartosz, G.; Sadowska-Bartosz, I. Generation of hydrogen peroxide in beer and strong alcoholic beverages. *Processes* **2025**, *13*, 277. [CrossRef]
36. Javanbakht, T.; Laurent, S.; Stanicki, D.; Frenette, M. Correlation between physicochemical properties of superparamagnetic iron oxide nanoparticles and their reactivity with hydrogen peroxide. *Can. J. Chem.* **2022**, *98*, 601–608. [CrossRef]
37. Wang, J.-Q.; Gao, Y.; Long, D.; Yin, J.-F.; Zeng, L.; Xu, Y.-Q.; Xu, Y.-Q. Effects of hydrogen peroxide produced by catechins on the aroma of tea beverages. *Foods* **2022**, *11*, 1273. [CrossRef]
38. Nelson, A.L.; Porter, L. Hydrogen peroxide toxicity. In *StatPearls*; StatPearls Publishing: Treasure Island, FL, USA, 2025.
39. Cairns, A.M.; Watson, M.; Creanor, S.L.; Foye, R.H. The pH and titratable acidity of a range of diluting drinks and their potential effect on dental erosion. *J. Dent.* **2002**, *30*, 313–317. [CrossRef]
40. Reddy, A.; Norris, D.F.; Momeni, S.S.; Waldo, B.; Ruby, J.D. The pH of beverages in the United States. *Am. Dent. Assoc.* **2016**, *147*, 255–263. [CrossRef]
41. Bitencourt, S.B.; Catanoze, I.A.; de Silva, E.V.F.; dos Santos, P.H.; dos Santos, D.M.; Turcio, K.H.L.; Guiotti, A.M. Effect of acidic beverages on surface roughness and color stability of artificial teeth and acrylic resin. *J. Adv. Prosthodont.* **2020**, *12*, 55–60. [CrossRef]
42. Javanbakht, T.; Sokolowski, W. Thiol-ene/acrylate systems for biomedical shape-memory polymers. In *Shape Memory Polymers for Biomedical Applications*; Woodhead Publishing: Sawston, UK, 2015; pp. 157–166. [CrossRef]

43. Touyz, L.Z.G.; Touyz, S.J.J.; Nassani, L.M. Acidic and alcoholic beverages and teeth, with clinical advisories. *EC Dent. Sci.* **2018**, *17*, 1759–1765.
44. Kumar, N.; Amin, F.; Hashem, D.; Khan, S.; Zaidi, H.; Rahman, S.; Farhan, T.; Mahmood, S.J.; Asghar, M.A.; Zafar, M.S. Evaluating the pH of various commercially available beverages in Pakistan: Impact of highly acidic beverages on the surface hardness and weight loss of human teeth. *Biomimetics* **2022**, *7*, 102. [CrossRef]

foods

MDPI

Article

Phytochemicals, Organic Acid, and Vitamins in Red Rhapsody Strawberry—Content and Storage Stability

Hung Trieu Hong [1,2,*], Julius Rami [3], Michael Rychlik [3], Tim J. O'Hare [1] and Michael E. Netzel [1,*]

[1] Queensland Alliance for Agriculture and Food Innovation, The University of Queensland, Coopers Plains, Brisbane, QLD 4108, Australia; t.ohare@uq.edu.au
[2] School of Agriculture and Food Sustainability, The University of Queensland, St Lucia, Brisbane, QLD 4072, Australia
[3] Department of Analytical Food Chemistry, Technical University of Munich, 85354 Freising, Germany; julius.rami@gmail.com (J.R.); michael.rychlik@tum.de (M.R.)
* Correspondence: h.trieu@uq.edu.au (H.T.H.); m.netzel@uq.edu.au (M.E.N.)

Abstract: Strawberries are highly perishable fruits harvested at full ripeness, and their nutritional quality together with their phytochemical composition can be significantly affected by storage duration and temperature. This study investigated the changes in key bioactive compounds, including folate, vitamin C, anthocyanins, quercetin-3-glucoside, ellagic acid, and organic acids, in "Red Rhapsody" strawberries stored at two typical household temperatures (4 °C and 23 °C). While storage duration and temperature did not have a significant impact ($p > 0.05$) on folate content, significant changes in other phytochemicals were observed. The total anthocyanin content increased significantly ($p < 0.05$), from 30.0 mg/100 g fresh weight (FW) at Day 0 to 84.4 mg/100 g FW at Day 7 at 23 °C, a 2.8-fold increase. Conversely, the vitamin C content was significantly reduced ($p < 0.05$), from 54.1 mg/100 g FW at Day 0 to 28.4 mg/100 g FW at Day 7 at 23 °C, while it remained stable at 4 °C. Additionally, the concentrations of quercetin-3-glucoside, ellagic acid, and organic acids underwent significant changes during the storage period. The total folate content fluctuated between 73.2 and 81.6 µg/100 g FW at both temperatures. These results suggest that storage temperature and duration influence the individual phytochemicals and nutrients of strawberries differently, with potential implications for their nutritional value and bioactive compound content.

Keywords: strawberry; folates; vitamin C; anthocyanins; ellagic acid; organic acids; storage stability; nutrition

1. Introduction

Strawberries (*Fragaria ananassa*) are widely consumed for their taste and health benefits. The global market value of strawberries was approximately USD 20.22 billion in 2023 and is expected to reach USD 32.35 billion by 2032 [1]. Strawberries are an important dietary source of natural bioactive compounds, such as anthocyanins, quercetin, ascorbic acid, ellagic acid, and folates, which are associated with various health benefits. Strawberries are particularly recognized as an anthocyanin-rich fruit, with total anthocyanin contents ranging from 24.8 to 101.0 mg/100 g fresh weight (FW) [2]. Anthocyanins in strawberries have been associated with various health-promoting properties, including enhancing heart health [3], possessing antioxidant functions [2], preventing high blood pressure [4], and having positive effects on degenerative diseases and cardiovascular disorders [5]. Strawberries are also a valuable natural source of folates, with folate content ranging from

https://doi.org/10.3390/foods14030379

59 to 153 μg/100 g FW [6,7]. Folates, a group of B vitamins, play crucial roles in various metabolic pathways. Tetrahydrofolate (H4folate) polyglutamates in particular are involved in one-carbon metabolism, which supports nucleotide biosynthesis, the re-methylation of homocysteine to methionine, amino acid production in mitochondria, and DNA replication and repair in the nucleus [8–10]. As the human body cannot synthesize folates, a deficiency can result in neural tube defects in infants [11], as well as decreased methionine levels and increased homocysteine levels, both of which are linked to cardiovascular diseases [12,13] and Alzheimer's disease [14].

Ellagic acid, quercetin, and vitamin C (ascorbic acid) are other important bioactive compounds found in strawberries [15,16]. Ellagic acid is known for its antioxidant properties [17], and its potential in preventing cancer and cardiovascular diseases [18]. Ascorbic acid is essential for collagen biosynthesis [19], acts as an antioxidant [20], and has protective effects against Alzheimer's disease [21] and in cancer treatment [22]. Quercetin is widely recognized for its anti-inflammatory, antioxidant, antiviral, and anticancer properties, as well as its ability to ease cardiovascular diseases [23].

The primary bioactive compounds in strawberries, such as anthocyanins, folates, ascorbic acid, and organic acids, are well known to be highly sensitive to environmental factors, such as temperature, light exposure, and storage [24–26]. Elevated temperatures and extended storage times can result in the degradation of these previously reported health-benefitting compounds. For instance, anthocyanins are particularly unstable under prolonged exposure to heat and light [27]. Similarly, ascorbic acid content in guava juice decreased significantly when stored at higher temperatures [28]. Folates also undergo significant degradation during storage, particularly at higher temperatures or prolonged storage periods [24]. These changes in the phytochemical and micronutrient composition can potentially lead to a loss of the nutritional value of strawberries and their health-promoting potential. Understanding how common storage temperatures affect the stability of these compounds is critical for optimizing postharvest handling and prolonging the shelf life of strawberries without compromising their nutritional quality.

Strawberries are typically harvested at a ripe eating stage and stored under optimal conditions (0 °C) to extend their postharvest lives, as they are highly perishable and delicate fruits [29]. However, strawberries are often stored at refrigerated (4 °C) or room temperature (23 °C) for a few days in retail settings and at home before consumption. Extended cold storage can cause phytochemical changes, as recent studies have reported a significant increase in anthocyanin content (a 1.5-fold increase) when strawberries are stored at 2 °C [30]. How domestic storage conditions affect key phytochemicals and micronutrients, such as anthocyanins, ellagic acid, quercetin-3-glucoside, ascorbic acid and folates, is not well documented. The aim of this study was to evaluate the effect of two common household storage temperatures (4 °C and 23 °C) on key phytochemicals and micronutrients (anthocyanins, ellagic acid, quercetin-3-glucoside, ascorbic acid and folate) in "Red Rhapsody" strawberries, a commercially significant cultivar in Australia, over a 14-day storage period.

2. Material and Methods

2.1. Plant Material

Fresh "Red Rhapsody" strawberries (7 kg) were obtained from a commercial farm in Brisbane, Queensland, Australia (Figure 1). Samples were transported (1 h) to the laboratory at the Health and Food Sciences Precinct, Coopers Plains. Samples were selectively divided into two equal batches (based on the size of fruits, the observation of pigment intensity, and red pigment coverage to make sure all fruits used for the current trial are identical) for storage at 4 °C and 23 °C, respectively. For the control group of samples (0 day), five strawberries were then collected and rapidly frozen using liquid nitrogen to halt any further metabolic processes. All samples were stored at −35 °C until analysis. The remaining samples were placed in commercial plastic punnets, which were loosely sealed in plastic barrier bags and stored in two separate batches: batch 1 at 4 °C and batch 2 at 23 °C, with a humidity level of 90% ± 3%. The humidity level was set based on the moisture content of the strawberry fruit and previous publications [27]. The humidity was monitored using a digital hygrometer thermometer (ThermoPro TP55, Guangdong, China).

Figure 1. Fresh "Red Rhapsody" strawberries were collected from a commercial farm located in Brisbane, Queensland, Australia.

Five strawberries were removed at 1, 7, and 14 days of storage at 4 °C and at 1, 3, and 7 days of storage at 23 °C. These samples were immediately frozen at −35 °C for later analysis. The frozen strawberries were freeze-dried at −50 °C (CSK Climateck, CSK Scientific, Brisbane, QLD, Australia), then ground into a powder using a Waring Laboratory Blender (Australian Scientific, Kotara, NSW, Australia). Powdered samples were stored at −80 °C for later analysis of individual and total anthocyanins, moisture, ascorbic acid, organic acids, folates, ellagic acid, and quercetin-3-glucoside.

2.2. Chemicals

Cyanidin-3-glucoside (Cy3G), pelargonidin-3-glucoside (Pg3G), quercetin-3-glucoside, citric acid, succinic acid, ellagic acid, and ascorbic acid (AA) were sourced from Sigma–Aldrich (Sydney, NSW, Australia) and Extrasynthese (Genay, France). Both the unlabeled and isotope-labeled internal standards were sourced from Professor Michael Rychlik (Analytical Food Chemistry, Technical University of Munich, Freising, Germany). All other chemicals were acquired from Merck (Darmstadt, Germany) or Sigma–Aldrich. Milli-Q water (Millipore Australia Pty Ltd., Kilsyth, VIC, Australia) was used throughout the study unless otherwise stated.

2.3. Methods

2.3.1. Folate Extraction and Determination

A stable isotope dilution assay (SIDA) method was utilized to analyze folate concentrations in strawberry samples stored at different temperatures and durations using a UHPLC-PDA-MS/MS system [7] with some modifications. Briefly, 50–100 mg of freeze-dried strawberry was suspended in 10 mL of MES-buffer (pH 5). A magnetic stirrer bar was placed in the mixtures and left to equilibrate on a magnetic stirrer (IKA C-MAG HS 10, IKA, Breisgau, Germany) for 15 min. The ^{13}C-labeled internal standard (IS), including [^{13}C5]-PteGlu (^{13}C-THF), [$^{13}C_5$]-H_4folate (^{13}C-FA), [$^{13}C_5$]-5-CH_3-H_4folate (^{13}C-5mTHF), and [$^{13}C_5$]-5-CHO-H_4folate (^{13}C-5fTHF), were dissolved in MES-buffer to achieve a final concentration of 60–70 μg/L. These stock solutions were added to the sample mixtures to achieve a final concentration of ~1 μg/L for ^{13}C-5mTHF, 0.05 μg/L for ^{13}C-5fTHF and ^{13}C-THF, and 0.015 μg/L for ^{13}C-FA in the final extraction solutions (2 mL). The resulting mixtures after adding IS were equilibrated for a further 15 min.

The samples were cooked at a boiling temperature of 100 °C for 10 min (Buechi Heating Bath B-491, Buechi AG, Aargau, Switzerland) and cooled down in an ice bath. After reaching room temperature, chicken pancreas (2 mL) and rat serum (300 μL) were added to each sample. The samples were then incubated in a water bath (Thermoline Scientific, Wetherill Park, NSW, Australia) at 37 °C overnight. The samples were subsequently reheated again at 100 °C for 10 min and then cooled in an ice bath. The suspension was transferred to a 50 mL centrifuge tube, diluted with 10 mL acetonitrile and centrifuged at 3900 rpm (Eppendorf centrifuge 5810, Eppendorf AG, Hamburg, Germany) for 20 min. The supernatant was decanted into a fresh 50 mL centrifuge tube and the pellet discarded.

The supernatant (approximate 20 mL) was purified by solid-phase extraction (SPE, Anion-exchange SAX-cartridges, Phenomenex Inc., Lane Cove, NSW, Australia). Folates were eluted from the SPE by using 2 mL buffer including 5% sodium chloride, 100 mM sodium acetate, 0.1 g/L dithiothreitol and 1 g/L ascorbic acid. The final eluate was filtered through a 0.20 μm hydrophilic PTFE membrane for folate analysis by a Shimadzu UHPLC-ESI-MS/MS following the previously instrumental method [6]. Folates were identified and then quantified following the previous study of Striegel, Chebib [6].

2.3.2. Ascorbic Acid Extraction and Determination

Ascorbic acid (AA) in strawberry samples was extracted following the previous report of Campos, Ribeiro [31], with some modifications. Briefly, powdered samples (from 50 to 100 mg) were homogenized with 15 mL of a buffer containing 8% acetic acid, 3% metaphosphoric acid, 1 mM and ethylenediamine tetraacetic acid (EDTA). Dehydroascorbic acid (DH-AA) in the extracts/samples was also converted to AA following the previous method [32].

AA was determined using a Shimadzu Nexera X2 UHPLC system coupled with a Shimadzu MS-8045 TQMS, triple quadrupole mass spectrometer (Shimadzu, Kyoto, Japan), controlled by Lab Solutions software (Version 5.91). The column oven (CTO-20AC) was maintained at 25 °C. The electrospray ionization (ESI) source was operated in the negative ion mode to ionize AA. Multiple reaction monitoring (MRM) transitions for AA (Table 1) were m/z 175.1 → 115.2 at 14 eV(quantifier) and m/z 175.1 → 87.1 at 20 eV(qualifier). The ESI source included a nebulizer gas flow of 3 L/min, a desolvation line (DL) temperature of 250 °C, a drying gas flow of 10 L/min, and a heat block temperature of 400 °C. Data collection was carried out using Shimadzu LabSolutions Insight software for LC-MS (Version 3.2).

Chromatographic separation of AA and other organic acids was carried out on an UPLC column (BEH AcclaimTM C30, 250 × 2.1 mm i.d., 3.0 μm particle size; Thermo Scientific, Waltham, MA, USA), with the column temperature being kept at 25 °C. Two mobile phases were used to elute AA and organic acids, including mobile phase A (MQ-water) and mobile phase B (acetonitrile). The column flow rate was maintained at 0.2 mL/min throughout the analysis. The elution program started with 100% mobile phase A for 5.8 min, followed by a linear increase to 95% mobile phase B for over 6.5 min, purging for 1 min, conditioning for 1 min, and re-equilibration for 3.5 min.

Table 1. MRM scan parameters for citric acid, succinic acid and ascorbic acid (AA).

Compounds	Precursor (m/z)	Product (m/z)	Dwell Time (ms)	Q1 Pre Bias (V)	CE (V)	Q3 Pre Bias (V)
Citric acid	191.0	147.2	58.8	11.0	9.0	14.0
	191.0	110.9	58.8	12.0	9.0	14.0
	191.0	103	58.8	14.0	9.0	14.0
	191.0	86.9	58.8	16.0	13.0	19.0
Succinic acid	117.3	99.1	6.0	14.0	13.0	16.0
	117.3	73.05	6.0	12.0	13.0	26.0
AA	175.1	115.2	47.0	24.0	14.0	28.0
	175.1	87.1	47.0	24.0	20.0	28.0

The linearity was obtained by using concentrations of seven-external standard (AA, citric acid and succinic acid) concentrations over the range 1–40 μg/mL plotted against their peak-area measurements by using LC-MS. The initial AA concentration was subtracted by the total AA (TAA) obtained after a reduction in AA extract with dithiothreitol (DTT) to account for DH-AA contents in all samples.

2.3.3. Organic Acid Extraction and Determination

Organic acids were extracted following the method previously described by Hernandez, Lobo [33,34] with some modifications. In brief, freeze-dried powder strawberry (0.1 g) was extracted with 30 mL of MQ-water using an ultra-sonication bath at room temperature (RT) for 15 min and then a shaker (RP1812, Paton Scientific, Victor Harbor, SA, Australia) at 200 rpm at RT. Samples were then centrifuged (Eppendorf Centrifuge 5804) at 3900 rpm for 10 min at RT. Supernatant was collected, while the sample pellet was re-extracted using the same procedure outlined above. Supernatants were combined and filtered through a 0.20 μm hydrophilic PTFE membrane for organic acid analysis using a Shimadzu UHPLC-ESI-MS/MS.

Citric acid and succinic acid were determined by using the above Shimadzu LC-MS/MS system. The electrospray ionization (ESI) source was employed in the negative mode. The MRM transitions were used at optimal collision energies for succinic acid and citric acid (Table 1).

2.3.4. Ellagic Acid Anthocyanins and Quercetin-3-Glucoside Extraction and Determination

Ellagic acid, anthocyanins, and quercetin-3-glucoside extraction was extracted following the previous method of Hong, Netzel [25], with some modifications. Briefly, approximately 0.2 g of freeze-dried powder was homogenized with 3 mL of cold extraction solution (80% methanol in water, 0.1 M HCl) at 4 °C in a 15 mL Falcon tube. The tubes were sonicated at 4 °C, for 10 min. The mixture was then shaken on a horizontal reciprocating shaker RP 1812 (Paton Scientific, Victor Harbor, SA, Australia) at 250 rpm, 4 °C for 10 min under dim light. The tubes were centrifuged at 4000 rpm, 4 °C for 10 min (Eppendorf Centrifuge 5804). The supernatants were collected, and the pellets were extracted again using the same procedure as outlined above. Supernatants were combined and filtered through a 0.20 μm hydrophilic PTFE membrane for chemical analysis. The extraction was carried out in triplicate (n = 3).

2.3.5. Moisture Content

The moisture contents of strawberry samples were determined in five replicates following the AOAC method 934.01 (AOAC, 1990) described previously [27].

2.4. *Chemical Determination*

Anthocyanins in strawberry samples were identified by LC-DAD-MS following the method previously reported Hong, Netzel [35]. Other compounds were detected by spiking external standards of AA, ellagic acid, citric acid, succinic acid, quercetin-3-glucoside, and folates to determine elution times and to confirm maximum absorbances on a DAD detector, as well as molecular masses and fragment patterns on the mass spectrometer (TQMS, MS-8045) and (MRM) transitions (Table 1).

2.5. Statistical Analysis

The obtained data from the current study were analyzed using a one-way analysis of variance (ANOVA) for each temperature performed by Minitab 17 software for Windows (Minitab Inc, State College, PA, USA). A Tukey's test was employed to evaluate the results of the current study. Statistical differences between means were assessed using the least significant difference (LSD) method, with a significance level set at $p < 0.05$.

3. Results and Discussion

3.1. The Effect of Storage Durations on Folate Content

The total folate content in "Red Rhapsody" strawberries ranged from 74.3 to 81.6 μg/100 g fresh weight (FW). This range is consistent with previous reports on folate content in Australian strawberry cultivars, which ranged from similar values [6,7], and is notably higher than the 4.0–28.0 μg/100 g FW reported in other studies using different methods [36]. The predominant form of folate in strawberries was 5-methyltetrahydrofolate (5mTHF), which accounted for over 90% of the total folate content, with 5mTHF measured at approximately 74.8 μg/100 g FW. Other folate vitamers included tetrahydrofolate (THF), which was present at 2.4 μg/100 g FW, and 5-formyltetrahydrofolate (5fTHF) at 10 μg/100 g FW (Figure 2). Smaller amounts of folic acid (FA) and 10-formyl-folic acid (10fFA) were detected, with concentrations of 0.3 μg/100 g FW for each.

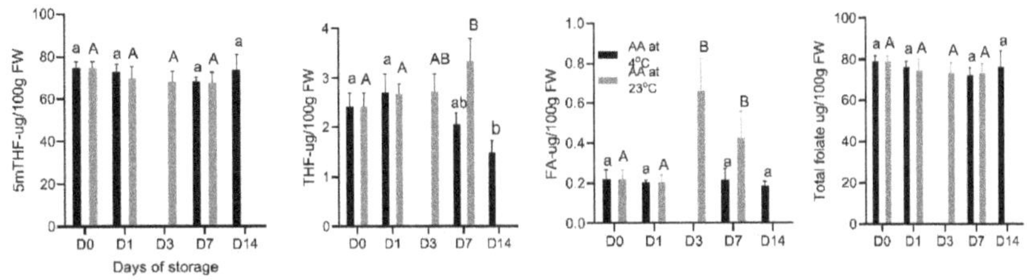

Figure 2. Individual and total folate contents in "Red Rhapsody" strawberries throughout the 14-day storage period at 4 °C and the 7-day storage at 23 °C. Data are presented as mean ± SD (n = 5). Means within columns that are followed by different letters denote significant differences ($p < 0.05$).

Storage conditions, including temperature and duration, did not significantly affect the total folate content in strawberries. At room temperature (23 °C), folate levels showed a slight, non-significant decrease over the 7-day storage period. The initial folate content was 78.7 μg/100 g FW on day 0 (D0), which decreased slightly to 73.2 μg/100 g FW on Day 3 (D3) and to 73.7 μg/100 g FW by Day 7 (D7). At lower temperatures (4 °C), even with storage durations extended up to two weeks, the folate content remained relatively stable at around 73.8 μg/100 g FW (Figure 2). The major folate component, 5mTHF, remained unchanged throughout the storage period regardless of temperature.

However, some changes were observed in other folate vitamers during storage. THF and FA concentrations increased significantly ($p < 0.05$) during the 7-day storage period at 23 °C. THF, in particular, increased from 2.4 μg/100 g FW at D0 to 3.4 μg/100 g FW at D7, while FA increased slightly from 0.2 to 0.5 μg/100 g FW. Notably, THF was the only folate form that showed a significant decrease at 4 °C and a significant increase at 23 °C, suggesting a potential temperature-dependent conversion or degradation process.

These findings highlight the fact that, despite minor fluctuations in individual folate components, the total folate content in strawberries is generally stable across the storage conditions. The increase in THF and FA content during storage at 23 °C suggests the possibility of conversions or interconversions among folate vitamers under high temperatures [37,38], which may be of interest for subsequent studies investigating folate metabolism in postharvest fruits.

3.2. The Effect of Storage Durations on Anthocyanin Content

Storage duration and temperature did not alter the anthocyanin profile in "Red Rhapsody" strawberries. Four anthocyanin components were identified: pelargonidin-3-glucoside (Pg-3-G), pelargonidin-3-rutinoside (Pg-3-R), cyanidin-3-glucoside (Cy-3-G), and pelargonidin-3-malonylglucoside (Pg-3-MG). Pg-3-G was the predominant anthocyanin, followed by Pg-3-R, with Cy-3-G and Pg-3-MG present in all samples regardless of storage period or temperature. The anthocyanin profile observed in this study was consistent with previous reports [2].

Although the anthocyanin components remained unchanged, significant increases in total anthocyanin content were observed with increasing storage durations at both temperatures ($p < 0.05$). At 23 °C, the total anthocyanin content was approximately 30 mg/100 g FW on D0. This value increased significantly ($p < 0.05$) to 51.2 mg/100 g FW on D1 and continued to rise to 76.9 mg/100 g FW on D3, followed by 84.4 mg/100 g FW on D7. At 4 °C, the increase in anthocyanin content was slower, with a significant ($p < 0.05$) rise observed only after 7 days of storage, reaching 41.1 mg/100 g FW on D7 and 44.9 mg/100 g FW on D14 (Figure 3A). The continued accumulation of anthocyanins throughout the storage period suggests that the enzymes involved in anthocyanin synthesis remained active in strawberry fruit tissues at both temperatures. These findings align with previous studies showing ongoing anthocyanin accumulation during the storage of strawberry [39], purple sweetcorn [27], and five primocane raspberry genotypes [40].

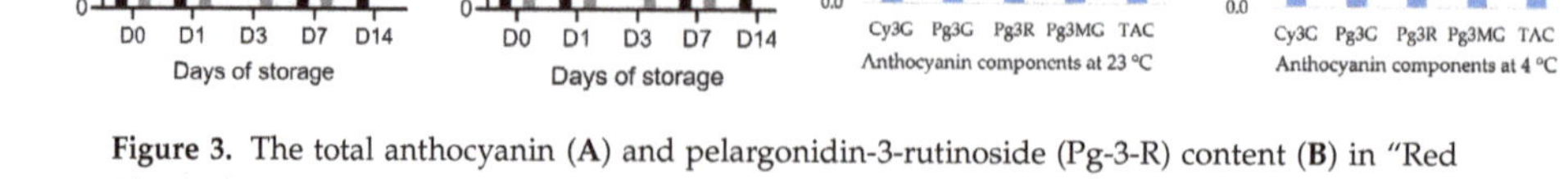

Figure 3. The total anthocyanin (**A**) and pelargonidin-3-rutinoside (Pg-3-R) content (**B**) in "Red Rhapsody" strawberries together with the increasing rates of different anthocyanin components (**C**,**D**) during a 14-day storage period at 4 °C and a 7-day storage period at 23 °C. Data are presented as mean ± SD (n = 5). Means within columns that are followed by different letters denote significant differences ($p < 0.05$).

Similarly to TAC, the concentrations of individual anthocyanin components increased over the storage period, with a faster accumulation rate at 23 °C compared to 4 °C (Figure 3C,D). Among the components, Pg-3-R accumulated at the fastest rate. At 23 °C, its concentration on D7 was approximately 3.6 times higher than on D0 (Figure 3B). At 4 °C, Pg-3-R showed a two-fold increase by D14. In contrast, Cy-3-G exhibited the slowest accumulation rate at 23 °C, with a 1.8-fold increase on D7 compared to its concentrationont D0, while Pg-3-G had the slowest increase at 4 °C, showing only a 1.4-fold increase by D14

(Figure 3C,D). These results indicate that individual anthocyanin components accumulate at different rates depending on the storage temperature and duration.

3.3. The Effect of Storage Durations on Quercetin-3-Glucoside and Free Ellagic Acid Content

The increase in the concentration of free ellagic acid during the storage of strawberry fruits at both temperatures indicates that storage duration significantly affects free ellagic acid content (Figure 4). The free ellagic acid content was approximately 1.8 mg/100 g FW at D0, a value consistent with previous studies about free ellagic acid content in strawberries [15]. This concentration increased significantly to 2.2 at D1, 2.4 at D3, and 4.2 mg/100 g FW on D7 at 23 °C. At 4 °C, however, only a significant increase in the concentration of free ellagic acid was observed at D14, reaching about 3.0 mg/100 g FW. To the best of our knowledge, this is the first report documenting an increase in free ellagic acid in strawberry fruits during storage at both temperatures. The reason for this increase in free ellagic acid is currently unclear. However, it is noteworthy that the total ellagic acid content in strawberries is more than 10 times higher than the free ellagic acid content [15]. Therefore, it is possible that storage conditions led to the conversion of other forms of ellagic acid into free ellagic acid.

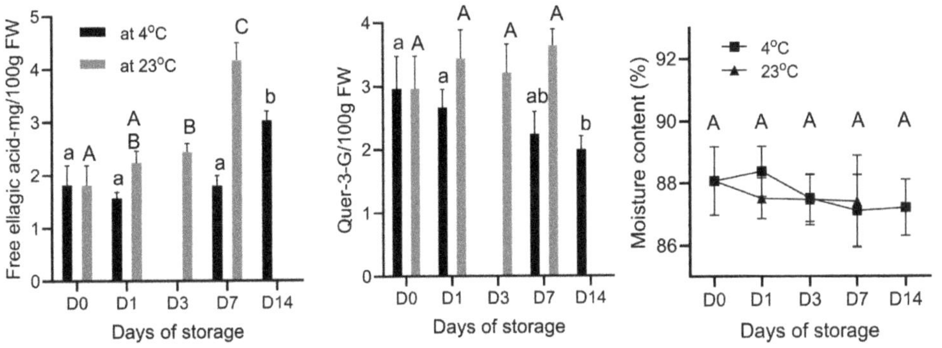

Figure 4. Free ellagic acid, quercetin-3-glucoside, and moisture content in "Red Rhapsody" strawberry fruits during a 14-day storage period at 4 °C and a 7-day storage period at 23 °C. Data are presented as mean ± SD (n = 5). Means within columns that are followed by different letters denote significant differences ($p < 0.05$).

In contrast to free ellagic acid, the concentration of quercetin-3-glucoside (Quer-3-G) showed differing trends depending on storage temperature (Figure 4). While the concentration of Quer-3-G did not change significantly at 23 °C, a significant decrease was observed at 4 °C. Specifically, the concentration of Quer-3-G was 2.9 mg/100 g FW at D0 (consistent with previous findings by da Silva Pinto, Lajolo [15]), but it decreased to 2.0 mg/100 FW on D14. This decrease may be attributed to the continued activity of enzymes responsible for the biosynthesis of anthocyanins from quercetin-3-glucoside at 4 °C, which could lead to the utilization of Quer-3-G in the production of anthocyanins. However, the biosynthesis of quercetin-3-glucoside appears to be slower or inactive at this lower temperature.

The moisture content of strawberry fruit at both storage temperatures was not statistically significant ($p > 0.05$). These results suggest that the storage conditions effectively maintained the moisture content in the strawberries. Given the stability of moisture content, the observed changes in phytochemicals, organic acids, and vitamins in the strawberries reflect the physiological processes occurring in the fruit during storage rather than being influenced by moisture changes.

3.4. *The Effect of Storage Durations on Ascorbic Acid Content*

Storage duration had significant effects on the free ascorbic acid (AA), total ascorbic acid (TAA), and dehydroascorbic acid (DH-AA) content in strawberry fruit at both storage temperatures. Free AA concentration remained stable over the two-week storage period at 4 °C. However, at 23 °C, the concentration of free AA decreased significantly from 54.1 mg/100 g FW at D0 to 37.3 mg/100 g FW on D3, and it further declined to 28.4 mg/100 g FW on D7 (Figure 5A).

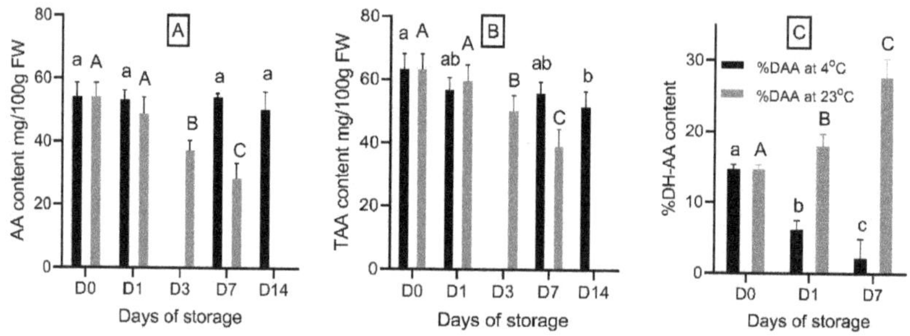

Figure 5. Free AA (**A**), TAA (**B**) content and percentage of DH-AA relative to total AA (**C**) in "Red Rhapsody" strawberry fruits during a 14-day storage period at 4 °C and a 7-day storage period at 23 °C. Data are presented as mean ± SD (n = 5). Means within columns that are followed by different letters denote significant differences ($p < 0.05$).

TAA also decreased at a faster rate at 23 °C than at 4 °C. In fresh strawberry fruit, TAA was measured at 63.3 mg/100 g FW, but it underwent a significant decline to 50.2 mg/100 g FW by D3 and 39.1 mg/100 g FW by D7 at 23 °C. In contrast, at 4 °C, only a significant reduction in TAA was observed after 14 days of storage, reaching 51.8 mg/100 g FW (Figure 5B). These findings are consistent with previous fruit studies that have reported a reduction in AA content during the storage of various fruits, such as oranges and pineapples stored at 10 °C and 20 °C [41], as well as strawberry fruit [42] and pears [43]. The degradation of AA during storage may help protect folates and other unstable compounds in strawberry fruit.

Figure 5C further supports the significant decrease in free AA and TAA content (Figure 5A,B). The percentage of DH-AA in strawberry fruit increased significantly ($p < 0.05$) from 14.6% at D0 to 17.9% on D1 and 27.4% on D7 at 23 °C, indicating that a greater proportion of free AA converted to DH-AA during storage at this temperature. This conversion was reflected in the significant decrease in free AA content observed in Figure 5A at 23 °C. In contrast, at 4 °C, the percentage of DH-AA decreased over the storage period (Figure 5C), suggesting that DH-AA is less stable than free AA in strawberry fruit and undergoes degradation over time, even at lower temperatures. Meanwhile, the free AA concentration remained relatively stable at 4 °C (Figure 4). To our knowledge, this study is the first to show that the relative amount of DH-AA increased during storage at 23 °C and decreased at 4 °C in strawberries. These results indicate that cold storage inhibits the conversion of free AA to DH-AA.

3.5. *The Effect of Storage Durations on Organic Acid Content*

Similarly to AA, storage duration had different effects on the citric acid (the principle organic acid) and succinic acid in strawberry fruits at both temperatures (Figure 6). The concentration of citric acid decreased significantly after three days of storage at 23 °C, from 743.7 at D0 to 618.1 at D3, followed by 484.3 mg/100 g FW at D7. In contrast, citric acid concentration remained stable over the two-week storage period at 4 °C. This finding is consistent with previous studies that observed a decrease in citric acid content as strawberries ripen [44].

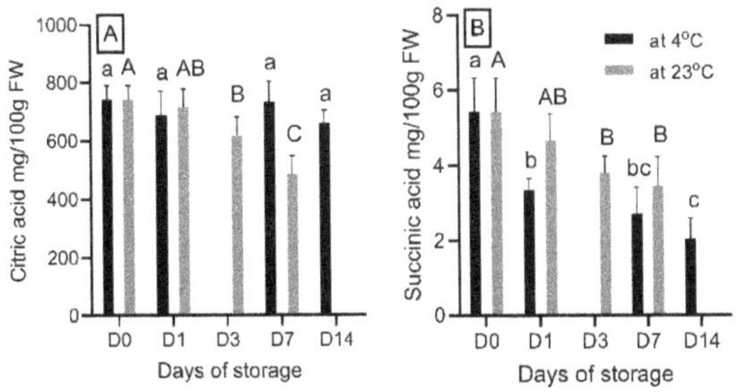

Figure 6. Citric acid (**A**), and succinic acid (**B**) content in "Red Rhapsody" strawberry fruits during a 14-day storage period at 4 °C and a 7-day storage period at 23 °C. Data are presented as mean ± SD (n = 5). Means within columns that are followed by different letters denote significant differences ($p < 0.05$).

Succinic acid, a metabolic product of citric acid in ripening fruits [45], decreased in concentration at both storage temperatures, although the decrease occurred at a faster rate at 4 °C. At 23 °C, the succinic acid content significantly decreased from approximately 5.4 mg/100 g FW at D0 to 3.8 mg/100 g FW at D3, followed by 3.4 mg/100 g FW at D7. At 4 °C, the succinic acid contents were much lower, with values of 3.3 mg/100 g FW at D1, 2.7 mg/100 g FW at D7, and 2.0 mg/100 g FW at D14. These results suggest that succinic acid levels decreased over the storage period; however, the loss was partly compensated by the degradation of citric acid into succinic acid at 23 °C.

4. Conclusions

The results of the current study demonstrated that refrigerated storage (4 °C) helped preserve the concentrations of key phytochemicals and micronutrients in strawberries, maintaining the nutritional quality over a two-week storage period. In contrast, storage at 23 °C led to significant increases in certain beneficial compounds, such as total anthocyanins and ellagic acid, which may enhance the nutritional value and antioxidant properties of strawberries. However, the higher temperatures also led to significant losses of ascorbic acid and organic acids, both of which play crucial roles in the fruit's sensory attributes and overall nutritional quality. These findings suggest that, while refrigeration effectively preserves the nutritional profile of strawberries, room temperature storage may improve certain bioactive compounds, though it may also cause nutrient degradation that could negatively impact the overall quality. Therefore, the optimal storage condition for strawberries depends on the desired balance between maintaining overall nutritional quality and enhancing specific bioactive compounds.

Author Contributions: Conceptualization, H.T.H., M.E.N., M.R. and T.J.O.; methodology, H.T.H., M.R. and M.E.N.; validation, H.T.H.; formal analysis, J.R and H.T.H.; investigation, J.R and H.T.H.; resources, M.E.N.; data curation, J.R. and H.T.H.; writing—original draft preparation, H.T.H.; writing—review and editing, J.R., H.T.H., T.J.O., M.R. and M.E.N.; supervision, T.J.O. and M.E.N.; project administration, M.E.N. and T.J.O.; funding acquisition, M.E.N. and T.J.O. All authors have read and agreed to the published version of the manuscript. author contributions.

Funding: This research was funded by Horticulture Innovation Australia Ltd., "Naturally Nutritious" project HN15001.

Institutional Review Board Statement: Not applicable.

Informed Consent Statement: Not applicable.

Data Availability Statement: The original contributions presented in this study are included in the article. Further inquiries can be directed to the corresponding authors.

Conflicts of Interest: The authors declare no conflicts of interest.

References

1. Zion Market Research. Strawberry Industry Prospective. 2024. Available online: https://www.zionmarketresearch.com/report/strawberry-market (accessed on 28 December 2024).
2. Fredericks, C.H.; Fanning, K.J.; Gidley, M.J.; Netzel, G.; Zabaras, D.; Herrington, M.; Netzel, M. High-anthocyanin strawberries through cultivar selection. *J. Sci. Food Agric.* **2013**, *93*, 846–852. [CrossRef]
3. Basu, A.; Rhone, M.; Lyons, T.J. Berries: Emerging impact on cardiovascular health. *Nutr. Rev.* **2010**, *68*, 168–177. [CrossRef]
4. Cassidy, A.; O'Reilly, E.J.; Kay, C.; Sampson, L.; Franz, M.; Forman, J.P.; Curhan, G.; Rimm, E.B. Habitual intake of flavonoid subclasses and incident hypertension in adults. *Am. J. Clin. Nutr.* **2011**, *93*, 338–347. [CrossRef] [PubMed]
5. de Pascual-Teresa, S.; Moreno, D.A.; Garcia-Viguera, C. Flavanols and anthocyanins in cardiovascular health: A review of current evidence. *Int. J. Mol. Sci.* **2010**, *11*, 1679–1703. [CrossRef]
6. Striegel, L.; Chebib, S.; Netzel, M.E.; Rychlik, M. Improved Stable Isotope Dilution Assay for Dietary Folates Using LC-MS/MS and Its Application to Strawberries. *Front. Chem.* **2018**, *6*, 11. [CrossRef]
7. Rami, J.; Dumler, C.; Weber, N.; Rychlik, M.; Netzel, G.; Hong, H.T.; Wright, O.; O'Hare, T.J.; Netzel, M.E. Folate in Red Rhapsody Strawberry—Content and Storage Stability. *Proceedings* **2021**, *70*, 47.
8. Schirch, V.; Strong, W.B. Interaction of folylpolyglutamates with enzymes in one-carbon metabolism. *Arch. Biochem. Biophys.* **1989**, *269*, 371–380. [CrossRef]
9. Selhub, J. Folate, vitamin B12 and vitamin B6 and one carbon metabolism. *J. Nutr. Health Aging* **2002**, *6*, 39–42. [PubMed]
10. Shane, B. Folylpolyglutamate synthesis and role in the regulation of one-carbon metabolism. *Vitam. Horm.* **1989**, *45*, 263–335. [CrossRef] [PubMed]
11. Van der Put, N.M.; Blom, H.J. Neural tube defects and a disturbed folate dependent homocysteine metabolism. *Eur. J. Obs. Gynecol. Reprod. Biol.* **2000**, *92*, 57–61. [CrossRef] [PubMed]
12. Steegers-Theunissen, R.P.; Wathen, N.C.; Eskes, T.K.; van Raaij-Selten, B.; Chard, T. Maternal and fetal levels of methionine and homocysteine in early human pregnancy. *Br. J. Obs. Gynaecol.* **1997**, *104*, 20–24. [CrossRef]
13. Robinson, K. Homocysteine, B vitamins, and risk of cardiovascular disease. *Heart* **2000**, *83*, 127–130. [CrossRef]
14. Clarke, R.; Smith, A.D.; Jobst, K.A.; Refsum, H.; Sutton, L.; Ueland, P.M. Folate, vitamin B12, and serum total homocysteine levels in confirmed Alzheimer disease. *Arch. Neurol.* **1998**, *55*, 1449–1455. [CrossRef]
15. da Silva Pinto, M.; Lajolo, F.M.; Genovese, M.I. Bioactive compounds and quantification of total ellagic acid in strawberries (*Fragaria x ananassa* Duch.). *Food Chem.* **2008**, *107*, 1629–1635. [CrossRef]
16. Cruz-Rus, E.; Amaya, I.; Sánchez-Sevilla, J.F.; Botella, M.A.; Valpuesta, V. Regulation of L-ascorbic acid content in strawberry fruits. *J. Exp. Bot.* **2011**, *62*, 4191–4201. [CrossRef]
17. Priyadarsini, K.I.; Khopde, S.M.; Kumar, S.S.; Mohan, H. Free radical studies of ellagic acid, a natural phenolic antioxidant. *J. Agric. Food Chem.* **2002**, *50*, 2200–2206. [CrossRef] [PubMed]
18. Vattem, D.A.; Shetty, K. Biological functionality of ellagic acid: A review. *J. Food Biochem.* **2005**, *29*, 234–266. [CrossRef]
19. Choi, H.I.; Park, J.I.; Kim, H.J.; Kim, D.W.; Kim, S.S. A novel L-ascorbic acid and peptide conjugate with increased stability and collagen biosynthesis. *BMB Rep.* **2009**, *42*, 743–746. [CrossRef] [PubMed]
20. Yen, G.C.; Duh, P.D.; Tsai, H.L. Antioxidant and pro-oxidant properties of ascorbic acid and gallic acid. *Food Chem.* **2002**, *79*, 307–313. [CrossRef]
21. Bowman, G.L. Ascorbic acid, cognitive function, and Alzheimer's disease: A current review and future direction. *Biofactors* **2012**, *38*, 114–122. [CrossRef]
22. Du, J.; Cullen, J.J.; Buettner, G.R. Ascorbic acid: Chemistry, biology and the treatment of cancer. *Biochim. Biophys. Acta* **2012**, *1826*, 443–457. [CrossRef]
23. Wang, W.Y.; Sun, C.X.; Mao, L.K.; Ma, P.H.; Liu, F.G.; Yang, J.; Gao, Y.X. The biological activities, chemical stability, metabolism and delivery systems of quercetin: A review. *Trends Food Sci. Technol.* **2016**, *56*, 21–38. [CrossRef]
24. Gazzali, A.M.; Lobry, M.; Colombeau, L.; Acherar, S.; Azaïs, H.; Mordon, S.; Arnoux, P.; Baros, F.; Vanderesse, R.; Frochot, C. Stability of folic acid under several parameters. *Eur. J. Pharm. Sci.* **2016**, *93*, 419–430. [CrossRef] [PubMed]
25. Hong, H.T.; Netzel, M.E.; O'Hare, T.J. Optimisation of extraction procedure and development of LC-DAD-MS methodology for anthocyanin analysis in anthocyanin-pigmented corn kernels. *Food Chem.* **2020**, *319*, 126515. [CrossRef] [PubMed]
26. Yin, X.; Chen, K.; Cheng, H.; Chen, X.; Feng, S.; Song, Y.; Liang, L. Chemical stability of ascorbic acid integrated into commercial products: A review on bioactivity and delivery technology. *Antioxidants* **2022**, *11*, 153. [CrossRef]

27. Hong, H.T.; Phan, A.D.T.; O'Hare, T.J. Temperature and Maturity Stages Affect Anthocyanin Development and Phenolic and Sugar Content of Purple-Pericarp Supersweet Sweetcorn during Storage. *J. Agric. Food Chem.* **2021**, *69*, 922–931. [CrossRef]
28. Siriwoharn, T.; Surawang, S. Protective effect of sweet basil extracts against vitamin C degradation in a model solution and in guava juice. *J. Food Process. Preserv.* **2018**, *42*, e13646. [CrossRef]
29. Milena Guajardo, C.M. Marisa Bunning. Strawberries. 2021. Available online: https://fsi.colostate.edu/strawberries (accessed on 19 November 2021).
30. Virgen-Ortiz, J.J.; Morales-Ventura, J.M.; Colin-Chavez, C.; Esquivel-Chavez, F.; Vargas-Arispuro, I.; Aispuro-Hernandez, E.; Martinez-Tellez, M.A. Postharvest application of pectic-oligosaccharides on quality attributes, activities of defense-related enzymes, and anthocyanin accumulation in strawberry. *J. Sci. Food Agric.* **2020**, *100*, 1949–1961. [CrossRef]
31. Campos, F.M.; Ribeiro, S.M.R.; Della Lucia, C.M.; Pinheiro-Sant'Ana, H.M.; Stringheta, P.C. Optimization of Methodology to Analyze Ascorbic and Dehydroascorbic Acid in Vegetables. *Quim. Nova* **2009**, *32*, 87–91. [CrossRef]
32. Spinola, V.; Mendes, B.; Camara, J.S.; Castilho, P.C. An improved and fast UHPLC-PDA methodology for determination of L-ascorbic and dehydroascorbic acids in fruits and vegetables. Evaluation of degradation rate during storage. *Anal. Bioanal. Chem.* **2012**, *403*, 1049–1058. [CrossRef]
33. Hernandez, Y.; Lobo, M.G.; Gonzalez, M. Factors affecting sample extraction in the liquid chromatographic determination of organic acids in papaya and pineapple. *Food Chem.* **2009**, *114*, 734–741. [CrossRef]
34. Moldoveanu, S.C.; Poole, T.; Scott, W.A. An LC-MS Method for the Analysis of Some Organic Acids in Tobacco Leaf, Snus, and Wet Snuff. *Beiträge Zur. Tab. Int./Contrib. Tob. Res.* **2018**, *28*, 30–41. [CrossRef]
35. Hong, H.T.; Netzel, M.E.; O'Hare, T.J. Anthocyanin composition and changes during kernel development in purple-pericarp supersweet sweetcorn. *Food Chem.* **2020**, *315*, 126284. [CrossRef] [PubMed]
36. Basu, A.; Nguyen, A.; Betts, N.M.; Lyons, T.J. Strawberry as a functional food: An evidence-based review. *Crit. Rev. Food Sci. Nutr.* **2014**, *54*, 790–806. [CrossRef] [PubMed]
37. Gorelova, V.; Bastien, O.; De Clerck, O.; Lespinats, S.; Rébeillé, F.; Van Der Straeten, D. Evolution of folate biosynthesis and metabolism across algae and land plant lineages. *Sci. Rep.* **2019**, *9*, 5731. [CrossRef] [PubMed]
38. Revuelta, J.L.; Serrano-Amatriain, C.; Ledesma-Amaro, R.; Jiménez, A. Formation of folates by microorganisms: Towards the biotechnological production of this vitamin. *Appl. Microbiol. Biotechnol.* **2018**, *102*, 8613–8620. [CrossRef]
39. Kalt, W.; Prange, R.K.; Lidster, P.D. Postharvest Color Development of Strawberries—Influence of Maturity, Temperature and Light. *Can. J. Plant Sci.* **1993**, *73*, 541–548. [CrossRef]
40. Palonen, P.; Weber, C. Fruit color stability, anthocyanin content, and shelf life were not correlated with ethylene production rate in five primocane raspberry genotypes. *Sci. Hortic.* **2019**, *247*, 9–16. [CrossRef]
41. Adisa, V.A. The Influence of Molds and Some Storage Factors on the Ascorbic-Acid Content of Orange and Pineapple Fruits. *Food Chem.* **1986**, *22*, 139–146. [CrossRef]
42. Nunes, M.C.N.; Brecht, J.K.; Morais, A.M.M.B.; Sargent, S.A. Controlling temperature and water loss to maintain ascorbic acid levels in strawberries during postharvest handling. *J. Food Sci.* **1998**, *63*, 1033–1036. [CrossRef]
43. Lee, S.K.; Kader, A.A. Preharvest and postharvest factors influencing vitamin C content of horticultural crops. *Postharvest Biol. Technol.* **2000**, *20*, 207–220. [CrossRef]
44. Fecka, I.; Nowicka, A.; Kucharska, A.Z.; Sokol-Letowska, A. The effect of strawberry ripeness on the content of polyphenols, cinnamates, L-ascorbic and carboxylic acids. *J. Food Compos. Anal.* **2021**, *95*, 103669. [CrossRef]
45. Cercos, M.; Soler, G.; Iglesias, D.J.; Gadea, J.; Forment, J.; Talon, M. Global analysis of gene expression during development and ripening of citrus fruit flesh. A proposed mechanism for citric Acid utilization. *Plant Mol. Biol.* **2006**, *62*, 513–527. [CrossRef]

beverages

MDPI

Article

Untargeted Metabolomic Analysis Using High-Resolution Orbitrap Mass Spectrometry for the Comparison of Volatile and Non-Volatile Compounds in Hot and Cold Brew Coffee

Seongeung Lee *, Eunmee Han, Jisun Kang, Seohee Kwon, Minkyung Sung, Minkyoung Kim, Hyeokjun Cho and Gyeonghweon Lee

Lotte R&D Center, 201, Magokjungang-ro, Gangseo-gu, Seoul 07594, Republic of Korea; eunmee.han@lotte.net (E.H.); jisun.kang@lotte.net (J.K.); shkwon8x@lotte.net (S.K.); minkyung.sung@lotte.net (M.S.); minkyoung.kim@lotte.net (M.K.); hyeokjun_cho@lotte.net (H.C.); ghlee1@lotte.net (G.L.)

* Correspondence: sungoung.lee@lotte.net

Abstract: Coffee contains several bioactive compounds, such as alkaloids and phenolic compounds, which contribute to its flavor and are influenced by the brewing method. The differences in coffee compounds based on brewing conditions have been studied in previous research, but no studies have yet utilized orbitrap mass spectrometry for this purpose. This study compared non-volatile and aromatic compounds in hot and cold brew coffee using high-resolution orbitrap mass spectrometry, followed by multivariate statistical analysis including principal component analysis and volcano plotting. A total of 163 non-volatile compounds and 93 volatile compounds were identified and annotated, with 18 non-volatile and 13 aroma-active compounds indicating differences between the brewing methods. Notably, certain quinic acids, such as 4,5-dicaffeoylquinic acid, and coumarin derivatives were more abundant in hot brew coffee, indicating that non-volatile compounds are significantly affected by extraction temperature. However, the major non-volatile compounds, including chlorogenic acid and trigonelline, are not affected by brewing conditions. For volatile compounds, phenolic compounds and indole were sensitive to temperature, while pyrazine and furan compounds were more influenced by extraction time. Additionally, in our results, several previously unreported bioactive compounds were detected in coffee, suggesting a need for further research to understand their potential functions and benefits.

Keywords: hot brew coffee; cold brew coffee; volatile; non-volatile; untargeted; high-resolution; orbitrap

1. Introduction

Coffee, a beverage that is well-known and easily accessible around the world, has experienced steady growth in popularity and production over recent decades. In the 2024/25 period, coffee bean production reached 176.2 million 60 kg bags, consistently remaining within the range of 165 to 175 million bags [1]. Coffee contains several bioactive compounds, such as alkaloids and phenolic compounds, including caffeine, chlorogenic acid, and trigonelline. Some of these compounds are associated with various health benefits, such as antihypertensive, hypoglycemic, and anticancer effects [2–4]. These compounds are also related to the flavor and taste of coffee. The flavor profile of coffee is mainly influenced by its chemical components, where non-volatile compounds affect its taste, while volatile compounds affect its aroma. The flavor profile of coffee can be affected by various factors, such as geographical characteristics, processing methods, roasting degree, decaffeination, and brewing methods, all of which ultimately influence consumer choice [5]. For example, variations in diterpene composition have been observed among different *Coffea arabica* cultivars, and sensory analyses have indicated that the characteristics of brewed coffee differed based on cultivation altitude [6,7]. The brewing process, recognized as the final step before coffee consumption, has garnered increasing attention from food manufacturers. Coffee can be categorized based on brewing temperature into traditional hot brew and cold brew coffee, with the latter being extracted using water at approximately 25 degrees Celsius. In recent years, the coffee industry has experienced notable shifts. As consumer demand for premium coffee continues to grow, cold brew has gained significant popularity. According to market analysis, the cold brew coffee segment was valued at USD 321 million in 2017 and is projected to reach USD 1.37 billion by 2024 [1,8]. Brewing parameters, such as grind size, extraction time, pressure, and temperature, are crucial factors influencing the sensory quality and flavor perception of coffee [8]. With the expansion of the cold brew coffee market and the growing interest in this beverage, scholarly research comparing the characteristics and differences between cold brew and hot brew coffee has progressively increased. These studies focus on aspects such as flavor, aroma, and chemical compounds like metabolites. According to Gloess et al. [9] and Batali et al. [10], hot brew coffee exhibited a roasted and smoky aroma, whereas cold brew coffee demonstrated fruity, floral, and sweet notes. The brewing method significantly influences the levels of non-volatile bioactive compounds in coffee, such as caffeine and phenolic acids, which are associated with the purported health benefits of coffee beans, including their antioxidant and anti-tumor effects [11–13]. Additionally, a previous study analyzing volatile compounds in medium-roasted Colombian Arabica coffee prepared through both cold brewing and hot brewing methods found that cold brew coffee exhibited a greater concentration of pyrazines and furans compared to hot brew coffee [14].

Metabolomics is a scientific approach used to quantitatively and qualitatively describe the overall levels of endogenous biological metabolites and their responses to external and internal factors. This method is renowned for its high efficiency, precision, and comprehensiveness, making it widely applicable in fields such as plant studies [15], pharmaceuticals [16], and food science [17–19]. This approach is frequently employed across various disciplines for both qualitative and quantitative analysis of metabolites from whole organisms, tissues, and cells, with the aim of identifying key differential metabolites. Untargeted metabolomics, coupled with high-resolution mass spectrometry (HR-MS), is a powerful tool for comparative analyses, providing comprehensive insights into the metabolic profiles of different samples, such as cold and hot brew coffee.

Orbitrap mass spectrometry (OMS), a type of HR-MS, is especially valuable for food authentication due to its improved dynamic range, simplified mass calibration, and tandem mass spectrometry capabilities [20]. Also, OMS exhibits high mass accuracy, with performance levels of 5 ppm under standard calibration and 3 ppm when using an internal standard. It allows for streamlined sample pretreatment, facilitating quicker and more efficient identification and quantification in complex matrices [21]. Thus, OMS has proven to be a highly sensitive and selective method, significantly boosting analytical efficiency, and to be especially valuable for analyzing substances, such as metabolites, newly emerging contaminants, and transformation products [22].

Several studies have analyzed untargeted chemical compounds in coffee using HR-MS, including time-of-flight mass spectrometry (TOF) and other mass spectrometry techniques, including triple quadrupole (QQQ) with liquid (LC) and gas chromatography (GC). For instance, Vezzulli et al. [23] reported on the metabolite differences between coffee species such as *Coffea arabica* and *Coffea canephora* var. *Robusta* using LC-QTOF analysis. Pérez Míguez et al. [24] examined the variation in metabolic compounds in coffee depending on the roasting process, also using LC-QTOF analysis. Additionally, Cai et al. [25] presented comparative profiling data of both volatile and non-volatile compounds in hot and cold brew coffee using LC-Q-ion trap mass spectrometry and GC-MS, respectively. However, there is a lack of comprehensive metabolomic studies utilizing HR-OMS technology to investigate the effects of different brewing methods on coffee.

Therefore, the objective of this study was to compare non-volatile compounds, such as alkaloids and phenolic compounds, in hot and cold brew coffee through untargeted metabolomics analysis using LC-HR-OMS. Additionally, aromatic compounds, such as furans and pyrazines, were analyzed using GC-HR-OMS, followed by multivariate statistical analysis. Also, unlike other studies, this research standardized the extraction time across all conditions to accurately assess the impact of extraction temperatures. The results of this study are expected to provide a clearer understanding of the differences between hot brew and cold brew coffee and facilitate a comparison of analytical data obtained from orbitrap and other HR-MS techniques.

2. Materials and Methods

2.1. Materials

Green coffee beans produced in Ethiopia were provided for our research by Lotte Chilsung Beverage Co., Ltd. (Seoul, Republic of Korea). To analyze untargeted metabolites in hot and cold brew coffee using HR-OMS, high-performance liquid chromatography (HPLC)-grade acetonitrile and methanol were purchased from J.T. Baker (Phillipsburg, NJ, USA). Other reagents, including ammonium formate and formic acid, were purchased from Sigma-Aldrich Chemical Co. (St. Louis, MO, USA).

2.2. Preparing Coffee

The green coffee beans were roasted at 190–210 °C for 30 min using a roasting machine (Model Probat BRZ 2, Probat, Emmerich am Rhein, Germany) and then ground with a semi-automatic grinder (Model K32S20, Mahlkonig GmbH & Co. KG, Hamburg, Germany) with the dial set to 4. The ground coffee was then sieved to obtain particles sized between 50 and 100 mesh, which were used for both hot and cold brew coffee preparation. For hot brew coffee, a total of 24 g of ground coffee was placed in a coffee filter paper. Approximately 300 g of purified water was then poured directly over the filter, and heated to brew the coffee. The coffee extract was released at a temperature of 95 °C, and the brewing process took approximately 3 min until the extraction volume reached 240 mL. For cold brew coffee, a total of 24 g of ground coffee was placed in a coffee filter. About 300 g of purified water was then poured directly over the filter. The coffee extract was released at a temperature of 25 °C, and the brewing process took approximately 3 min, until the extraction volume reached 240 mL.

2.3. Analysis of Non-Volatile Compounds

For analyzing non-volatile compounds in hot and cold brew coffee, a modified version of the method by Cai et al. and Li et al. [25,26] was followed. Briefly, the hot and cold brew coffee was vigorously shaken for 1 min before sampling. A 500 μL aliquot of coffee was mixed with 500 μL of 80% methanol solution, and then the mixture was vortexed for 10 min. Following centrifugation at 4000 rpm for 5 min using a centrifuge (Model Combi R515, Hanil Scientific, Gimpo, Republic of Korea), the supernatant was filtered through a 0.2 μm syringe filter (Sartorius, Göttingen, Germany). Finally, 1 μL of the sample was injected into the LC-HR-OMS system (Model Q-Exactive, Thermo Scientific, Waltham, MA,

USA) featuring an electrospray ionization (ESI) source. Chromatographic separation was performed using a C18 column (Zorbax Eclipse Plus, 2.1 × 100 mm, particle size 100 μm, Agilent, Santa Clara, CA, USA) maintained at 30 °C. The separation was conducted using mobile phases comprising water with 0.1% formic acid and 5 mM ammonium formate (solvent A) and 90% acetonitrile with 0.1% formic acid and 5 mM ammonium formate (solvent B), which were eluted following a specified gradient profile: 5–60% solvent B over 10 min, 60–100% solvent B over 5 min, 100–5% solvent B over 2 min, and then maintaining 5% solvent B for 3 min, resulting in a total analysis time of 20 min. The ESI probe temperature was set to 250 °C, the spray voltage to 3 kV, the ion transfer line temperature to 300 °C, the sheath gas pressure to 35 psi, and the auxiliary gas flow to 10 (arbitrary units). Full MS and data-dependent MS^2 ($ddMS^2$)/top N (N = 4) scan modes were conducted in both positive and negative ion monitoring modes, respectively, with the following parameters: a resolution of 80,000 for full MS in the m/z range of 100 to 1000, a maximum trap filling time of 80 milliseconds, a resolution of 35,000 for fragment scans, and collision energies of 20, 40, and 60 eV with nitrogen as the collision gas.

2.4. Analysis of Volatile Compounds

For analyzing volatile compounds in hot and cold brew coffee using solid-phase microextraction (SPME), a modified version of the method by Pan et al. [27]. was followed. A 4 mL coffee sample was placed into a 20 mL headspace vial containing 2.5 g of NaCl. The vial was closed with a PTFE-silicone septum and equilibrated at 60 °C for 10 min with continuous agitation. Subsequently, a 50/30 μm divinylbenzene/carboxen™/polydimethylsiloxane (DVB/CAR/PDMS) SPME fiber was exposed to the coffee headspace for 45 min at the same temperature. The fiber was then removed and inserted into the gas chromatography injector at 250 °C for 5 min.

The GC-HR-OMS (Model Q-Exactive GC, Thermo Scientific, Waltham, MA, USA) analysis was conducted following the procedure outlined by Heo et al. [28]. Volatile compounds were separated using DB-WAX fused silica capillary columns (30 m × 0.25 mm I.D., 0.25 μm film thickness, Agilent, Santa Clara, CA, USA). Helium was used as the carrier gas at a constant flow rate of 1 mL/min. The oven temperature was initially maintained at 40 °C for 10 min, then increased at a rate of 8 °C/min to 260 °C, and then maintained at that temperature for 10 min. Mass spectrometry was performed in the electron impact mode at 70 eV, scanning a range of m/z 35–500 at a resolution of 60,000.

2.5. Data Processing and Multivariate Statistical Analysis

The raw data signals (n = 5) from HR-OMS were processed using Compound Discoverer, software version 3.3 (Thermo Scientific, Waltham, MA, USA). All statistical analyses were conducted based on the intensity area under the peaks for each qualitatively identified molecular mass. To conduct qualitative analysis, LC-HR-OMS raw data were identified by matching the mass spectrum and tandem mass spectrum with the mzCloud and mzVault databases in Compound Discoverer. Additionally, GC-HR-OMS raw data were identified by matching the mass spectrum with the NIST 17 standard database in Compound Discoverer after deconvolution. To visualize the differences between hot and cold brew coffee, principal component analysis (PCA) and volcano plotting were conducted using Compound Discoverer software. The PCA reduces the dimension of data by transforming it into a smaller set of uncorrelated variables to highlight key patterns. Volcano plotting is a graphical representation that shows the relationship between the fold change in log scale and statistical significance calculated by Student's t-test with the data.

3. Results

3.1. *Non-Volatile Compounds in Hot and Cold Brew Coffee*

Metabolic profiling using LC-HR-OMS was performed on hot and cold brew coffee. Compound Discoverer conducted peak alignment and baseline correction, resulting in the identification of approximately 1400 peaks in the positive ion mode and 1300 peaks in the negative ion mode in brewed coffee samples. The process of annotation followed the methods of Souza et al. [29] and Sukor et al. [30] with some modifications The analysis commenced with the exclusion of peaks that lacked MS/MS fragmentation data, exhibited a ΔMass (ppm) exceeding 2.0, had a peak rating of 8.0 or below, or presented a CV (%) value of less than or equal to 10%. The remaining data were subjected to further analysis using MS/MS fragmentation patterns, which were compared against several spectral databases to enhance the accuracy of compound identification. Key databases referenced in this process included the mzCloud spectral library, mzVault, ChemSpider™, and a mass list of natural and flavonoid compounds. Subsequently, the peaks exhibiting a matching rate of 80% or higher with these databases were annotated (Figure 1), ensuring that only highly confident matches were retained. As a result of this annotation process, a total of 163 compounds were identified as metabolites present in both hot and cold brew coffee. Specifically, 93 compounds were annotated as being in the positive ion mode, while 70 compounds were identified as being in a negative ion mode, offering a comprehensive metabolic profile of the coffee samples. A detailed list of these annotated compounds, along with their respective data, is provided in Table S1.

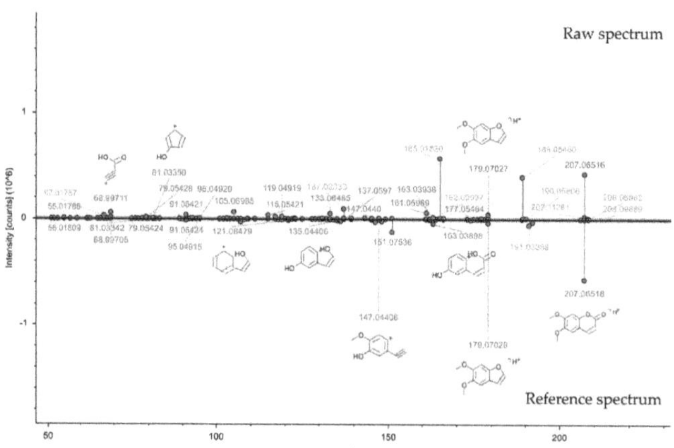

Figure 1. Mirror match of fragmentation of scoparone by mzCloud spectral library: best matching rate, 86.2%. The green color indicates matching, while the red color indicates non-matching.

Among these compounds, to evaluate the variation and distribution of hot and cold brew coffee based on its non-volatile chemical compound profile, a PCA score plot was examined. As illustrated in Figure 2, the first two principal components (PC1 and PC2) accounted for a combined total of 91.5% (positive ion mode) and 94.4% (negative ion mode) of the variance across the groups. This high percentage of explained variance highlights the model's strong capability to differentiate between the samples, effectively distinguishing the brewing conditions used in our study. The clear separation observed in the PCA plot demonstrates the robustness of the statistical model in capturing the chemical differences associated with each method of extraction. Also, volcano plot analysis was employed, integrating Student's t-test (p-value < 0.05) with Fold Change analysis (Log_2 Fold Change > 1.0), to examine the chemical variations among the hot and cold brew coffee samples in positive and negative ion modes, respectively (Figure 3). As shown in the results, the volcano plot comparing hot and cold brew coffee revealed 18 significant compounds (in both positive and negative ion modes), with 17 compounds exhibiting significant up-regulation in hot brew coffee, having the exception of 3-(3-pyridinyl)propanoic acid (detected in the positive ion mode). Detailed up-regulated compound information, including regarding molecule weight, formula, and CAS number, is provided in Table 1.

Table 1. Differential non-volatile compounds in hot and cold brew coffee using high-resolution orbitrap mass spectrometry with volcano plotting up-regulation (p-value < 0.05, Log_2 Fold Change > 1.0).

Compounds	CAS [1]	Formula	Molecule Weight	Reference Ion	Peak Rating
6-Methyl-2-pyridinemethanol	1122-71-0	C7H9NO	123.0683	[M + H] + 1	9.2
2-Methylbenzoic acid	118-90-1	C8H8O2	136.0524	[M − H] − 1	9.2
Tyrosol	501-94-0	C8H10O2	138.0681	[M − H] − 1	9
3-(3-pyridinyl)propanoic acid [2]	3724-19-4	C8H9NO2	151.0632	[M + H] + 1	8.3
6-Methoxyquinoline	5263-87-6	C10H9NO	159.0683	[M + H] + 1	8.8
4-Coumaric acid	7400-08-0	C9H8O3	164.0473	[M − H] − 1	8.7
Norharman	244-63-3	C11H8N2	168.0687	[M + H] + 1	9.6
3-Dehydroshikimate	2922-42-1	C7H8O5	172.0371	[M − H] − 1	9.2
Indole-3-acetic acid	87-51-4	C10H9NO2	175.0633	[M − H] − 1	8.8
1-(4-Methylphenyl)pyrrolidine-2,5-dione	–	C11H11NO2	189.0790	[M + H] + 1	8.7
Scoparone	120-08-1	C11H10O4	206.0579	[M + H] + 1	8.5
Euparin	532-48-9	C13H12O3	216.0787	[M − H] − 1	9
Resveratrol	510-36-0	C14H12O3	228.0786	[M − H] − 1	10
Naringenin	480-41-1	C15H12O5	272.0686	[M − H] − 1	8.6
N-Caffeoyltryptophan	109163-69-1	C20H18N2O5	366.1215	[M − H] − 1	9.2
Mascaroside II	–	C26H34O9	490.2206	[M + H] + 1	10
Mascaroside I	–	C26H36O10	508.2311	[M + H] + 1	9.4
4,5-Dicaffeoylquinic acid	57378-72-0	C25H24O12	516.1268	[M − H] − 1	9.6
		C25H24O12	516.1272	[M + H] + 1	8.8

[1] Minus sign '–' indicates the CAS was not available in the literature. [2] This compound was only up-regulated in cold brew coffee.

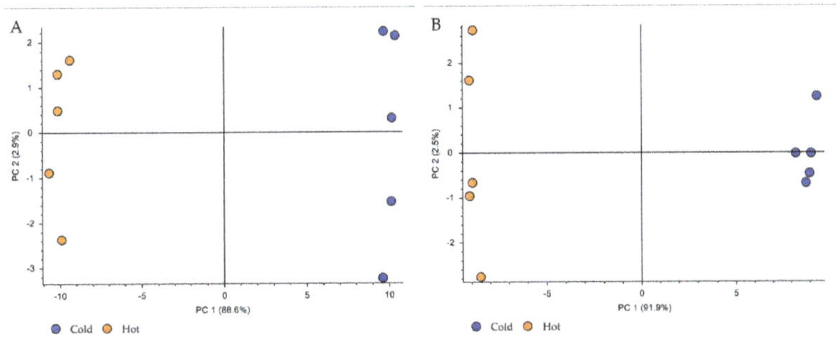

Figure 2. PCA score plot for non-volatile compounds in hot and cold brew coffee: (**A**) positive ion mode, (**B**) negative ion mode.

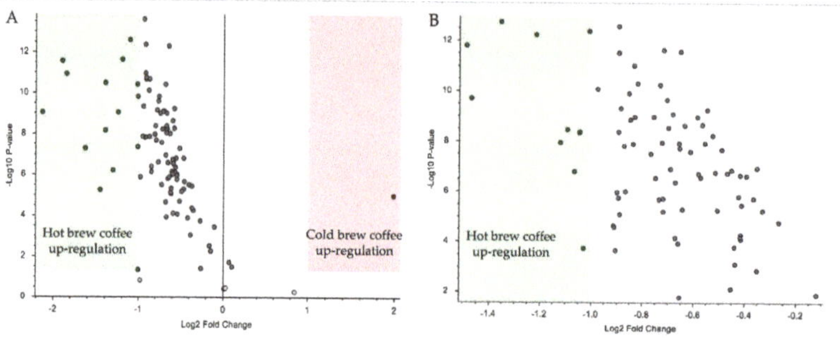

Figure 3. Volcano plotting for non-volatile compounds in hot and cold brew coffee: (**A**) positive ion mode; (**B**) negative ion mode.

3.2. Volatile Compounds in Hot and Cold Brew Coffee

Metabolic profiling using GC-HR-OMS was conducted on hot and cold brew coffee. Compound Discoverer performed peak alignment, baseline correction, and deconvolution, resulting in the identification of approximately 1400 peaks in hot brew coffee and 1300 peaks in cold brew coffee. This process followed the method of Sapozhnikova [31] with some modifications. Firstly, peak detection is carried out using high-resolution deconvolution with a mass tolerance of less than 5 ppm. The deconvolution of raw data are particularly advantageous for reducing and eliminating interferences. After deconvolution, peaks with a CV (%) below 10% were identified, with an S/N threshold greater than 3 and a total ion chromatogram (TIC) threshold of 5,000,000, where TIC threshold refers to the total intensity area of all ions at a specific retention time. And then, a library search was conducted using the NIST 17 standard database in Compound Discoverer. Among these 1300–1400 peaks, 93 compounds were annotated as metabolic compounds in hot and cold brew coffee (Table S2). The annotated metabolic compounds in hot and cold brew coffee were classified into alcohol, aldehyde, ester, fatty acid, furan, hydroquinone, indole, ketone, pyrazine, phenol, and others.

In order to assess the differences and distribution patterns between hot and cold brew coffee based on their respective profiles of volatile chemical compounds, a PCA score plot was employed. As shown in Figure 4A, the first two principal components (PC1 and PC2) together explained a significant portion of the overall variance, accounting for 93.0% of the total variation observed among the different groups. These two components were instrumental in distinguishing the coffee samples, providing insight into how the volatile profiles differ between hot and cold brew coffees. By capturing the majority of the variance within the first two components, the PCA score plot effectively demonstrates the underlying chemical distinctions that contribute to the sensory and flavor characteristics of the respective coffee types. In addition, a volcano plot analysis was conducted to investigate the volatile compound differences between the hot and cold brew coffee samples. This analysis combined Student's t-test (p-value < 0.05) with Fold Change analysis (Log_2 Fold Change > 0.9) to identify significant variations (Figure 4B). As shown in the results, the volcano plot comparing hot and cold brew revealed 13 significant compounds, with 10 compounds exhibiting significant up-regulation in hot brew coffee. Detailed information on the up-regulated compounds, including molecular weight, odor, and CAS number, is provided in Table 2.

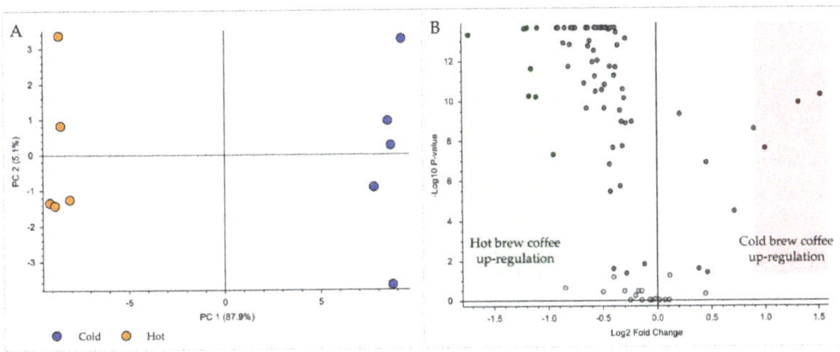

Figure 4. PCA score plot and volcano plotting for volatile compounds in hot and cold brew coffee: (**A**) PCA score plot; (**B**) volcano plotting.

Table 2. Differential volatile compounds in hot and cold brew using high-resolution orbitrap mass spectrometry with volcano plotting of up-regulation (p-value < 0.05, Log_2 Fold Change > 0.9).

Up-Regulation	Compounds	Molecule Weight	CAS [1]	Odor
Cold brew	Cedrenol	220.1827	28231-03-0	Woody, sweet [32]
	7,9-Di-tert-butyl-1-oxaspiro(4,5)deca-6,9-diene-2,8-dione	276.1725	82304-66-3	
	Dicyclohexyl adipate	310.2144	849-99-0	
Hot brew	Catechol	110.0368	120-80-9	
	Indole	117.0578	120-72-9	Animal, floral, moth ball [25]
	3-Methylindole	131.0735	83-34-1	Animal [32]
	1,3-Benzenediol, 4-ethyl-	138.0681	2896-60-8	
	3-Methoxy-1,2-benzenediol	140.0473	934-00-9	
	2-Naphthalenol	144.0575	135-19-3	Slight phenolic odor [32]
	2-Methoxy-4-vinylphenol	150.0681	7786-61-0	Smoky, woody, powdery [32]
	2-Amino-5-methoxybenzaldehyde	151.0633	26831-52-7	
	1-(2,3-Dihydroxyphenyl)ethanone	152.0473	–	
	4-Vinylphenol	162.0681	2628-16-2	Sweet, vanilla [32]

[1] Minus sign '–' indicates the CAS was not available in the literature.

4. Discussion

In our results, 163 non-volatile compounds were identified and annotated in hot and cold brew coffee (Table S1). These compounds were classified into alkaloids, including caffeine, theobromine, and trigonelline; phenolic compounds such as chlorogenic acid, caffeic acid, and quinic acid; and organic acids, amino acids, coumarins, furans, vitamins, pyrazines, and others. These compounds play a crucial role in determining the sensory profile of coffee, contributing significantly to its flavor characteristics [33]. Additionally, they provide various bioactive functions, including antioxidant properties and stimulatory effects. Bitterness was a major sensory attribute that impacted consumer acceptability. Both caffeine and chlorogenic acid played a role in contributing to this bitterness, while chlorogenic acid additionally contributed to astringency [34]. In our results, the major non-volatile compounds in coffee, including caffeine, chlorogenic acid, and trigonelline, showed no significant differences between hot and cold brew coffee, suggesting these compounds are less influenced by the coffee brewing condition. According to Fuller and Rao [35], the caffeine and 3-chlorogenic acid levels in coffee vary with different roast degrees and particle sizes. Their findings indicated that roasting temperature influenced the 3-chlorogenic acid content. However, no significant difference was observed between cold and hot brewed coffee, which is in agreement with our results. Intragranular pore diffusion limits the extraction of chlorogenic acid, despite its high solubility in water. This characteristic allows for efficient extraction of chlorogenic acid at both high and low temperatures, enhancing its extraction under various brewing conditions [36]. Trigonelline plays a role in the overall aroma profile of brewed coffee. While it typically has a mild, bitter taste, it breaks down quickly during the roasting process, forming potent volatile compounds like pyridines and pyrroles. Córdoba et al. and Stankek et al. [14,37] investigated the variation in trigonelline content in coffee as influenced by different brewing methods. In their results, most coffee varieties showed no significant variation, though some exceptions were observed. Specifically, Brazilian coffee beans exhibited higher trigonelline levels in cold brew extraction, whereas Guatemalan beans demonstrated higher trigonelline concentrations in hot brew extraction.

Our findings indicate that 17 non-volatile compounds exhibited significant differences between hot brew and cold brew coffee (Table 1 and Figure 3). Among these compounds, one of the most prominent in coffee is 4,5-dicaffeoylquinic acid, which is a derivative of quinic acid. Silveira et al. [38] reported that the extraction efficiency of quinic acids tends to increase proportionally with rising temperatures. Notably, a sharp enhancement in extraction was observed when the temperature exceeded 95 °C, suggesting that higher thermal conditions significantly promote the solubility and release of quinic acids. Additionally, coumarin derivatives, including 4-coumaric acid and scoparone, showed relatively higher extraction efficiencies in hot brew coffee compared to cold brew coffee. Similarly, Doctor et al. [39] reported findings that coumarin derivatives exhibit increased extraction efficiency as temperature rises. Other compounds, such as alkaloids, including norharman and polyphenol, tyrosol, and shikimic acid, exhibited relatively higher extraction efficiencies in hot brew coffee. This finding underscores the critical role of temperature in optimizing the extraction of bioactive compounds during the brewing process, particularly at elevated temperatures where extraction kinetics are more favorable [40]. 3-(3-Pyridinyl)propanoic acid, as a derivative of pyridine, was shown to be present in relatively high amounts in cold brew coffee. However, no other literature demonstrating similar results could be found. Therefore, future research should aim to verify the reproducibility of this substance in cold brew coffee and investigate the underlying causes.

In our results, 93 volatile compounds were identified and annotated in hot and cold brew coffee (Table S2). These compounds were classified into pyrazines which are associated with nutty and roasted aromas, such as 2-methylpyrazine, 2-ethylpyrazine, and 2-ethyl-5-methylpyrazine; phenolic compounds which are associated with smoky aromas, such as cresol; furans which are associated with sweet and nutty aromas, such as 2-methylfuran and furfural; and pyridines; alcohols; and others [32].

Pyrazines and furans are among the most abundant volatile compounds in coffee, primarily generated through the Maillard reaction during the roasting process of coffee beans. In previous studies, pyrazines, including 2-methylpyrazine and 2-ethylpyrazine, were typically found to exhibit higher extraction efficiencies in cold brew coffee compared to hot brew coffee [25,36]. However, in our results, most pyrazines showed no significant difference between cold brew and hot brew extractions (Table 2). This discrepancy suggests that extraction time, rather than temperature, may play a more critical role in the extraction of pyrazines. While previous studies utilized relatively longer extraction times for cold brew, our study maintained consistent extraction times across both methods, focusing specifically on temperature variations. Additionally, Maksimowski et al. [41] reported that the content of pyrazines increases as the extraction time is extended under conditions of 25 °C. A similar trend was observed with furans such as furfural and 2-methylfuran.

On the other hand, phenolic compounds, including catechol, 4-ehtly-1,3-benzenediol, 2-methoxy-4-vinylphenol (smoky and woody aroma), and 4-vinylphenol (sweet and vanilla aroma), as well as indole and 3-methylindole (animalic aroma), were present in relatively higher amounts in hot brew coffee compared to cold brew coffee in our results [32]. According to Cai et al. [25], 4-vinylguaiacol (2-methoxy-4-vinylphenol) was present in relatively higher amounts in hot brew coffee. Additionally, Pan et al. [36] reported that 4-ethylguaiacol and indole were present in higher concentrations in hot brew coffee, which is in agreement with our results. This finding suggests that extraction temperature, rather than time, may play a more critical role in the extraction of phenolic compounds and indole. In particular, indole was not detected in some other studies, suggesting that its presence may be influenced by factors such as coffee variety and roasting conditions [36].

Cedrenol, a compound classified as a terpenoid, is known for its woody and sweet flavor. In our study, it was found to be present in relatively higher concentrations in cold brew coffee compared to hot brew coffee. In general, elevated temperatures increase the saturated vapor pressure, leading to greater losses of some volatile compounds [42].

In this study, we utilized HR-OMS and advanced software, such as Compound Discoverer, to identify and annotate a total of 163 non-volatile compounds and 93 volatile compounds in both hot brew and cold brew coffee, significantly more than in previous research. Among the annotated compounds, in addition to those previously mentioned, we identified several chemicals, such as esculetin, scoparone, scopoletin, and resveratrol, which have not yet been reported in relation to coffee. Given that many of these compounds exhibit potent antioxidant properties and hold potential health benefits, further in-depth studies are warranted to explore their biological relevance and possible applications.

5. Conclusions

In conclusion, we identified and annotated a total of 163 non-volatile compounds and 93 volatile compounds in both hot brew and cold brew coffee using HR-OMS. The major classes of non-volatile compounds were alkaloids, including caffeine and trigonelline, and phenolic compounds, including chlorogenic acid. The main classes of volatile compounds were pyrazines, phenolic compounds, and furans. The influence of hot and cold brewing on the coffee's non-volatile and volatile compound profiles showed notable differences, with 18 distinct non-volatile compounds and 13 aroma-active compounds identified. Our analysis revealed that certain quinic acids, such as 4,5-dicaffeoylquinic acid, and coumarin derivatives were present in higher amounts in hot brew coffee. Based on this, we can conclude that some non-volatile compounds are significantly influenced by temperature during the extraction process. Regarding volatile compounds, phenolic compounds and indole are significantly influenced by temperature during the extraction process. On the other hand, a comparison with other studies revealed that pyrazine and furan compounds are more affected by extraction time than by temperature. Finally, among the annotated compounds, several previously unreported bioactive compounds in coffee were detected. Although our study qualitatively identified compounds not previously reported in previous research, the quantification of each compound was not assessed. Therefore, further research should focus on the quantification of these compounds, while concurrently evaluating the functionality and toxicity of the major compounds.

Supplementary Materials: The following supporting information can be downloaded at: https://www.mdpi.com/article/10.3390/beverages11010010/s1: Table S1: Non-Volatile compounds identified in hot and cold brew coffee using liquid chromatography high-resolution orbitrap mass spectrometry; Table S2: Volatile compounds identified in hot and cold brew coffee using gas chromatography high-resolution orbitrap mass spectrometry.

Author Contributions: Conceptualization, S.L. and E.H.; methodology, S.L.; software, S.L.; validation, S.L., E.H., J.K. and S.K.; formal analysis, S.L. and M.S.; investigation, S.L., E.H., J.K. and S.K.; resources, M.S.; data curation, S.L. and S.K.; writing—original draft preparation, S.L.; writing—review and editing, S.L. and M.K.; visualization, S.L.; supervision, H.C. and G.L.; project administration, M.K. and H.C.; funding acquisition, G.L. All authors have read and agreed to the published version of the manuscript.

Funding: This research received no external funding.

Data Availability Statement: The data presented in this study are available on request from the corresponding author.

Conflicts of Interest: All authors, including S.L., E.H., J.K., S.K., M.S., M.K., H.C. and G.L., are employees of the Lotte R&D Center, and the materials used in the experiment were provided by our affiliate, Lotte Chilsung. The employment relationship did not affect the results or conclusions of the experiment.

References

1. Coffee: World Markets and Trade. Available online: https://www.fas.usda.gov/data/coffee-world-markets-and-trade (accessed on 25 September 2024).
2. Ludwig, I.A.; Clifford, M.N.; Lean, M.E.; Ashihara, H.; Crozier, A. Coffee: Biochemistry and potential impact on health. *Food Funct.* **2014**, *5*, 1695–1717. [CrossRef]
3. van Dam, R.M.; Hu, F.B.; Willett, W.C. Coffee, caffeine, and health. *N. Engl. J. Med.* **2020**, *383*, 369–378. [CrossRef]
4. Socała, K.; Szopa, A.; Serefko, A.; Poleszak, E.; Wlaź, P. Neuroprotective Effects of Coffee Bioactive Compounds: A Review. *Int. J. Mol. Sci.* **2021**, *22*, 107. [CrossRef]
5. Sunarharum, W.B.; Williams, D.J.; Smyth, H.E. Complexity of coffee flavor: A compositional and sensory perspective. *Food Res. Int.* **2014**, *62*, 315–325. [CrossRef]
6. Kitzberger, C.S.G.; dos Santos Scholz, M.B.; de Toledo Benassi, M. Bioactive compounds content in roasted coffee from traditional and modern Coffea arabica cultivars grown under the same edapho-climatic conditions. *Food Res. Int.* **2014**, *61*, 61–66. [CrossRef]
7. Martins, P.M.M.; Batista, N.N.; Miguel, M.G.d.C.P.; Simão, J.B.P.; Soares, J.R.; Schwan, R.F. Coffee growing altitude influences the microbiota, chemical compounds and the quality of fermented coffees. *Food Res. Int.* **2020**, *129*, 108872. [CrossRef] [PubMed]
8. Cordoba, N.; Fernandez-Alduenda, M.; Moreno, F.L.; Ruiz, Y. Coffee extraction: A review of parameters and their influence on the physicochemical characteristics and flavour of coffee brews. *Trends Food Sci. Technol.* **2020**, *96*, 45–60. [CrossRef]
9. Gloess, A.N.; Schönbächler, B.; Klopprogge, B.; DAmbrosio, L.; Chatelain, K.; Bongartz, A.; Strittmatter, A.; Rast, M.; Yeretzian, C. Comparison of nine common coffee extraction methods: Instrumental and sensory analysis. *Eur. Food Res. Technol.* **2013**, *236*, 607–627. [CrossRef]
10. Batali, M.E.; Lim, L.X.; Liang, J.; Yeager, S.E.; Thompson, A.N.; Han, J.; Ristenpart, W.D.; Guinard, J.-X. Sensory analysis of full immersion coffee: Cold brew is more floral, and less bitter, sour, and rubbery than hot brew. *Foods* **2022**, *11*, 2440. [CrossRef]
11. Gavrilyuk, O.; Braaten, T.; Skeie, G.; Weiderpass, E.; Dumeaux, V.; Lund, E. High coffee consumption and different brewing methods in relation to postmenopausal endometrial cancer risk in the Norwegian women and cancer study: A population-based prospective study. *BMC Women's Health* **2014**, *14*, 48. [CrossRef]
12. Rothwell, J.A.; Loftfield, E.; Wedekind, R.; Freedman, N.; Kambanis, C.; Scalbert, A.; Sinha, R. A metabolomic study of the variability of the chemical composition of commonly consumed coffee brews. *Metabolites* **2019**, *9*, 17. [CrossRef]
13. Muzykiewicz-Szymańska, A.; Nowak, A.; Wira, D.; Klimowicz, A. The effect of brewing process parameters on antioxidant activity and caffeine content in infusions of roasted and unroasted Arabica coffee beans originated from different countries. *Molecules* **2021**, *26*, 3681. [CrossRef] [PubMed]
14. Córdoba, N.; Moreno, F.L.; Osorio, C.; Velásquez, S.; Ruiz, Y. Chemical and sensory evaluation of cold brew coffees using different roasting profiles and brewing methods. *Food Res. Int.* **2021**, *141*, 110141. [CrossRef] [PubMed]
15. Yang, J.; Wen, H.; Zhang, L.; Zhang, X.; Fu, Z.; Li, J. The influence of ripening stage and region on the chemical compounds in mulberry fruits (*Morus atropurpurea* Roxb.) based on UPLC-QTOF-MS. *Food Res. Int.* **2017**, *100*, 159–165. [CrossRef] [PubMed]
16. Lu, Y.; Li, N.; Gao, L.; Xu, Y.-J.; Huang, C.; Yu, K.; Ling, Q.; Cheng, Q.; Chen, S.; Zhu, M. Acetylcarnitine is a candidate diagnostic and prognostic biomarker of hepatocellular carcinoma. *Cancer Res.* **2016**, *76*, 2912–2920. [CrossRef] [PubMed]

17. Wang, Y.; Li, C.; Li, L.; Yang, X.; Chen, S.; Wu, Y.; Zhao, Y.; Wang, J.; Wei, Y.; Yang, D. Application of UHPLC-Q/TOF-MS-based metabolomics in the evaluation of metabolites and taste quality of Chinese fish sauce (Yu-lu) during fermentation. *Food Chem.* **2019**, *296*, 132–141. [CrossRef] [PubMed]
18. Anderson, H.E.; Liden, T.; Berger, B.K.; Zanella, D.; Manh, L.H.; Wang, S.; Schug, K.A. Profiling of contemporary beer styles using liquid chromatography quadrupole time-of-flight mass spectrometry, multivariate analysis, and machine learning techniques. *Anal. Chim. Acta* **2021**, *1172*, 338668. [CrossRef]
19. Chang, W.C.-W.; Wu, H.-Y.; Kan, H.-L.; Lin, Y.-C.; Tsai, P.-J.; Chen, Y.-C.; Pan, Y.-Y.; Liao, P.-C. Discovery of spoilage markers for chicken eggs using liquid chromatography-high resolution mass spectrometry-based untargeted and targeted foodomics. *J. Agric. Food Chem.* **2021**, *69*, 4331–4341. [CrossRef]
20. Li, Y.; Zhang, J.; Jin, Y.; Wang, L.; Zhao, W.; Zhang, W.; Zhai, L.; Zhang, Y.; Zhang, Y.; Zhou, J. Hybrid quadrupole-orbitrap mass spectrometry analysis with accurate-mass database and parallel reaction monitoring for high-throughput screening and quantification of multi-xenobiotics in honey. *J. Chromatogr. A* **2016**, *1429*, 119–126. [CrossRef] [PubMed]
21. Chiesa, L.M.; Pavlovic, R.; Panseri, S.; Arioli, F. Evaluation of parabens and their metabolites in fish and fish products: A comprehensive analytical approach using LC-HRMS. *Food Addit. Contam.* **2018**, *35*, 2400–2413. [CrossRef]
22. Senyuva, H.Z.; Gökmen, V.; Sarikaya, E.A. Future perspectives in Orbitrap™-high-resolution mass spectrometry in food analysis: A review. *Food Addit. Contam.* **2015**, *32*, 1568–1606. [CrossRef] [PubMed]
23. Vezzulli, F.; Rocchetti, G.; Lambri, M.; Lucini, L. Metabolomics combined with sensory analysis reveals the impact of different extraction methods on coffee beverages from coffea arabica and coffea canephora var. robusta. *Foods* **2022**, *11*, 807. [CrossRef] [PubMed]
24. Pérez-Míguez, R.; Sánchez-López, E.; Plaza, M.; Castro-Puyana, M.; Marina, M.L. A non-targeted metabolomic approach based on reversed-phase liquid chromatography–mass spectrometry to evaluate coffee roasting process. *Anal. Bioanal. Chem.* **2018**, *410*, 7859–7870. [CrossRef] [PubMed]
25. Cai, Y.; Xu, Z.; Pan, X.; Gao, M.; Wu, M.; Wu, J.; Lao, F. Comparative profiling of hot and cold brew coffee flavor using chromatographic and sensory approaches. *Foods* **2022**, *11*, 2968. [CrossRef] [PubMed]
26. Li, J.; Zhao, W.; Pan, X.; Lao, F.; Liao, X.; Shi, Y.; Wu, J. Improvement of antioxidant properties of jujube puree by biotransformation of polyphenols via Streptococcus thermophilus fermentation. *Food Chem.* **2022**, *13*, 100214. [CrossRef] [PubMed]
27. Pan, X.; Wu, J.; Zhang, W.; Liu, J.; Yang, X.; Liao, X.; Hu, X.; Lao, F. Effects of sugar matrices on the release of key aroma compounds in fresh and high hydrostatic pressure processed Tainong mango juices. *Food Chem.* **2021**, *338*, 128117. [CrossRef]
28. Heo, J.; Adhikari, K.; Choi, K.S.; Lee, J. Analysis of caffeine, chlorogenic acid, trigonelline, and volatile compounds in cold brew coffee using high-performance liquid chromatography and solid-phase microextraction—Gas chromatography-mass spectrometry. *Foods* **2020**, *9*, 1746. [CrossRef]
29. Souza, A.; Ntai, I.; Tautenhahn, R. Accelerated unknown compound annotation with confidence: From spectra to structure in untargeted metabolomics experiments. *Thermo Fish. Sci. Appl. Note* **2018**, 65362.
30. Sukor, N.S.M.; Zakri, Z.H.M.; Rasol, N.E.; Salim, F. Annotation and Identification of Phytochemicals from Eleusine indica Using High-Performance Liquid Chromatography Tandem Mass Spectrometry: Databases-Driven Approach. *Molecules* **2023**, *28*, 3111. [CrossRef] [PubMed]
31. Sapozhnikova, Y. Non-targeted screening of chemicals migrating from paper-based food packaging by GC-Orbitrap mass spectrometry. *Talanta* **2021**, *226*, 122120. [CrossRef]
32. The Good Scents Company Search Page. Available online: http://www.thegoodscentscompany.com/search2.html (accessed on 7 October 2024).
33. Buffo, R.A.; Cardelli-Freire, C. Coffee flavour: An overview. *Flavour Fragr. J.* **2004**, *19*, 99–104. [CrossRef]
34. Masi, C.; Dinnella, C.; Monteleone, E.; Prescott, J. The impact of individual variations in taste sensitivity on coffee perceptions and preferences. *Physiol. Behav.* **2015**, *138*, 219–226. [CrossRef] [PubMed]
35. Fuller, M.; Rao, N.Z. The effect of time, roasting temperature, and grind size on caffeine and chlorogenic acid concentrations in cold brew coffee. *Sci. Rep.* **2017**, *7*, 17979. [CrossRef] [PubMed]
36. Pan, L.; Xiao, Y.; Jiang, F.; Jiang, T.; Zhu, J.; Tang, W.; Liu, X.; Zhou, Y.; Yu, L. Comparison of characterization of cold brew and hot brew coffee prepared at various roasting degrees. *J. Food Process. Preserv.* **2023**, *2023*, 3175570. [CrossRef]
37. Stanek, N.; Zarębska, M.; Biłos, Ł.; Barabosz, K.; Nowakowska-Bogdan, E.; Semeniuk, I.; Błaszkiewicz, J.; Kulesza, R.; Matejuk, R.; Szkutnik, K. Influence of coffee brewing methods on the chromatographic and spectroscopic profiles, antioxidant and sensory properties. *Sci. Rep.* **2021**, *11*, 21377. [CrossRef] [PubMed]
38. Silveira, T.F.F.d.; Meinhart, A.D.; Souza, T.C.L.d.; Cunha, E.C.E.; Moraes, M.R.d.; Lorini, A.; Teixeira Filho, J.; Godoy, H.T. Impact of water temperature of chimarrão on phenolic compounds extraction. *Food Sci. Technol.* **2020**, *41*, 856–863. [CrossRef]
39. Doctor, N.; Parker, G.; Vang, K.; Smith, M.; Kayan, B.; Yang, Y. Stability and extraction of vanillin and coumarin under subcritical water conditions. *Molecules* **2020**, *25*, 1061. [CrossRef]

40. Mestdagh, F.; Glabasnia, A.; Giuliano, P. The brew—Extracting for excellence. In *The Craft and Science of Coffee*; Elsevier: Amsterdam, The Netherlands, 2017; pp. 355–380.
41. Maksimowski, D.; Pachura, N.; Oziembłowski, M.; Nawirska-Olszańska, A.; Szumny, A. Coffee roasting and extraction as a factor in cold brew coffee quality. *Appl. Sci.* **2022**, *12*, 2582. [CrossRef]
42. López, J.A.S.; Wellinger, M.; Gloess, A.N.; Zimmermann, R.; Yeretzian, C. Extraction kinetics of coffee aroma compounds using a semi-automatic machine: On-line analysis by PTR-ToF-MS. *Int. J. Mass Spectrom.* **2016**, *401*, 22–30. [CrossRef]

foods MDPI

Article

Heat Treatment of Hazelnut Allergens Monitored by Polyclonal Sera and Epitope Fingerprinting

Karolin Kern [1], Suttinee Santa-Ardharnpreecha [1], Nicolas Delaroque [1], Sabine Dölle-Bierke [2], Regina Treudler [3], Eva Ehrentreich-Förster [4], Isabell Rothkopf [5], Margitta Worm [2] and Michael Szardenings [1,6,*]

1 Department Diagnostics, Fraunhofer Institute for Cell Therapy and Immunology (IZI), Perlickstr. 1, 04103 Leipzig, Germany; karolin.kern@izi.fraunhofer.de (K.K.); nicolas.delaroque@izi.fraunhofer.de (N.D.)
2 Division of Allergy and Immunology, Department of Dermatology, Venerology and Allergy, Charité-Universitätsmedizin, 10117 Berlin, Germany; margitta.worm@charite.de (M.W.)
3 Institute of Allergology (IFA), Charité-Universitätsmedizin, 12203 Berlin, Germany
4 Bioanalytics and Bioprocesses Branch, Fraunhofer Institute for Cell Therapy and Immunology (IZI), Am Mühlenberg 13, 14476 Potsdam-Golm, Germany; eva.ehrentreich@izi-bb.fraunhofer.de
5 Fraunhofer Institute for Process Engineering and Packaging IVV, 85354 Freising, Germany; isabell.rothkopf@ivv.fraunhofer.de
6 Epitopic GmbH, Deutscher Platz 5e, 04103 Leipzig, Germany
* Correspondence: michael.szardenings@izi.fraunhofer.de

Abstract: Hazelnuts are frequently involved in IgE-mediated reactions and are the main cause of nut allergies in Europe. Most food products are processed before human consumption. Food processing can modify the structure, properties, and function of proteins, and as a result, the IgE-binding capacity of allergens can be affected. In this study, we aimed to investigate epitope changes caused by the roasting of hazelnuts using epitope fingerprinting. Rabbit sera were raised against hazelnut proteins, and their epitopes were characterized. Immunoassays using specific polyclonal antibodies from rabbits targeting the main allergens in hazelnuts revealed marked reductions in the levels of Cor a 1 (PR-10), Cor a 11 (7S globulin), and Cor a 14 (2S albumin). However, rabbit antibodies can recognize different epitopes. Using antibodies that are different and characterized could help establish reliable methods for estimating the effects of treatments on the allergenicity of foods. In this work, we provide the first practical application that could lead to sets of peptide epitopes to compare and standardize immune diagnostics, even for complex protein preparations.

Keywords: epitope fingerprints; thermal processing; rabbit sera

1. Introduction

Tree nuts are popular worldwide. They are beneficial food sources because they are rich in unsaturated fats and have comparatively high levels of phenolics, phytosterols, tocopherols, minerals, and fiber [1]. This may explain why the consumption of both raw and processed tree nuts is increasing. In recent years, there has been an increase in tree nut consumption because of its favorable health effects [2]. For example, a diet rich in tree nuts has been shown to positively affect cardiovascular-risk biomarkers [3]. In Europe, tree nuts are among the predominant foods involved in allergic reactions to plants and may cause anaphylaxis [4]. The prevalence of nut allergy in Central and Northern Europe ranges from 1–2% [5], and hazelnuts in particular are one of the most common triggers of food-induced anaphylaxis in children and adults [6]. Most nut allergens are highly resistant to enzymatic degradation and denaturation. Several allergenic proteins of hazelnut have been registered in the WHO-IUS list of allergenic proteins: Cor a 1 (Bet v1 homolog), Cor a 2 (profilin), Cor a 8 (Lipid Transfer Protein LTP), Cor a 9 (11S legumin), Cor a 11 (7S vicilin), Cor a 14 (2S albumin), Cor a 12, Cor a 13, and Cor a 15 (oleosins) [7]. In general, hazelnut allergies can be of two forms: primary allergic sensitization to heat-stable hazelnut proteins or secondary food allergy due to immunological cross-reactivity between the structurally related Bet v 1 (a major birch pollen allergen) and members of the pathogenesis-related protein 10 (PR-10) family [8].

Most food products must be processed before human consumption to improve safety, functionality, and organoleptic properties. Unit operations involving the use of heat, such as pasteurization, sterilization, boiling, frying, roasting, and drying, are among the most common operations in food processing. Food processing can modify the structure, properties, and function of proteins, and as a result, the IgE-binding capacity of allergens can be affected. Some processing methods can effectively reduce the content of specific allergens but do not completely abolish them. As some processing treatments have been shown to decrease the allergenicity of certain foods, food processing may play an important role in the development of hypoallergenic foods and the induction of food tolerance [1,9]. Other processes can also increase the allergenicity of certain foods [9]. Roasting, defined as dry heat treatment, is one of the most important processes that provides the product with the required alterations to become value-added nuts. Hazelnut roasting times of 30–50 min at 142 °C are the industrial standard, commonly referred to as low-temperature–long-time (LTLT) processes. In contrast, modern food processing increasingly employs high-temperature–short-time (HTST) methods to preserve raw material quality, including nutritional properties. HTST roasting systems, such as infrared, fluidized bed, or accelerated drum roasting, achieve roasting times of approximately 10 min. This study contrasts LTLT and HTST roasting processes, ensuring sufficient differences in conditions to observe distinct effects [10,11].

These treatments can induce changes in the chemical properties of proteins or biochemical reactions inside food matrix components [12,13]. Numerous studies have been conducted on the modification of the allergenicity of tree nuts using different heat treatments [12,14,15]. Using oral provocation tests, the heat processing of hazelnuts has been shown to reduce allergenicity [16].

Rabbit serum is used to characterize food allergenicity [17–19]. Rabbit serum has been used, amongst others, for the investigation of changes in allergenicity after heat processing [17], cross-reactivity between almond and *Prunus mahaleb* [18], and cross-reactivity between animal IgG and human IgE antibodies [19]. These examples demonstrate that rabbit serum can probably replace human serum in investigations of food allergenicity; however, previous studies have not characterized immune responses at the epitope level.

Sensitization against antigens is currently determined by IgE measurement of the antigenic protein, although it may be assessed much better at the level of individual allergy-relevant peptide epitopes. In addition, IgE-binding epitopes that are often less than 20 amino acids long are more suitable [20]. Several methods have been reported for mapping IgE-binding epitopes [21]. Arrays of overlapping peptides synthesized on a nitrocellulose membrane (SPOTmembrane) [22] can be used to determine sequential epitopes [23,24]. Overlapping peptides can also be used in an enzyme-linked immunosorbent assay (ELISA) format [25]. However, this approach is more effective and frequently used in microarray formats because many more peptides can be used with less serum [26–28]. However, the determination of conformational epitopes requires sophisticated techniques such as peptide derivatives [29], nuclear magnetic resonance, mass spectrometry, mutant generation, and in silico analysis [30,31]. Phage displays can be used for linear or conformational epitope mapping [32–34].

In this study, we aimed to investigate epitope changes caused by the roasting of hazelnuts using epitope fingerprinting. Rabbit sera were raised against hazelnut proteins and their epitopes were characterized [33]. Epitopes similar to those previously identified in serum samples from hazelnut-sensitized patients were also identified. These peptide epitopes were confirmed using peptide microarrays with rabbit anti-hazelnut polyclonal antisera. Most epitope peptides were bound to IgG from rabbit sera. Western blot analysis was performed to investigate the cross-reactivity with other proteins from legumes and tree nuts.

2. Materials and Methods

2.1. Hazelnut

Forty-eight raw and undamaged hazelnuts with a diameter of 11–13 mm were roasted in a convection oven with a Pyrotherm EEBP 6400.8 MX from Küppersbusch, Gelsenkirchen, Germany, under the following four conditions: 142 °C for 30 min, 142 °C for 50 min, 200 °C for 8 min, and 200 °C for 10 min.

To ensure that the roasts were comparable, the nuts were roasted at the same roasting temperature in the same oven and height. The roasted hazelnuts were ground using a food processor (Moulinette DPA1X; Tefal, 73312 Geislingen/Steige, Germany) to obtain a fine powder.

Roasting conditions were selected as two typical ways to achieve the same roasting color: low-temperature–long-time vs. high-temperature–short-time. Shorter times at low temperatures and longer times at high temperatures were assumed to cause less or more damage to the nuts, respectively.

2.2. Animal Sera

Immunization of rabbits and preparation of sera were carried out by Seramun Diagnostica GmbH (Spreenhagener Str. 1, 15754 Heidesee, Germany; permit 2347-15-2023-20-E, LAVG Brandenburg). Rabbits were immunized with de-oiled hazelnut flour and complete Freund's adjuvant. A boost immunization with incomplete Freund's adjuvant was performed on day 21. Rabbits were bled after 28 days, and sera were used for phage display and array experiments.

2.3. Patient Sera

The human blood samples from patients allergic to tree nuts used in this study were provided by Charité-Universitätsmedizin Berlin (Department of Dermatology, Venerology, and Allergy) and Universitätsklinikum Leipzig (Department of Dermatology, Venerology, and Allergy). The use of samples for this study was authorized by the responsible ethics committee (Ethics Committee of the Medical Faculty of Leipzig University (AZ202/16-ek) and Charité's Ethics Committee (EA01/022/20). Donors were informed that the blood samples would only be used for research purposes. Each donor signed a statement regarding the informed consent.

2.4. Peptide Phage Display Experiments

A special naïve peptide phage display was used in the selection experiments [32], and only two selection rounds were performed, as previously described [32,33]. Briefly, a selection from the ENTE-1 peptide phage display library was performed using Dynabeads Protein A (Thermo Fisher Scientific, Waltham, MA, USA). Two rounds of selections were performed. Pooled DNA of the recovered phagemids from the first and second rounds of selection was subjected to next-generation sequencing (NGS) on an Illumina MiSeq, as previously described. The back and forward runs were combined using PEAR (Version 0.9.6 [35])and processed using Trimmomatic (Version 0.33 [36])). Finally, datasets from each sequencing run were cured of sequencing errors and other artifacts, as described in the latest version of LibDB software (Version 1.34, Epitopic GmbH, Leipzig, Germany). Any sequences deviating from the library codon structure were sorted using this procedure because they potentially contained additional reading errors from the sequencer.

Identification of Epitopes by Motifs

Sequences statistically enriched by binding to serum antibodies and sharing motifs resembling antigen sequences were selected from NGS datasets, as previously described [32,37]. Sequences with motifs matching the antigens were retrieved from sequence pools obtained after the first or second round of selection. The alignment of sequences identified longer stretches of similarity to the antigen as well as structural features such as Cys-constrained peptides in a few hundred sequences. Using a naïve library allows for the search for epitopes that match multiple antigens in the dataset of a single selection experiment.

2.5. IgG-Binding Measurements Using Peptide Microarrays

2.5.1. Selection of Peptide Epitopes/Mimotopes for Spotting on Microarray Slides

Peptides were derived from either native hazelnut antigen sequences (epitopes) or mimotopes obtained from phage-displayed sequences in selection experiments. Selected peptides and their origins are listed in Supplementary Information.

2.5.2. Peptide Microarrays

The peptides were purchased from Peptides & Elephants GmbH (Hennigsdorf, Germany). All peptides had a C-terminal ebes-ϵ-azido-Lys linker; therefore, they were printed and immobilized in triplicate on DBCO-coated glass slides using click chemistry, as previously described [32,37].

The slides contained two clusters of 287 peptide epitopes, spotted in a quadruple. We measured the binding of IgG to peptide epitopes.

Each slide was blocked for 1 h at 4 °C in array buffer (PBS containing 0.1% *v/v* Tween® 20 and 1% *w/v* casein, pH 7.4). The slides were incubated for 1 h at RT with rabbit anti-hazelnut polyclonal antiserum diluted 1:3000 in array buffer, washed twice (PBS containing 0.1% *v/v* Tween® 20, pH 7.4), and incubated for 1 h at RT in a solution of Cy5-labeled goat anti-mouse antibody diluted 1:5000 in array buffer (ThermoFisher Scientific, Waltham, MA, USA). Finally, the slides were washed twice, and fluorescence was measured in a microarray reader at 10 µm resolution using a laser at 532 nm with 25% power/PMT Gain 600 and 635 nm with 25% power/PMT Gain 600 (Genepix 4300; Molecular Devices, San Jose, CA, USA).

2.5.3. Determination of Spot Intensities

This method has been described previously [32,37]. The first step is the detection of the a priori known grid (GAL file) in the image, as the grid may be rotated or shifted. Using the known position of each spot, we used a segmentation approach that combines a seeding threshold and masking threshold with geodesic dilation to distinguish between the foreground and background signals. In the last step of image analysis, we extracted information regarding the fluorescence intensity and shape of the segmentation. As a reliable and robust method to decide on a positive or negative spot result, we used the total count of all pixels in the spot after subtracting the background per pixel, summarizing all intensities in a spot segment, and subtracting the median of the block background for every pixel.

The raw total fluorescence signal intensities of the quadruples were used to calculate the upper and lower quartiles. The difference between the upper and lower quartiles was multiplied by 1.5. This number was added to the upper quartile and subtracted from that of the lower quartile. All raw fluorescence signal intensities that were not within the calculated range were excluded. The adjusted raw fluorescence signal intensity was calculated as a multiple of the background (water) signal.

2.6. SDS-PAGE and Western Blot

2.6.1. Sample Extraction and SDS-PAGE

The extraction protocol was performed as described by Dooper et al. [38], and ground hazelnut (50 mg) was suspended in 5 mL 0.05 M Tris-HCl buffer (pH 7.5) by vortexing for 5 min at room temperature Then, it was left at room temperature overnight, yielding three layers the next day: the fat, water, and ground nut layers. Only water layers were collected. The remaining extract was discarded.

Then, 21 µL of the sample was mixed with 7 µL SDS sample buffer. The mixture was heated for 5 min at 99 °C. The treated mixture (20 µL) was loaded onto 10% SDS gel. The gel system was applied using Schägger and Tricine SDS-PAGE [39]. The gel was run at 90 V for stacking, followed by 120 V for 40 min to separate the gel.

After protein separation, the gel was briefly rinsed with tap water. It was then fixed in fixing buffer (50% methanol, 10% acetic acid, and 100 mM ammonium acetate) for 15–30 min at room temperature. After 2–3 times washing with tap water, the gel was stained with Coomassie for 30 min. It was then washed thrice with tap water and destained in a destaining solution (10% acetic acid) until the background was clear.

2.6.2. Sample Transfer and Blotting

For Western blotting, the gel was cast after protein separation without staining with Coomassie dye in an electrophoretic transfer cell (Mini Trans-Blot® Cell, Bio-Rad Laboratories, Hercules, CA, USA). This procedure was performed following the manufacturer's instructions. The sample was transferred to nitrocellulose blotting paper (Mini Trans-Blot® Cell, Bio-Rad Laboratories, Hercules, CA, USA) in Towbin buffer [40] by applying 40 mA overnight at 4 °C.

The following day, the free surface on the blotting paper was blocked by incubation with 5% bovine serum albumin (BSA) in PBS for 1 h at room temperature with agitation. It was then washed once with washing buffer and incubated with serum diluted in blocking buffer (1:2000) for 1 h at 4 °C. After washing with washing buffer, the blotting paper was incubated with a diluted secondary antibody (mouse monoclonal anti-rabbit IgG (γ-chain specific)-peroxidase) (1:5000 Abcam) for 1 h at 4 °C. The sections were then washed three times with PBS and incubated with 3,3′-diaminobenzidine (DAB) peroxidase substrate solution for 5 min at room temperature with agitation. The reaction was stopped using 2N sulfuric acid. Finally, the blot paper was dried and photographed.

2.7. Protein Structure Comparison

The Protean 3D® software of the DNAStar (Version 17.4.3. DNASTAR, Inc., Madison, WI, USA.) was used to align and compare the protein structures cited in the results section. The structures were aligned using an automated algorithm applied to the entire structures and the RMSD values were calculated with the software's routines.

3. Results

3.1. Epitope Fingerprinting (In Vitro)

Data from selection experiments using a random peptide library with five rabbit samples immunized with hazelnut powder were matched to known allergenic hazelnut proteins. Epitopes were identified using a previously described statistical peptide phage-display method [33]. Seventy-five potential epitopes were identified using rabbit anti-hazelnut polyclonal antisera. Overall, most epitopes were recognized by antibodies from at least two rabbits. Seventeen epitopes were detected using antibodies from both rabbits. Antibodies from rabbit sera recognized motifs for epitopes of 11 allergenic hazelnut proteins (Table 1). Comparing the results from different sera allowed for the selection of peptides with a minimal size recognized by different sera.

Table 1. Overview of the recognition patterns of different hazelnut proteins and identified epitopes using rabbit serum (R1 and R2), Western blotting (WB), and peptide phage display with next-generation sequencing (NGS). Peptides based on the epitopes were also tested using sera from individuals sensitized with hazelnuts and/or peanuts. Numbers before "/" mean positive-tested and total numbers of tested patients. The peptide sequences are not listed in the table. Unroasted (UR) samples are shown for Western blot. See supplement for signal intensities and peptide codes.

		WB		NGS			Patient		
Protein	**Mass (Da)**	**R1 (UR)**	**R2 (UR)**	**R1**	**R2**	**Epitope-Rabbits**	**IgE**	**IgG**	**IgE and IgG**
Cor a 1	17.5	x	x	x	x	3-vfnYEVETps			
				x	x	32-pkVAPQa	39/230	62/235	27/229
				x	x	59-fgEGSRyky			
				x	x	94-dklEKVCSelk			
				x	x	124-kgdHEINaee	111/333	110/340	63/331
				x		9-eTPSVIsa			
				x		67-yvkERVDe			
				x		10-etPSVISa			
				x		143-llraVETYll	8/230	30/235	3/229
				x		141-aEKLLRav	43/329	39/336	13/327
						13-VIPPARLF	32/333	61/340	16/331
						38-aiTSVenvgg	25/285	54/294	12/284
Cor a 2	14.2	x	x	x	x	30-hdGSVWaqsssf	114/333	117/340	69/331
				x	x	11-lmcdiDGQGQq			
				x	x	1-swqaYVDEhl			
				x	x	104-giyeEPVTPgqc			
				x		25-aSAIVGhd			
						45-kPEEItg			
						51-IKDFDEPGSLA	72/329	104/340	48/331
						58-epGHLAPtg			
Cor a 6	34.2			x	x	19-fivEASLkag			
				x	x	238-lekiHLTEEKl			
				x	x	298-tvEEYLqqf			
				x		248-ilKDIQEspi			
				x		281-eESFGVe			
				x		290-qLYPDVky			
					x	39-tVSDPvk			
				x		63-dlyDHGSlv			
				x		119-drVHAVEp			
				x		131-atKVEIRrk			
				x		141-EAEGIPYTY			
				x		202-vDDPRtl			
				x		72-lvkaIKHVDVv			
Cor a 8	9			x	x	73-nCLKDtak			
					x	24-slTCPQik	80/333	109/340	59/284
				x		33-nlTPCVLy			
				x		104-kispsTNCNnv	46/329	75/336	22/327
						5-kLVCAvllc	7/329	8/173	6/169
						49-PSCCKGVRA	68/333	74/340	36/331
Cor a 9	58.8			x	x	63-dhnDQQFqc			
basic subunits	22	x	x	x	x	289-rqewERQErqere			
acidic subunit	40	x	x	x	x	106-itgVLFPgcp	75/285	108/294	60/284
				x	x	119-EDPQQQs			
				x	x	160-agVAHWCyndg			
					x	59-ieSWDHn			
				x		262-rlQSNQdk	66/333	90/340	49/331
					x	302-seQERERqrrQGGRg	39/167	62/340	25/331
				x		131-qGQGQSq			
				x		141-QDRHQk	25/285	34/294	10/284
				x		185-YANQLDe			
					x	201-npddEHQrQGQQQFgqr			
					x	237-nVFSGfd_ef			
				x		277-egRLQVVRPer			
				x		340-r_diYTEQVgr			
				x		411-VFDDelr	71/171	90/336	35/327
						354-vnsnTLPVLrwlql	23/171	35/173	13/169

Table 1. *Cont.*

		WB				NGS	Patient		
Protein	Mass (Da)	R1 (UR)	R2 (UR)	R1	R2	Epitope-Rabbits	IgE	IgG	IgE and IgG
Cor a 10	73.5	x		x	x	16-ilFGCLfai			
				x	x	55-ngHVELia			
				x	x	111-edKEVQkd			
				x	x	236-tfDVSILTIDNgvf			
				x	x	296-rREAEra			
				x	x	305-iSSQHQVRvries			
				x	x	361-nQIDEIvLVGGs			
				x	x	382-iKDYFdgk			
				x	x	460-ftTYQDQqtv			
				x	x	468-tvSIQVFege			
					x	70-sWVGFtdg			
				x		205-iiNEPTaa			
						257-dtHLGGedf			
				x		322-GVDFSepltr			
				x		397-pdEAVAYgaa			
				x		433-lGIETvgg			
						615-dDNQSAeke			
						629-lkEVEAVCnp			
Cor a 11	48	x	x	x	x	421-fkNQDQAff			
				x		46-gnSSEESyg	71/333	87/336	37/327
				x		74-kteeGRVQVLENftk			
				x		139-kreSFNVEhgd			
				x		192-gGEDPeSfY			
				x		338-ssSGSYQki			
						14-kcRDERQf	50/333	33/340	17/331
						40-ERQQEE			
						56-eqeeNPYVF	47/333	84/340	31/331
Cor a 12	17	x	x	x		149-iqSRAQegr			
						7-QLQVHPQRGHG	94/333	109/340	64/331
						121-EMKDRAEQFGQQHV	64/333	73/340	27/331
Cor a 13	14–16				x	14-qpRSHQvvka			
Cor a 14	12	x	x	x	x	30-vdvDEDivn			
				x	x	45-eSCREQAQRQqnl	19/285	52/294	9/284
				x	x	55-qnlNQCQry			
						134-rlspQRCEirsARf	91/333	86/340	13/331
						38-NQQGRR	49/333	69/336	27/327

3.2. Results from Array Measurements

Hazelnut peptide epitopes were synthesized and spotted as microarrays on glass slides [33]. Sera from three immunized rabbits were used for the analysis (Table 1). Rabbit IgGs bound to several epitopes that were previously identified in human serum samples. In the array experiments, at least 50% of these were found in all three rabbits, and 80% were found in at least one. Seven of these peptides showed a high signal intensity (Table 1). All peptides recognized by the rabbit serum samples were identified as IgE and/or IgG in human samples.

3.3. Changing the Thermal Processing to Roasting

The hazelnuts were roasted under different conditions to simulate low-temperature–long-time (LTLT) and high-temperature–short-time (HTST) processes. Two trials involved moderately long roasting times (30 and 50 min) at 142 °C, while the other two involved fast roasting (8 and 10 min) at 200 °C.

Table 2 shows whether heat treatment of a hazelnut allergen resulted in a change in the protein (heat labile) or not (heat stable).

Coomassie-stained SDS-PAGE (Figure 1A) showed no significant alterations in protein patterns after different treatments. All major hazelnut proteins could still be found in different amounts in the soluble form of the roasted hazelnuts.

Table 2. Hazelnut allergens and their behavior to heat treatment are described in the literature and Western blot (R1 = rabbit 1 and R2 = rabbit2).

Protein	Characteristic	Protein Class	Western Blot
Cor a 1	Heat-labile [41]		Heat-labile
Cor a 10	Heat-stable [42]		Heat-stable
Cor a 11	Heat-stable [43]; heat-labile [44]		Heat-stable (R1); heat-labile (R2)
Cor a 12	Heat-stable	Oleosin [45]	
Cor a 13	Heat-stable	Oleosin [45]	
Cor a 14	Heat-stable [43]		Heat-stable (R2); heat-labile (R1)
Cor a 15	Heat-stable	Oleosin [45]	
Cor a 2	Heat-stable [46]; heat-labile [47]		
Cor a 6	n.d.	Isoflavone Reductase	
Cor a 8	Heat-stable [41]; heat-labile [43,44]		
Cor a 9	Heat-stable [43]; heat-labile [44]		Heat-stable (R1, R2)

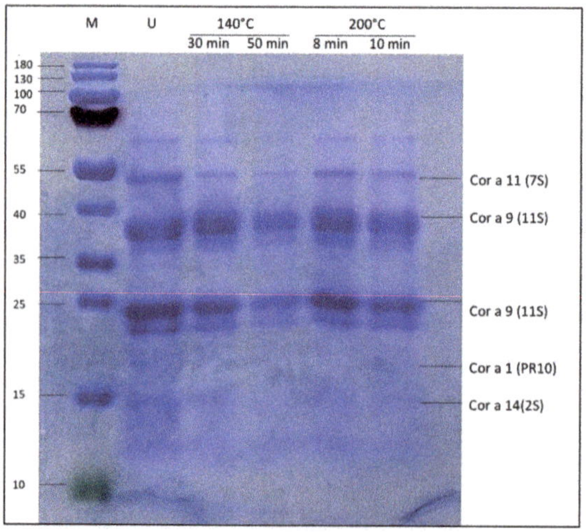

(A)

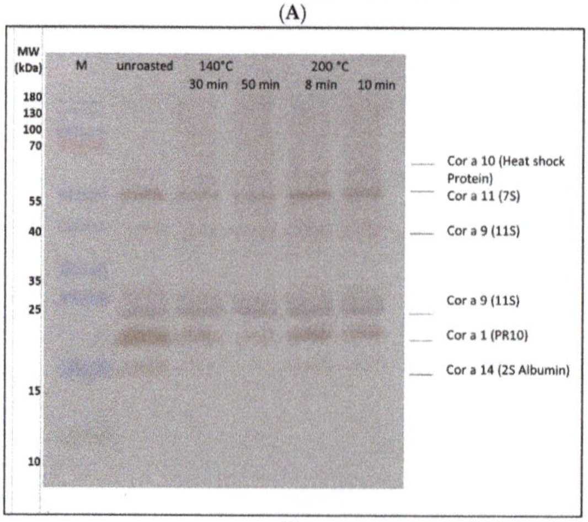

(B)

Figure 1. *Cont.*

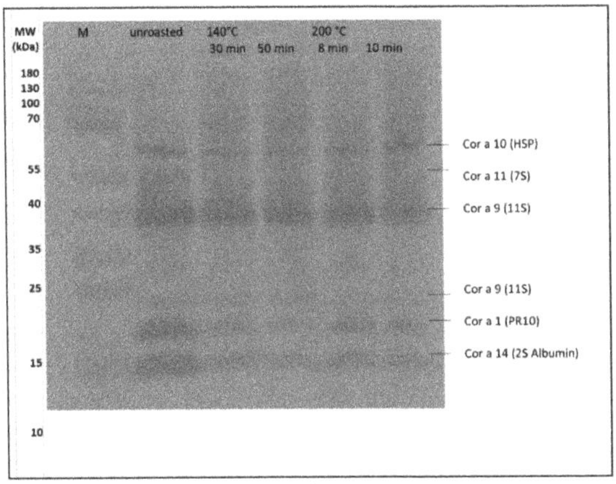

(C)

Figure 1. Extracts from hazelnut roasted under different conditions: Coomassie-stained SDS-PAGE (**A**) and Western blot with sera from two rabbits (**B**,**C**). (**B**) serum from rabbit 1 after immunization (diluted 1:2000); (**C**) serum from rabbit 2 after immunization (diluted 1:2000).

Rabbit IgG binding to extracts from roasted ground hazelnuts was further investigated using Western blotting. A band at 22 kDa was observed in all hazelnut extracts tested with the sera before immunization from both rabbits as well as when no serum was introduced into the test (see Supplementary Materials). This indicates that the secondary antibody may also bind to a hazelnut protein because the intensity of this protein band decreases upon heat treatment. Before immunization, no further bands were observed, indicating no binding to the hazelnut protein. Rabbit IgG showed strong binding to hazelnut proteins with molecular weights of 65 and 55 kDa (Cor a 11) and 22 kDa (Cor a 9) (Figure 1B,C). Faint bands were visible at 17 and 15 kDa (Cor a 1, Cor a 2, and/or Cor a 12). IgG binding to the 40 kDa (Cor a 9) hazelnut protein can be observed by immunoblotting with the serum from rabbit 2. Roasting at 140 °C for 50 min slightly decreased the amount of protein detected in the Western blot for rabbit 1, which may be due to the denaturation of epitopes or entire proteins.

3.4. *Cross-Reactivity with Different Plant Extracts*

Western blotting of different plant extracts used in the skin-prick test was performed to study cross-reactivity with other plant proteins. Antibodies from both rabbit sera recognized the same proteins but with different signal intensities, and both rabbit sera showed binding not only to several hazelnut proteins but also to one Brazil nut and almond 22 kDa protein (11S seed storage globulin). Rabbit 1 serum strongly recognized the three lupin proteins, whereas rabbit 2 serum strongly recognized a distinct lupin protein of approximately 30 kDa. Moreover, rabbit 1 serum antibodies bound to several soy proteins, and a slight 22 kDa (walnut protein) band was observed. Rabbit 2 serum weakly recognizes these proteins (Figure 2).

The cross-reactivity observed in the Western blot was verified by the alignment of the Cor a 9 protein (band 22 kDa). Epitope 63-DHNDQQFQCA and 183-NNYANQLDENPR shared high sequence similarities among the different plant species (Figure 3A). The structural similarities between the proteins of different plant species are shown in Figure 3B–F and their RMDS values in Table 3.

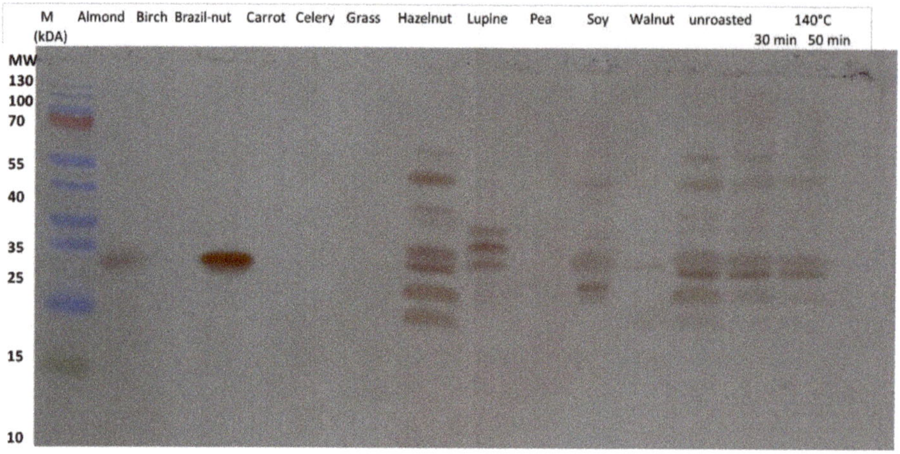

(A)

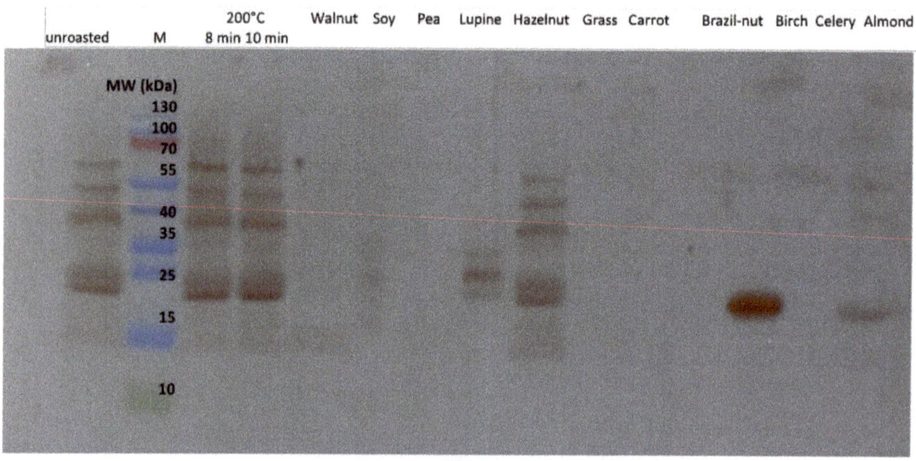

(B)

Figure 2. Western blots with plant extracts used in skin-prick tests of the FoodAllergen project with two rabbit sera after immunization. (**A**) Serum from rabbit 1; (**B**) serum from rabbit 2.

Table 3. RMSD values of structure alignments.

	2D5H	3C3V	3EHK	3KSC	6B4S
2D5H		1386	1309	1397	1283
3C3V	1386		1623	1627	
3EHK	1309	1623			
3KSC	1397	1627			
6B4S	1283				

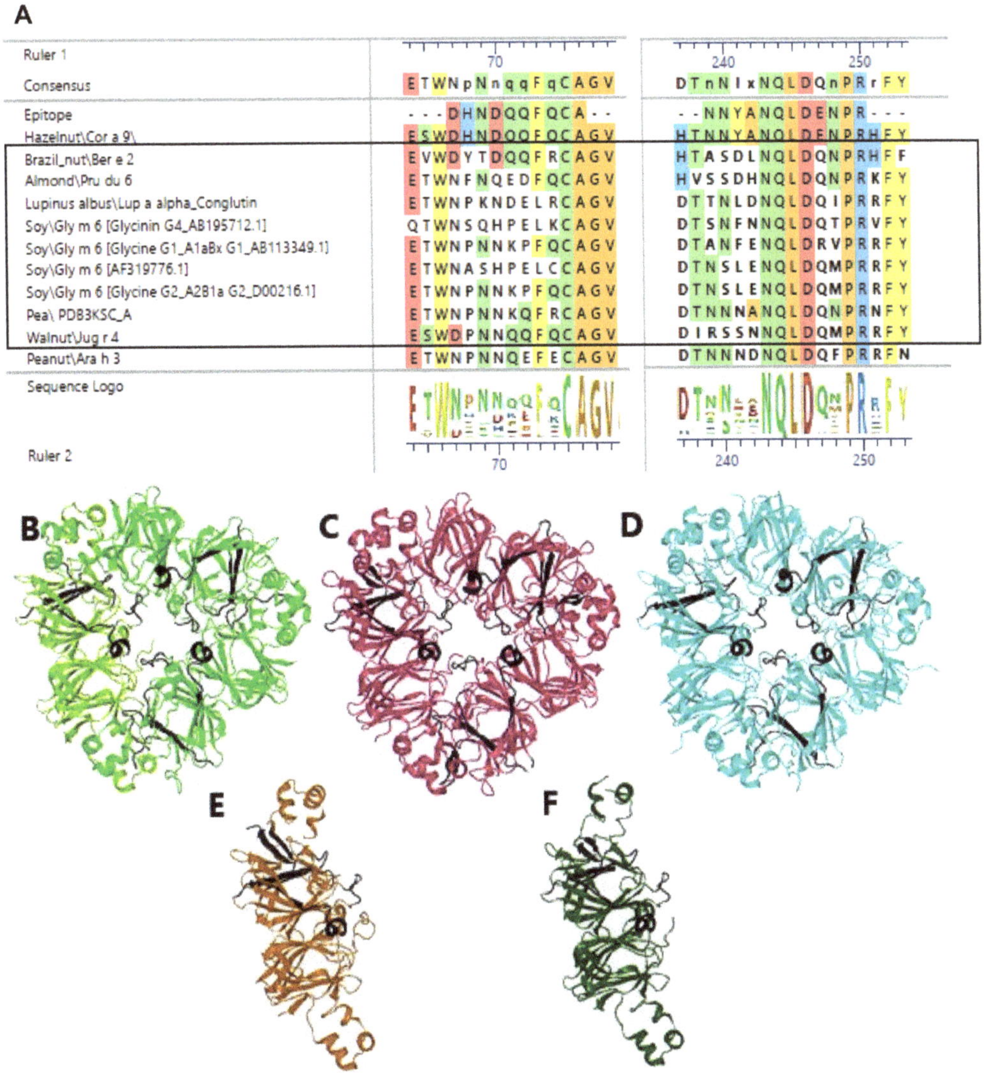

Figure 3. Alignment (**A**) of epitope 63-DHNDQQFQCA and 183-NNYANQLDENPR of the Cor a 9 protein with different plant species. Species detected by rabbit antibodies are marked in boxes. Structure of 11S globulins of different species soybean Gly m 6 2D5H ((**B**), green [48]), almond 3EHK ((**C**), pink [49]), pea 3KSC ((**D**), light blue [48]), peanut Ara h 3 3C3V ((**E**), orange [50]), and Brazil nut 6B4S ((**F**), dark green [51]). Identified epitopes are colored in black.

4. Discussion

Heat treatment or other methods to reduce the allergenicity of food by modifying allergens cannot be measured or quantified using uncharacterized polyclonal antibody mixtures. A large number of individual mAbs cannot cover the variety of individual immune responses. Basophile activation tests require patient material and provide only patient-specific information [52–54]. In a recent study, polyclonal rabbit serum was used for the characterization of the immune response to processed pea proteins [55]. Here, we compared the immune responses of allergic patients with those of rabbits at the epitope level. The similarity of the epitopes could be used to develop more reliable methods, perhaps even standards, to better understand the beneficial effects of food processing on allergenicity reduction.

Our data highlight the advantages of short, highly specific peptide epitopes for detailed analysis of immune responses. This enabled the characterization and comparison of antibodies against hazelnut proteins in humans and immunized animals. Several epitopes in different hazelnut allergens can be identified using rabbit anti-hazelnut polyclonal antisera. Overall, the microarray with peptides based on epitopes identified using human sera revealed that the immune system of vaccinated rabbits recognizes the same epitopes as antibodies present in the serum of individuals. Immunized rabbits have been used several times to detect the immunogenicity of food allergies [17–19]. Those studies showed that epitopes recognized by antibodies from human serum were also recognized by antibodies from rabbit serum. The data presented herein demonstrated that rabbit serum can be used as a surrogate for human serum. Rabbit serum offers several advantages. The B-cell response is better controlled in humans, where cross-reactivity with other plant proteins may obscure the results. It is known what rabbits have been fed on and the resulting potential food-induced immune responses. Proteins from other species detected by antibodies resulting from immunization with a given antigen were most likely due to cross-reactivity (Figure 2). The animals were unintentionally exposed to antigens.

In this study, we investigated the influence of thermal treatment on the in vitro allergenic reactivity of hazelnut proteins at the epitope level. Roasting at temperatures above 180 °C significantly changes the secondary structure. The α-helix content increases with increasing temperature, whereas the β-sheet content decreases [56]. Different heat treatment procedures result in different protein degradation products. The results of this study are in agreement with previous findings that thermal treatment reduces the IgE immunoreactivity of hazelnut proteins [13,57]. Reduced signal intensity for Cor a 1 (PR10) and Cor a 14 (2S albumin) after roasting was reported by Pastorello et al. and Lopez et al. [41,44]. Compared to Lamberti et al., discrepancies were observed for Cor a 11 (7S globulin) and Cor a 14. Lamberti et al. showed that the immune reactivity of Cor a 11 and Cor a 14 was stable after roasting at 140 °C but was reduced after roasting at 170 °C [43]. Denaturation of 2S albumins was observed at temperatures above 120 °C [58]. However, the epitopes in 2S albumins are highly resistant to severe heat treatment [59]. The immunogenicity of other 2S albumins could also be increased [60]. 7S globulin forms aggregates after heating [61], which might lead to an increase in immunoreactivity. The application of different batches of polyclonal sera characterized for recognizing different epitopes, such as R1 and R2 sera in this study, can help monitor different epitopes in the analysis of specific proteins and individual heat-labile epitopes, for example, Cor a 14.

The observed cross-reactivity in the Western blot could be well explained by the sequence alignment (Figure 3) and the highly conserved structure (Figure 3B,C). Surprisingly, denatured pea protein was not detected by Western blotting, although the structure and sequence of pea 7S globulin are very similar to those of hazelnut 7S globulin. Minor changes in the pea protein sequence are sufficient for rabbit antibodies to not detect the pea protein in the Western blot.

5. Conclusions

The allergenicity of food is usually assessed by measuring the total amount of allergenic proteins using monoclonal or polyclonal antibodies. Changes in immunogenicity are regarded as an increase or decrease in the allergenicity of a protein upon treatment. However, allergenic proteins contain multiple epitopes. Variations in individual patients' antibodies can be of individual relevance to the allergenicity of a protein, even when not taking into account cross-reactive antibodies or remaining protein fragments. Overall, the reproducibility of the experiments depends on the antibodies used. This explains the different published results in Table 1. Using polyclonal antibodies that are well characterized at the epitope level could help establish reliable methods. Each of the sera used in this study would lead to different outcomes on the allergenicity of foods after treatments. In this work, we show that a combination of different sera characterized by peptide epitopes (Table 2) should help to compare and standardize immune diagnostics. We show that this is even possible for complex protein preparations. Being able to predict potential cross-reactivity, these analytes could even be mixtures from multiple allergen sources (Figure 2).

Supplementary Materials: The following supporting information can be downloaded at https://www.mdpi.com/article/10.3390/foods13233932/s1: Table S1 showing the measured intensities with a peptide microarray covering the epitopes (Table S1) for different rabbit sera; Table S2 shows a complete version of Table 2. Figure S1 shows the extracts from hazelnut roasted under different conditions at timepoint T0.

Author Contributions: Conceptualization, I.R. and M.S.; methodology, K.K., S.S.-A., N.D. and M.S.; software, K.K. and M.S.; validation, K.K.; formal analysis, K.K.; investigation, K.K., S.S.-A., N.D. and I.R.; resources, N.D., S.D.-B., R.T., E.E.-F., I.R. and M.S.; data curation, K.K. and S.D.-B.; writing—original draft preparation, K.K., R.T. and M.S.; writing—review and editing, K.K., S.S.-A., N.D., S.D.-B., R.T., E.E.-F., I.R., M.W. and M.S.; visualization, K.K. and M.S.; supervision, I.R. and M.S.; project administration, M.S.; funding acquisition, N.D., S.D.-B. and M.S. All authors have read and agreed to the published version of the manuscript.

Funding: This work involved allergen epitopes and peptides from the project "FoodAllergen", funded by Fraunhofer Zukunftsstiftung and the POC4allergies project (Grant Agreement No. 768686), which was funded by ERA PerMed ERA-NET. The contributions of M.W. and S.D.-B. were funded by the German Federal Ministry of Education and Research (01KU2005).

Institutional Review Board Statement: The study was conducted in accordance with the Declaration of Helsinki, and approved by the Institutional Review Board of the Medical Faculty of Leipzig University (AZ202/16-ek 05072016 and 2016/07/05) and Charité (EA01/022/20).

Informed Consent Statement: Not applicable.

Data Availability Statement: The original contributions presented in the study are included in the article and Supplementary Materials, further inquiries can be directed to the corresponding author.

Acknowledgments: The authors thank Lisbeth Ramirez-Caballero for support in the experiment and reading earlier texts and Dorothee Wehrmann for help with epitope identification in human sera.

Conflicts of Interest: M.S. is managing the director and shareholder of Epitopic GmbH (Leipzig, Germany), the company providing the software LibDB used in this manuscript. M.S. and N.D. are inventors of the patented library used in this study. The remaining authors declare no conflicts of interest.

References

1. Verhoeckx, K.C.; Vissers, Y.M.; Baumert, J.L.; Faludi, R.; Feys, M.; Flanagan, S.; Herouet-Guicheney, C.; Holzhauser, T.; Shimojo, R.; van der Bolt, N.; et al. Food processing and allergenicity. *Food Chem. Toxicol.* **2015**, *80*, 223–240. [CrossRef] [PubMed]
2. Liu, X.; Li, Y.; Guasch-Ferré, M.; Willett, W.C.; Drouin-Chartier, J.-P.; Bhupathiraju, S.N.; Tobias, D.K. Changes in nut consumption influence long-term weight change in US men and women. *BMJ Nutr. Prev. Health* **2019**, *2*, 90–99. [CrossRef] [PubMed]
3. Kendall, C.W.C.; Josse, A.R.; Esfahani, A.; Jenkins, D.J.A. Nuts, metabolic syndrome and Diabetes. *Br. J. Nutr.* **2010**, *104*, 465–473. [CrossRef] [PubMed]
4. Spolidoro, G.C.I.; Lisik, D.; Nyassi, S.; Ioannidou, A.; Ali, M.M.; Amera, Y.T.; Rovner, G.; Khaleva, E.; Venter, C.; van Ree, R.; et al. Prevalence of tree nut allergy in Europe: A systematic review and meta-analysis. *Allergy* **2024**, *79*, 302–323. [CrossRef]
5. Lyons, S.A.; Clausen, M.; Knulst, A.C.; Ballmer-Weber, B.K.; Fernandez-Rivas, M.; Barreales, L.; Bieli, C.; Dubakiene, R.; Fernandez-Perez, C.; Jedrzejczak-Czechowicz, M.; et al. Prevalence of Food Sensitization and Food Allergy in Children Across Europe. *J. Allergy Clin. Immunol. Pract.* **2020**, *8*, 2736–2746.e9. [CrossRef]
6. Dölle-Bierke, S.; Höfer, V.; Francuzik, W.; Näher, A.-F.; Bilo, M.B.; Cichocka-Jarosz, E.; de Oliveira, L.C.L.; Fernandez-Rivas, M.; García, B.E.; Hartmann, K.; et al. Food-Induced Anaphylaxis: Data From the European Anaphylaxis Registry. *J. Allergy Clin. Immunology. Pract.* **2023**, *11*, 2069–2079.e7. [CrossRef]
7. Costa, J.; Mafra, I.; Carrapatoso, I.; Oliveira, M.B.P.P. Hazelnut Allergens: Molecular Characterization, Detection, and Clinical Relevance. *Crit. Rev. Food Sci. Nutr.* **2016**, *56*, 2579–2605. [CrossRef]

8. Geroldinger-Simic, M.; Zelniker, T.; Aberer, W.; Ebner, C.; Egger, C.; Greiderer, A.; Nicole, P.; Lidholm, J.; Ballmer-Weber, B.K.; Vieths, S.; et al. Birch pollen-related food allergy: Clinical aspects and the role of allergen-specific IgE and IgG4 Antibodies. *J. Allergy Clin. Immunol.* **2011**, *127*, 616–622.e1. [CrossRef]
9. Cabanillas, B.; Novak, N. Effects of daily food processing on allergenicity. *Crit. Rev. Food Sci. Nutr.* **2019**, *59*, 31–42. [CrossRef]
10. Demir, A.D.; Celayeta, J.M.F.; Cronin, K.; Abodayeh, K. Modelling of the kinetics of colour change in hazelnuts during air roasting. *J. Food Eng.* **2002**, *55*, 283–292. [CrossRef]
11. Sruthi, N.; Premjit, Y.; Pandiselvam, R.; Kothakota, A.; Ramesh, S. An overview of conventional and emerging techniques of roasting: Effect on food bioactive Signatures. *Food Chem.* **2021**, *348*, 129088. [CrossRef] [PubMed]
12. Masthoff, L.J.; Hoff, R.; Verhoeckx, K.C.M.; van Os-Medendorp, H.; Michelsen-Huisman, A.; Baumert, J.L.; Pasmans, S.G.; Meijer, Y.; Knulst, A.C. A systematic review of the effect of thermal processing on the allergenicity of tree nuts. *Allergy* **2013**, *68*, 983–993. [CrossRef] [PubMed]
13. Cuadrado, C.; Sanchiz, A.; Vicente, F.; Ballesteros, I.; Linacero, R. Changes Induced by Pressure Processing on Immunoreactive Proteins of Tree Nut. *Molecules* **2020**, *25*, 954. [CrossRef] [PubMed]
14. Hansen, K.S.; Ballmer-Weber, B.K.; Lüttkopf, D.; Skov, P.S.; Wüthrich, B.; Bindslev-Jensen, C.; Vieths, S.; Poulsen, L.K. Roasted hazelnuts—Allergenic activity evaluated by double-blind, placebo-controlled food challenge. *Allergy* **2003**, *58*, 132–138. [CrossRef]
15. Mondoulet, L.; Paty, E.; Drumare, M.F.; Ah-Leung, S.; Scheinmann, P.; Willemot, R.M.; Wal, J.M.; Bernard, H. Influence of thermal processing on the allergenicity of peanut Proteins. *J. Agric. Food Chem.* **2005**, *53*, 4547–4553. [CrossRef]
16. Worm, M.; Hompes, S.; Fiedler, E.; Illner, A.; Zuberbier, T.; Vieths, S. Impact of native, heat-processed and encapsulated hazelnuts on the allergic response in hazelnut-allergic Patients. *Clin. Exp. Allergy* **2009**, *39*, 159–166. [CrossRef]
17. Mattison, C.P.; Bren-Mattison, Y.; Vant-Hull, B.; Vargas, A.M.; Wasserman, R.L.; Grimm, C.C. Heat-induced alterations in cashew allergen solubility and IgE binding. *Toxicol. Rep.* **2016**, *3*, 244–251. [CrossRef] [PubMed]
18. Noble, K.A.; Liu, C.; Sathe, S.K.; Roux, K.H. A Cherry Seed-Derived Spice, Mahleb, is Recognized by Anti-Almond Antibodies Including Almond-Allergic Patient IgE. *J. Food Sci.* **2017**, *82*, 1786–1791. [CrossRef]
19. Hofer, H.; Asam, C.; Hauser, M.; Nagl, B.; Laimer, J.; Himly, M.; Briza, P.; Ebner, C.; Lang, R.; Hawranek, T.; et al. Tackling Bet v 1 and associated food allergies with a single hybrid protein. *J. Allergy Clin. Immunol.* **2017**, *140*, 525–533.e10. [CrossRef]
20. Matsuo, H.; Yokooji, T.; Taogoshi, T. Common food allergens and their IgE-binding epitopes. *Allergol. Int.* **2015**, *64*, 332–343. [CrossRef]
21. Hensen, S.M.M.; Derksen, M.; Pruijn, G.J.M. Multiplex peptide-based B cell epitope mapping. *Methods Mol. Biol.* **2014**, *1184*, 295–308. [CrossRef] [PubMed]
22. Frank, R. The SPOT-synthesis technique. Synthetic peptide arrays on membrane supports—Principles and applications. *J. Immunol. Methods* **2002**, *267*, 13–26. [CrossRef] [PubMed]
23. Matsuo, H.; Morita, E.; Tatham, A.S.; Morimoto, K.; Horikawa, T.; Osuna, H.; Ikezawa, Z.; Kaneko, S.; Kohno, K.; Dekio, S. Identification of the IgE-binding epitope in omega-5 gliadin, a major allergen in wheat-dependent exercise-induced anaphylaxis. *J. Biol. Chem.* **2004**, *279*, 12135–12140. [CrossRef]
24. Chatchatee, P.; Järvinen, K.M.; Bardina, L.; Vila, L.; Beyer, K.; Sampson, H.A. Identification of IgE and IgG binding epitopes on beta- and kappa-casein in cow's milk allergic patients. *Clin. Exp. Allergy* **2001**, *31*, 1256–1262. [CrossRef]
25. Li, Y.; Gao, Y.; Wang, Y.; Duan, Y.; Fu, Y.; Yang, H.; Xi, J. Localization of an IgE epitope of glycinin A2 peptide chain. *J. Sci. Food Agric.* **2024**, *104*, 3697–3704. [CrossRef]
26. Perez-Gordo, M.; Lin, J.; Bardina, L.; Pastor-Vargas, C.; Cases, B.; Vivanco, F.; Cuesta-Herranz, J.; Sampson, H.A. Epitope mapping of Atlantic salmon major allergen by peptide microarray immunoassay. *Int. Arch. Allergy Immunol.* **2012**, *157*, 31–40. [CrossRef]
27. Desmet, C.; Coelho-Cruz, B.; Mehn, D.; Colpo, P.; Ruiz-Moreno, A. ASFV epitope mapping by high density peptides microarrays. *Virus Res.* **2024**, *339*, 199287. [CrossRef] [PubMed]
28. Havenith, H.; Kern, K.; Rautenberger, P.; Spiegel, H.; Szardenings, M.; Ueberham, E.; Lehmann, J.; Buntru, M.; Vogel, S.; Treudler, R.; et al. Combination of two epitope identification techniques enables the rational design of soy allergen Gly m 4 mutants. *Biotechnol. J.* **2017**, *12*. [CrossRef]
29. Heidarinia, H.; Tajbakhsh, E.; Rostamian, M.; Momtaz, H. Epitope mapping of Acinetobacter baumannii outer membrane protein W (OmpW) and laboratory study of an OmpW-derivative peptide. *Heliyon* **2023**, *9*, e18614. [CrossRef]
30. Pomés, A. Relevant B cell epitopes in allergic disease. *Int. Arch. Allergy Immunol.* **2010**, *152*, 1–11. [CrossRef]
31. Sharma, P.; Gaur, S.N.; Arora, N. In silico identification of IgE-binding epitopes of osmotin protein. *PLoS ONE* **2013**, *8*, e54755. [CrossRef] [PubMed]
32. Szardenings, M.; Delaroque, N.; Kern, K.; Ramirez-Caballero, L.; Puder, M.; Ehrentreich-Förster, E.; Beige, J.; Zürner, S.; Popp, G.; Wolf, J.; et al. Detection of Antibodies against Endemic and SARS-CoV-2 Coronaviruses with Short Peptide Epitopes. *Vaccines* **2023**, *11*, 1403. [CrossRef] [PubMed]
33. Kern, K.; Havenith, H.; Delaroque, N.; Rautenberger, P.; Lehmann, J.; Fischer, M.; Spiegel, H.; Schillberg, S.; Ehrentreich-Foerster, E.; Aurich, S.; et al. The immunome of soy bean allergy: Comprehensive identification and characterization of epitope. *Clin. Exp. Allergy.* **2019**, *49*, 239–251. [CrossRef]
34. Cariccio, V.L.; Domina, M.; Benfatto, S.; Venza, M.; Venza, I.; Faleri, A.; Bruttini, M.; Bartolini, E.; Giuliani, M.M.; Santini, L.; et al. Phage display revisited: Epitope mapping of a monoclonal antibody directed against Neisseria meningitidis adhesin A using the PROFILER technology. *mAbs* **2016**, *8*, 741–750. [CrossRef]

35. Comeau, A.M.; Douglas, G.M.; Langille, M.G.I. Microbiome Helper: A Custom and Streamlined Workflow for Microbiome Research. *mSystems* **2017**, *2*. [CrossRef] [PubMed]
36. Bolger, A.M.; Lohse, M.; Usadel, B. Trimmomatic: A flexible trimmer for Illumina sequence data. *Bioinformatics* **2014**, *30*, 2114–2120. [CrossRef]
37. Kern, K.; Delaroque, N.; Boysen, A.; Puder, M.; Wendt, R.; Kölsch, A.; Ehrentreich-Förster, E.; Stærk, K.; Andersen, T.E.; Andersen, K.; et al. Glycosylation of bacterial antigens changes epitope pattern. *Front. Immunol.* **2023**, *14*, 1258136. [CrossRef]
38. Dooper, M.M.; Plassen, C.; Holden, L.; Moen, L.H.; Namork, E.; Egaas, E. Antibody binding to hazelnut (*Corylus avellana*) proteins: The effects of extraction procedure and hazelnut source. *Food Agric. Immunol.* **2008**, *19*, 229–240. [CrossRef]
39. Schägger, H. Tricine–SDS-PAGE. *Nat. Protoc.* **2006**, *1*, 16–22. [CrossRef]
40. Towbin, H.; Staehelin, T.; Gordon, J. Electrophoretic transfer of proteins from polyacrylamide gels to nitrocellulose sheets: Procedure and some applications. *Proc. Natl. Acad. Sci. USA* **1979**, *76*, 4350–4354. [CrossRef]
41. Pastorello, E.A.; Vieths, S.; Pravettoni, V.; Farioli, L.; Trambaioli, C.; Fortunato, D.; Lüttkopf, D.; Calamari, M.; Ansaloni, R.; Scibilia, J.; et al. Identification of hazelnut major allergens in sensitive patients with positive double-blind, placebo-controlled food challenge Results. *J. Allergy Clin. Immunol.* **2002**, *109*, 563–570. [CrossRef] [PubMed]
42. Wigotzki, M.; Steinhart, H.; Paschke, A. Influence of Varieties, Storage and Heat Treatment on IgE-Binding Proteins in Hazelnuts (*Corylus avellana*). *Food Agric. Immunol.* **2000**, *12*, 217–229. [CrossRef]
43. Lamberti, C.; Nebbia, S.; Antoniazzi, S.; Cirrincione, S.; Marengo, E.; Manfredi, M.; Smorgon, D.; Monti, G.; Faccio, A.; Giuffrida, M.G.; et al. Effect of hot air and infrared roasting on hazelnut allergenicity. *Food Chem.* **2021**, *342*, 128174. [CrossRef] [PubMed]
44. López, E.; Cuadrado, C.; Burbano, C.; Jiménez, M.A.; Rodríguez, J.; Crespo, J.F. Effects of autoclaving and high pressure on allergenicity of hazelnut Proteins. *J. Clin. Bioinform.* **2012**, *2*, 12. [CrossRef]
45. Fu, L.; He, Z.; Zeng, M.; Qin, F.; Chen, J. Effects of preheat treatments on the composition, rheological properties, and physical stability of soybean oil Bodies. *J. Food Sci.* **2020**, *85*, 3150–3159. [CrossRef]
46. Barazorda-Ccahuana, H.L.; Theiss-De-Rosso, V.; Valencia, D.E.; Gómez, B. Heat-Stable. Hazelnut Profilin: Molecular Dynamics Simulations and Immunoinformatics Analysis. *Polymers* **2020**, *12*, 1742. [CrossRef]
47. Müller, U.; Lüttkopf, D.; Hoffmann, A.; Petersen, A.; Becker, W.M.; Schocker, F.; Niggemann, B.; Altmann, F.; Kolarich, D.; Haustein, D.; et al. Allergens in raw and roasted hazelnuts (*Corylus avellana*) and their cross-reactivity to pollen. *Eur. Food Res. Technol.* **2000**, *212*, 2–12. [CrossRef]
48. Tandang-Silvas, M.R.G.; Fukuda, T.; Fukuda, C.; Prak, K.; Cabanos, C.; Kimura, A.; Itoh, T.; Mikami, B.; Utsumi, S.; Maruyama, N. Conservation and divergence on plant seed 11S globulins based on crystal structures. *Biochim. Biophys. Acta* **2010**, *1804*, 1432–1442. [CrossRef]
49. Gaur, V.; Salunke, D.M. 3EHK: Crystal Structure of Pru du Amandin, an Allergenic Protein from Prunus Dulcis. 2009. Available online: www.rcsb.org (accessed on 27 November 2024).
50. Jin, T.; Guo, F.; Chen, Y.-W.; Howard, A.; Zhang, Y.-Z. Crystal structure of Ara h 3, a major allergen in peanut. *Mol. Immunol.* **2009**, *46*, 1796–1804. [CrossRef]
51. Zhang, Y.Z.; Guo, F. 6B4S: Crystal Structure of Brazil nut (Bertholletia excelsa) Allergen Ber e 2. 2018. Available online: www.rcsb.org (accessed on 27 November 2024).
52. Fernández-Lozano, C.; Rita, C.G.; Barra-Castro, A.; Caballer, L.H.; Roldán, E.; López, C.P.; Martinez-Botas, J.; Berges-Gimeno, M.P. Specific IgE and Basophil Activation Test by Microarray: A Promising Tool for Diagnosis of Platinum Compound Hypersensitivity Reaction. *Int. J. Mol. Sci.* **2024**, *25*, 3890. [CrossRef]
53. Briceno Noriega, D.; Teodorowicz, M.; Savelkoul, H.; Ruinemans-Koerts, J. The Basophil Activation Test for Clinical Management of Food Allergies: Recent Advances and Future Directions. *J. Asthma Allergy* **2021**, *14*, 1335–1348. [CrossRef] [PubMed]
54. Santos, A.F.; Alpan, O.; Hoffmann, H. Basophil activation test: Mechanisms and considerations for use in clinical trials and clinical practice. *Allergy* **2021**, *76*, 2420–2432. [CrossRef] [PubMed]
55. García Arteaga, V.; Demand, V.; Kern, K.; Strube, A.; Szardenings, M.; Muranyi, I.; Eisner, P.; Schweiggert-Weisz, U. Enzymatic Hydrolysis and Fermentation of Pea Protein Isolate and Its Effects on Antigenic Proteins, Functional Properties, and Sensory Profile. *Foods* **2022**, *11*, 118. [CrossRef]
56. Ma, X.; Li, H.; Zhang, J.; Ge, Y.; He, L.; Kang, W.; Huang, W.; Sun, J.-L.; Chen, Y. Effect of Roasting on the Conformational Structure and IgE Binding of Sesame Allergens. *J. Agric. Food Chem.* **2022**, *70*, 9442–9450. [CrossRef]
57. Cuadrado, C.; Arribas, C.; Sanchiz, A.; Pedrosa, M.M.; Gamboa, P.; Betancor, D.; Blanco, C.; Cabanillas, B.; Linacero, R. Effects of enzymatic hydrolysis combined with pressured heating on tree nut allergenicity. *Food Chem.* **2024**, *451*, 139433. [CrossRef] [PubMed]
58. Moreno, F.J.; Clemente, A. 2S Albumin Storage Proteins: What Makes them Food Allergens? *Open Biochem. J.* **2008**, *2*, 16–28. [CrossRef]
59. Clemente, A.; Chambers, S.J.; Lodi, F.; Nicoletti, C.; Brett, G.M. Use of the indirect competitive ELISA for the detection of Brazil nut in food products. *Food Control* **2004**, *15*, 65–69. [CrossRef]

60. Starkl, P.; Krishnamurthy, D.; Szalai, K.; Felix, F.; Lukschal, A.; Oberthuer, D.; Sampson, H.A.; Swoboda, I.; Betzel, C.; Untersmayr, E.; et al. Heating Affects Structure, Enterocyte Adsorption and Signalling, As Well as Immunogenicity of the Peanut Allergen Ara h 2. *Open Allergy J.* **2011**, *4*, 24–34. [CrossRef]
61. Chihi, M.-L.; Mession, J.-L.; Sok, N.; Saurel, R. Heat-Induced Soluble Protein Aggregates from Mixed Pea Globulins and β-Lactoglobulin. *J. Agric. Food Chem.* **2016**, *64*, 2780–2791. [CrossRef]

beverages

MDPI

Article

Changes in the Profile of Phenolic Compounds and in the Antioxidant, Hypoglycemic, and Antidiabetic Activities of a Beverage Based on Sesame By-Product Caused by the Simulated Gastrointestinal Digestion Process

Jonhatan Carrasco-Chávez [1], María Fernanda Quintero-Soto [1,2,*], Rosalio Velarde-Barraza [3], Ignacio Alfredo Rivero-Espejel [3], Ismael Díaz-Peña [4], Martha Elena Vázquez-Ontiveros [1], Ramona Julieta Espinoza-Moreno [1], Luz Adriana Ontiveros-García [1], Jesús Mateo Amillano-Cisneros [1], Janitzio Xiomara K. Perales-Sánchez [5], Oscar Daniel Argüelles-López [5], Fernando Salas-López [6] and Jennifer Vianey Félix-Medina [2,4]

1 Maestría en Biotecnología Agropecuaria, Universidad Politécnica del Mar y la Sierra, Carretera a Potrerillos del Norote Km 3, La Cruz, Elota 82700, Sinaloa, Mexico; 2023030003@upmys.edu.mx (J.C.-C.); mvazquez@upmys.edu.mx (M.E.V.-O.); respinoza@upmys.edu.mx (R.J.E.-M.); lontiveros@upmys.edu.mx (L.A.O.-G.); jamillano@upmys.edu.mx (J.M.A.-C.)

2 Posdoc CONAHCYT, Universidad Politécnica del Mar y la Sierra, Carretera a Potrerillos del Norote Km 3, La Cruz, Elota 82700, Sinaloa, Mexico; jevianey9@gmail.com

3 Centro de Graduados e Investigación en Química, Instituto Tecnológico de Tijuana Unidad Otay-Tecnológico Nacional de México, Blvd. Industrial s/n, Tijuana 22000, Baja California, Mexico; rosaliovb@gmail.com (R.V.-B.); irivero@tectijuana.mx (I.A.R.-E.)

4 Instituto Tecnológico de Culiacán-Tecnológico Nacional de México, Calle Juan de Dios Bátiz 310 pte, Colonia Guadalupe, Culiacán de Rosales 80220, Sinaloa, Mexico; diaz_1986@hotmail.es

5 Facultad de Ciencias Químicas Biológicas, Universidad Autónoma de Sinaloa, Calle Josefa Ortiz de Domínguez, esq. Blvd de las Américas, Ciudad Universitaria, Culiacán 80010, Sinaloa, Mexico; janitzio.perales@uas.edu.mx (J.X.K.P.-S.); oscararguelles@uas.edu.mx (O.D.A.-L.)

6 Facultad de Ciencias Biomédicas, Universidad Autónoma de Occidente, UR Culiacán. Blvd. Lola Beltrán y Blvd. Rolando Arjona, 4 de marzo, Culiacán 80020, Sinaloa, Mexico; fernando.salas@uadeo.mx

* Correspondence: mquintero@upmys.edu.mx

Abstract: The by-product of extracting oil from sesame has good functional properties for use in the development of new food products. In this study, the effect of simulated gastrointestinal digestion on the bioaccessibility of phenolic compounds, as well as the antioxidant, antidiabetic, and hypoglycemic properties of a beverage from sesame by-products was analyzed. Oral digestion significantly decreased the total phenolic content of the beverage, while the total flavonoid content increased. Both phytochemicals increased after gastric and intestinal digestion. Twenty-five phenolic compounds were characterized in the sesame by-product beverage, including feruloylquinic acid, which was present in all the digestive fractions analyzed. The most abundant free phenolic was quercetin, which was also present in all the digested fractions. On the other hand, the most abundant phenolic in the digested fractions was isorhamnetin. During gastrointestinal digestion, the antioxidant activity and inhibitory effects on the α-glucosidase and DPP-IV of the beverage significantly improved, while the ability to inhibit the α-amylase significantly increased during oral digestion, remaining constant throughout digestion. Correlation analysis indicated that flavonoids, including quercetin, may be the compounds with the greatest effect on the evaluated activities. The results of this study not only improve the understanding of the impact of gastrointestinal digestion on the bioaccessibility of phenolic compounds but also suggest potential applications in formulating functional foods with enhanced antioxidant, hypoglycemic, and antidiabetic properties, contributing to the development of health-promoting food products.

Keywords: α-amylase inhibition; α-glucosidase inhibition; dipeptidyl peptidase-IV inhibition; phenolic acids; flavonoids; HPLC-DAD-MS/MS

1. Introduction

The food industry is one of the main generators of waste or by-products worldwide, which in some cases go directly to landfills, generating a negative environmental impact [1]. The production of food, where conventional raw materials are insufficient to meet global demand, has led to the overexploitation of crops, causing an increase in food costs and poor distribution [2,3]. Therefore, it is important to seek appropriate solutions to establish economically and environmentally sustainable parameters that support the transformation of the food industry towards a circular bioeconomy [3,4]. By-products, with proper processing, can be used as raw materials in the creation of new products or foods. As a result, there has been significant interest in valorizing food waste for the development of new value-added products [1,2]. These by-products represent alternative sources to conventional raw materials, helping to reduce the excessive consumption of the latter, which has been gradually depleting [1].

The by-product of sesame seed oil extraction is often discarded as a secondary raw material despite having a good content of nutritional compounds and phytochemicals [5,6]. When the oil is extracted from the seed (which represents more than 50% of the original components), the remaining compounds increase, mainly protein, fiber, minerals, and phenolic compounds [7,8]. These latter compounds are responsible for the various biological activities of the by-product [9].

Unlike other by-products from the food industry, the by-product of sesame oil extraction retains its techno-functional properties, as well as its nutritional and nutraceutical qualities [10]; this makes it suitable for use as a non-conventional raw material in the production of various food products such as cookies and beverages [5,6]. In this regard, Lucini Mas et al. [5] made wheat cookies enriched with sesame by-product, finding that the addition did not alter the functional parameters of the cookies but significantly increased the antioxidant activity and phenolic compound content by three times the expected values. This was attributed to the generation of Maillard reaction compounds formed during the baking process and the increased bioavailability of phenolic compounds. In a previous study, it was observed that heating combined with the shear force generated by the extrusion process modified the phenolic compound profile of a beverage based on sesame by-product, enhancing its antioxidant and antidiabetic activity [6]. Despite observing an increase in antioxidant and antidiabetic activities due to the extrusion process, it is important to understand the bioavailability of the phenolic compounds present in the beverage during gastrointestinal digestion, as well as the effect that digestion has on biological activity.

Generally, the digestibility of food has a negative impact on the bioavailability of compounds and their biological activities [11]. Most digestibility studies focus on understanding changes in the profile of compounds and antioxidant activity. Chen et al. [8] indicated that the thermal processing applied to six varieties of sesame increased the content of phenolic compounds, flavonoids, and antioxidant activity after roasting. However, they observed a decrease in antioxidant activity when the samples were subjected to the in vitro gastrointestinal digestion process, but depending on the sesame variety, this characteristic either decreased or improved compared to the raw sample. This is due to the type of phenolic compound and the food matrix subjected to the process. On the other hand, Tomé-Sánchez et al. [12] observed that for different wheat-based matrices (germinated wheat, spray-dried wheat bran, and microencapsulated wheat bran), the effects of gastrointestinal digestion were different. In the case of wheat brans, antioxidant activity either decreased or increased depending on the evaluation method (ABTS, DPPH, FRAP, ORAC), while for germinated wheat, an increase in activity was observed during gastric and intestinal digestion. Luo et al. [13] evaluated the effect of gastrointestinal digestion on the antioxidant activity and phenolic compounds of black, brown, and white sesame seeds. It was observed that black and brown sesame seeds presented better stability to antioxidant activity during the digestion process, but the phenolic compounds of beige sesame seeds were the ones that presented the best bioavailability. Among the 12 compounds identified by these authors, syringic acid was the one that presented the best bioavailability during the different phases of digestion (oral, gastric, and intestinal) of the three types of sesame.

Only one report was found where the hypoglycemic activity of sesame by-product was evaluated after the in vitro gastrointestinal digestion process. Doungwichitrkul et al. [14] isolated and identified 29 phytochemical compounds, mainly furofuran lignans, in the black sesame by-product and evaluated its inhibitory effect against the rat intestinal α-glucosidase. The sesaminol triglucoside compound did not present inhibitory effects against the said enzyme. However, when subjected to the gastrointestinal digestion process, its potential to inhibit the α-glucosidase enzyme increased; this is due to the transformations it underwent during digestion (elimination of a glucose group and two methylene groups).

Other studies focus on examining the importance of phenolic compounds in plant extracts, foods, or beverages and are limited to analyzing their content and their bioactivity. This study is one of the few that focuses on analyzing the effects of in vitro gastrointestinal digestion on the phenolic compound profile and their bioactivities, especially on the phenolic compounds of the sesame by-product-based beverage. No reports were found that evaluate the effect of in vitro gastrointestinal digestion on the antidiabetic activity of sesame or sesame by-products. Therefore, the objective of this research is to determine the changes caused by the gastrointestinal digestion process in the phenolic compound profile, as well as the antioxidant activity, hypoglycemic activity, and antidiabetic activity of a novel beverage made from extruded sesame by-products.

2. Materials and Methods

2.1. Preparation of the Functional Beverage

The methodology reported by Quintero-Soto et al. [6] was followed for the formulation of the beverage, where a 10% concentration of the by-product previously selected as the sensory acceptance was highest. Twenty grams of previously extruded by-product flour [10] was mixed with 5 g of sugar substitute (BC Metco® Sugar) and 200 mL of purified water. The mixture was stirred (500 rpm/20 min) and stored at 8 °C until further use.

2.2. Gastrointestinal Digestion Process

The digestion process was carried out in three phases, as described by Cárdenas-Castro et al. [15] and Lucini Mas et al. [5], with some modifications. For the oral phase, 200 mg of lyophilized beverage was mixed and digested with a human salivary α-amylase solution (A0521, Sigma-Aldrich, St. Louis, MO, USA, 1000–3000 units/mg protein; 100 μL of a 0.049 mg/mL solution in 0.05 M phosphate buffer, pH 6.9, 37 °C, for 2 min). For the gastric phase, a pepsin solution was added to the samples (P-7000, Sigma-Aldrich, ≥250 units/mg solid; 400 μL of a 300 mg/mL solution in 0.2 M HCl-KCl buffer, pH 1.5, 40 °C) and placed in a water bath with agitation for 2 h. For the intestinal digestion, a pancreatin solution (P-1750, Sigma-Aldrich, 4 × USP specifications; 3 mL of a 5 mg/mL solution in 0.1 M $NaHCO_3$, pH = 7.5) and bile salts (1,071,304, US Pharmacopeia, USP, Rockville, Maryland, USA; 2 mL of a 25 mg/mL solution in 0.1 M $NaHCO_3$, pH = 7.5) were added. The final mixture was placed in a cellulose membrane (6000–8000 Da, 25.5 mm × 30 m, Fisherbrand™, Waltham, MA, USA) to simulate the small intestine and dialyzed in 0.1 M $NaHCO_3$ (pH = 7.5) with agitation (150 rpm, 3 h, 37 °C). The fraction that passed through the membrane was considered the fraction of compounds available for passive diffusion into the circulatory system in the small intestine. The fraction remaining inside the membrane was considered the indigestible fraction, which was mixed with 10 mL of methanol (400 rpm/30 min) and centrifuged to recover the supernatant containing the non-digestible phenolic compounds.

During each phase of digestion, aliquots were collected, adjusted to pH 7.0, and washed with ethyl acetate to recover phenolic compounds. The recovered ethyl acetate was evaporated to dryness, and the samples were reconstituted in 2 mL of 80% methanol for analysis by HPLC-DAD-MS/MS. These same aliquots were used to determine biological activities.

2.3. Determination of In Vitro Bioaccessibility of Released Phenolic Compounds

Bioaccessibility expressed as a percentage was calculated as the ratio of phenolic compounds released from the beverage during digestion and the initial content of phenolic compounds in the beverage before digestion [11], using the following formula:

$$\%\ \text{bioaccessibility} = \left(\frac{\text{Phenolics after the digestion process}}{\text{Phenolics before the digestion process}} \right) \times 100$$

Values over 100% are indicative of an increase in the bioaccessible amount of phenolic compounds in each digestion phase. The bioaccessibility of phenolic compounds that were not initially found in the undigested beverage (UD) was considered to be 100%.

2.4. Characterization of the Profile of Phenolic Compounds

The phenolic compound profile of the undigested, digested (oral, gastric, and intestinal phases), and indigestible samples was performed using HPLC-DAD (1100 Series, Agilent Technologies, Santa Clara, CA, USA) coupled to a mass detector (1100 Series LC/MSD Trap, Agilent Technologies, USA) as reported by Quintero-Soto et al. [6]. Twenty-five microliters of the sample were separated on a Zorbax SB-C18 column (150 × 3 mm, 5 μm, Agilent Technologies, Santa Clara, CA, USA) using water-formic acid (99:1 v/v) and acetonitrile as mobile phases (A and B, respectively). A linear gradient from 1 to 60% of B over 60 min, with a flow rate of 0.4 mL/min and detection wavelengths of 280 nm (hydroxybenzoic acids and flavan-3-ols), 320 nm (hydroxycinnamic acids), and 350 nm (flavonoids) were used. The identification of phenolic compounds was based on retention time, UV spectra, and mass spectra. The mass spectra were acquired on a mass spectrometer with an ion trap analyzer and an electrospray ionization source. The mass spectrometer parameters for compound identification were as follows: positive and negative ion scanning at 35 V and 300 °C, collision-induced dissociation fragmentation (10–45 V), m/z = 100 to 2000, nitrogen for drying, and helium for collision. Data was analyzed with Agilent ChemStation for LC 3D software (Rev. A. 09. 01 [1206] Agilent Technologies, USA) and LC/MSD Trap software 5.2 (Build 63.8, Bruker Daltonik GmbH). The individual phenolic compound content was calculated using commercial standard curves.

2.5. Quantification of Total Phenolic Compounds and Flavonoids

The total phenolic content (TPC) was determined by mixing 0.2 mL of the sample and 2.2 mL of Folin-Ciocalteu reagent in the dark for 3 min. Then, 0.6 mL of sodium carbonate (7%) was added, and the samples were kept in the dark for 90 min. Finally, the absorbance of the samples was measured at 750 nm and compared to a gallic acid standard curve [16]. Total flavonoid content was determined as described by Xu and Chang [17] by mixing 0.1 mL of the sample, 0.5 mL of water, and 0.25 mL of sodium nitrate (5%) in the dark for 6 min. Subsequently, 0.5 mL of silver chloride (5%) and 0.25 mL of sodium hydroxide (1 M) were added to the mixture. The absorbance of the samples was measured at 510 nm and compared to a quercetin standard curve.

2.6. Determination of Antioxidant Activity

The in vitro antioxidant activity was determined using the ABTS [18] and DPPH [19] methods. For ABTS, three milliliters of the ABTS radical previously activated in potassium persulfate (2.6 mM) and adjusted with phosphate buffer (10 mM, pH 7.4) to an absorbance of 0.70 at 734 nm were mixed with 0.75 mL of the sample and incubated for 6 min at room temperature. The samples were then monitored at 734 nm. For DPPH, 20 μL of the sample was mixed with 180 μL of DPPH radical solution (0.1 mM) and incubated for 30 min at 37 °C. The absorbance of the samples was measured at 510 nm. A trolox standard curve was used in both assays to calculate the results.

2.7. Determination of Hypoglycemic Activity

The hypoglycemic activity was determined by the inhibition of the α-amylase and α-glucosidase enzymes [6]. Ten microliters of the sample and 10 μL of α-amylase solution (2 U/mL) were mixed and incubated for 10 min at 25 °C. Then, 10 μL of soluble starch (1%) was added, and the mixture was incubated again (10 min at 25 °C). Subsequently, 20 μL of a 3,5-dinitrosalicylic acid solution was added and incubated for 5 min at 25 °C. Finally, 200 μL of distilled water was added, and the absorbance was read at 540 nm. The results were expressed as the percentage inhibition of the α-amylase enzyme (AA) and IC_{50}.

For the determination of α-glucosidase enzyme inhibition, 50 μL of the sample and 100 μL of α-glucosidase solution (1 U/mL) were mixed and incubated for 10 min at 37 °C. Subsequently, 50 μL of *p*-nitrophenyl-*α*-D-glucopyranoside (5 mM) was added, followed by incubation (10 min at 37 °C). Finally, the absorbance was read at 405 nm. The results were expressed as the percentage inhibition of the α-glucosidase enzyme (AG) and IC_{50}.

2.8. Determination of Antidiabetic Activity

The antidiabetic activity was determined by the samples' potential to inhibit the dipeptidyl-peptidase IV (DPP-IV) enzyme using the MAK203 kit (Sigma-Aldrich, St. Louis, MO, USA). The results were expressed as the percentage inhibition of the DPPIV enzyme (%DPP-IV) and IC_{50}.

2.9. Statistical Analysis

The statistical analyses were performed in triplicate (n = 3), and the data were analyzed through a mean comparison using Fisher's least significant difference test ($p < 0.05$) with the STATGRAPHIC plus version 5.1 program (Statistical Graphics Corporation™, Rockville, MD, USA). Pearson correlation analysis and principal component analysis were conducted using the MINITAB 17 program (Minitab, LLC., PA, USA).

3. Results and Discussion

3.1. Changes in the Profile and Content of Phenolic Compounds After the Gastrointestinal Digestion Process

Table 1 shows the profile of phenolic compounds in the sesame seed oil by-product beverage before and during gastrointestinal digestion, as well as the compounds available for the microbiota. In total, 25 phenolic compounds were identified, 13 of which were phenolic acids and 12 flavonoids. Some of these had been previously identified in sesame and its products [6–8].

Table 1. Tentative identification of phenolic compounds from sesame by-product beverage before and after in vitro gastrointestinal digestion.

#	TR (min)	*m/z* [M-H]⁻	Error (ppm)	MS/MS (Relative Intensity, %)	Proposed Compounds *
1	2.20	169.45	−1.95	125 (100), 97 (8)	Gallic acid [20]
2	6.11	153.35	−1.50	109 (100)	Protocatechuic acid [21]
3	6.53	341.35	−0.56	162 (100), 135 (25)	Caffeic acid hexoside [22]
4	7.01	167.35	−1.26	151 (36), 108 (100)	Vanillic acid [6]
5	8.22	179.59	−2.40	145 (20), 135 (100), 117 (36)	Caffeic acid [6]
6	9.05	353.89	−1.64	191 (100), 179 (10), 173 (37)	Caffeoylquinic acid [23]
7	9.32	325.11	−0.22	163 (100), 119 (41)	*p*-Coumaric acid hexoside [24]
8	10.56	289.89	−2.18	245 (100), 221 (7), 203 (3)	Catechin [20]
9	18.11	163.55	−3.07	146 (62), 119 (100)	p-Coumaric acid [24]
10	19.23	353.68	−1.05	191 (100), 177 (25), 145 (59)	Chlorogenic acid [6]
11	23.06	355.71	−1.38	193 (28), 117 (69)	Ferulic acid hexoside [20]
12	25.63	147.66	−3.47	103 (100), 122 (23), 102 (13)	Cinnamic acid [21]
13	25.98	193.22	−0.21	175 (100), 115 (42), 143 (49)	Ferulic acid [6]
14	26.15	771.89	−0.47	609 (56), 301 (100)	Rutin hexoside [25]

Table 1. *Cont.*

#	TR (min)	*m/z* [M-H]-	Error (ppm)	MS/MS (Relative Intensity, %)	Proposed Compounds *
15	27.46	447.96	−1.63	285 (100), 229 (23), 216 (12)	Kaempferol hexoside [24]
16	27.68	623.85	−0.35	463 (34), 301 (100)	Quercetin dihexoside [26]
17	27.99	609.96	−0.74	301 (100), 300 (88), 151 (10)	Rutin [20]
18	30.12	463.37	−0.30	301 (100), 179 (12)	Quercetin hexoside [27]
19	30.56	367.87	−2.07	134 (30), 193 (100), 278 (10)	Feruloilquinic acid [12]
20	31.62	301.87	−2.12	179 (100), 151 (7)	Quercetin [20]
21	32.13	285.67	−1.51	193 (16), 151 (100)	Luteolin [6]
22	34.19	431.78	−1.55	269 (100), 117 (56)	Apigenin hexoside [28]
23	35.01	285.35	−0.46	257 (100), 229 (13), 216 (19)	Kaempferol [20]
24	35.26	268.98	0.26	161 (13), 152 (100), 117 (52)	Apigenin [6]
25	36.97	315.43	−0.54	301 (10), 300 (100), 285 (49)	Isorhamnetin [6]

TR: retention time. * The numbers in brackets correspond to the reference number.

Of the 25 phenolic compounds identified, only 12 were observed in the undigested sample (UD), while the other 13 were generated by the degradation and transformation of the initial phenolic compounds. Quercetin was present in the highest concentration in UD, while isorhamnetin was found in the three digestive phases and in the fraction available for the intestinal microbiota (Table 2).

The phenolic compounds identified in this study have shown various health effects. Gallic acid, in addition to its antioxidant activity, has been reported to attenuate ulcerative colitis in vivo [29]. Protocatechuic acid promotes the lysosomal autophagy pathway and alleviates neuronal damage in in vivo and in vitro models of Alzheimer's disease [30]. Caffeic acid mitigates myocardial fibrosis and improves cardiac function after myocardial infarction by inhibiting the signaling pathways of the transforming growth factor-β receptor 1 [31]; additionally, caffeic acid and its derivatives, such as chlorogenic acid (3-caffeoylquinic acid) and 5-caffeoylquinic acid, have been observed to have anti-inflammatory activity by inhibiting proinflammatory cytokines and increasing anti-inflammatory cytokine secretions [32].

On the other hand, flavonoids like catechin demonstrated antioxidant, antidiabetic, anti-infectious, and anticancer effects, specifically against gastric cancer cells [33]. Quercetin and rutin have also demonstrated anti-inflammatory activity [34,35]. Kaempferol has shown positive effects on metabolic diseases like diabetes mellitus, obesity, non-alcoholic fatty liver disease, steatohepatitis, and atherosclerosis [36]. Apigenin and its derivatives have demonstrated antimicrobial, anticancer, anti-inflammatory, and antioxidant effects [37]. Isorhamnetin has shown various pharmacological properties associated with the prevention of neurodegenerative disorders like Alzheimer's disease [38], bladder cancer [39], breast cancer [40], colon cancer [41], and ulcerative colitis [42].

Changes in the profile of phenolic compounds during gastrointestinal digestion have been reported by various authors and are dependent on the type of matrix, the concentration of the compound, and whether the compound is in its free, bound, or conjugated form [5,8,11,12,15,43,44]. The changes observed in the phenolic profile of the beverage are described below.

Gallic acid first appeared at the beginning of digestion in the oral phase (OP) (Table 2), with a significant increase observed during the gastric phase (GP) ($p < 0.05$). Subsequently, there was a marked decrease caused by the dehydroxylation of gallic acid, leading to the formation of protocatechuic acid in the intestinal phase (IP).

Table 2. Quantification of phenolic compounds from sesame by-product beverage at different stages of gastrointestinal digestion.

Phytochemical	UD mg/100 g (mg/100 mL)	OP mg/100 g (mg/100 mL)	GP mg/100 g (mg/100 mL)	IP mg/100 g (mg/100 mL)	ID mg/100 g (mg/100 mL)	LOD mg/mL
Gallic acid	ND ND	7.28 ± 0.06^{b} (0.91 ± 0.01^{b})	9.46 ± 0.37^{a} (1.18 ± 0.05^{a})	4.78 ± 0.21^{c} (0.60 ± 0.03^{c})	0.25 ± 0.02^{d} (0.03 ± 0.00^{d})	0.0012
Protocatechuic acid	16.73 ± 0.60^{a} (2.09 ± 0.07^{a})	ND ND	ND ND	3.06 ± 0.05^{c} (0.38 ± 0.01^{c})	10.06 ± 0.66^{b} (1.26 ± 0.08^{b})	0.0025
Caffeic acid hexoside	0.76 ± 0.11^{b} (0.09 ± 0.01^{b})	ND ND	1.91 ± 0.07^{a} (0.24 ± 0.01^{a})	1.88 ± 0.07^{a} (0.24 ± 0.01^{a})	ND ND	0.0006
Vanillic acid	ND ND	ND ND	ND ND	64.35 ± 4.50 (8.04 ± 0.56)	ND ND	0.0750
Caffeic acid	ND ND	0.71 ± 0.03^{b} (0.09 ± 0.00^{b})	4.58 ± 0.33^{a} (0.57 ± 0.04^{a})	ND ND	ND ND	0.0006
Caffeoylquinic acid	4.44 ± 0.14^{b} (0.55 ± 0.02^{b})	4.31 ± 0.35^{b} (0.54 ± 0.04^{b})	ND ND	14.09 ± 0.97^{a} (1.76 ± 0.12^{a})	2.17 ± 0.03^{c} (0.27 ± 0.00^{c})	0.0004
p-Coumaric acid hexoside	ND ND	ND ND	ND ND	ND ND	11.38 ± 0.81 (1.42 ± 0.10)	0.0099
Catechin	19.85 ± 0.08^{b} (2.48 ± 0.01^{b})	4.77 ± 0.11^{d} (0.60 ± 0.01^{d})	9.86 ± 0.67^{c} (1.23 ± 0.08^{c})	10.49 ± 0.14^{c} (1.31 ± 0.02^{c})	39.42 ± 3.42^{a} (4.93 ± 0.43^{a})	0.0029
p-Coumaric acid	ND ND	ND ND	ND ND	3.18 ± 0.04^{a} (0.40 ± 0.01^{a})	1.86 ± 0.03^{b} (0.23 ± 0.00^{b})	0.0099
Chlorogenic acid	30.18 ± 1.33^{a} (3.77 ± 0.17^{a})	22.30 ± 0.33^{b} (2.79 ± 0.04^{b})	ND ND	ND ND	15.65 ± 1.78^{c} (1.96 ± 0.22^{c})	0.0241
Ferulic acid hexoside	34.59 ± 0.58^{a} (4.32 ± 0.11^{a})	21.10 ± 1.04^{b} (2.64 ± 0.13^{b})	19.05 ± 0.40^{c} (2.38 ± 0.05^{c})	19.09 ± 0.10^{c} (2.39 ± 0.01^{c})	11.00 ± 0.47^{d} (1.37 ± 0.06^{d})	0.0150
Cinnamic acid	14.25 ± 0.13^{b} (1.78 ± 0.02^{b})	6.96 ± 0.43^{e} (0.87 ± 0.05^{e})	12.27 ± 0.50^{c} (1.53 ± 0.06^{c})	24.82 ± 0.62^{a} (3.10 ± 0.08^{a})	8.86 ± 0.13^{d} (1.11 ± 0.02^{d})	0.0099
Ferulic acid	ND ND	12.54 ± 0.038^{c} (1.57 ± 0.05^{c})	26.42 ± 1.61^{b} (3.30 ± 0.20^{b})	ND ND	49.55 ± 1.62^{a} (6.19 ± 3.27^{a})	0.0151
Rutin hexoside	12.02 ± 0.14^{a} (1.50 ± 0.02^{a})	ND ND	ND ND	ND ND	11.79 ± 0.49^{a} (1.47 ± 0.06^{a})	0.0019
Kaempferol hexoside	ND ND	ND ND	ND ND	ND ND	72.97 ± 0.49 (9.12 ± 0.06)	0.0094
Quercetin dihexoside	41.39 ± 0.62 (5.17 ± 0.08)	ND ND	ND ND	ND ND	ND ND	0.0016
Rutin	ND ND	9.71 ± 0.16^{b} (1.21 ± 0.02^{b})	10.10 ± 0.23^{a} (1.26 ± 0.03^{a})	ND ND	4.97 ± 0.09^{c} (0.62 ± 0.01^{c})	0.0005
Quercetin hexoside	25.07 ± 0.84 (3.13 ± 0.10)	ND ND	ND ND	ND ND	ND ND	0.0016
Feruloilquinic acid	22.30 ± 2.03^{b} (2.79 ± 0.25^{b})	21.38 ± 1.09^{b} (2.67 ± 0.14^{b})	14.53 ± 0.64^{c} (1.82 ± 0.08^{c})	14.50 ± 0.13^{c} (1.81 ± 0.02^{c})	32.40 ± 0.54^{a} (4.05 ± 0.07^{a})	0.0150

Table 2. *Cont.*

Phytochemical	UD mg/100 g (mg/100 mL)	OP mg/100 g (mg/100 mL)	GP mg/100 g (mg/100 mL)	IP mg/100 g (mg/100 mL)	ID mg/100 g (mg/100 mL)	LOD mg/mL
Quercetin	295.98 ± 5.69^{c} (37.00 ± 0.71^{c})	404.79 ± 12.35^{b} (50.60 ± 1.54^{b})	414.13 ± 5.28^{b} (51.77 ± 0.66^{b})	448.96 ± 1.49^{a} (56.12 ± 0.19^{a})	ND ND	0.0015
Luteolin	ND ND	11.93 ± 0.42^{a} (1.49 ± 0.05^{a})	ND ND	8.19 ± 0.18^{b} (1.02 ± 0.02^{b})	ND ND	0.0023
Apigenin hexoside	25.80 ± 1.60^{a} (3.23 ± 0.20^{a})	ND ND	ND ND	16.80 ± 1.32^{b} (2.10 ± 0.16^{b})	ND ND	0.0057
Kaempferol	ND ND	ND ND	99.02 ± 5.52^{b} (12.38 ± 0.69^{b})	110.88 ± 0.51^{a} (13.86 ± 0.06^{a})	ND ND	0.0094
Apigenin	ND ND	16.10 ± 0.57^{a} (2.01 ± 0.07^{a})	15.91 ± 0.59^{a} (1.99 ± 0.07^{a})	16.81 ± 0.22^{a} (2.11 ± 0.03^{a})	ND ND	0.0057
Isorhamnetin	ND ND	710.48 ± 15.39^{a} (88.81 ± 0.20^{a})	696.68 ± 1.63^{a} (87.08 ± 0.20^{a})	496.02 ± 7.97^{b} (62.00 ± 1.00^{b})	708.92 ± 14.58^{a} (88.61 ± 1.82^{a})	0.0709

UD: Undigested (free extract), OP: Oral phase, GP: Gastric phase, IP: Intestinal phase, ID: Indigestible fraction (available to the microbiota), LOD: Limit of detection (mg/mL), ND: Not detected. The results are expressed in mg/100 g dw and mg/100 mL of beverage (values in parentheses). Values (means ± SD) with different letters within a row are significantly different at $p < 0.05$ among the means according to Fisher's test.

Caffeic acid hexoside was one of the phenolics observed for the first time in UD and then disappeared in OP as a result of the hydrolysis of the hexose molecule, generating caffeic acid. This acid appeared for the first time in OP. During GP, a significant increase in caffeic acid ($p < 0.05$) was observed, resulting from the hydrolysis of the quinic group of caffeoylquinic acid and chlorogenic acid. During IP, caffeic acid is transformed into p-coumaric acid by the dehydroxylation of the hydroxyl group. Both caffeic acid hexoside and caffeoylquinic acid reappeared in IP, likely as a result of hydrolysis caused by the mixture of enzymes. Phenolic compounds bound to the food matrix can be released under the acidic conditions of the stomach, where proteins, lipids, and other macromolecules begin to denature, releasing phenolic compounds bound to the food matrix such as caffeic acid hexoside and caffeoylquinic acid or other precursors involved in their synthesis. Ali et al. [45] demonstrated the binding of caffeoylquinic acid to coffee bean storage protein.

Ferulic acid hexoside and feruloylquinic acid were present both in UD and in all phases of digestion. A significant decrease in their concentration was observed as digestion progressed (Table 2). This decrease is reflected in the generation of ferulic acid in OP due to the hydrolysis of the hexose from some ferulic acid hexoside molecules. Ferulic acid more than doubled during GP as a result of the release of the quinic group from feruloylquinic acid.

As for flavonoids, rutin hexoside was observed in UD, which, during OP, lost the hexoside group and was transformed into rutin. Rutin appeared for the first time in this phase and remained stable in GP. During IP, the rutinoside group was hydrolyzed, leading to the formation of quercetin. On the other hand, quercetin dihexoside and quercetin hexoside were present in UD, and both flavonoids were transformed into quercetin during OP through the hydrolysis of their hexose molecules. These transformations were reflected in the significant increase ($p < 0.05$) in quercetin concentration during OP. With regard to apigenin, it appeared for the first time in OP due to the hydrolysis of the hexose group from apigenin hexoside, which was initially identified in UP. The concentration of apigenin remained stable throughout the gastrointestinal digestion process. Throughout the three phases of digestion, the presence of isorhamnetin was observed. This flavonoid was not among those identified in UD. It first appeared in OP, suggesting that it was bound to some component of the beverage and was likely released as a result of hydrolysis caused by the enzyme α-amylase. Isorhamnetin remained stable during OP and GP, but a significant decrease ($p > 0.05$) of around 30% was observed in IP.

In the fraction available to the intestinal microbiota, that is, the indigestible (ID) fraction, 15 compounds were identified. Among them were p-coumaric acid hexoside and kaempferol hexoside, which were only observed in this fraction. The presence of isorhamnetin also stands out. This is relevant since this phenolic compound has been shown to have positive effects on colon cancer [41]. Unabsorbed flavonoids play an important role in the intestinal environment by modulating the diversity and microbial composition of the gut microbiota [46]. Aglycones, such as isorhamnetin, are transformed by intestinal microbial enzymes to produce a variety of ring-fission products, including 3,4-dihydroxyphenylacetic acid, 3-hydroxyphenylacetic acid, protocatechuic acid, and hippuric acid [47,48]. Isorhamnetin has been reported to enhance the abundance and diversity of gut microbiota in mice, with dominant genera mainly including Bacteroidales S24-7 group_norank, Erysipelotrichaceae_uncultured, Lactobacillus, Lachnospiraceae_uncultured, Allobaculum, Lachnospiraceae NK4A136 group, Turicibacter, and Lachnoclostridium [49]. It also improves lipid metabolism in patients with type 2 diabetes mellitus due to the reduction in Firmicutes and Bacteroidetes ratios [49].

The total phytochemical content and its behavior during gastrointestinal digestion are shown in Table 3. In UD, a total phenolic compound (TPC) content of 178.19 mg GAE/100 g was observed, similar to the values reported by Lucini Mas et al. [5] for a cookie made from sesame by-products. During OP, a 50% decrease in TPC was observed, followed by an increase of approximately 65% and 50% in GP and IP, respectively. Regarding the total flavonoid content (TFC), a significant increase ($p < 0.05$) of 100%, 130%, and 150% was observed in OP, GP, and IP, respectively, compared to the initial concentration (61.64 mg QE/100 g). The behavior observed for TPC and TFC is consistent with the findings reported by Chen et al. [8] for six varieties of sesame subjected to an in vitro gastrointestinal digestion process. Luo et al. [13] reported a gradual increase in the TPC content of beige, brown, and black sesame seeds during all phases of digestion.

Table 3. Total phenolic and flavonoid content from sesame by-product beverage at different stages of gastrointestinal digestion.

Sample	Total Phenolic		Total Flavonoid	
	(mg GAE/100 g)	(mg GAE/100 mL)	(mg QE/100 g)	(mg QE/100 mL)
UD	178.19 ± 2.31 [b]	22.27 ± 2.31 [b]	61.64 ± 2.58 [e]	7.71 ± 0.32 [e]
OP	88.19 ± 6.96 [e]	11.02 ± 0.87 [e]	124.03 ± 8.57 [c]	15.50 ± 1.07 [c]
GP	153.35 ± 3.43 [c]	19.17 ± 0.43 [c]	159.15 ± 1.50 [b]	19.89 ± 0.19 [b]
IP	230.10 ± 4.76 [a]	28.76 ± 0.63 [a]	187.12 ± 3.33 [a]	23.39 ± 0.42 [a]
ID	107.86 ± 4.76 [d]	13.48 ± 0.60 [d]	86.91 ± 4.93 [d]	10.86 ± 0.62 [d]

UD: Undigested (free extract, OP: Oral phase, GP: Gastric phase, IP: Intestinal phase, ID: Indigestible fraction (available to the microbiota). The results are expressed in mg/100 g freeze-dried beverage dw and mg/100 mL of beverage. Values (means ± SD) with different letters within a column are significantly different at $p < 0.05$ among the means according to Fisher's test.

Regarding bioaccessibility, TPC values of 50%, 86%, and 129% were observed for OP, GP, and IP, respectively (Table 4). These results are contrary to those reported by Jara-Palacios et al. [50], who found higher TPC bioaccessibility in GP (128%) than in IP (75%) for white grape vinification by-products (pulp). For TFC, values of 201%, 258%, and 304% were observed for OP, GP, and IP, respectively. This behavior aligns with the findings reported by Jara-Palacios et al. [50], (30% in GP and 47% in IP) and Liu et al. [51] in Prinsepia utilis Royle fruits (30% in GP and 47% in IP). Among individual phenolics, quercetin showed the highest bioaccessibility (137%) in OP, while caffeic acid hexoside was the most bioaccessible in GP (251%) and IP (247%). These values of bioaccessibility are higher than those reported by Sęczyk et al. [11] for phenolic compounds in different types of wheat (21% to 95%) and similar to those reported by Tomé-Sánchez et al. [12] for germinated whole wheat (11% to 300%), spray-dried wheat bran (51% to 583%), and microencapsulated wheat bran (6% to 120%) subjected to gastrointestinal digestion.

Table 4. Bioaccessibility percentage of the phenolic compounds from sesame by-product beverage at different stages of gastrointestinal digestion.

Compounds	Bioaccessibility *		
	OP	GP	IP
Gallic acid	100	100	100
Protocatechuic acid	ND	ND	18
Caffeic acid hexoside	ND	251	247
Vanillic acid	ND	ND	100
Caffeic acid	100	100	ND
Caffeoylquinic acid	97	ND	317
p-Coumaric acid hexoside	100	100	100
Catechin	24	50	53
p-Coumaric acid	ND	ND	100
Chlorogenic acid	74	ND	ND
Ferulic acid hexoside	61	56	55
Cinnamic acid	49	86	174
Ferulic acid	100	100	ND
Rutin hexoside	ND	ND	ND
Kaempferol hexoside	ND	ND	ND

Table 4. Cont.

Compounds	Bioaccessibility *		
	OP	GP	IP
Quercetin dihexoside	ND	ND	ND
Rutin	100	100	ND
Quercetin hexoside	ND	ND	ND
Feruloilquinic acid	96	65	65
Quercetin	137	140	152
Luteolin	100	ND	100
Apigenin hexoside	ND	ND	65
Kaempferol	ND	100	100
Apigenin	100	100	100
Isorhamnetin	100	100	100
Total phenolic	50	86	129
Total flavonoid	201	258	304

OP: Oral phase, GP: Gastric phase, IP: Intestinal phase, ND: Not detected. * Bioaccessibility expressed as a percentage was calculated as the ratio of phenolic compounds released from the beverage during digestion and the initial content of phenolic compounds in the beverage before digestion, values over 100% are indicative of an increase in the bioaccessible amount of phenolic compounds in each digestion phase.

3.2. *Effect of the Gastrointestinal Digestion Process on Antioxidant, Hypoglycemic, and Antidiabetic Properties*

In Figure 1, the antioxidant activity (AAOX) values measured by ABTS (Figure 1A,B) and DPPH (Figure 1C,D) for the beverage before and after gastrointestinal digestion are shown. The AAOX by ABTS for UD was 1913 µmol TE/100 g, higher than the value reported by Lucini Mas et al. [5] for a sesame-based cookie (1080 µmol TE/100 g). Five hundred milliliters of the beverage provide 1195 µmol TE of AAOX, equivalent to consuming a nixtamalized blue corn tortilla (1066 µmol TE/30 g) [52]. During gastrointestinal digestion, the AAOX by ABTS experienced a significant decrease ($p < 0.05$) in OP, followed by an increase in GP and IP. This behavior is similar to that of TPC. For DPPH, the AAOX was 1388 µmol TE/100 g, values higher than those observed by Chen et al. [8] for six varieties of sesame (650 to 800 µmol TE/100 g). On the other hand, 500 mL of the beverage contains 865 µmol TE for AAOX, values within the range reported by Mainente et al. [53] for 24 types of cider (390 to 973.5 µmol TE/500 mL). During the gastrointestinal digestion process, a gradual increase was observed, reaching an AAOX of 2191.70 µmol TE/100 g in IP, increasing by around 40%.

In Figure 2, the IC_{50} values (µg/mL) for the inhibition of α-amylase (Figure 2A) and α-glucosidase (Figure 2B) are shown. The ability of phenolic compounds to inhibit the enzyme α-amylase (AA) increased at the beginning of gastrointestinal digestion, remaining statistically unchanged throughout the three phases of digestion (IC_{50}: 836.13 to 859.15 µg/mL). These values are higher than those reported by Kartelias et al. [54] for kombucha tea supplemented with different herbs, spices, and fruits (158 to 243 µg/mL).

The same authors also evaluated the inhibition of α-glucosidase (AG), reporting better IC_{50} values (0.85 to 22.24 µg/mL) than those observed in this study, where an IC_{50} of 219.78 µg/mL was obtained in UD. These values improved during gastrointestinal digestion, with IC_{50} values of 202.60, 172.29, and 125.69 µg/mL for OP, GP, and IP, respectively. This behavior is similar to that observed in fruits of Rhus chinensis by Fu et al. [55], who reported an increase in AG enzyme inhibition as digestion progressed.

The antidiabetic activity of the beverage was determined by the inhibition of the DPP-IV enzyme (Figure 3). In this study, a similar behavior was observed among the inhibition of DPP-IV, α-glucosidase, and TFC. In UD, an IC_{50} value of 93.36 µg/mL was obtained, a value similar to that reported for the phytochemical extract of a beverage made from extruded sesame by-products (110 µg/mL) [6] and for fruits of *Prinsepia utilis* Royle (150 µg/mL) [51]. During gastrointestinal digestion, antidiabetic activity increased (82.02,

70.24, and 52.97 µg/mL for OG, GP, and IP, respectively). This is similar to what was reported by Liu et al. [51] for gastrointestinally digested fruits of *Prinsepia utilis* Royle.

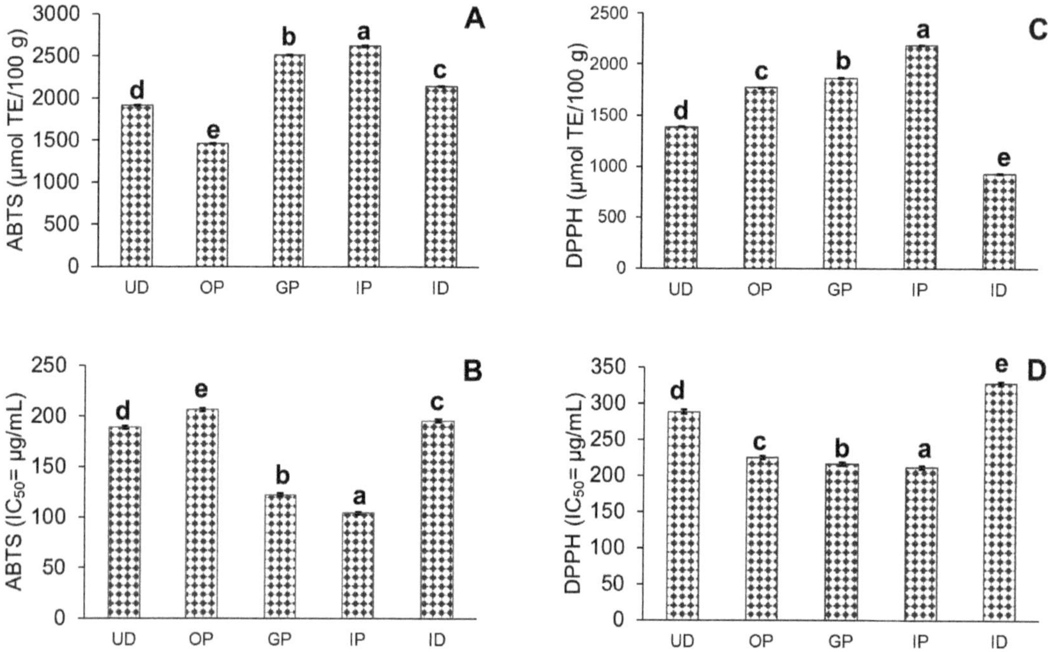

Figure 1. Effect of in vitro digestion on the antioxidant activity of the beverage of by-product sesame seeds. The results of ABTS are expressed in µmol TE/100 g (**A**) and IC_{50} (**B**). The results of DPPH are expressed in µmol TE/100 g (**C**) and IC_{50} (**D**). UD: Undigested (free extract), OP: Oral phase, GP: Gastric phase, IP: Intestinal phase, ID: Indigestible fraction (available to the microbiota). Different letters in the same graph indicate differences ($p < 0.05$) among the means according to Fisher's test.

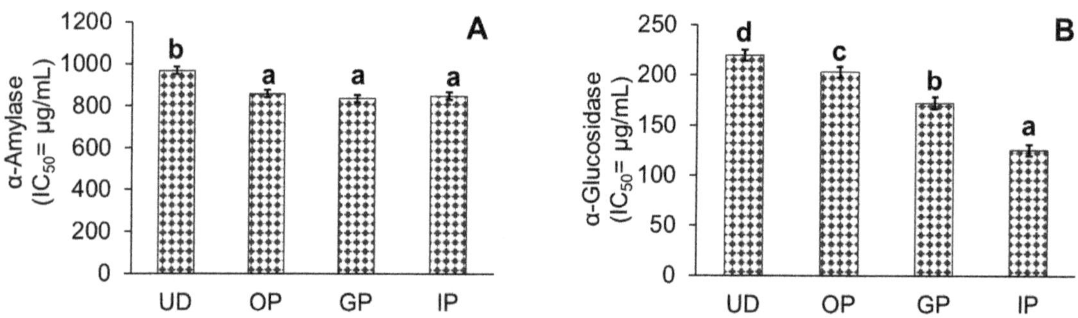

Figure 2. Effect of in vitro digestion on the hypoglycemic activity of the beverage of by-product sesame seeds. Inhibition of α-amylase (**A**) and Inhibition of α-glucosidase (**B**). UD: Undigested (free extract), OP: Oral phase, GP: Gastric phase, IP: Intestinal phase. Different letters in the same graph indicate differences ($p < 0.05$) among the means according to Fisher's test.

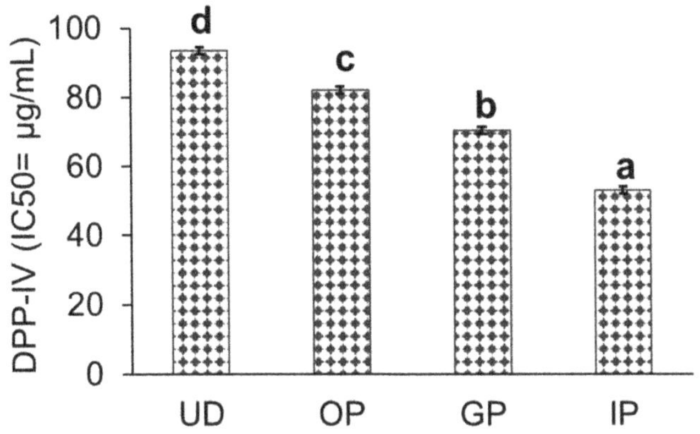

Figure 3. Effect of in vitro digestion on the inhibition of dipeptidyl-Peptidase IV (DPP-IV) of the beverage of by-product sesame seeds. UD: Undigested (free extract), OP: Oral phase, GP: Gastric phase, IP: Intestinal phase. Different letters in the graph indicate differences ($p < 0.05$) among the means according to Fisher's test.

3.3. Relationship Between the Phenolic Compound Profile and the Evaluated Biological Activities

To establish the relationships and similarities between the identified phytochemicals and the evaluated activities, a Pearson correlation analysis (Figure 4) and a principal component analysis (PCA) (Figure 5) were performed.

The positive and highly linear correlations between quercetin and the inhibition of DPPH (r = 0.94; $p > 0.01$), AA (r = 0.98; $p > 0.05$), AG (r = 0.95; $p > 0.05$), and DPP-IV (r = 0.96; $p > 0.01$) suggest that this flavonoid is responsible for these activities. This is also reflected in the PCA, where the vectors are oriented in the same direction. In the case of AAOX, a positive correlation with ABTS vs. TPC (r = 0.79; $p > 0.05$) was observed, while DPPH showed positive correlations with TFC (r = 0.082; $p > 0.05$) and apigenin (r = 0.89; $p > 0.05$). The TPC also showed a positive correlation with DPP-IV (r = 0.72; $p > 0.05$) and α-glucosidase inhibition (r = 0.71; $p > 0.05$). These correlations have been previously reported by various authors [6,20,24].

In addition to positive correlations, negative correlations were observed between bioactive activities and the evaluated compounds. Inhibition of the DPP-IV enzyme showed a negative correlation with feruloylquinic acid (r = −0.096; $p > 0.05$), kaempferol (r = −0.89; $p > 0.05$), and catechin (r = −0.87; $p > 0.05$). Inhibition of α-glucosidases showed a negative correlation with feruloylquinic acid (r = −0.95; $p > 0.05$), kaempferol hexoside (r = −0.89; $p > 0.05$), and ferulic acid (r = −0.80; $p > 0.01$).

The inhibition of α-amylase showed a negative correlation with feruloylquinic acid (r = −0.90; $p > 0.05$), kaempferol hexoside (r = −0.99; $p > 0.001$), catechin (r = −0.92; $p > 0.05$), and p-coumaric acid (r = −0.99; $p > 0.05$). AAOX by DPPH correlated negatively with feruloylquinic acid (r = −0.94; $p > 0.05$), rutin hexoside (r = −0.88; $p > 0.05$), and catechin (r = −0.88; $p > 0.05$).

The PCA showed the differences and similarities between UD, OP, GP, IP, and ID (Figure 5). Together, the first two components explain approximately 73% of the observed variation in the data. The first component (PC1) accounted for 46.6%, while the second component (PC2) accounted for 26.14%. Regarding the distribution of parameters, 22 of the 32 analyzed characteristics were located in the positive quadrant of PC1, including ABTS, DPPH, AA, AG, and DPP-IV. The PCA revealed a separation of the samples into three main clusters: one for the digested samples (purple oval), one for the sample before digestion (green oval), and another for the sample available for the intestinal microbiota (blue oval). IP was positioned in the positive quadrant of the two principal components, indicating that the in vitro gastrointestinal digestion process had a significant positive impact on the samples.

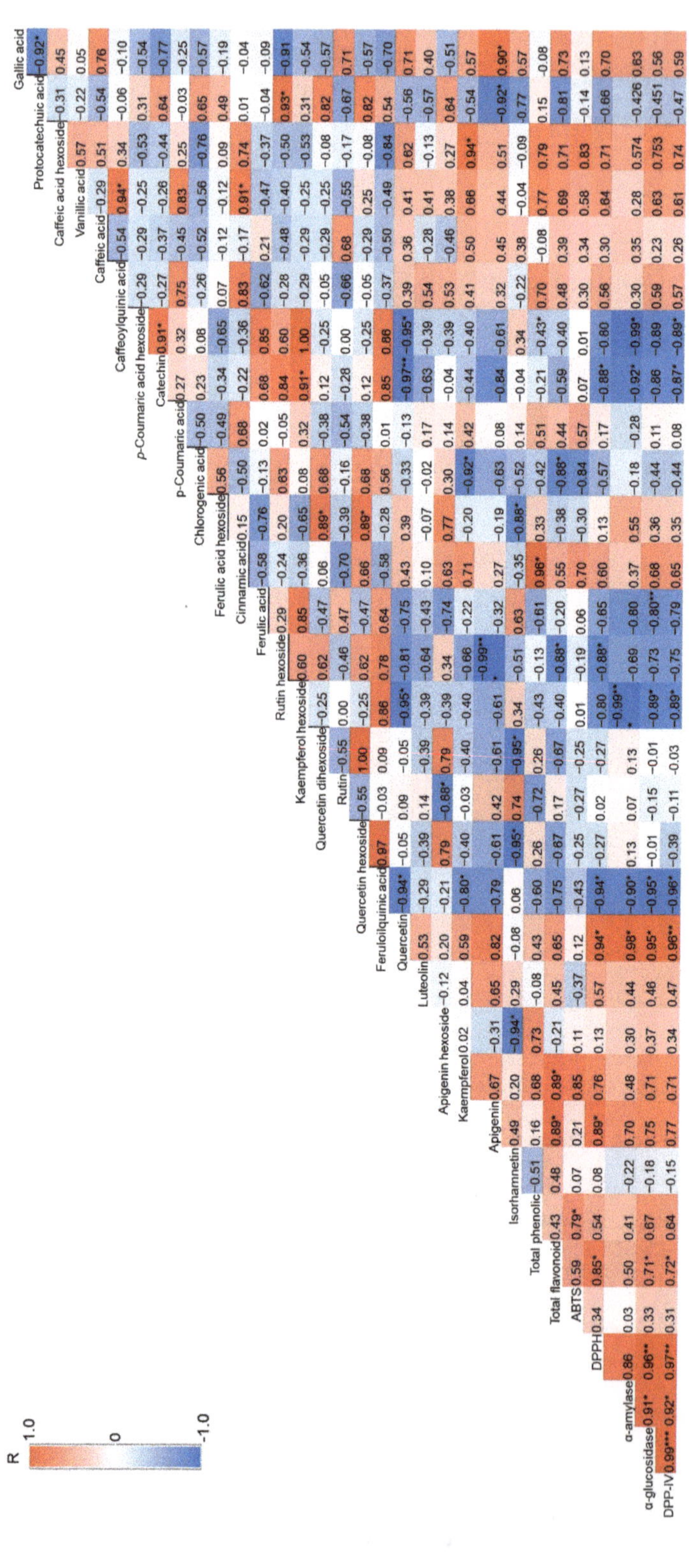

Figure 4. Correlation analysis between identified phytochemicals and evaluated biological activities. *** Correlation is significant at the 0.001 level, ** Correlation is significant at the 0.01 level, * Correlation is significant at the 0.05 level.

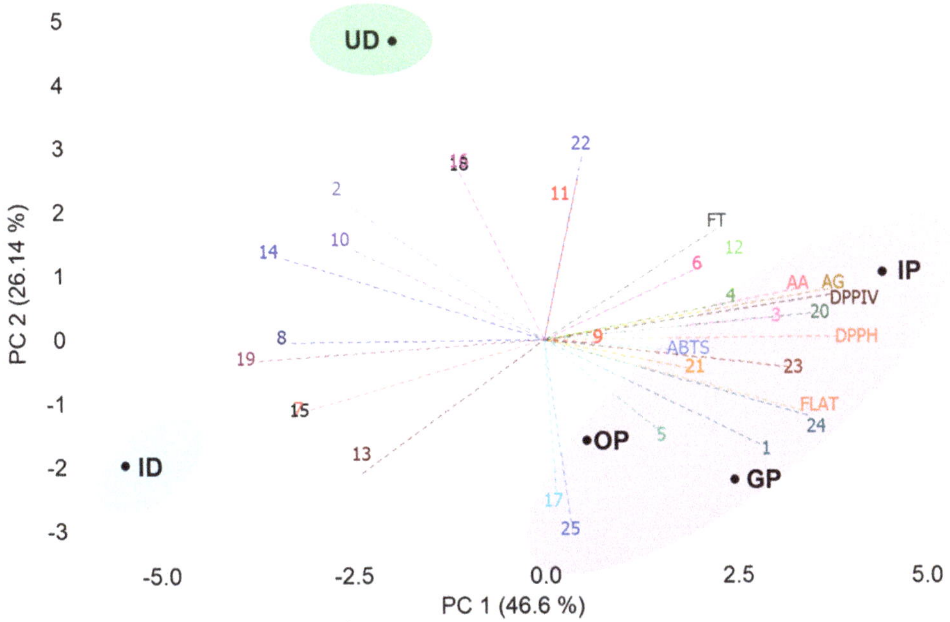

Figure 5. Principal component analysis describing the relationship among different phytochemicals and evaluated biological activities. 1: Gallic acid, 2: Protocatechuic acid, 3: Caffeic acid hexoside, 4: Vanillic acid, 5: Caffeic acid, 6: Caffeoylquinic acid, 7: p-Coumaric acid hexoside, 8: Catechin, 9: p-Coumaric acid, 10: Chlorogenic acid, 11: Ferulic acid hexoside, 12: Cinnamic acid, 13: Ferulic acid, 14: Rutin hexoside, 15: Kaempferol hexoside, 16: Quercetin dihexoside, 17: Rutin, 18: Quercetin hexoside, 19: Feruloilquinic acid, 20: Quercetin, 21: Luteolin, 22: Apigenin hexoside, 23: Kaempferol, 24: Apigenin, 25: Isorhamnetin, FT: Total phenolic, FLAT: Total flavonoid, AA: α-Amylase, α-glucosidase, DPPIV: Dipeptidyl peptidase IV, UD: Undigested (free extract), OP: Oral phase, GP: Gastric phase, IP: Intestinal phase.

4. Conclusions

This study demonstrates the significant impact of gastrointestinal digestion on the compound profile and biological activities of a beverage based on sesame by-products. Notably, compounds like quercetin, isorhamnetin, and caffeic acid hexoside showed strong bioaccessibility after digestion and are linked to enhanced antioxidant, hypoglycemic, and antidiabetic activities. These findings suggest that the digestion process not only improves the bioaccessibility of these bioactive compounds but also potentially enhances their health benefits. Compounds like quercetin showed strong correlations with these activities, suggesting their critical role in mediating health benefits. The principal component analysis highlighted distinct clusters among the samples, reflecting the changes brought about by digestion. Overall, the findings support the hypothesis that gastrointestinal digestion transforms and optimizes phenolic compounds, potentially leading to enhanced bioactivity. Although various studies have evaluated the antioxidant activity and bioaccessibility of phenolic compounds in sesame-based products, this research is the first report reporting the antidiabetic activity of the beverage based on sesame by-products after the in vitro gastrointestinal digestion process. This research contributes valuable insights into the bioaccessibility and bioactivity of phytochemicals, providing a foundation for further exploration of their health benefits and applications in functional foods. Future research, including in vivo studies and clinical trials, is recommended to validate these findings and further explore how this enhanced bioactivity can be harnessed in functional food formulations, contributing to the development of innovative products for health improvement.

Author Contributions: Conceptualization, M.F.Q.-S. and J.V.F.-M.; methodology, J.C.-C., M.F.Q.-S., L.A.O.-G., R.J.E.-M. and J.M.A.-C.; software, I.D.-P., R.V.-B. and O.D.A.-L.; formal analysis, J.C.-C. and M.F.Q.-S.; investigation, J.C.-C., M.F.Q.-S. and J.V.F.-M.; resources, R.V.-B., I.A.R.-E. and J.X.K.P.-S.; data curation, J.C-C., M.E.V.-O. and F.S.-L.; writing—original draft preparation, J.C.-C.; writing—review and editing, M.F.Q.-S.; visualization, M.F.Q.-S. and J.V.F.-M.; supervision, M.F.Q.-S. and J.V.F.-M.; funding acquisition, M.F.Q.-S., M.E.V.-O., J.V.F.-M. and J.X.K.P.-S. All authors have read and agreed to the published version of the manuscript.

Funding: This research received no external funding.

Data Availability Statement: The datasets generated and/or analyzed during this study are available from the corresponding authors upon reasonable request.

Acknowledgments: The authors are grateful to the Consejo Nacional de Humanidades, Ciencia y Tecnología for the postdoctoral fellowship awarded to M.F.Q.-S.

Conflicts of Interest: The authors declare no conflicts of interest.

References

1. Coman, V.; Teleky, B.-E.; Mitrea, L.; Martău, G.A.; Szabo, K.; Călinoiu, L.-F.; Vodnar, D.C. Chapter Five—Bioactive potential of fruit and vegetable wastes. In *Advances in Food and Nutrition Research;* Toldrá, F., Ed.; Academic Press: Cambridge, MA, USA, 2020; Volume 91, pp. 157–225.
2. Mandha, J.; Shumoy, H.; Matemu, A.O.; Raes, K. Valorization of Mango By-Products to Enhance the Nutritional Content of Maize Complementary Porridges. *Foods* **2021**, *10*, 1635. [CrossRef] [PubMed]
3. Bondam, A.F.; Diolinda da Silveira, D.; Pozzada dos Santos, J.; Hoffmann, J.F. Phenolic compounds from coffee by-products: Extraction and application in the food and pharmaceutical industries. *Trends Food Sci. Technol.* **2022**, *123*, 172–186. [CrossRef]
4. Oliver-Simancas, R.; Labrador-Fernández, L.; Díaz-Maroto, M.C.; Pérez-Coello, M.S.; Alañón, M.E. Comprehensive research on mango by-products applications in food industry. *Trends Food Sci. Technol.* **2021**, *118*, 179–188. [CrossRef]
5. Lucini Mas, A.; Brigante, F.I.; Salvucci, E.; Ribotta, P.; Martinez, M.L.; Wunderlin, D.A.; Baroni, M.V. Novel cookie formulation with defatted sesame flour: Evaluation of its technological and sensory properties. Changes in phenolic profile, antioxidant activity, and gut microbiota after simulated gastrointestinal digestion. *Food Chem.* **2022**, *389*, 133122. [CrossRef] [PubMed]
6. Quintero-Soto, M.F.; Espinoza-Moreno, R.J.; Félix-Medina, J.V.; Salas-López, F.; López-Carrera, C.F.; Argüelles-López, O.D.; Vazquez-Ontiveros, M.E.; Gómez-Favela, M.A. Comparison of Phytochemical Profile and In Vitro Bioactivity of Beverages Based on the Unprocessed and Extruded Sesame (*Sesamum indicum* L.) Seed Byproduct. *Foods* **2022**, *11*, 3175. [CrossRef]
7. Chen, Y.; Lin, H.; Lin, M.; Lin, P.; Chen, J. Effects of thermal preparation and in vitro digestion on lignan profiles and antioxidant activity in defatted-sesame meal. *Food Chem. Toxicol.* **2019**, *128*, 89–96. [CrossRef]
8. Chen, Y.; Lin, H.; Lin, M.; Zheng, Y.; Chen, J. Effect of roasting and in vitro digestion on phenolic profiles and antioxidant activity of water-soluble extracts from sesame. *Food Chem. Toxicol.* **2020**, *139*, 111239. [CrossRef]
9. Morsy, M.K.; Sami, R.; Algarni, E.; Al-Mushhin, A.A.M.; Benajiba, N.A.A.; Almasoudi, A.G.; Mekawi, E. Phytochemical Profile and Antioxidant Activity of Sesame Seed (*Sesamum indicum*) By-Products for Stability and Shelf Life Improvement of Refined Olive Oil. *Antioxidants* **2022**, *11*, 338. [CrossRef]
10. Ruiz-Armenta, X.A.; Ruiz-Armenta, J.E.; Espinoza-Moreno, R.J.; Gutiérrez-Dorado, R.; Aguilar-Palazuelos, E.; Zazueta-Morales, J.d.J.; Gómez-Favela, M.A. Use of sesame by-product and optimized extrusion to obtain a functional flour with improved techno-functional, nutritional and antioxidant properties. *Acta Univ.* **2022**, *32*, e3494. [CrossRef]
11. Sęczyk, Ł.; Sugier, D.; Świeca, M.; Gawlik-Dziki, U. The effect of in vitro digestion, food matrix, and hydrothermal treatment on the potential bioaccessibility of selected phenolic compounds. *Food Chem.* **2021**, *344*, 128581. [CrossRef]
12. Tomé-Sánchez, I.; Martín-Diana, A.B.; Peñas, E.; Frias, J.; Rico, D.; Jiménez-Pulido, I.; Martínez-Villaluenga, C. Bioprocessed Wheat Ingredients: Characterization, Bioaccessibility of Phenolic Compounds, and Bioactivity During in vitro Digestion. *Front. Plant Sci.* **2021**, *12*, 790898. [CrossRef] [PubMed]
13. Luo, J.; Li, M.; Wu, H.; Liu, Z.; Barrow, C.; Dunshea, F.; Suleria, H.A.R. Bioaccessibility of phenolic compounds from sesame seeds (*Sesamum indicum* L.) during in vitro gastrointestinal digestion and colonic fermentation. *J. Food Process. Preserv.* **2022**, *46*, e16669. [CrossRef]
14. Doungwichitrkul, T.; Damsud, T.; Phuwapraisirisan, P. α-Glucosidase Inhibitors from Cold-Pressed Black Sesame (Sesamum indicum) Meal: Characterization of New Furofuran Lignans, Kinetic Study, and In Vitro Gastrointestinal Digestion. *J. Agric. Food Chem.* **2024**, *72*, 1044–1054. [CrossRef] [PubMed]
15. Cárdenas-Castro, A.P.; Zamora-Gasga, V.M.; Alvarez-Parrilla, E.; Ruíz-Valdiviezo, V.M.; Venema, K.; Sáyago-Ayerdi, S.G. In vitro gastrointestinal digestion and colonic fermentation of tomato (*Solanum lycopersicum* L.) and husk tomato (*Physalis ixocarpa* Brot.): Phenolic compounds released and bioconverted by gut microbiota. *Food Chem.* **2021**, *360*, 130051. [CrossRef] [PubMed]
16. Singleton, V.L.; Orthofer, R.; Lamuela-Raventós, R.M. Analysis of Total Phenols and Other Oxidation Substrates and Antioxidants by Means of Folin-Ciocalteu Reagent. *Methods Enzymol.* **1999**, *299*, 152–178. [CrossRef]

17. Xu, B.J.; Chang, S.K. A comparative study on phenolic profiles and antioxidant activities of legumes as affected by extraction solvents. *J. Food Sci.* **2007**, *72*, S159–S166. [CrossRef]
18. Re, R.; Pellegrini, N.; Proteggente, A.; Pannala, A.; Yang, M.; Rice-Evans, C. Antioxidant activity applying an improved ABTS radical cation decolorization assay. *Free Radic. Biol. Med.* **1999**, *26*, 1231–1237. [CrossRef]
19. Brand-Williams, W.; Cuvelier, M.E.; Berset, C. Use of a Free Radical Method to Evaluate Antioxidant Activity. *LWT-Food Sci. Technol.* **1995**, *28*, 25–30. [CrossRef]
20. Quintero-Soto, M.F.; Saracho-Peña, A.G.; Chavez-Ontiveros, J.; Garzon-Tiznado, J.A.; Pineda-Hidalgo, K.V.; Delgado-Vargas, F.; Lopez-Valenzuela, J.A. Phenolic profiles and their contribution to the antioxidant activity of selected chickpea genotypes from Mexico and ICRISAT collections. *Plant Foods Hum. Nutr.* **2018**, *73*, 122–129. [CrossRef]
21. Garduño-Félix, K.G.; Ramirez, K.; Salazar-Salas, N.Y.; Amabilis-Sosa, L.E.; Rochín-Medina, J.J. Phenolic profile in black sesame sprouts biostimulated with Bacillus clausii. *J. Food Meas. Charact.* **2021**, *15*, 5418–5426. [CrossRef]
22. Rochín-Medina, J.J.; Sotelo-Castro, J.A.; Salazar-Salas, N.Y.; López-Valenzuela, J.A.; Ramírez, K. Antioxidant and anti-Salmonella activities of eggplant peel compounds obtained by solvent-free calcium-based extraction. *CyTA-J. Food* **2019**, *17*, 873–881. [CrossRef]
23. Delgado-Vargas, F.; Vega-Álvarez, M.; Landeros Sánchez, A.; López-Angulo, G.; Salazar-Salas, N.Y.; Quintero-Soto, M.F.; Pineda-Hidalgo, K.V.; López-Valenzuela, J.A. Metabolic changes associated with chilling injury tolerance in tomato fruit with hot water pretreatment. *J. Food Biochem.* **2022**, *46*, e14056. [CrossRef] [PubMed]
24. Félix-Medina, J.V.; Gutiérrez-Dorado, R.; López-Valenzuela, J.A.; López-Ángulo, G.; Quintero-Soto, M.F.; Perales-Sánchez, J.X.K.; Montes-Ávila, J. Nutritional, antioxidant and phytochemical characterization of healthy ready-to-eat expanded snack produced from maize/common bean mixture by extrusion. *LWT-Food Sci. Technol.* **2021**, *142*, 111053. [CrossRef]
25. Gómez-Romero, M.; Segura-Carretero, A.; Fernández-Gutiérrez, A. Metabolite profiling and quantification of phenolic compounds in methanol extracts of tomato fruit. *Phytochemistry* **2010**, *71*, 1848–1864. [CrossRef] [PubMed]
26. Romdhane, M.H.; Chahdoura, H.; Barros, L.; Dias, M.I.; Corrêa, R.C.; Morales, P.; Ciudad-Mulero, M.; Flamini, G.; Majdoub, H.; Ferreira, I.C.F.R. Chemical Composition, Nutritional Value, and Biological Evaluation of Tunisian Okra Pods (*Abelmoschus esculentus* L. Moench). *Molecules* **2020**, *25*, 4739. [CrossRef]
27. Pereira-Feitosa, L.G.; Monge, M.; Lopes, N.P.; Rodrigues de Oliveira, D.C. Distribution of flavonoids and other phenolics in Mikania species (Compositae) of Brazil. *Biochem. Syst. Ecol.* **2021**, *97*, 104273. [CrossRef]
28. Abad-García, B.; Berrueta, L.A.; Garmón-Lobato, S.; Gallo, B.; Vicente, F. A general analytical strategy for the characterization of phenolic compounds in fruit juices by high-performance liquid chromatography with diode array detection coupled to electrospray ionization and triple quadrupole mass spectrometry. *J. Chromatogr. A* **2009**, *1216*, 5398–5415. [CrossRef]
29. Leng, Y.; Zhang, X.; Zhang, Q.; Xia, J.; Zhang, Y.; Ma, C.; Liu, K.; Li, H.; Hong, Y.; Xie, Z. Gallic acid attenuates murine ulcerative colitis by promoting group 3 innate lymphocytes, affecting gut microbiota, and bile acid metabolism. *J. Nutr. Biochem.* **2024**, *131*, 109677. [CrossRef]
30. Li, H.; Zheng, T.; Lian, F.; Xu, T.; Yin, W.; Jiang, Y. Anthocyanin-rich blueberry extracts and anthocyanin metabolite protocatechuic acid promote autophagy-lysosomal pathway and alleviate neurons damage in in vivo and in vitro models of Alzheimer's disease. *Nutrition* **2022**, *93*, 111473. [CrossRef]
31. Jiang, W.; Deng, B.; Xie, M.; Feng, Y.; Jiang, X.; Yang, B.; Tan, Z.; Ou, H.; Tan, Y.; Liu, S.; et al. Caffeic acid mitigates myocardial fibrosis and improves heart function in post-myocardial infarction by inhibiting transforming growth factor-β receptor 1 signaling pathways. *Biomed. Pharmacother.* **2024**, *177*, 117012. [CrossRef]
32. Wu, T.-Y.; Chen, C.-C.; Lin, J.-Y. Anti-inflammatory in vitro activities of eleven selected caffeic acid derivatives based on a combination of pro−/anti-inflammatory cytokine secretions and principal component analysis—A comprehensive evaluation. *Food Chem.* **2024**, *458*, 140201. [CrossRef] [PubMed]
33. Ding, Y.; Li, H.; Cao, S.; Yu, Y. Effects of catechin on the malignant biological behavior of gastric cancer cells through the PI3K/Akt signaling pathway. *Toxicol. Appl. Pharmacol.* **2024**, *490*, 117036. [CrossRef] [PubMed]
34. Li, F.; Zhang, L.; Zhang, X.; Fang, Q.; Xu, Y.; Wang, H. Rutin alleviates Pb-induced oxidative stress, inflammation and cell death via activating Nrf2/ARE system in SH-SY5Y cells. *NeuroToxicology* **2024**, *104*, 1–10. [CrossRef] [PubMed]
35. Jin, F.; Li, J.; Zhao, C.; Gu, L.; Pu, M.; Jiang, S.; Liang, M.; Zhao, Y.; Shen, J.; Agabuwei, A.; et al. Quercetin alleviates kidney damage caused by mercury Chloride: The protective effects of quercetin on autophagy and inflammation were studied based on TRIM32/TLR4/LC3 pathway. *Toxicon* **2024**, *248*, 108031. [CrossRef] [PubMed]
36. Yao, Y.-X.; Yu, Y.-J.; Dai, S.; Zhang, C.-Y.; Xue, X.-Y.; Zhou, M.-L.; Yao, C.-H.; Li, Y.-X. Kaempferol efficacy in metabolic diseases: Molecular mechanisms of action in diabetes mellitus, obesity, non-alcoholic fatty liver disease, steatohepatitis, and atherosclerosis. *Biomed. Pharmacother.* **2024**, *175*, 116694. [CrossRef]
37. Liu, S.; Zheng, X.; Luo, Z.; Tang, C.; Hu, Y.; Peng, Q.; Mi, P.; Chen, H.; Yao, X. The synthesis and bioactivity of apigenin derivatives. *Fitoterapia* **2024**, *179*, 106228. [CrossRef]
38. Ishola, I.O.; Osele, M.O.; Chijioke, M.C.; Adeyemi, O.O. Isorhamnetin enhanced cortico-hippocampal learning and memory capability in mice with scopolamine-induced amnesia: Role of antioxidant defense, cholinergic and BDNF signaling. *Brain Res.* **2019**, *1712*, 188–196. [CrossRef]
39. Zhang, P.; Sun, Y.; Shi, L.; Sun, D.; Wang, L.; Feng, D.; Ding, C. Effect of isorhamnetin on carbonic anhydrase IX expression and tumorigenesis of bladder cancer by activating PPARγ/PTEN/AKT pathway. *Tissue Cell* **2023**, *82*, 102048. [CrossRef]

40. Yang, T.; Xiao, Y.; Liu, S.; Luo, F.; Tang, D.; Yu, Y.; Xie, Y. Isorhamnetin induces cell cycle arrest and apoptosis by triggering DNA damage and regulating the AMPK/mTOR/p70S6K signaling pathway in doxorubicin-resistant breast cancer. *Phytomedicine* **2023**, *114*, 154780. [CrossRef]
41. Luan, Y.; Luan, Y.; Zhao, Y.; Xiong, F.; Li, Y.; Liu, L.; Cao, Y.; Dai, F. Isorhamnetin in Tsoong blocks Hsp70 expression to promote apoptosis of colon cancer cells. *Saudi J. Biol. Sci.* **2019**, *26*, 1011–1022. [CrossRef]
42. Ru, Y.; Luo, Y.; Liu, D.; Huang, Q.; Zhou, X.; Linghu, M.; Luo, X.; Lv, Z.; Wu, Y.; Zhang, H.; et al. Isorhamnetin alleviates ferroptosis-mediated colitis by activating the NRF2/HO-1 pathway and chelating iron. *Int. Immunopharmacol.* **2024**, *135*, 112318. [CrossRef] [PubMed]
43. Cárdenas-Castro, A.P.; Rochín-Medina, J.J.; Ramírez, K.; Tovar, J.; Sáyago-Ayerdi, S.G. In Vitro Intestinal Bioaccessibility and Colonic Biotransformation of Polyphenols from Mini Bell Peppers (*Capsicum annuum* L.). *Plant Foods Hum. Nutr.* **2022**, *77*, 77–82. [CrossRef] [PubMed]
44. Picos-Salas, M.A.; Leyva-López, N.; Bastidas-Bastidas, P.d.J.; Antunes-Ricardo, M.; Cabanillas-Bojórquez, L.A.; Angulo-Escalante, M.A.; Heredia, J.B.; Gutiérrez-Grijalva, E.P. Supercritical CO_2 extraction of naringenin from Mexican oregano (*Lippia graveolens*): Its antioxidant capacity under simulated gastrointestinal digestion. *Sci. Rep.* **2024**, *14*, 1146. [CrossRef] [PubMed]
45. Ali, M.; Homann, T.; Kreisel, J.; Khalil, M.; Puhlmann, R.; Kruse, H.-P.; Rawel, H. Characterization and Modeling of the Interactions between Coffee Storage Proteins and Phenolic Compounds. *J. Agric. Food Chem.* **2012**, *60*, 11601–11608. [CrossRef] [PubMed]
46. Xiong, H.-H.; Lin, S.-Y.; Chen, L.-L.; Ouyang, K.-H.; Wang, W.-J. The Interaction between Flavonoids and Intestinal Microbes: A Review. *Foods* **2023**, *12*, 320. [CrossRef]
47. Murota, K.; Nakamura, Y.; Uehara, M. Flavonoid metabolism: The interaction of metabolites and gut microbiota. *Biosci. Biotechnol. Biochem.* **2018**, *82*, 600–610. [CrossRef]
48. Lin, W.; Wang, W.; Yang, H.; Wang, D.; Ling, W. Influence of Intestinal Microbiota on the Catabolism of Flavonoids in Mice. *J Food Sci.* **2016**, *81*, H3026–H3034. [CrossRef]
49. Li, J.; Yuan, H.; Zhao, Z.; Li, L.; Li, X.; Zhu, L.; Wang, X.; Sun, P.; Xiao, Y. The mitigative effect of isorhamnetin against type 2 diabetes via gut microbiota regulation in mice. *Front. Nutr.* **2022**, *9*, 1070908. [CrossRef]
50. Jara-Palacios, M.J.; Gonçalves, S.; Hernanz, D.; Heredia, F.J.; Romano, A. Effects of in vitro gastrointestinal digestion on phenolic compounds and antioxidant activity of different white winemaking byproducts extracts. *Food Res. Int.* **2018**, *109*, 433–439. [CrossRef]
51. Liu, X.; Shi, J.; Yi, J.; Zhang, X.; Ma, Q.; Cai, S. The effect of in vitro simulated gastrointestinal digestion on phenolic bioaccessibility and bioactivities of Prinsepia utilis Royle fruits. *LWT-Food Sci. Technol.* **2021**, *138*, 110782. [CrossRef]
52. Gámez-Valdez, L.C.; Gutiérrez-Dorado, R.; Gómez-Aldapa, C.A.; Perales-Sánchez, J.X.K.; Milán-Carrillo, J.; Cuevas-Rodríguez, E.O.; Mora-Rochín, S.; Reyes-Moreno, C. Effect of the extruded amaranth flour addition on the nutritional, nutraceutical and sensory quality of tortillas produced from extruded creole blue maize flour. *Biotecnia* **2021**, *23*, 103–112. [CrossRef]
53. Mainente, F.; Vincenzi, S.; Rizzi, C.; Pasini, G. Polyphenols Profile and Antioxidant Activity Characterization of Commercial North Italian Ciders in Relation to Their Geographical Area of Production and Cidermaking Procedures. *Beverages* **2023**, *9*, 54. [CrossRef]
54. Kartelias, I.G.; Panagiotakopoulos, I.; Nasopoulou, C.; Karantonis, H.C. Evaluating the Effect of Adding Selected Herbs, Spices, and Fruits to Fermented Olympus Mountain Tea (*Sideritis scardica*) Kombucha Sweetened with Thyme Honey: Assessment of Physicochemical and Functional Properties. *Beverages* **2024**, *10*, 9. [CrossRef]
55. Fu, Y.; Liu, X.; Ma, Q.; Yi, J.; Cai, S. Phytochemical bioaccessibility and in vitro antidiabetic effects of Chinese sumac (*Rhus chinensis* Mill.) fruits after a simulated digestion: Insights into the mechanisms with molecular docking analysis. *Int. J. Food Sci. Technol.* **2022**, *57*, 2656–2669. [CrossRef]

nutrients

MDPI

Article

The Consumption of Non-Sugar Sweetened and Ready-to-Drink Beverages as Emerging Types of Beverages in Shanghai

Zhengyuan Wang [1], Liping Shen [1], Jinpeng Ning [2], Zhuo Sun [1], Yiwen Xu [3], Zehuan Shi [1], Qi Song [1], Wei Lu [1], Wenqing Ma [1], Shupeng Mai [1] and Jiajie Zang [1,*]

[1] Department of Nutrition and Health, Division of Health Risk Factors Monitoring and Control, Shanghai Municipal Center for Disease Control and Prevention, Shanghai 200336, China; wangzhengyuan@scdc.sh.cn (Z.W.); shenliping@scdc.sh.cn (L.S.); sunzhuo@scdc.sh.cn (Z.S.); shizehuan@scdc.sh.cn (Z.S.); songqi@scdc.sh.cn (Q.S.); luwei@scdc.sh.cn (W.L.); mawenqing@scdc.sh.cn (W.M.); maishupeng@scdc.sh.cn (S.M.)

[2] School of Public Health, Shanghai University of Traditional Chinese Medicine, Shanghai 201203, China; 17505564961@163.com

[3] The College of Medical Technology, Shanghai University of Medicine and Health Sciences, Shanghai 201318, China; xu_yw1021@163.com

* Correspondence: zangjiajie@scdc.sh.cn

Abstract: Background: The Chinese beverage industry is experiencing rapid growth, particularly in the popularity of non-sugar sweetened beverages (NSSs) and ready-to-drink beverages (RSBs). This study aimed to assess current consumption patterns and determinants of various beverage types among retail visitors. Methods: A total of 44 observation points, including 22 supermarkets and 22 convenience stores, were randomly selected across Shanghai. At each location, at least 100 individuals were recruited to participate. Data were collected using an electronic self-administered questionnaire. Results: The consumption rates of total beverages, sugar-sweetened beverages, NSSs, and RSBs were 57.70%, 56.94%, 19.60%, and 29.50%, respectively; the median consumption amounts among the drinking population were 162.57 mL/day, 137.98 mL/day, 32.85 mL/day, and 32.85 mL/day, respectively. The consumption proportions of NSSs and RSBs ranked 2nd and 3rd. The multifactorial analyses showed that people aged 6–18 years consumed more beverages ($p < 0.05$). Males were more likely to consume sugar-sweetened beverages and NSSs, but females were more likely to consume RSBs ($p < 0.05$). Higher educated people and bachelors were more likely to consume beverages ($p < 0.05$). Conclusions: The emerging beverage categories, NSSs and RSBs, warrant attention due to their significant consumption rates. Tailored intervention strategies should be considered for demographic groups varying by age, gender, and educational attainment.

Keywords: beverage; non-sugar sweetened beverages; ready-to-drink beverages; retail visitors; Shanghai

1. Introduction

In recent years, the global beverage industry has experienced rapid expansion, with a corresponding surge in beverage consumption. According to data from the National Bureau of Statistics [1], China's beverage industry output reached 175 million tons in 2023, marking a staggering 37-fold increase over the past three decades, with an impressive average annual growth rate of 13%. Per capita beverage consumption has soared to 30 times its level of 30 years ago, a growth trajectory that merits significant attention. Sugar-sweetened beverages (SSBs), favored for their diversity and convenience, dominate the market, with a notable uptick in consumption. A global analysis of SSB intake among adults across 185 countries revealed a 91.76 g/week increase from 1990 to 2018, with sub-Saharan Africa witnessing the most substantial rise of 741.52 g/week [2]. While China's

SSB consumption does not lead globally, the upward trend is unmistakable [2]. A 2016 survey across 27 Chinese cities indicated that a substantial majority—74% of children (4–9 years), 85% of adolescents (10–17 years), and 83% of adults (18–55 years)—consumed at least 500 mL of SSBs weekly [3].

An umbrella review [4] had shown that high dietary sugar consumption was associated with a range of health risks. SSBs are the largest source of dietary sugars, so the intake of SSBs has become an urgent public health problem. Studies have shown that for every additional 355 mL of SSB consumed per day, the body weight of adults would increase by 0.22 kg and the body mass index of children would increase by 0.06 kg/m^2 in 1 year [5]. To add insult to injury, SSBs are also associated with an increased risk of cardiovascular disease. A meta-analysis of seven prospective cohort studies [6] found that consuming SSBs one or more times per day increased the risk of cardiovascular disease by approximately 9% compared with consuming no SSBs or less than one SSB per month.

In addition to various chronic diseases, SSBs increase the risk of death. A 2010 analysis of data from the Global Dietary Surveys estimated that 184,000 people die each year due to excessive intake of SSBs [7]. The Nurses' Health Study and Health Professionals Follow-up Study also found [8] that there was a positive association between SSBs and mortality, which was mainly correlated with mortality through cardiovascular disease. This implies that long-term excessive SSB intake may increase the risk of mortality.

High SSB consumption is linked to a host of health issues. Consequently, many have switched to non-sugar sweetened options to indulge their sweet tooth without the caloric intake. In China, NSS sales hit $0.21 billion in 2019, growing at a CAGR of 42.84%, and are projected to reach $0.57 billion by 2027 [9]. The WHO guidelines flag potential NSS health risks such as cardiovascular disease and metabolic issues, calling for more research on their safety and health impact. They also note that NSS benefits generally exceed the drawbacks, underscoring the importance of mindful NSS consumption [10]. On the other hand, beyond traditionally packaged SSBs, the market for ready-to-drink beverages (RSBs) is burgeoning in China. Data from China Report Hall [11] indicates that China's RSB sales surpassed $14.36 billion in 2022, with projections suggesting a leap to over $28.72 billion by 2025, reflecting a swift growth trajectory. Prior research on beverage consumption has predominantly centered on packaged SSBs, with scant attention to RSBs, potentially leading to an underestimation of overall beverage intake. Consequently, our study aimed to assess the current landscape of beverage consumption alongside potential influencing factors among Shanghai's residents in 2023.

2. Method

2.1. Participants

In this study, we utilized the shopping mall intercept methodology commonly recognized in market research to randomly engage visitors at the retail location for our survey, taking into account both those who made purchases and those who did not, thereby providing a comprehensive dataset for analysis. This study first randomly selected eight districts with different economic levels and geographical locations (namely Jiading, Pudong, Putuo, Changning, Xuhui, Jing'an, Songjiang, Fengxian) from the 16 districts in Shanghai. Then, 22 supermarkets and 22 convenience stores were randomly selected as observation points within these selected districts. For each observation point, at least 40 people were randomly recruited from the age group of 6–18 years old, and at least 20 people were randomly recruited from each of the age groups of 18–40 years old, 40–60 years old, and >60 years old, with an equal distribution between gender with each age group.

2.2. Data Collection

Our survey was conducted from June to July in 2023. We used a self-administered electronic questionnaire, which took approximately 8–10 min to complete. The uniformly trained surveyors guided and assisted the participants in filling out the questionnaires after obtaining informed consent. Demographic information (e.g., gender, age, personal income, etc.), Noncommunicable Chronic Disease status (NCD), height, weight, and beverages consumption status were collected through the questionnaire. The beverages included nine types such as NSSs, carbonated beverages (CBs), fruit-flavored beverages (FFBs, including 100% NFC), sugar-sweetened tea beverages (STBs), vegetable protein beverages (VPBs), lactic acid bacteria beverages (LABs), RSBs, sports functional beverages (SFBs), and sugar-sweetened coffee beverages (SCBs). The term "beverages" did not include water, and no information related to it was surveyed. The participants were asked about the consumption frequency and average amount of each type of beverage in the past month. The consumption frequency was divided into five groups: no drink, 1–3 times/month, 1–3 times/week, 4–7 times/week, and >1 time/day. We showed the survey participants the common sizes of bottled beverages to help them estimate their average consumption.

2.3. Covariates and Categorization

The education level was classified into three groups: low education (junior high school and below), medium education (senior high school/secondary school/technical school/college), and high education (bachelor's degree and above). Marriage status analysis did not include people aged 6–18 years. Personal income, after excluding those aged 6–18 years old, was classified into low income (<$500/month), medium income ($500–1000/month), and high income (>$1000/month). Obesity was defined according to the standards of the National Health and Wellness Commission of the People's Republic of China [12–14]. We divided the beverages into two categories: SSBs and NSSs, based on whether they contain added sugar. The SSBs included CBs, FFBs, STBs, VPBs, LABs, RSBs, SFBs, and SCBs. Total beverages (TBs) equal NSSs plus SSBs. The beverage consumption population was defined as those who had drank beverages at least once in the past month.

2.4. Statistical Analysis

Statistical analyses were completed with SPSS 25.0. All statistical significance tests were two-sided, and the level of $p < 0.05$ was considered statistically significant. The chi-square test was used to analyze qualitative variables, and quantitative variables were compared using non-parametric tests. The binary logistic regression analyses were used to analyze the factors influencing the intake of beverages.

3. Results

3.1. Basic Information of the Survey Population

The total number of respondents for this study was 6549. After excluding the failed questionnaires, the final sample consisted of 6394 respondents, with a pass rate of 98.0%. Of these, 3135 (49.03%) were males and 3259 (50.97%) were females, with an average age of 35.8 ± 21.79 years. There were significant statistical differences in the composition ratios for personal income, occupational status, marriage status, and obesity between males and females ($p < 0.05$), as shown in Table 1. The composition ratios of other survey factors did not show statistically significant differences.

Table 1. Basic information of the survey population.

	Male	Female	Total	χ^{2} (1)	p (2)
Total	3135 (49.03%)	3259 (50.97%)	6394 (100%)		
Age					
6–18	1053 (33.59%)	1034 (31.73%)	2087 (32.64%)	3.987	0.263
18–40	748 (23.86%)	770 (23.63%)	1518 (23.74%)		
40–60	670 (21.37%)	753 (23.11%)	1423 (22.26%)		
>60	664 (21.18%)	702 (21.54%)	1366 (21.36%)		
Personal income					
low	458 (22.00%)	610 (27.42%)	1068 (24.80%)	52.02	<0.001
medium	942 (45.24%)	1099 (49.39%)	2041 (47.39%)		
high	682 (32.76%)	516 (23.19%)	1198 (27.82%)		

Table 1. *Cont.*

	Male	Female	Total	$\chi^{2\,(1)}$	$p^{\,(2)}$
Education level					
low	1263 (40.29%)	1344 (41.24%)	2607 (40.77%)	4.67	0.099
medium	1310 (41.79%)	1282 (39.34%)	2592 (40.54%)		
high	562 (17.93%)	633 (19.42%)	1195 (18.690%)		
Occupational status					
health-related industries	607 (19.36%)	710 (21.79%)	1317 (20.60%)	5.74	0.017
others	2528 (80.64%)	2549 (78.21%)	5077 (79.40%)		
Marriage status					
bachelor	1502 (47.91%)	1423 (43.67%)	2925 (45.75%)	16.372	<0.001
married	1569 (50.05%)	1735 (53.24%)	3304 (51.67%)		
divorced/widowed	64 (2.04%)	101 (3.10%)	165 (2.58%)		
Obesity					
yes	443 (14.13%)	325 (9.97%)	768 (12.01%)	26.15	<0.001
no	2692 (85.87%)	2934 (90.03%)	5626 (88.99%)		
NCD [(3)]					
yes	633 (20.19%)	602 (18.47%)	1235 (19.31%)	3.03	0.082
no	2502 (79.81%)	2657 (81.53%)	5159 (80.69%)		
Weight loss/shaping in the past month					
yes	249 (7.94%)	299 (9.18%)	548 (8.57%)	3.10	0.079
no	2886 (92.06%)	2960 (90.83%)	5846 (91.43%)		
Mood swing in the past month					
yes	72 (2.30%)	81 (2.49%)	153 (2.39%)	0.24	0.621
no	3063 (97.70%)	3178 (97.52%)	6241 (97.61%)		

[(1)] Chi-square test statistics; [(2)] significance of the test, considering a level of 5%; [(3)] NCD = Noncommunicable Chronic Diseases.

3.2. Consumption Rate and Consumption Amount of Different Types of Beverages

The consumption rates of TBs, SSBs, NSSs, and RSBs were 57.70%, 56.94%, 19.60%, and 29.90%, respectively, in Table 2. The consumption rates of the four types of beverages among different age, personal income, education level, marriage status, NCD, and weight loss/shaping groups showed statistically significant differences ($p < 0.05$). In addition to these variables, the four types of beverages had their own characteristics. There were significant statistical differences in the consumption rates of NSSs and RSBs across the gender index, in the consumption rates of TBs, SSBs, and NSSs across the occupational status and obesity indices, and in the consumption rates of SSBs and NSSs across the mood swing index ($p < 0.05$).

The median consumption amounts of TBs, SSBs, NSSs, and RSBs in the drinking population ware 162.57 mL/day, 137.98 mL/day, 32.85 mL/day, and 32.85 mL/day, respectively, in Table 2. There were significant statistical differences in the consumption amounts of TBs and SSBs among different gender, age, personal income, education level, occupational status, marriage status, obesity, as well as weight loss/shaping and mood swing groups ($p < 0.05$). Moreover, there were significant statistical differences in the consumption amount of SSBs across the NCD index, in the consumption amounts of NSSs across the gender, occupational status, NCD, obesity, and mood swing indices, and in the consumption amounts of RSBs across the age, occupational status, and marriage status, NCD, obesity, weight loss/shaping and mood swing indices ($p < 0.05$).

3.3. Consumption Amount and Frequency Distribution of Beverages

Of TB consumption amount, SSB consumption accounted for 84.5% and NSS consumption accounted for 9.5%. The largest share of SSBs was CBs with 15.6% and the smallest was LABs with 7.9%. RSBs had a slightly lower share than CBs, coming in second with 13.6%, in Figure 1a. Moreover, the consumption amount distribution of various types of beverages across different genders, ages, and education level groups are shown in Figure 2b–d.

Table 2. Consumption rate and consumption amounts of beverages.

Variables	TB (1)				SSB (2)				NSS (3)				RSB (4)			
	Drink Rate (%)	p (5)	Consumption Amount Medium (P25, P75)	p	Drink Rate (%)	p	Consumption Amount Medium (P25, P75)	p	Drink Rate (%)	p	Consumption Amount Medium (P25, P75)	p	Drink Rate (%)	p	Consumption Amount Medium (P25, P75)	p
Gender		0.23		<0.001		0.192		<0.001		<0.001		0.001		0.015		0.940
male	58.30		175.71 (65.70, 396.56)		57.80		160.97 (65.70, 359.83)		20.50		32.85 (32.85, 85.71)		28.50		32.85 (32.85, 142.86)	
female	56.90		142.86 (52.56, 302.14)		56.20		133.38 (49.28, 285.72)		16.70		32.85 (21.68, 85.71)		31.30		32.85 (32.85, 142.86)	
Age		<0.001		<0.001		<0.001		<0.001		<0.001		0.056		<0.001		0.001
6–18	73.70		174.11 (65.70, 379.98)		75.80		131.41 (39.42, 292.10)		28.20		32.85 (32.85, 142.86)		39.30		32.85 (32.85, 142.86)	
18–40	70.70		176.14 (65.70, 384.27)		69.40		165.85 (65.70, 357.14)		25.00		32.85 (32.85, 142.86)		40.00		32.85 (32.85, 142.86)	
40–60	46.40		142.86 (45.99, 357.14)		45.80		132.85 (45.99, 320.21)		14.10		32.85 (32.85, 142.86)		23.10		32.85 (32.85, 65.70)	
>60	28.80		77.43 (32.85, 215.13)		28.70		72.92 (32.85, 214.29)		5.20		32.85 (19.71, 96.42)		8.30		32.85 (21.68, 131.41)	
Personal income		<0.001		<0.001		<0.001		<0.001		<0.001		0.568		<0.001		0.518
low	40.20		134.00 (42.86, 301.42)		39.50		129.99 (45.01, 282.85)		11.00		32.85 (23.00, 94.29)		17.50		32.85 (32.85, 137.13)	
medium	45.50		134.00 (49.28, 318.57)		44.90		128.27 (45.99, 289.89)		12.60		32.85 (32.85, 131.41)		22.10		32.85 (32.85, 98.55)	
high	64.20		178.57 (65.70, 419.28)		63.40		171.43 (65.70, 374.46)		23.00		32.85 (32.85, 142.86)		34.30		32.85 (32.85, 142.86)	
Education level		<0.001		<0.001		<0.001		<0.001		<0.001		0.094		<0.001		0.374
low	53.40		131.41 (49.28, 294.24)		53.80		101.99 (32.85, 241.77)		15.90		32.85 (21.68, 142.86)		22.80		32.85 (32.85, 142.86)	
medium	57.20		177.40 (65.70, 416.55)		57.70		156.00 (54.53, 357.14)		20.90		32.85 (32.85, 142.86)		31.50		32.85 (32.85, 142.86)	
high	68.60		175.71 (65.70, 381.51)		68.00		164.26 (65.70, 334.10)		25.00		32.85 (26.28, 142.86)		40.00		32.85 (32.85, 131.41)	
Occupational status		<0.001		0.004		<0.001		<0.001		<0.001		<0.001		0.102		0.003
health-related industries	63.20		175.71 (65.70, 423.21)		66.10		98.55 (32.85, 274.26)		25.90		65.70 (32.85, 228.57)		31.20		32.85 (29.07, 122.85)	
others	56.10		154.40 (57.14, 342.14)		55.60		142.86 (54.53, 314.98)		17.80		32.85 (26.28, 100.00)		29.00		32.85 (32.85, 142.86)	
Marriage status		<0.001		<0.001		<0.001		<0.001		<0.001		0.324		<0.001		<0.001
bachelor	74.70		175.71 (65.70, 366.06)		73.70		161.54 (65.70, 339.20)		26.40		32.85 (29.57, 131.41)		42.40		32.85 (32.85, 142.86)	
married	43.40		131.41 (41.71, 307.11)		42.80		118.27 (39.42, 285.71)		12.00		32.85 (26.28, 98.55)		19.30		32.85 (32.85, 98.55)	
divorced/widowed	45.20		146.99 (46.98, 368.42)		44.60		142.86 (44.02, 282.85)		12.50		52.56 (21.68, 142.86)		19.60		32.85 (26.28, 142.86)	
Obesity		0.002		<0.001		0.003		0.001		0.001		0.004		0.123		0.003
yes	62.80		177.40 (65.70, 463.11)		62.00		165.06 (65.70, 425.70)		22.80		32.85 (32.85, 142.86)		32.30		39.42 (32.85, 142.86)	
no	56.90		151.76 (55.85, 328.68)		56.30		142.86 (52.56, 305.43)		18.00		32.85 (24.97, 98.55)		29.60		32.85 (32.85, 142.86)	
NCD (6)		<0.001		0.850		<0.001		<0.001		<0.001		<0.001		<0.001		0.014
yes	41.00		161.10 (52.56, 405.67)		43.90		65.70 (13.14, 236.81)		15.30		124.84 (32.85, 357.14)		16.90		32.85 (26.28, 98.55)	
no	62.10		162.57 (61.42, 354.25)		61.70		142.86 (55.85, 318.57)		20.70		32.85 (27.43, 131.41)		32.80		32.85 (32.85, 142.86)	
Weight loss/shaping in the past month		<0.001		<0.001		<0.001		<0.001		<0.001		0.055		<0.001		0.034
yes	74.30		220.71 (98.55, 485.69)		73.50		197.71 (91.33, 423.69)		32.80		32.85 (32.85, 142.86)		46.90		32.85 (32.85, 142.86)	
no	56.20		151.12 (55.85, 342.86)		56.50		131.41 (39.42, 289.75)		18.40		32.85 (29.07, 142.86)		27.90		32.85 (32.85, 142.86)	
Mood swing in the past month		0.097		<0.001		<0.001		<0.001		<0.001		<0.001		0.117		<0.001
yes	62.30		295.66 (127.14, 571.43)		82.00		1.85 (0, 175.71)		37.30		274.26 (65.70, 714.29)		25.40		32.85 (19.71, 45.99)	
no	57.30		153.09 (57.14, 342.86)		56.70		142.86 (54.53, 318.57)		18.50		32.85 (26.28, 107.14)		30.70		32.85 (32.85, 142.86)	
All	57.70		162.57 (59.13, 357.14)		56.94		137.98 (45.99, 306.06)		19.60		32.85 (32.85, 142.86)		29.90		32.85 (32.85, 142.86)	

(1) Total beverages; (2) sugar-sweetened beverages; (3) non-sugar sweetened beverages; (4) ready-to-drink beverages; (5) significance of the test, considering a level of 5%; (6) NCD = Noncommunicable Chronic Diseases.

The proportion of people who drank TBs, SSBs, NSSs, and RSBs at least once a week were 47.7%, 41.6%, 6.0%, and 9.6%, respectively, in Figure 1a. Moreover, the consumption frequency distribution of various types of beverages across different genders, ages, and education level groups are shown in Figure 2b–d.

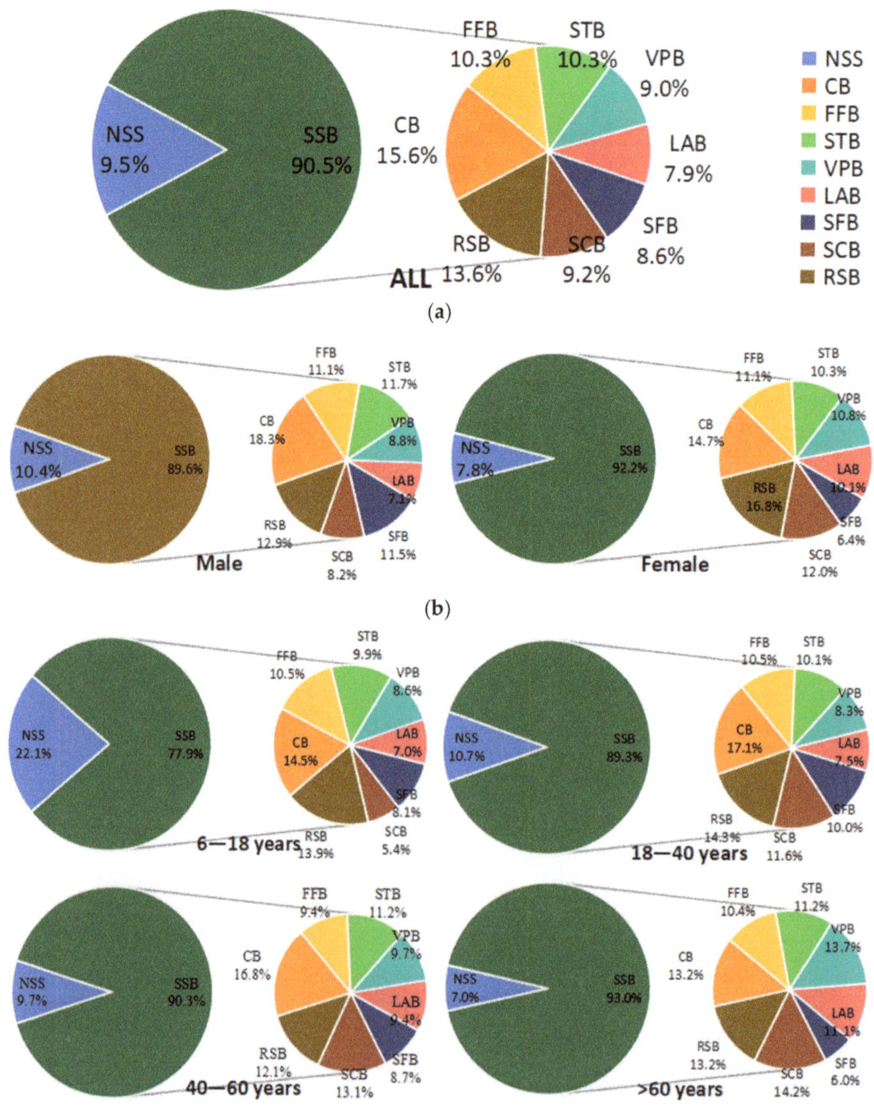

Figure 1. *Cont.*

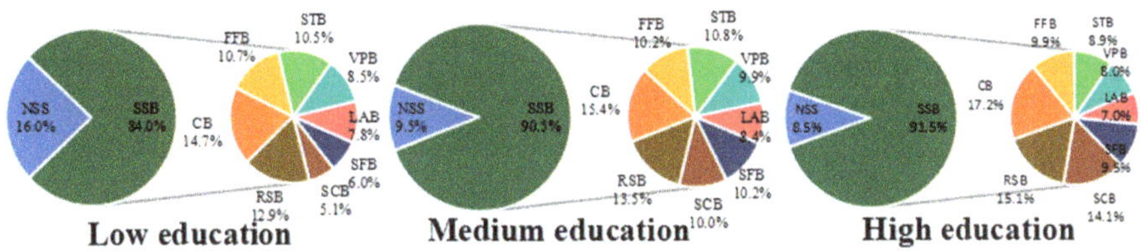

(d)

Figure 1. The consumption amount distribution of various types of beverages; (**a–d**), respectively, display the composition ratios of various beverage consumptions for all subjects, different genders, different age groups, and different levels of education. (NSSs = non-sugar sweetened beverages; SSBs = Sugar-sweetened beverages; CBs = carbonated beverages; FFBs = fruit-flavored beverages (including 100% NFC); STBs = sugar-sweetened tea beverages; VPBs = vegetable protein beverages; LABs = lactic acid bacteria beverages; SFBs = sports functional beverages; SCBs = sugar-sweetened coffee beverages; RSBs = ready-to-drink beverages).

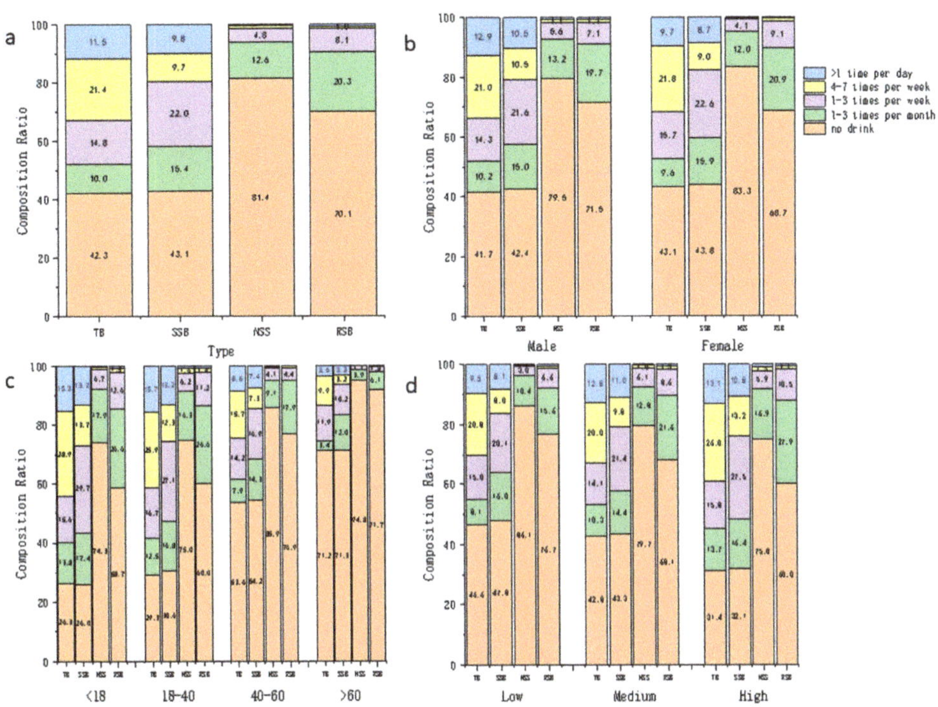

Figure 2. Consumption frequency distribution of beverages (%); (**a–d**), respectively, demonstrate the composition ratios of different consumption frequencies for four types of beverages in all subjects, different genders, different age groups, and different levels of education. (Total = total beverages; SSBs = sugar-sweetened beverages; NSSs = non-sugar sweetened beverages; RSBs = ready-to-drink beverages).

3.4. The Multifactorial Analyses Result of Influencing Factors Related to the Consumption of Beverages

Logistic regression analysis was conducted with gender, age, education level, occupational status, marriage status, obesity, NCD, weight loss/shaping, and mood swing as independent variables, and the consumption rate of different types of beverages as dependent variables in Table 3.

Table 3. Logistic regression analyses of influencing factors related to the consumption of beverages.

	TB (1)				SSB (2)				NSS (3)				RSB (4)			
	β (5)	*p*	OR (5)	95% CI (5)	β	*p*	OR	95% CI	β	*p*	OR	95% CI	β	*p*	OR	95% CI
Gender																
male			reference				reference				reference				reference	
female	−0.037	0.507	0.964	0.864–1.075	−0.043	0.432	0.957	0.859–1.067	−0.247	<0.001	0.781	0.684–0.892	0.180	0.002	1.197	1.067–1.343
Age																
6–18			reference				reference				reference				reference	
18–40	−0.536	<0.001	0.585	0.476–0.720	−0.574	<0.001	0.563	0.459–0.691	−0.537	<0.001	0.584	0.469–0.729	−0.509	<0.001	0.601	0.494–0.732
40–60	−1.171	<0.001	0.310	0.241–0.400	−1.163	<0.001	0.312	0.243–0.402	−0.866	<0.001	0.421	0.314–0.564	−0.859	<0.001	0.423	0.328–0.547
>60	−1.653	<0.001	0.191	0.147–0.249	−1.626	<0.001	0.197	0.152–0.255	−1.651	<0.001	0.192	0.134–0.274	−1.783	<0.001	0.168	0.124–0.228
Education level																
low			reference				reference				reference				reference	
medium	0.399	<0.001	1.490	1.306–1.700	0.400	<0.001	1.492	1.308–1.701	0.667	<0.001	1.949	1.648–2.304	0.693	<0.001	2.000	1.728–2.314
high	0.779	<0.001	2.180	1.809–2.626	0.796	<0.001	2.216	1.841–2.668	0.937	<0.001	2.553	2.021–3.225	1.025	<0.001	2.786	2.277–3.410
Occupational status																
health-related industries	0.261	<0.001	1.298	1.127–1.496	0.229	0.001	1.257	1.092–1.446	0.158	0.061	0.848	0.992–1.381	0.096	0.197	1.101	0.951–1.274
others			reference				reference				reference				reference	
Marriage status																
married			reference				reference				reference				reference	
bachelor	0.439	<0.001	1.551	1.266–1.899	0.433	<0.001	1.542	1.262–1.884	0.362	0.001	1.437	1.152–1.792	0.526	<0.001	1.692	1.393–2.055
divorced/widowed	0.300	0.079	1.350	0.965–1.888	0.291	0.088	1.338	0.957–1.871	0.268	0.305	1.307	0.783–2.181	0.345	0.108	1.412	0.927–2.151
Obesity																
no			reference				reference				reference				reference	
yes	0.132	0.132	1.141	0.961–1.354	0.128	0.143	1.136	0.958–1.347	0.175	0.078	1.191	0.981–1.447	0.039	0.666	1.039	0.872–1.238
NCD																
no			reference				reference				reference				reference	
yes	−0.026	0.746	0.975	0.834–1.139	−0.046	0.563	0.955	0.817–1.116	0.031	0.793	1.031	0.820–1.298	0.115	0.244	1.122	0.924–1.362
Weight loss/shaping in the past month																
no			reference				reference				reference				reference	
yes	0.533	<0.001	1.704	1.375–2.111	0.534	<0.001	1.706	1.380–2.110	0.618	<0.001	1.855	1.510–2.278	0.526	<0.001	1.692	1.396–2.051
Mood swing in the past month																
no			reference				reference				reference				reference	
yes	0.273	0.147	1.314	0.908–1.902	0.277	0.139	1.319	0.914–1.904	−0.024	0.908	0.976	0.649–1.469	0.192	0.287	1.212	0.851–1.727

(1) Total beverages; (2) sugar-sweetened beverages; (3) non-sugar sweetened beverages; (4) ready-to-drink beverages; (5) logistic regression analyses statistics.

Logistic regression analysis of TBs showed that, compared with the 6–18 age group, the risk decreased in the 18–40 age group (OR = 0.585, 95% CI: 0.476–0.720); the 40–60 age group (OR = 0.310, 95% CI: 0.241–0.400), and the >60 age group (OR = 0.191, 95% CI: 0.147–0.249). Compared with the low-education group, the risk increased by 49.0% (OR = 1.490, 95% CI: 1.306–1.700) in the middle-education group and 118% (OR = 2.180, 95% CI: 1.809–2.626) in the high-education group. Compared with the married group, the risk increased in the bachelor group (OR = 1.551, 95% CI: 1.266–1.899). Those with weight loss/shaping had a 70.4% increased risk (OR = 1.704, 95% CI: 1.375–2.111).

Similar results were found in the analyses of the other three types of beverages in the different age, education level, marriage status, and weight loss/shaping groups. In addition to these variables, the other three types of beverages had their own characteristics. Females had a lower risk of consuming NSSs (OR = 0.781, 95% CI: 0.684–0.892) but a higher risk of consuming RSBs (OR = 1.197, 95% CI: 1.067–1.343) than males. People working in health-related industries were at higher risk of consuming TBs and SSBs [(OR = 1.298, 95% CI: 1.127–1.496); OR = 1.257, 95% CI: 1.092–1.446)].

4. Discussion

In recent years, global production and consumption of beverages have been on the rise. At the same time, numerous studies [15,16] have shown that excessive intake of SSBs and NSSs can cause various health problems. According to our study, the consumption rates of TBs, SSBs, NSSs, and RSBs in Shanghai were 57.70%, 56.94%, 19.60%, and 29.50%, respectively. The 2017–2018 U.S. National Health and Nutrition Examination Survey (NHANES) [17] showed that the percentage of adults aged 20–39 years consuming any amount of SSBs on a given day was 72.6% across all races and/or ethnicity groups. A national survey of Australians [18] aged 18 years and older showed that almost half (47.3%) of participants had consumed an SSB in the past week, with 13.6% consuming it at least once a day. In our survey, about 34.3% of participants aged 18 years and older had consumed an SSB in the past week, with 8.1% consuming it at least once a day on average. Although the SSB consumption level in Shanghai is lower than in developed countries such as the United States and Australia, the sales data of two major brands, the Coca-Cola Company (Atlanta, GA, USA) and PepsiCo Inc. (New York, NY, USA). [19], and studies [20,21], all showed that the consumption of SSBs in China is on a growing trend. The rate of SSB consumption at least once every 3 days among urban residents was 11.4% in 18 provinces of China, and the average SSB consumption amount in the drinking population was 130.0 g/d [20]. Corresponding to our survey, the rate of SSB consumption at least 4 times/week was 19.5%, nearly twice the national level, and the average consumption in the drinking population was 412.0 mL/day, which is over three times the national level. The high consumption rate of beverages in Shanghai may be related to the fact that the level of economic development in Shanghai is among the highest in the country. The article about global nutrition dynamics published by the American Journal of Clinical Nutrition [22] showed that an increase in the level of economic development may bring about changes in the eating habits and lifestyles of the residents, such as purchasing more SSBs, etc.

Regarding TB consumption, CBs still hold the largest share. Additionally, emerging types of beverages were also very much worth noting. RSBs and NSSs accounted for 12.3% and 9.5%, respectively, ranking them second and third. They have become the main types of beverages, and a lot of analyses [9,10] showed that the consumption proportion in Chinese consumers is still expanding. RSB ingredients are opaque, and there is a misconception about NSSs that many people think they are healthier than pre-packaged SSBs because of their low-calorie characteristic. Therefore, it is necessary to guide people to recognize and learn about new types of beverages and strengthen health education in them.

In our study, both consumption rate and consumption amount showed a downward trend with increasing age. In a systematic assessment of beverage intake in 187 countries/regions worldwide, it was shown that the consumption of SSBs usually followed an inverse age gradient [23], a finding that was also confirmed in our study. The NHANES showed that children (aged 2 to 19) always consume SSBs at a higher rate than adults (aged 20+) [24]. A study investigating the consumption of SSBs in São Paulo found that adolescents had the highest consumption of SSBs [25]. These data showed that adolescents are the main consumers of beverages. The excessive consumption of SSBs poses a threat to the healthy development of children and adolescents, especially by increasing the risk of obesity [6,26] and dental caries [15,16]. Therefore, for the adolescent population, we need to strengthen guidance and regulation of beverage consumption to control their intake of SSBs.

There were differences in beverage consumption preferences between males and females. Our study found that although the difference in SSB consumption rates between males and females was not statistically significant ($p = 0.192$), the consumption amount was higher among males than females, which is consistent with the results for Chinese children and adolescents [26] and for the U.S. in all age groups [16]. Similar results had been found in São Paulo, Taiwan, and several other regions around the world [23,25,27]. The reasons females consume less SSBs may include a lower energy intake requirement resulting in a lower total food intake, and that females have growing concern for their appearances and increasing health awareness compared with males [28]. In addition, this study also found that the consumption rate of RSBs in females was higher than in males. The Report of Development Prospect Prediction and Investment Strategy Planning on China New Tea Drinking Industry (2023–2028) described that women's consumption rate of RSBs was higher than men's, with women's share reaching 65.09% in 2022 [29]. These data further proved the difference in beverage consumption preferences between different gender groups.

The beverage consumption was also closely related to economic level. When analyzing the association between income and educational level with the beverage consumption, we found that the highest consumption rates and amounts of TBs, SSBs, NSSs, and RSBs were in the groups with the highest incomes and the highest educational level. This is similar to some other national studies. In Mexico, two nationally representative dietary intake surveys reported higher average energy intake of SSBs for children from higher socioeconomic status families and from urban families compared with rural families [30]. The Korean National Health and Nutrition Examination Survey showed that people in the highest income group were more likely to consume SSBs (OR = 1.18 for females and 1.25 for males) [31]. However, there are also some inconsistent studies. The result of a survey on the consumption patterns of SSBs in the United States showed that lower-income and lower educational attainment people consumed more SSBs [32]. Australia also found higher odds of SSB consumption and higher caloric intake among people of low income [33]. This research inconsistency may be related to the development of economic levels in different countries. People with higher incomes in economically developed countries have additional nutritional knowledge and preventive information about nutrition and tend to choose healthy foods [34], whereas in countries with relatively weaker economies, people with low incomes decline purchasing higher-price foods. China is still a developing country, but the price of beverages is often relatively high, for example, a cup of RSB costs about $2–4. This may be the reason why people with low incomes tend not to consume beverages. It suggests that we also should strengthen the health education of high-income people to raise awareness of the potential negative effects of beverages consumption.

We also found an interesting phenomenon: the bachelor consumed more beverages than married person. This phenomenon may be due to the fact that there is a positive relationship between being married and being healthy [35,36], as partners are bound to each other, which helps to form healthy habits, and married persons may consume healthier foods and fewer beverages on a daily basis. Research on the relationship between marriage and health showed that emerging adults with an unhealthy BMI have an improved BMI after entering marriage, further favoring chronic disease management and improving future health [37–39]. This is similar to our findings.

In our study, we delved into the beverage consumption patterns across various demographic groups in Shanghai. Our comprehensive inclusion of different beverage types, especially emerging types of beverages such as NSSs and RSBs, which were the beverage categories that have rarely been studied separately in the past but have increasing significant consumption, coupled with a rational sampling strategy and a robust sample size, has effectively mitigated the shortcomings observed in prior research. This has allowed us to present more representative insights into the consumption of beverages in the region.

However, there are some limitations in our study. While our work provided a temporal snapshot and reduced recall bias, it may not fully capture the complexities of long-term beverage consumption habits. Our survey focused on retail visitors, which may limit the generalizability of our findings to the broader population. Additionally, household income indeed may affect the beverage consumption of kids, but our survey did not cover this aspect.

5. Conclusions

In summary, the beverage consumption in Shanghai, although lower than internationally, is still high and suggests that timely interventions are needed. Firstly, we need to strengthen health education, especially for adolescents, who are the main consumers of SSBs. Secondly, health education should not only focus on CBs, which were the most significant contributors to TB consumption, but also pay attention to emerging types of beverages, such as NSSs and RSBs. Thirdly, there should be different intervention strategies for different genders, e.g., males should focus on controlling SSB and NSS intake, while females should focus on preventing excessive RSB intake. Finally, health awareness needs to be raised even among high-income and highly educated populations.

Author Contributions: Z.W., L.S., Z.S. (Zehuan Shi), Z.S. (Zhuo Sun), Q.S., W.L., W.M., S.M. and J.Z.: Investigation and Review; Z.W., J.N. and Y.X.: Formal analysis, Writing—original draft, Writing—reviewing and editing; Z.W., L.S. and J.Z.: Project administration. All authors have read and agreed to the published version of the manuscript.

Funding: The current study was supported by the key projects in the three-year plan of Shanghai municipal public health system (2023–2025) (GWVI-4), key disciplines in the three-year plan of Shanghai municipal public health system (2023–2025) (GWVI-11.1-42, GWVI-11.1-45), academic leader in the three-year plan of Shanghai municipal public health system (2023–2025) (GWVI-11.2-XD21), Young Talents in Shanghai Health Science Popularization (JKKPYC-2023-B12), and Shanghai Sailing Program (No. 23YF1437000).

Institutional Review Board Statement: The Shanghai Municipal Center for Disease Control and Prevention Ethical Review Committee approved the present study, KY-2024-16, 13 March 2024.

Informed Consent Statement: All participants provided informed consent before the investigation.

Data Availability Statement: The original contributions presented in this study are included in the article, further inquiries can be directed to the corresponding author.

Acknowledgments: We are grateful to all subjects who participated in our study and the healthcare professionals at the Centers for Disease Control and Prevention of Shanghai.

Conflicts of Interest: The authors declare no conflicts of interest.

References

1. National Bureau of Statistics of China. [EB/OL]. (2023-12-31). Available online: https://www.stats.gov.cn/ (accessed on 8 April 2024).
2. Lara-Castor, L.; Micha, R.; Cudhea, F.; Miller, V.; Shi, P.; Zhang, J.; Sharib, J.R.; Erndt-Marino, J.; Cash, S.B.; Mozaffarian, D. Sugar-sweetened beverage intakes among adults between 1990 and 2018 in 185 countries. *Nat. Commun.* **2023**, *14*, 5957. [CrossRef] [PubMed]
3. Zhang, N.; Morin, C.; Guelinckx, I.; Moreno, L.A.; Kavouras, S.A.; Gandy, J.; Martinez, H.; Salas-Salvadó, J.; Ma, G. Fluid intake in urban China: Results of the 2016 Liq.In (7) national cross-sectional surveys. *Eur. J. Nutr.* **2018**, *57* (Suppl. S3), 77–88. [CrossRef] [PubMed]

4. Huang, Y.; Chen, Z.; Chen, B.; Li, J.; Yuan, X.; Li, J.; Wang, W.; Dai, T.; Chen, H.; Wang, Y.; et al. Dietary sugar consumption and health: Umbrella review. *BMJ* **2023**, *381*, e071609. [CrossRef]
5. Malik, V.S.; Pan, A.; Willett, W.C.; Hu, F.B. Sugar-sweetened beverages and weight gain in children and adults: A systematic review and meta-analysis. *Am. J. Clin. Nutr.* **2013**, *98*, 1084–1102. [CrossRef]
6. Malik, V.S.; Hu, F.B. The role of sugar-sweetened beverages in the global epidemics of obesity and chronic diseases. *Nat. Rev. Endocrinol.* **2022**, *18*, 205–218. [CrossRef] [PubMed]
7. Singh, G.M.; Micha, R.; Khatibzadeh, S.; Lim, S.; Ezzati, M.; Mozaffarian, D. Estimated Global, Regional, and National Disease Burdens Related to Sugar-Sweetened Beverage Consumption in 2010. *Circulation* **2015**, *132*, 639–666. [CrossRef]
8. Malik, V.S.; Li, Y.; Pan, A.; De Koning, L.; Schernhammer, E.; Willett, W.C.; Hu, F.B. Long-Term Consumption of Sugar-Sweetened and Artificially Sweetened Beverages and Risk of Mortality in US Adults. *Circulation* **2019**, *139*, 2113–2125. [CrossRef] [PubMed]
9. Report Hall. Milk Tea Market Analysis 2023: Asia Is the Major Consuming Region of Milk Tea Market [EB/OL]. (2023-10-13). Available online: https://www.chinabgao.com/freereport/90251.html (accessed on 5 March 2024).
10. TIMON. China's Sugar-Free Beverage Industry Research Report in 2022 [EB/OL]. (2022-05-26). Available online: https://data.eastmoney.com/report/zw_industry.jshtml?infocode=AP202205261568020049 (accessed on 25 April 2024).
11. World Health Organization. Use of Non-Sugar Sweeteners: WHO Guideline [R]. (2023.5.15). Available online: https://www.who.int/publications/i/item/9789240073616 (accessed on 25 April 2024).
12. National Health Commission of the People's Republic of China. Criteria of Weight for Adults [R/OL]. (2013-04-18). Available online: http://www.nhc.gov.cn/zwgkzt/yingyang/201308/a233d450fdbc47c5ad4f08b7e394d1e8.shtml (accessed on 8 April 2024).
13. National Health Commission of the People's Republic of China. Screening for Overweight and Obesity Among School-Age Children and Adolescents [R/OL]. (2018-02-23). Available online: http://www.nhc.gov.cn/fzs/s7852d/201803/7595325d43bb4417945edd570b1b12fa.shtml (accessed on 8 April 2024).
14. National Health Commission of the People's Republic of China. Growth Standard for Children Under 7 Years of Age [R/OL]. (2022-09-19). Available online: http://www.nhc.gov.cn/wjw/fyjk/202211/16d8b049fdf547978a910911c19bf389.shtml (accessed on 8 April 2024).
15. Hill, C.M.; Chi, D.L.; Mancl, L.A.; Jones-Smith, J.C.; Chan, N.; Saelens, B.E.; McKinney, C.M. Sugar-sweetened beverage intake and convenience store shopping as mediators of the food insecurity—Tooth decay relationship among low-income children in Washington state. *PLoS ONE* **2023**, *18*, e0290287. [CrossRef]
16. Laniado, N.; Sanders, A.E.; Godfrey, E.M.; Salazar, C.R.; Badner, V.M. Sugar-sweetened beverage consumption and caries experience: An examination of children and adults in the United States, National Health and Nutrition Examination Survey 2011–2014. *J. Am. Dent. Assoc.* **2020**, *151*, 782–789. [CrossRef]
17. Dai, J.; Soto, M.J.; Dunn, C.G.; Bleich, S.N. Trends and patterns in sugar-sweetened beverage consumption among children and adults by race and/or ethnicity, 2003–2018. *Public Health Nutr.* **2021**, *24*, 2405–2410. [CrossRef]
18. Miller, C.; Ettridge, K.; Wakefield, M.; Pettigrew, S.; Coveney, J.; Roder, D.; Durkin, S.; Wittert, G.; Martin, J.; Dono, J. Consumption of Sugar-Sweetened Beverages, Juice, Artificially-Sweetened Soda and Bottled Water: An Australian Population Study. *Nutrients* **2020**, *12*, 817. [CrossRef] [PubMed]
19. Kleiman, S.; Ng, S.W.; Popkin, B. Drinking to our health: Can beverage companies cut calories while maintaining profits? *Obes. Rev.* **2012**, *13*, 258–274. [CrossRef] [PubMed]
20. Pan, F.; Luan, D.C.; Zhang, T.W.; Mao, W.F.; Liang, D.; Liu, A.D.; Li, J.W. Assessment of sugar-sweetened beverages consumption and free sugar intake among urban residents aged 3 and above in China. *Chin. J. Food Hygiene* **2022**, *34*, 126–130.
21. Guo, H.; Zhao, L.; Xu, X.; Yu, W.; Ju, L.; Yu, D. Consumption of sugar-sweetened beverages among 18 years old and over adults in 2010–2012 in China. *J. Hyg. Res. [Wei Sheng Yan Jiu]* **2018**, *47*, 22–26. [PubMed]
22. Popkin, B.M. Global nutrition dynamics: The world is shifting rapidly toward a diet linked with noncommunicable diseases. *Am. J. Clin. Nutr.* **2006**, *84*, 289–298. [CrossRef]
23. Singh, G.M.; Micha, R.; Khatibzadeh, S.; Shi, P.; Lim, S.; Andrews, K.G.; Engell, R.E.; Ezzati, M.; Mozaffarian, D.; Global Burden of Diseases Nutrition and Chronic Diseases Expert Group (NutriCoDE). Global, Regional, and National Consumption of Sugar-Sweetened Beverages, Fruit Juices, and Milk: A Systematic Assessment of Beverage Intake in 187 Countries. *PLoS ONE* **2015**, *10*, e0124845. [CrossRef]
24. Bleich, S.N.; Vercammen, K.A.; Koma, J.W.; Li, Z. Trends in Beverage Consumption among Children and Adults, 2003–2014. *Obesity* **2018**, *26*, 432–441. [CrossRef]
25. Fontes, A.S.; Pallottini, A.C.; Vieira, D.A.d.S.; Fontanelli, M.d.M.; Marchioni, D.M.; Cesar, C.L.G.; Alves, M.C.G.P.; Goldbaum, M.; Fisberg, R.M. Demographic, socioeconomic and lifestyle factors associated with sugar-sweetened beverage intake: A population-based study. *Rev. Bras. Epidemiol.* **2020**, *23*, e200003. [CrossRef]
26. Gui, Z.H.; Zhu, Y.N.; Cai, L.; Sun, F.H.; Ma, Y.H.; Jing, J.; Chen, Y.J. Sugar-Sweetened Beverage Consumption and Risks of Obesity and Hy-pertension in Chinese Children and Adolescents: A National Cross-Sectional Analysis. *Nutrients* **2017**, *9*, 1302. [CrossRef]
27. Chan, T.-F.; Lin, W.-T.; Huang, H.-L.; Lee, C.-Y.; Wu, P.-W.; Chiu, Y.-W.; Huang, C.-C.; Tsai, S.; Lin, C.-L.; Lee, C.-H. Consumption of sugar-sweetened beverages is associated with components of the metabolic syndrome in adolescents. *Nutrients* **2014**, *6*, 2088–2103. [CrossRef]
28. Nathanson, C.A. Sex roles as variables in preventive health behavior. *J. Community Health* **1977**, *3*, 142–155. [CrossRef] [PubMed]

29. Foresight Industry Research Institute. Report of Development Prospect Prediction and Investment Strategy Planning on China New Tea Drinking Industry (2023–2028) [EB/OL]. (2023-06-13). Available online: https://bg.qianzhan.com/report/detail/1905081728257184.html (accessed on 2 May 2024).
30. Barquera, S.; Campirano, F.; Bonvecchio, A.; Hernández-Barrera, L.; Rivera, J.A.; Popkin, B.M. Caloric beverage consumption patterns in Mexican children. *Nutr. J.* **2010**, *9*, 47. [CrossRef] [PubMed]
31. Han, E.; Kim, T.H.; Powell, L.M. Beverage consumption and individual-level associations in South Korea. *BMC Public Health* **2013**, *13*, 195. [CrossRef] [PubMed]
32. Han, E.; Powell, L.M. Consumption Patterns of Sugar-Sweetened Beverages in the United States. *J. Acad. Nutr. Diet.* **2013**, *113*, 43–53. [CrossRef]
33. Hafekost, K.; Mitrou, F.; Lawrence, D.; Zubrick, S.R. Sugar sweetened beverage consumption by Australian children: Implications for public health strategy. *BMC Public Health* **2011**, *11*, 950. [CrossRef]
34. Darmon, N.; Drewnowski, A. Contribution of food prices and diet cost to socioeconomic disparities in diet quality and health: A systematic review and analysis. *Nutr. Rev.* **2015**, *73*, 643–660. [CrossRef]
35. Tatangelo, G.; McCabe, M.; Campbell, S.; Szoeke, C. Gender, marital status and longevity. *Maturitas* **2017**, *100*, 64–69. [CrossRef]
36. Robles, T.F. Marital quality and health: Implications for marriage in the 21st century. *Curr. Dir. Psychol. Sci.* **2014**, *23*, 427–432. [CrossRef]
37. Cortez, G.R.; Lee, S.; Roberson, P.N.E. Get healthy to marry or marry to get healthy? *Pers. Relatsh.* **2020**, *27*, 613–629. [CrossRef]
38. Quan, S.; Zhang, H. The relationship between Marriage and Body Mass Index in China: Evidence from the China Health and Nutrition Survey. *Econ. Hum. Biol.* **2024**, *53*, 101368. [CrossRef]
39. Turnic, T.N.; Jakovljevic, V.; Strizhkova, Z.; Polukhin, N.; Ryaboy, D.; Kartashova, M.; Korenkova, M.; Kolchina, V.; Reshetnikov, V. The Association between Marital Status and Obesity: A Systematic Review and Meta-Analysis. *Diseases* **2024**, *12*, 146. [CrossRef] [PubMed]

foods

MDPI

Article

Green Synthesis of Molecularly Imprinted Polymers for Selective Extraction of Protocatechuic Acid from Mango Juice

Liping Zhang *, Xin Song, Yuxiao Dong and Xiyan Zhao

School of Basic Medicine and Forensic Medicine, Henan University of Science and Technology, Luoyang 471000, China; s2036558171@163.com (X.S.); yxd2020871071@163.com (Y.D.); xyz20040816@163.com (X.Z.)

* Correspondence: 9906023@haust.edu.cn

Abstract: A novel and environmentally friendly molecularly imprinted polymer (PCA-MIP) was successfully synthesized in an aqueous solution for the selective extraction of protocatechuic acid (PCA). In this study, a deep eutectic solvent (DES, choline chloride/methacrylic acid, 1:2, mol/mol) and chitosan were employed as the eco-friendly functional monomers. These two components interacted with PCA through hydrogen bonding, integrating a multitude of recognition sites within the PCA-MIP. Thus, the resulting PCA-MIP exhibited outstanding adsorption performance, rapid adsorption rate, and better selectivity, with a maximum binding capacity of 30.56 mg/g and an equilibrium time of 30 min. The scanning electron microscope (SEM) and Brunauer–Emmett–Teller (BET) analyses revealed that the synthesized polymers possessed a uniform morphology and substantial surface areas, which were conducive to their adsorption properties. Moreover, the PCA-MIP integrated with HPLC demonstrated its efficacy as an adsorbent for the selective extraction of PCA from mango juice. The PCA-MIP presented itself as an exemplary adsorbent, offering a highly effective and eco-friendly method for the enrichment of PCA from complex matrices.

Keywords: molecularly imprinted polymer; deep eutectic solvent; chitosan; green synthesis; protocatechuic acid

1. Introduction

Protocatechuic acid (PCA), a member of natural phenolic acids, is extensively found in a variety of vegetables and fruits. Extensive pharmacological research has demonstrated that PCA exhibits significant anti-inflammatory, antioxidant, tumor-inhibitory, and neuroprotective properties [1–3]. Nowadays, PCA is predominantly extracted from foods and natural products [4,5]. However, due to the intricate composition of food matrices and interference from other compounds, conventional enrichment techniques for PCA are cumbersome, solvent-consuming, and make it difficult to achieve high-purity products [6]. Consequently, it is essential to develop novel adsorbent materials for the selective enrichment and extraction of PCA from complex crude extracts prior to chromatographic analysis.

Molecular imprinting is an effective method for the fabrication of selective adsorbents [7–9]. The molecularly imprinted polymer (MIP) is prepared through a polymerization process that incorporates a template, functional monomer, and crosslinker agent. After removal of template molecules via an appropriate method, the resultant polymer is endowed with imprinting cavities that are complementary to the spatial structure of the template [10]. These "cryo-memory sites" could again selectively recognize templates from complex matrices through hydrogen bonding, hydrophobic effects, or electrostatic interactions [11,12]. The main advantage of MIPs lies in their exceptional selectivity and affinity for the target molecules, similar to the well-known "lock-and-key" model. Furthermore, MIPs are characterized by their minimal preparation cost and higher physicochemical stability.

Currently, a variety of MIPs have been reported for their efficacy in the extraction and detection of PCA from complex samples. Xie et al. (2015) synthesized magnetic MIPs, with acetonitrile utilized as the dispersion medium, acrylamide serving as the functional monomer, and ethylene glycol dimethacrylate acting as the crosslinker, which exhibited impressive recoveries and minimal detection limits for the selective extraction of PCA from plant extracts [13]. Denderz et al. (2014) introduced two MIPs for gallic acid (GA) and PCA, crafted with methacrylic acid as the monomer and methanol as the porogen agent. These MIPs had been successfully applied as sorbents, facilitating the precise quantification of targeted phenolic compounds within red wine matrices [14]. Li et al. (2015) first fabricated the magnetic hollow porous MIPs (HPMIPs) for PCA in the presence of 4-vinylpyridine and glycidilmethacrylate as a co-monomer [15]. However, the constituent components such as functional monomers (acrylic acid, acrylamide, vinyl pyridine, etc.), crosslinking agents (ethylene glycol dimethacrylate and methylene-bis-acrylamide), and porogen agents (methanol, acetonitrile, chloroform) all have some toxicity, limiting the application of MIPs in the food, therapeutic, and pharmaceutical industries. Consequently, the synthesis of MIPs based on green chemistry strategies has garnered significant interest [16,17]. The core principle entails striving to minimize or avoid the utilization of toxic, hazardous, and volatile organic solvents as well as monomers, and instead adopting some green, safe, and biodegradable materials and methods, thereby mitigating the detrimental impacts on the environment and human health [18].

Deep eutectic solvents (DESs) constitute a class of self-assembling eutectic fluids composed of hydrogen bond acceptors and hydrogen bond donors in a predefined ratio [19]. DES, as an emerging eco-friendly solvent, has been successfully applied in the preparation of MIPs [20,21]. Compared with conventional functional monomers, the abundant functional groups inherent in DESs could bind to the template through hydrogen bonding interactions, thus augmenting the affinity and selectivity of MIPs [22,23]. Furthermore, DESs have been demonstrated to serve as crosslinking agents, thereby conferring enhanced rigidity to the polymeric structure [24]. By leveraging the malleable properties of DESs, poorly soluble template molecules can be transformed into DESs, which can then be used as both a monomer and a template for the preparation of MIPs [25,26]. Moreover, DESs-based MIPs exhibit commendable compatibility with aqueous media and provide excellent recognition performance for templates in aqueous solution [27,28]. Therefore, the combination of DESs and MIPs has provided a new idea for the development of eco-friendly materials.

Moreover, chitosan (CS) is widely recognized as an environmentally friendly natural polysaccharide endowed with the virtues of nontoxic, biocompatible, and biodegradable advantages [29]. Recently, CS has demonstrated significant potential application value in MIP synthesis [30]. Typically, CS inherently contains a multitude of functional groups, such as glycosides, hydroxyl, and amino, rendering it an ideal functional monomer for the binding of templates in ion-imprinted polymers [31]. Additionally, chitosan can serve as a crosslinker or a supportive matrix, thereby reinforcing the rigid structure of MIPs and augmenting their specific surface area [32,33]. The chitosan-based MIPs have received considerable attention for the micro-solid phase extraction target from complex matrix [34–36]. However, it is noteworthy that several MIPs incorporate some toxic reagents during their synthesis. For instance, Laskar et al. (2021) synthesized chitosan-based magnetic molecularly imprinted polymers utilizing an acetonitrile/toluene (75/25, v/v) mixture as a porogenic solvent for separation and enrichment of tricyclazole from rice and water samples [37]. Despite organic solvents like acetonitrile and toluene being commonly utilized in the synthesis of MIPs, their potential hazards to health and the environment cannot be overlooked. This study presents a novel approach for PCA-MIP synthesis using environmentally friendly DESs and chitosan as functional monomers to improve adsorption performance while minimizing environmental impact.

In this study, we designed and synthesized an innovative PCA-MIP via green synthesis strategies in an aqueous solution, using DESs and CS as the co-functional monomers and glutaraldehyde as the crosslinking agent. The structural features of PCA-MIP were also meticulously examined. The preparation conditions, adsorption performance, selectivity, and reusability of PCA-MIP were further validated through a series of adsorption experiments. Moreover, the applicability of PCA-MIP in combination with HPLC for the targeted extraction and precise detection of PCA from the crude extracts of mango juice is also demonstrated. This integration showcases the practical utility and selectivity of PCA-MIP in real samples.

2. Materials and Methods

2.1. Materials and Reagents

Protocatechuic acid (PCA, ≥97.0%), gallic acid (GA, ≥98%), ethyl gallate (EG, ≥99%), naringin (NRG, ≥98%), chitosan (CS), and glutaraldehyde (Glu, 50%) were obtained from Macklin (Shanghai, China). Methacrylic acid (MAA, AR) was obtained from the Damao Chemical Reagent Factory (Tianjin, China). Choline chloride (ChCl, ≥98%) was obtained from Yuanye Biotechnology (Shanghai, China). Acetic acid and methanol were supplied by Yongda Chemical Reagent (Tianjin, China).

2.2. Preparation of PCA-MIP

Initially, the synthesis of DES was conducted by heating and mechanically stirring the ChCl and MAA (ChCl/MAA, 1/2, molar ratio) mixture for 2 h at 80 °C. This process culminated in the formation of a homogeneous and clarified liquid. Figure 1 illustrates the preparation process for PCA-MIP. The CS solution was obtained by dispersing 500 mg of CS into 50 mL of acetic acid solution (1%) and stirring to dissolve. Subsequently, 100 mg of PCA was dissolved in an appropriate amount of DES, and this solution was then gently immersed into 20 mL of prepared CS solution. The mixture was stirred for 1 h to ensure thorough mixing. An appropriate volume of Glu (0.5, 1.0, 2.0, and 3.0 mL), used as a crosslinking agent, was then added to the mixture and further stirred to ensure complete homogenization. Nitrogen gas was subsequently purged through the mixture to remove oxygen—a critical step to prevent unwanted side reactions during the polymerization process. The pre-polymerization mixture was then carefully transferred to a water bath at 65 °C and stirred continuously for 3 h to facilitate the polymerization reaction. The resulting polymeric matrix was evenly divided into centrifuge tubes and eluted with methanol and acetic acid mixture (9:1, v/v) until no trace of PCA was detectable by UV-vis spectrophotometry (UV-1000, Techcomp, Shanghai, China). Ultimately, the PCA-MIP was eluted three times with methanol and then dried under a vacuum. As a comparative control, the preparation process for non-imprinted polymer (PCA-NIP) was the same, but without PCA.

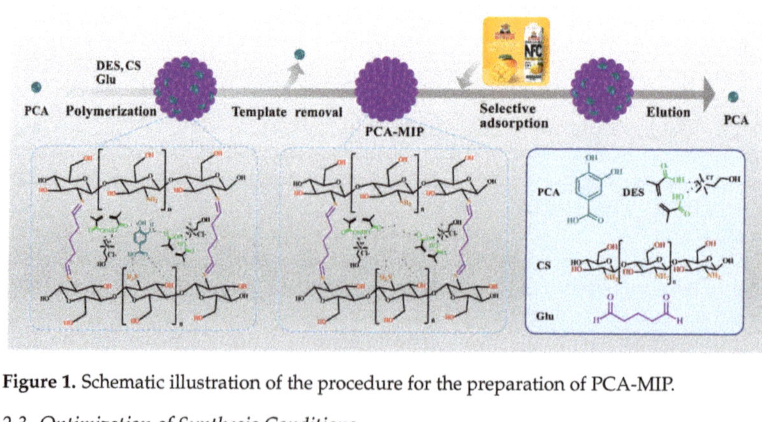

Figure 1. Schematic illustration of the procedure for the preparation of PCA-MIP.

2.3. Optimization of Synthesis Conditions

The synthesis conditions (content of DES and volume of Glu) for PCA-MIP and PCA-NIP were optimized using a series of binding experiments. Briefly, 10 mg of dried PCA-MIP and PCA-NIP, derived from different preparation conditions, were dispersed into 1.5 mL of

PCA aqueous solution at a fixed concentration of 500 μg/mL, with pH adjusted to 5.0. The mixtures were then subjected to a speed-regulated oscillator, ensuring uniform mixing for 4 h at room temperature. After centrifugal separation, the supernatant's PCA concentration was quantified using a UV-vis spectrophotometer, at an absorbance wavelength of 260 nm. All adsorption experiments were conducted in triplicate, and the amounts (Q, mg/g) of PCA adsorbed by PCA-MIP and PCA-NIP were calculated using Equation (1). The imprinting performance of polymers was rigorously assessed using the imprinting factor (IF), and the formula was presented as Equation (2).

$$Q = \frac{(C_0 - C_e) \times V}{M} \tag{1}$$

$$\mathrm{IF} = \frac{Q_{PCA-MIP}}{Q_{PCA-NIP}} \tag{2}$$

Here, C_0 (mg/mL) signifies the initial concentration of PCA in the solution, while C_e (mg/mL) is the concentration at equilibrium. The volume of the PCA solution is represented by V (mL) and M (g) is the mass of the polymers. $Q_{PCA-MIP}$ and $Q_{PCA-NIP}$ (mg/g) are the bonding amounts of polymers PCA-MIP and PCA-NIP toward PCA, respectively.

2.4. Characterization of PCA-MIP

The morphological characteristics of PCA-MIP and PCA-NIP were meticulously examined using an S-4800 SEM (Hitachi, Tokyo, Japan). FT-IR spectroscopy was employed to record the infrared spectra of PCA-MIP and PCA-NIP using a NICOLET380 Fourier transform infrared spectrophotometer (Thermo Fisher Scientific, Inc., Waltham, MA, USA). Furthermore, nitrogen adsorption–desorption of polymers was quantitatively assessed using an Autosorb-IQ Brunauer–Emmett–Teller analyzer (Quantachrome Instruments, Florida, USA). Thermal stability was measured using a Q50 Thermogravimetric analyzer (V20.8 Build 34, TA Instruments, Wilmington, DE, USA).

2.5. PCA-MIP Binding Study

The effect of incubation solution pH on the adsorption properties of PCA-MIP and PCA-NIP was initially investigated. A series of PCA aqueous solutions with an initial concentration of 500 μg/mL were prepared using water at pH values of 3.0, 4.0, 5.0, and 6.0. Subsequently, several portions of 10 mg of dried PCA-MIP and PCA-NIP were immersed in 1.5 mL of the PCA aqueous solutions at varying pH levels and mixed for 4 h. After centrifugation, the concentrations of PCA were measured using a UV-vis at 260 nm. The adsorption capacity (Q, mg/g) of PCA-MIP and PCA-NIP at different pH values was defined using Equation (1).

For equilibrium adsorption experiments, precise quantities of 10 mg each of the PCA-MIP and PCA-NIP were each suspended in 1.5 mL of PCA aqueous solutions with initial concentrations varying from 100 μg/mL to 1000 μg/mL, with the pH of the incubation medium standardized at 5.0. After shaking for 4 h, the mixtures were centrifuged to separate, and the equilibrium concentration of PCA was ascertained by a UV-vis spectrophotometer. The equilibrium adsorption capacity was then determined using Equation (1).

For kinetic adsorption experiments, 10 mg of PCA-MIP and PCA-NIP was introduced into 1.5 mL of PCA aqueous solutions with an initial concentration of 500 μg/mL, with the pH of the adsorption medium again set to 5.0. The mixtures were shaken at ambient temperature and the PCA concentrations were measured at certain intervals (10, 30, 60, 90, 120, 150, 180, and 240 min) using a UV-vis spectrophotometer. The dynamic adsorption capacity at time t, denoted as Q_t (mg/g), was determined using Equation (3).

$$Q_t = \frac{(C_0 - C_t) \times V}{M} \tag{3}$$

In this equation, C_t (μg/mL) are the concentrations of PCA at the respective time intervals.

Furthermore, selectivity adsorption experiments were performed by determining the adsorption capacity of polymers of PCA and its structural analogs, namely GA, NRG, and EG. Briefly, 10 mg of PCA-MIP and PCA-NIP were incubated in 1.5 mL of GA, NRG, and EG aqueous solutions (500 μg/mL). Following shaking for 4 h, the contents of the competitive molecules absorbed by polymers were quantified individually by a UV-Vis spectrophotometer at wavelengths of 272 nm, 280 nm, and 275 nm.

2.6. Study of PCA-MIP Reusability

The reusability of PCA-MIP was evaluated by determining the adsorption capacity with several rebinding–regeneration cycles. Initially, 10 mg of PCA-MIP was introduced into 1.5 mL of PCA aqueous solution (500 μg/mL, pH 5.0) and mixed for 4 h. Then, the PCA-MIP was collected by centrifugation, and the concentration of PCA in the supernatant was measured by UV-vis spectrophotometry. The sedimented PCA-MIP was then regenerated through three cycles of elution with a methanol/acetic acid mixture (9:1, v/v) to remove PCA. The following adsorption–desorption PCA-MIP experiments were repeated five times.

2.7. Extraction of PCA from Mango Juice

NFC mango juice (Anyang Jingshantang Beverage Co., Ltd., Anyang, China) was purchased from a local market in Luoyang. A 40 mL volume of the mango juice was first centrifuged for 20 min at 8000 rpm. Then, 25 mL of the supernatant was carefully extracted twice using an equivalent volume of ethyl acetate via liquid–liquid extraction. Subsequently, the ethyl acetate extract was evaporated to yield a dry residue. The mango juice crude extract was obtained by dissolving the dry residue in 20 mL of aqueous solution with a pH of 5.0. For the adsorption process, 1.5 mL of the mango juice crude extract was incubated with 30 mg of PCA-MIP for 4 h. The supernatant was harvested after separation, and adsorbents were then washed once with water and three times with a mixture of ethanol and acetic acid at a volume ratio of 9:1. The resultant crude extract, collected supernatant, eluate, and PCA standard were all filtered through 0.22 μm nylon membranes for HPLC analysis. HPLC detection was performed on a LaChrom Elite system (Hitachi, Tokyo, Japan) consisting of an LC 2130 pump and an LC 2030 detector that connected a Kromasii-C18 column (4.6 mm × 250 mm). The mobile phase was methanol/0.1% phosphoric acid (5/95, v/v) with a flow rate of 1.0 mL/min. The injected sample volume was 20 μL and the detection wavelength was 280 nm.

3. Results and Discussion

3.1. Preparation of PCA-MIP

Figure 1 illustrates the preparation route for PCA-MIP. Prior to polymerization, the green DES, rich in functional groups, engaged in hydrogen bonding interactions with PCA. In addition, CS, endowed with a number of amino and hydroxyl groups, formed a host–guest complex with PCA through electrostatic attraction and multi-site hydrogen bonding. In the presence of Glu, the free amino groups in CS were effectively crosslinked by the reactive aldehyde groups of Glu, thereby enhancing the mechanical properties of the co-polymers [38]. After the elution of PCA, PCA-MIP with selective recognition sites for PCA was fabricated. In order to obtain high adsorption capacity for efficient recognition, some important synthetic conditions need to be further optimized during the preparation process.

In this system, the quaternary ammonium salt of CH_3Cl mainly provided electrostatic interactions with PCA, and the HBD of MAA mainly produced hydrogen bonding interactions with PCA [20,39]. Therefore, the binding of CH_3Cl with MAA can enhance its interaction with the PCA template through multiple interaction modes, resulting in the formation of more imprinting sites. As a consequence, the amount of DES directly affected the number of recognition sites within the polymer; thus, the content of DES was first investigated. As depicted in Figure 2a, the adsorption capacity of PCA-MIP showed an upward trend with the increment in DES content. When the DES content reached 2.5 mL,

PCA-MIP achieved the highest adsorption capacity (Q = 25.39 mg/g) and the best imprinting performance (IF = 1.53). When the content of DES was less than 2.5 mL, there were few imprinting cavities in MIP to bind more PCA. However, as the content of DES increased over 2.5 mL, both the adsorption amount and IF experienced a decline. The reason may be that steric hindrance from larger-sized polymers reduced adsorption capacity. In addition, the excessive addition of DES decreased the cross-section of the polymer network, which led to loose imprinting and lower binding capacity [40,41]. As a result, a DES volume of 2.5 mL was selected for the synthesis of PCA-MIP.

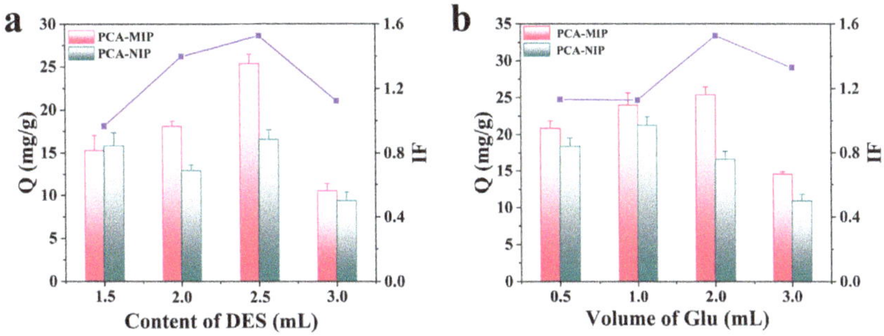

Figure 2. Effect of (**a**) content of DES and (**b**) volume of Glu on absorption capacity (Q) and imprinting factor (IF) of PCA-MIP and PCA-NIP. The pink bar represents PCA-MIP, the blue bar represents PCA-NIP, and the purple line is the imprinting factor.

The role of the crosslinker was to maintain the rigid structure of the polymer matrix; therefore, the volume of Glu employed was critical in determining the degree of crosslinking within the PCA-MIP. From Figure 2b, it was found that the binding capacity and IF of PCA-MIP showed the same trend as the volume of Glu increased. At a Glu volume of 2.0 mL, the PCA-MIP demonstrated its optimal adsorption performance. This enhancement is postulated to arise from the increased Glu content, which facilitated the polymer to maintain a larger rigid structure, thereby better retaining the imprinted cavities and enhancing the recognition capability [42]. Nonetheless, when the volume of Glu exceeds 2.0 mL, both Q and IF dropped dramatically. The reason is that an excessive degree of crosslinking can lead to the embedding of imprinted sites, impeding the complete elution of the template and reducing specific adsorption of the polymer. Therefore, 2.0 mL of Glu was identified as the optimal crosslinker concentration.

3.2. Morphological and Structural Characterization of PCA-MIP

The morphological features of PCA-MIP and PCA-NIP were characterized by SEM. As illustrated in Figure 3a, the PCA-MIP exhibited a distinctive cauliflower-like surface structure, with particle sizes of 10–12 μm. In contrast, the PCA-NIP particles were clustered together to form irregular microparticles of approximately 7–10 μm (Figure 3b). Similarly, the SEM images at an increased magnification of 30,000× revealed that PCA-MIP possessed a relatively smooth and porous structure (Figure 3a1), whereas the PCA-NIP showed an irregular shape with some degree of adhesion (Figure 3b1). These differences were likely attributed to the incorporation of the template, and PCA-MIP retained numerous imprinted cavities after template elution. Conversely, PCA-NIP without the template resulted in a more amorphous aggregation.

Figure 3. The SEM images of PCA-MIP (**a**,**a1**) and PCA-NIP (**b**,**b1**) with magnifications of 10,000× and 30,000×.

The FTIR spectra of PCA-MIP and PCA-NIP are illustrated in Figure 4a. It can be seen that both PCA-MIP and PCA-NIP present characteristic absorption peaks at 3300–3600 cm^{-1}, corresponding to the O-H stretching vibration of DES and CS. The broader band in PCA-MIP proved hydrogen bonding formation between PCA and DES. The peaks at approximately 2935 cm^{-1}, 2870 cm^{-1}, and 1457 cm^{-1}, 1380 cm^{-1} were assigned to the C-H stretching vibration and C-H bending vibration of methyl and methylene groups in DES or Glu. The 1033 cm^{-1} band belonged to the stretching vibration of C-N in ChCl. These observations suggested that DES actively participated in the polymerization. Additionally, the strong adsorption peak near 1638 cm^{-1} was in agreement with the C=N stretching vibrations, indicative of Schiff base formation between CS and Glu [43]. The peaks at 1715 cm^{-1} were characteristic of the aldehyde group, probably from unreacted Glu. In addition, the strong peak at 1067 cm^{-1} was attributed to the stretching vibration of -C-O- within the furan ring of CS.

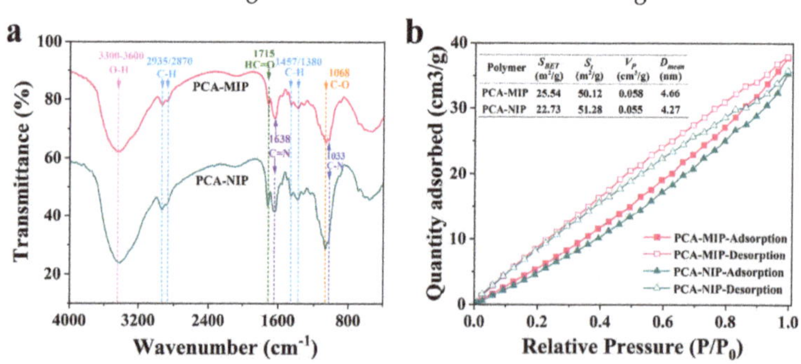

Figure 4. (**a**) FTIR spectra and (**b**) nitrogen adsorption–desorption isotherms of PCA-MIP and PCA-NIP. Inset in picture b are the pore parameters of PCA-MIP and PCA-NIP.

Nitrogen adsorption–desorption was carried out to assess the pore size distributions of PCA-MIP and PCA-NIP. As shown in Figure 4b, the BET isotherms exhibited a typical "type IV" pattern with H3 hysteretic loops within the relative pressure range of 0–1.0, indicating that PCA-MIP and PCA-NIP were mesoporous structures with narrow slit-like pores [44]. Further, the surface areas of the polymers were calculated using the Brunauer–Emmett–Teller (BET) equation. As shown in the inset in Figure 4b, PCA-MIP (25.54 m^2/g) exhibited a greater surface area than that of PCA-NIP (22.73 m^2/g), further substantiating the presence of imprinted sites within PCA-MIP. The pore size distribution (Figure S1) indicated that the pore volume of PCA-MIP was larger than that of PCA-NIP, with values of 0.058 cm^3/g and 0.055 cm^3/g, respectively. Moreover, the average pore sizes of the PCA-MIP and PCA-NIP were 4.66 nm and 4.27 nm, respectively, categorizing the synthesized polymers as mesoporous materials.

In order to investigate the thermal stability of PCA-MIP and PCA-NIP, thermogravimetric analysis (TGA) was performed. As shown in Figure S2, the two TGA curves were similar. Approximately 15.4% of mass loss was observed in the temperature range of 30 °C to 150 °C, probably due to evaporation of residual water within the adsorbent. There were two noticeable and drastic losses between 246 °C and 500 °C, which were mainly due to the decomposition of the polymer microsphere. As the temperature increased over 500 °C, the weight loss of both PCA-MIP and PCA-NIP remained relatively constant. It was worth noting that PCA-MIP had a weight loss of 63.5% at 500 °C, whereas PCA-NIP decomposed about 64.8% at the same temperature. These results indicate that PCA-MIP is more thermally stable than PCA-NIP.

3.3. *PCA-MIP Binding Study*

3.3.1. Effect of pH on the Adsorption Capacity

The pH value of the adsorption medium exerts a profound influence on both the existence state of PCA and the binding efficacy of PCA-MIP. To identify the optimal conditions for PCA adsorption, the pH of the adsorption medium was systematically investigated. Considering that the pKa value of PCA was 4.48, the incubation medium pH was selected from 3.0 to 6.0. As illustrated in Figure 5, the adsorption capacity of polymers gradually increased with pH increase from 3.0 to 5.0. Notably, the optimal imprinting performance of PCA-MIP with an imprinting factor of 1.40 was achieved at pH 5.0. This enhancement is primarily attributed to the neutrality of PCA at pH 5.0, allowing for the phenolic hydroxyl and carboxyl groups to effectively engage in multiple hydrogen bonding interactions with the complementary sites on PCA-MIP. Consequently, this alignment resulted in the high degree of adsorption of PCA-MIP for PCA [45]. However, upon surpassing pH 5.0, the adsorption capacity gradually decreased. This reduction may be attributed to the strong electrostatic repulsion or attraction between the ionized PCA and polymers, which weakened the adsorption process. Thus, the optimal pH of the incubation solution was 5.0 in the following experiments.

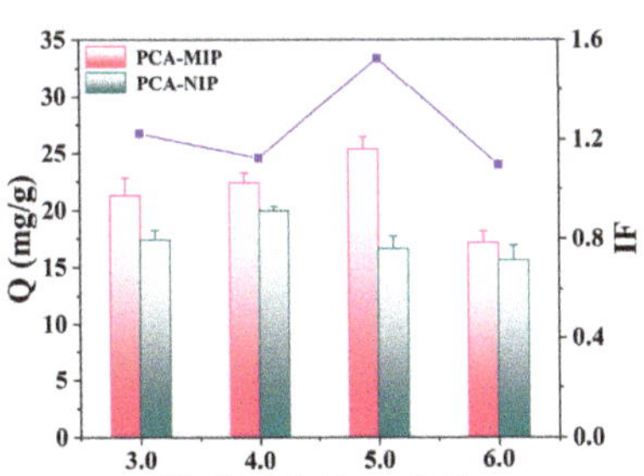

Figure 5. Effect of pH of incubation solution on absorption capacity (Q) and imprinting factor (IF) of PCA-MIP and PCA-NIP. The pink bar represents PCA-MIP, the blue bar represents PCA-NIP, and the purple line is the imprinting factor.

3.3.2. Equilibrium Adsorption Isotherms

The equilibrium adsorption experiments were investigated at a pH of 5.0, with initial concentrations of PCA from 100 to 1000 µg/mL. As illustrated in Figure 6a, the adsorption capacities of PCA-MIP and PCA-NIP exhibited a gradual increase with escalating PCA concentrations, reaching saturation at approximately 800 µg/mL. The amount of binding PCA-MIP was always higher than that of PCA-NIP under the same initial concentrations, and their maximum amounts were 38.07 mg/g and 31.70 mg/g, respectively. The results underscored that the numerous binding sites within PCA-MIP could effectively recognize PCA during the adsorption process. To elucidate the adsorption mechanism, the equilibrium adsorption isotherms were analyzed using Langmuir models, Freundlich models, and Scatchard models (Equations (4)–(6)).

$$\frac{C_e}{Q_e} = \frac{C_e}{Q_{max}} + \frac{1}{Q_{max}K_L} \tag{4}$$

$$InQ_e = InK_F + \frac{1}{n}InC_e \tag{5}$$

$$\frac{Q_e}{C_e} = \frac{Q_{max} - Q}{K_D} \tag{6}$$

In these equations, Q_e (mg/g) is the equilibrium adsorption capacity of PCA on polymers and Q_{max} (mg/g) represents the maximum adsorption capacity. C_e (mg/mL) is intended for the concentration of PCA at equilibrium. K_L (L/mg) and K_F (L/g) are the Langmuir and Freundlich constants, respectively, 1/n is the equilibrium adsorption index of the isotherm, and K_D (mg/mL) is the dissociation constant.

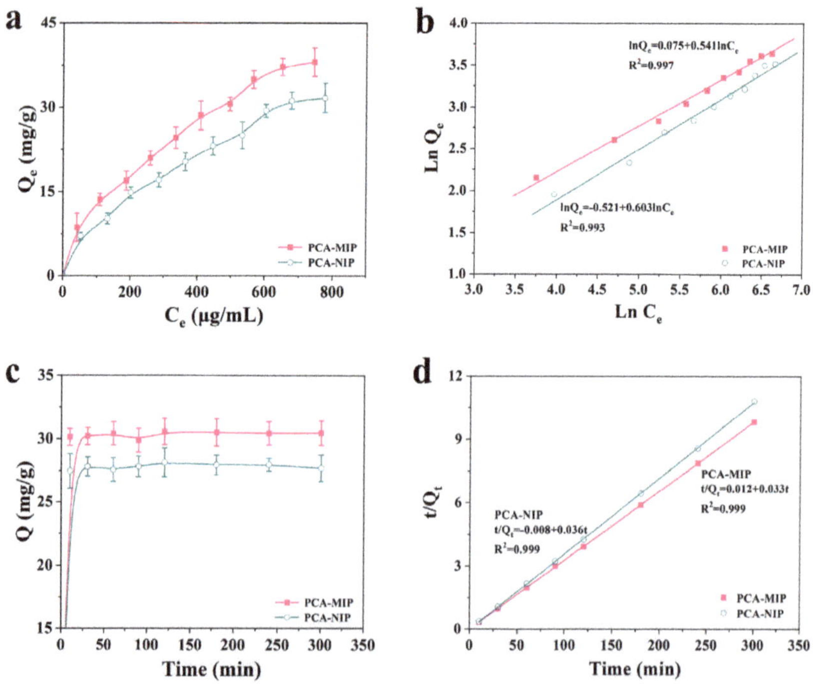

Figure 6. (**a**) Equilibrium adsorption curves, (**b**) linear fitting curves of Freundlich model, (**c**) adsorption kinetics curves, and (**d**) linear fitting curves of a pseudo-second-order kinetic model for PCA-MIP and PCA-NIP.

The fitting curves and calculated parameters corresponding to the adsorption isotherms are detailed in Figures 6b and S3 and Table 1. The Freundlich linear regression equations for PCA-MIP and PCA-NIP were expressed as $lnQ_e = 0.541lnC_e + 0.075$ ($R^2 = 0.997$) and $lnQ_e = 0.603\ lnC_e - 0.521$ ($R^2 = 0.993$), respectively. The Langmuir equations for these polymers were determined to be $C_e/Q_e = 0.018C_e + 6.439$ ($R^2 = 0.967$) and $C_e/Q_e = 0.019C_e + 9.766$ ($R^2 = 0.994$). Obviously, the correlation coefficient (R^2) values of the Freundlich model for PCA-MIP and PCA-NIP were greater than those of the Langmuir model, suggesting that the adsorption process of PCA-MIP was predominantly controlled by a multilayer adsorption mechanism [46]. Additionally, the Scatchard plot (Figure S4) for PCA-MIP was characterized by two distinct straight lines with varying slopes, suggesting the presence of both high-affinity specific binding sites and low-affinity non-specific binding sites for PCA. As detailed in Table 1, for the high-affinity imprinting sites for PCA-MIP, K_D and Q_{max} were calculated to be 0.074 mg/mL and 23.30 mg/g, respectively. However, the Scatchard plot for PCA-NIP displayed a single linear relationship, yielding K_D and Q_{max} values of 0.649 mg/mL and 58.83 mg/g, respectively, suggesting that PCA-NIP possessed only non-specific binding sites for PCA.

Table 1. The fitting parameters for the Freundlich model and Scatchard analysis for PCA-MIP and PCA-NIP.

Polymer	Q_{exp} (mg/g)	Freundlich Model			Low-Affinity Sites			High-Affinity Sites		
		K_F (L/g)	n	R^2	K_D (mg/mL)	Q_{max} (mg/g)	R^2	K_D (mg/mL)	Q_{max} (mg/g)	R^2
PCA-MIP	38.07	1.078	1.85	0.997	0.671	74.23	0.971	0.074	23.30	0.992
PCA-NIP	31.70	0.594	1.66	0.993	0.649	58.83	0.930	/	/	/

3.3.3. Kinetic Adsorption

The kinetic adsorption profiles of PCA-MIP and PCA-NIP are displayed in Figure 6c. Initially, both PCA-MIP and PCA-NIP exhibited a swift rise in binding capacity, with an especially rapid increase observed within the first 10 min. Subsequently, equilibrium was attained at approximately 30 min. Notably, PCA-MIP demonstrated a superior adsorption capacity of 30.56 mg/g compared to PCA-NIP (27.68 mg/g). The faster adsorption rate and higher binding amount were attributed to the existence of specific imprinting sites in PCA-MIP that facilitated the capture and mass transfer of PCA. Furthermore, the data were fitted with the pseudo-first-order kinetic model (Equation (7)) and the pseudo-second-order kinetic model (Equation (8)).

$$ln\left(Q_f - Q_t\right) = lnQ_f - k_1 t \tag{7}$$

$$\frac{t}{Q_t} = \frac{1}{k_2 Q_s^2} + \frac{t}{Q_s} \tag{8}$$

Here, Q_f and Q_s (mg/g) represent the adsorption amounts at equilibrium, and k_1 and k_2 denote the rate constants.

As depicted in Figures 6d and S5 alongside Table 2, the higher R^2 values demonstrated that the kinetic adsorption data for PCA-MIP exhibited a superior fit to the pseudo-second-order kinetic model ($R^2 = 0.999$), surpassing that of the pseudo-first-order model ($R^2 = 0.671$). In addition, the calculated adsorption amount from the pseudo-second-order kinetic model ($Q_s = 30.49$ mg/g) was in excellent agreement with the experimentally determined value of 30.56 mg/g. The results indicated that the process of PCA-MIP adsorption to PCA was a chemisorption process potentially involving a rate-limiting step [22].

Table 2. The fitting parameters for pseudo-first-order and pseudo-second-order models for PCA-MIP and PCA-NIP.

Polymer	Q_e (mg/g)	Pseudo-First-Order Model			Pseudo-Second-Order Model		
		Q_f (mg/g)	K_1 (mg/g/min)	R^2	Q_s (mg/g)	K_2 (g/mg/min)	R^2
PCA-MIP	30.56	2.55	0.005	0.671	30.49	0.089	0.999
PCA-NIP	27.68	1.77	0.005	0.867	27.80	0.161	0.999

3.3.4. Selectivity Adsorption

The selectivity of PCA-MIP was rigorously assessed by determining its adsorption capacity for PCA in comparison to its competitors, namely GA, NRG, and EG (structures shown in Figure 7a). From Figure 7a, it is notable that PCA-MIP presented a higher binding capacity for PCA (25.39 mg/g) compared to NRG and EG. This observation suggests that the specific imprinting sites within PCA-MIP conferred remarkable selectivity for PCA. Strangely, PCA-MIP also displayed higher adsorption capacity for GA, but with a slight imprinting factor. This is primarily attributed to the functional groups and structure of GA being similar to those of PCA and may enter the imprinted cavities of PCA-MIP during adsorption. Additionally, the presence of more phenolic hydroxyl groups in GA compared to PCA may enhance its nonspecific binding to the excess groups of CS. Therefore, the adsorption of GA onto PCA-MIP was predominantly a nonspecific adsorption process.

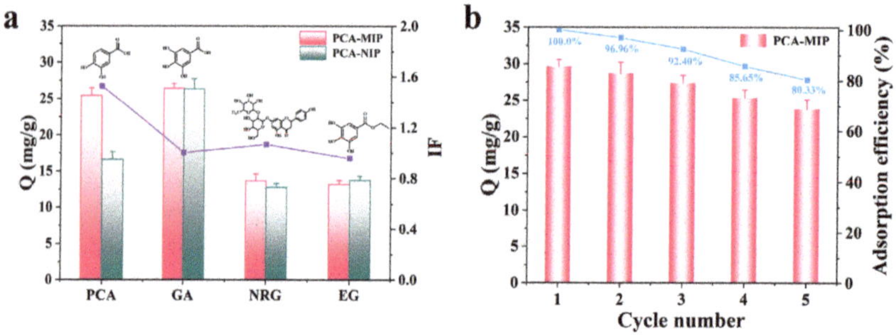

Figure 7. (**a**) Selective adsorption capacities of PCA-MIP and PCA-NIP for PCA and its analogs. (**b**) Reusability analysis of PCA-MIP via five sequential cycles of adsorption–desorption.

3.4. Reusability of PCA-MIP

To assess the reusability of PCA-MIP, a series of consecutive adsorption–desorption cycles were performed on the identical samples. As shown in Figure 7b, it was noted that a minor reduction in the adsorption capacity of PCA-MIP occurred, from 29.64 mg/g to 23.81 mg/g, while the adsorption efficiency remained consistently above 80.33% even after five cycles. The minor reduction in adsorption capacity might be ascribed to potential damage or blockage of the imprinting cavities during the elution process. By comparing with PCA-imprinted MIPs in Table 3, the preparation of PCA-MIP was facile, green, and time-efficient. In addition, PCA-MIP also demonstrated greater recognition ability than other MIPs [13,47–49]. Overall, the prepared PCA-MIP demonstrated satisfactory reusability and stability, making it a robust candidate for practical applications.

Table 3. Comparison of this work with other PCA-imprinted MIPs.

Polymer	Functional Monomer	Solvent	Synthesis Time (Hours)	Adsorption Capacity (mg/g)	Equilibrium Time (min)	Ref
MMIPs	4-vinylpyridine	Acetonitrile	24	7.5	40	[13]
Fe_3O_4@mSiO_2@MIP	4-vinylpyridine	Acetonitrile	24	17.2	140	[47]
MMIPs	acrylamide	Acetonitrile	24	-	30	[48]
MIPs@Fe_3O_4-NH_2	dopamine	Tris-HCl	3.5	46.48	50	[49]
PCA-MIP	DES and CS	Aqueous solution	3	30.56	30	This work

3.5. Extraction of PCA from Mango Juice

Recently, NFC mango juice, a beverage, has attracted more scientific attention due to its valuable health benefits as well as being rich in high levels of phenolic compounds [50]. It also acts as a major source of antioxidants such as PCA, phenolic acids, mango glycosides, quercetin, mango keto acid, isomangiferolic acid, and other biologically active substances [51]. However, due to its complicated matrix, the extraction and determination of the active components directly from mango juice are tedious and inefficient. Therefore, in order to reduce interference by other polar components, it is very necessary to employ a selective pretreatment step to separate the PCA from the mango juice.

PCA-MIP was expected to be an efficient adsorbent material and its ability to extract PCA from mango juice was evaluated. Initially, the HPLC method for the analysis of PCA was first established, encompassing linearity, correlation coefficients, limit of detection (LOD), and limit of quantification (LOQ). A linear calibration curve was obtained using the standard curve method. The HPLC peak area (A) for the corresponding PCA concentration (C) was found to be linear between 1 and 100 μg/mL. The linear equation was A = 23119C + 29874, with an R^2 value of 0.998. The LOD and LOQ were determined to be 0.81 μg/mL and 2.68 μg/mL, respectively, based on the content relative to 3 and 10 times the baseline noise.

Furthermore, the chromatograms of mango juice crude extract, the supernatant and eluent of PCA-MIP, and a PCA standard sample (5 μg/mL) were recorded. As shown in Figure 8d, the retention time of the PCA standard was identified as 5.60 min while the interference peaks in mango juice crude extract also occurred at nearly the same retention time as PCA (Figure 8a), which directly affected the accuracy of PCA content determination. Following selective extraction by PCA-MIP, there was a significant reduction in PCA in the supernatant (Figure 8b), suggesting specific adsorption of PCA by the polymer. In Figure 8c, the PCA peaks were significantly enhanced and the interference peaks were nearly eliminated, resulting in extraction recoveries of PCA from mango juice ranging from 52.99% to 61.08%. These results clearly highlight the good practical applicability of fabricated PCA-MIP in the separation and enrichment of PCA from complex samples.

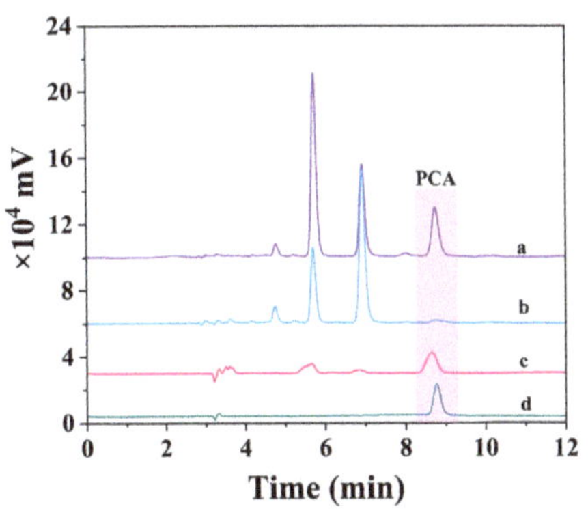

Figure 8. Chromatograms of (**a**) mango juice crude extract, (**b**) supernatant of mango juice crude extract from PCA-MIP, (**c**) elution of mango juice crude extract from DYM-MIP, and (**d**) PCA standard.

4. Conclusions

In summary, a novel type of PCA-MIP based on green synthesis strategies was prepared in an aqueous solution using DES and CS as the functional monomer and glutaraldehyde as the crosslinker. Compared with other MIPs, this work presented several distinct advantages: (1) the preparation strategy for PCA-MIP was facile, green, and time-efficient; (2) PCA-MIP exhibited excellent recognition ability and the adsorption capacity was greater than that of other MIPs; (3) PCA-MIP achieved a rapid mass transfer rate with an equilibrium time of 30 min. Thus, the resultant polymer presented outstanding adsorption performance, rapid adsorption rate, and better selectivity for PCA. Furthermore, PCA-MIP demonstrated great practical utility in the selective extraction of PCA from mango juice. Based on the research results and practicability potential, the following suggestions can be made: (1) Increasing the types of templates—more types of compounds can be selected, including proteins, peptides, and DNA, rather than only small molecule compounds. (2) Exploring more green functional monomers—more functional monomers containing specific functional groups should be explored to improve the adsorption capacity and imprinting performance of MIPs in aqueous environments. (3) Expanding the range of application—in addition to selective separation of PCA in the target juice, other complex products have been added that are more suitable for practical applications.

Supplementary Materials: The following supporting information can be downloaded at: https://www.mdpi.com/article/10.3390/foods13182955/s1, Figure S1: Pore volumes of PCA-MIP and PCA-NIP; Figure S2: Linear fitting curves of the Langmuir model for PCA-MIP and PCA-NIP; Figure S3: Linear fitting curves of the Scatchard model for PCA-MIP and PCA-NIP; Figure S4: Linear fitting curves of the pseudo-first-order model for PCA-MIP and PCA-NIP. Figure S5: The liner fitting curves of Pseudo-first-order model for PCA-MIP and PCA-NIP.

Author Contributions: Conceptualization, L.Z.; data curation, X.S., Y.D. and X.Z.; formal analysis, X.Z.; funding acquisition, L.Z.; investigation, Y.D.; methodology, X.S.; project administration, L.Z.; validation, Y.D.; writing—original draft, L.Z. and X.Z.; writing—review and editing, L.Z. All authors have read and agreed to the published version of the manuscript.

Funding: This research was funded by the Scientific and Technological Project of Henan Province (No. 242102310569), the Doctoral Startup Foundation of Henan University of Science and Technology (No. 13480055), and the National Key Laboratory of Cotton Bio-breeding and Integrated Utilization Open Fund (No. CB2024A28).

Institutional Review Board Statement: Not applicable.

Informed Consent Statement: Not applicable.

Data Availability Statement: The original contributions presented in the study are included in the article/Supplementary Materials, further inquiries can be directed to the corresponding author.

Conflicts of Interest: The authors declare no conflicts of interest.

Abbreviations

PCA	Protocatechuic acid
DES	Deep eutectic solvents
CS	Chitosan
Glu	Glutaraldehyde
MAA	Methacrylic acid
ChCl	Choline chloride
MIP	Molecularly imprinted polymer
NIP	Non-molecularly imprinted polymer
SEM	Scanning electron microscope
BET	Brunauer–Emmett–Teller
FT-IR	Fourier transform infrared spectroscopy
HPLC	High-performance liquid chromatography

References

1. Song, J.; He, Y.N.; Luo, C.H.; Feng, B.; Ran, F.; Xu, H.; Ci, Z.M.; Xu, R.C.; Han, L.; Zhang, D.K. New progress in the pharmacology of protocatechuic acid: A compound ingested in daily foods and herbs frequently and heavily. *Pharmacol. Res.* **2020**, *161*, 105109. [CrossRef] [PubMed]
2. Li, X.; Wang, X.; Chen, D.; Chen, S. Antioxidant activity and mechanism of protocatechuic acid in vitro. *Funct. Foods Health Dis.* **2011**, *1*, 232–244. [CrossRef]
3. Zhang, S.; Gai, Z.; Gui, T.; Chen, J.; Chen, Q.; Li, Y. Antioxidant effects of protocatechuic acid and protocatechuic aldehyde: Old wine in a new bottle. *Evid.-Based Complement. Altern. Med.* **2021**, *2021*, 6139308. [CrossRef]
4. Galvano, F.; Vitaglione, P.; Li Volti, G.; Di Giacomo, C.; Gazzolo, D.; Vanella, L.; La Fauci, L.; Fogliano, V. Protocatechuic acid: The missing human cyanidins' metabolite. *Mol. Nutr. Food Res.* **2008**, *52*, 386. [CrossRef] [PubMed]
5. Yang, Y.C.; Wei, M.C.; Huang, T.C.; Lee, S.Z. Extraction of protocatechuic acid from Scutellaria barbata D. Don using supercritical carbon dioxide. *J. Supercrit. Fluid* **2013**, *81*, 55–66. [CrossRef]
6. Li, H.; Nie, L.; Li, Y.; Zhang, Z.; Shi, H.; Hu, W.; Zhang, Y. Application of molecularly imprinted column for separation and purification of bioactive compound from cirsium segetum bunge. *Sep. Sci. Technol.* **2009**, *44*, 370–385. [CrossRef]
7. Zhou, T.; Ding, L.; Che, G.; Jiang, W.; Sang, L. Recent advances and trends of molecularly imprinted polymers for specific recognition in aqueous matrix: Preparation and application in sample pretreatment. *TrAC Trends Anal. Chem.* **2019**, *114*, 11–28. [CrossRef]
8. Musarurwa, H.; Tavengwa, N.T. Stimuli-responsive molecularly imprinted polymers as adsorbents of analytes in complex matrices. *Microchem. J.* **2022**, *181*, 107750. [CrossRef]
9. Kaya, S.I.; Cetinkaya, A.; Ozkan, S.A. Molecularly imprinted polymers as highly selective sorbents in sample preparation techniques and their applications in environmental water analysis. *Trends Environ. Anal. Chem.* **2023**, *37*, e00193. [CrossRef]
10. Lamaoui, A.; García-Guzmán, J.J.; Amine, A.; Palacios-Santander, J.M.; Cubillana-Aguilera, L. Synthesis techniques of molecularly imprinted polymer composites. In *Molecularly Imprinted Polymer Composites*; Elsevier: Amsterdam, The Netherlands, 2021; pp. 49–91.
11. Hasanah, A.N.; Safitri, N.; Zulfa, A.; Neli, N.; Rahayu, D.J.M. Factors affecting preparation of molecularly imprinted polymer and methods on finding template-monomer interaction as the key of selective properties of the materials. *Molecules* **2021**, *26*, 5612. [CrossRef]
12. Sobiech, M.; Lulinski, P. Molecularly imprinted solid phase extraction—Recent strategies, future prospects and forthcoming challenges in complex sample pretreatment process. *TrAC-Trends Anal. Chem.* **2024**, *174*, 117695. [CrossRef]
13. Xie, X.; Wei, F.; Chen, L.; Wang, S. Preparation of molecularly imprinted polymers based on magnetic nanoparticles for the selective extraction of protocatechuic acid from plant extracts. *J. Sep. Sci.* **2015**, *38*, 1046–1052. [CrossRef] [PubMed]
14. Denderz, N.; Lehotay, J. Using of molecularly imprinted polymers for determination of gallic and protocatechuic acids in red wines by high performance liquid chromatography. *J. Chromatogr. A* **2014**, *1372*, 72–80. [CrossRef] [PubMed]
15. Li, H.; Hu, X.; Zhang, Y.P.; Shi, S.Y.; Jiang, X.Y.; Chen, X.Q. High-capacity magnetic hollow porous molecularly imprinted polymers for specific extraction of protocatechuic acid. *J. Chromatogr. A* **2015**, *1404*, 21–27. [CrossRef] [PubMed]
16. Viveiros, R.; Rebocho, S.; Casimiro, T. Green Strategies for Molecularly Imprinted Polymer Development. *Polymers* **2018**, *10*, 306. [CrossRef]
17. Bagheri, A.R.; Arabi, M.; Ghaedi, M.; Ostovan, A.; Wang, X.Y.; Li, J.H.; Chen, L.X. Dummy molecularly imprinted polymers based on a green synthesis strategy for magnetic solid-phase extraction of acrylamide in food samples. *Talanta* **2019**, *195*, 390–400. [CrossRef]
18. Ostovan, A.; Arabi, M.; Wang, Y.Q.; Li, J.H.; Li, B.W.; Wang, X.Y.; Chen, L.X. Greenificated molecularly imprinted materials for advanced applications. *Adv. Mater.* **2022**, *34*, 2203154. [CrossRef]
19. Perna, F.M.; Vitale, P.; Capriati, V. Deep eutectic solvents and their applications as green solvents. *Curr. Opin. Green Sustain. Chem.* **2020**, *21*, 27–33. [CrossRef]
20. Marć, M.; Jatkowska, N.; Płotka-Wasylka, J.; Mateu, D.G.; Turrillas, F.A.E.; De la Guardia, M. Molecularly imprinted polymers based on deep eutectic solvents as a greenest materials for selective extraction of emerging contaminants from complex samples. *TrAC-Trends Anal. Chem.* **2024**, *178*, 117837. [CrossRef]
21. Madikizela, L.M.; Ncube, S.; Nomngongo, P.N.; Pakade, V.E. Molecular imprinting with deep eutectic solvents: Synthesis, applications, their significance, and benefits. *J. Mol. Liq.* **2022**, *362*, 119696. [CrossRef]
22. Zhang, L.P.; Hao, P.Z.; He, Y.F.; Li, S.J.; Li, T.; Wang, L.; He, S. Fabrication of magnetic molecularly imprinted polymer-based covalent-noncovalent synergistic imprinting strategies for the highly specific enrichment of luteolin from honeysuckle. *Front. Sustain. Food Syst.* **2024**, *8*, 1413458. [CrossRef]
23. Zhao, L.; Han, S.; Sun, R.A.; Yan, C. UiO66-based molecularly imprinted polymers with water-compatible deep eutectic solvent as functional monomer for purification of lysozyme from egg white. *Microchim. Acta* **2024**, *191*, 56. [CrossRef] [PubMed]
24. Liu, Z.W.; Wang, Y.Z.; Xu, F.T.; Wei, X.X.; Chen, J.; Li, H.Q.; He, X.Y.; Zhou, Y.G. A new magnetic molecularly imprinted polymer based on deep eutectic solvents as functional monomer and cross-linker for specific recognition of bovine hemoglobin. *Anal. Chim. Acta* **2020**, *1129*, 49–59. [CrossRef] [PubMed]
25. Fu, N.J.; Li, L.T.; Liu, X.; Fu, N.A.; Zhang, C.C.; Hu, L.D.; Li, D.H.; Tang, B.K.; Zhu, T. Specific recognition of polyphenols by molecularly imprinted polymers based on a ternary deep eutectic solvent. *J. Chromatogr. A* **2017**, *1530*, 23–34. [CrossRef]
26. Fu, N.J.; Liu, X.; Li, L.T.; Tang, B.K.; Row, K.H. Ternary choline chloride/caffeic acid/ethylene glycol deep eutectic solvent as both a monomer and template in a molecularly imprinted polymer. *J. Sep. Sci.* **2017**, *40*, 2286–2291. [CrossRef]
27. Cheng, G.H.; Chen, N.; Li, Z.; Zhao, K.X.; Duan, R.J.; Chen, Z.H.; Zhu, G.F. Fabrication of deep eutectic solvent-molecularly imprinted polymer in water: A green strategy for adsorption of bisphenol A. *J. Environ. Chem. Eng.* **2023**, *11*, 109651. [CrossRef]

28. Wang, M.W.; Qiao, F.X.; Yan, H.Y. A simple and benign protocol for the synthesis of a deep eutectic solvent-based hydrophilic molecularly imprinted resin in water for excellent selective molecular recognition in aqueous phase. *Green Chem.* **2021**, *23*, 5179–5188. [CrossRef]
29. Kou, S.; Peters, L.M.; Mucalo, M.R. Chitosan: A review of sources and preparation methods. *Int. J. Biol. Macromol.* **2021**, *169*, 85–94. [CrossRef]
30. Xu, L.; Huang, Y.A.; Zhu, Q.J.; Ye, C. Chitosan in Molecularly-Imprinted Polymers: Current and Future Prospects. *Int. J. Mol. Sci.* **2015**, *16*, 18328–18347. [CrossRef]
31. Zhang, J.W.; Tan, L.; Yuan, J.B.; Qiao, R.F.; Wang, C.Z.; Yang, F.Q.; Zhou, L.D.; Zhang, Q.H.; Xia, Z.N.; Yuan, C.S. Extraction of activated epimedium glycosides *in vivo* and *in vitro* by using bifunctional-monomer chitosan magnetic molecularly imprinted polymers and identification by UPLC-Q-TOF-MS. *Talanta* **2020**, *219*, 121350. [CrossRef]
32. Wang, Y.F.; Zhou, J.J.; Gu, H.M.; Jia, X.K.; Su, K.H.; Zhang, B.L.; Zhang, Q.Y. Preparation of anti-nonspecific adsorption chitosan-based bovine serum albumin imprinted polymers with outstanding adsorption capacity and selective recognition ability based on magnetic microspheres. *Macromol. Mater. Eng.* **2019**, *304*, 1800731. [CrossRef]
33. Ansari, S.; Masoum, S. Ultrasound-assisted dispersive solid-phase microextraction of capecitabine by multi-stimuli responsive molecularly imprinted polymer modified with chitosan nanoparticles followed by HPLC analysis. *Microchim. Acta* **2020**, *187*, 366. [CrossRef] [PubMed]
34. Li, G.Z.; Rwo, K.H. Hydrophilic molecularly imprinted chitosan based on deep eutectic solvents for the enrichment of gallic acid in red ginseng tea. *Polymers* **2019**, *11*, 1434. [CrossRef] [PubMed]
35. Ma, W.; Dai, Y.; Row, K.H. Molecular imprinted polymers based on magnetic chitosan with different deep eutectic solvent monomers for the selective separation of catechins in black tea. *Electrophoresis* **2018**, *39*, 2039–2046. [CrossRef]
36. Yuan, M.D.; Zhong, C.L.; Row, K.H.; Li, G.Z. Hydrophilic molecularly imprinted chitosan microspheres based on deep eutectic solvents for micro-solid phase extraction of catechin. *J. Liq. Chromatogr. Relat. Technol.* **2023**, *46*, 48–55. [CrossRef]
37. Laskar, N.; Ghoshal, D.; Gupta, S. Chitosan-based magnetic molecularly imprinted polymer: Synthesis and application in selective recognition of tricyclazole from rice and water samples. *Iran. Polym. J.* **2021**, *30*, 121–134. [CrossRef]
38. Xiao, X.D.; Li, Z.Q.; Liu, Y.; Jia, L. Preparation of chitosan-based molecularly imprinted material for enantioseparation of racemic mandelic acid in aqueous medium by solid phase extraction. *J. Sep. Sci.* **2019**, *42*, 3544–3552. [CrossRef]
39. Husin, N.A.; Muhamad, M.; Yahaya, N.; Miskam, M.; Kamal, N.; Asman, S.; Raoov, M.; Zain, N.N.M. Application of a new choline-imidazole based deep eutectic solvents in hybrid magnetic molecularly imprinted polymer for efficient and selective removal of naproxen from aqueous samples. *Mater. Chem. Phys.* **2021**, *261*, 124228. [CrossRef]
40. Xu, K.J.; Wang, Y.Z.; Wei, X.X.; Chen, J.; Xu, P.L.; Zhou, Y.G. Preparation of magnetic molecularly imprinted polymers based on a deep eutectic solvent as the functional monomer for specific recognition of lysozyme. *Microchim. Acta* **2018**, *185*, 146. [CrossRef]
41. Wang, Y.; Lu, Z.; Zhu, Z.; Zhao, X.; Gao, N.; Wang, D.; Hua, Z.; Yan, Y.; Huo, P.; Song, M. Enhanced selective photocatalytic properties of a novel magnetic retrievable imprinted $ZnFe_2O_4$/PPy composite with specific recognition ability. *RSC Adv.* **2016**, *6*, 51877–51887. [CrossRef]
42. Sun, Y.; Bai, L.; Wang, T.; Cao, S.; Han, C.; Sun, X. Effective scavenging and selective adsorption of salicylic acid from wastewater using a novel deep eutectic solvents-based chitosan-acrylamide surface molecularly imprinted hydrogel. *Appl. Surf. Sci.* **2023**, *608*, 155102. [CrossRef]
43. Khapre, M.A.; Pandey, S.; Jugade, R.M. Glutaraldehyde-cross-linked chitosan–alginate composite for organic dyes removal from aqueous solutions. *Int. J. Biol. Macromol.* **2021**, *190*, 862–875. [CrossRef] [PubMed]
44. Sing, K.; Everett, D.; Haul, R.; Moscou, L.; Pierotti, R.; Rouquerol, J.; Siemieniewska, T. Physical and biophysical chemistry division commission on colloid and surface chemistry including catalysis. *Pure Appl. Chem.* **1985**, *57*, 603–619. [CrossRef]
45. Zhang, L.P.; Zhu, F.H.; Guo, Y.; Yang, H.; Wang, L.; He, Y.F.; Li, S.J.; Li, T. Synthesis of molecularly imprinted polymer based on cooperative imprinting for enrichment of gallic acid in Puer tea. *J. Appl. Polym. Sci.* **2023**, *140*, e53712. [CrossRef]
46. Xie, L.W.; Guo, J.F.; Zhang, Y.P.; Shi, S.Y. Efficient Determination of Protocatechuic Acid in Fruit Juices by Selective and Rapid Magnetic Molecular Imprinted Solid Phase Extraction Coupled with HPLC. *J. Agric. Food Chem.* **2014**, *62*, 8221–8228. [CrossRef]
47. Xie, L.; Guo, J.; Zhang, Y.; Hu, Y.; You, Q.; Shi, S. Novel molecular imprinted polymers over magnetic mesoporous silica microspheres for selective and efficient determination of protocatechuic acid in Syzygium aromaticum. *Food Chem.* **2015**, *178*, 18–25. [CrossRef]
48. Zhou, Y.; Liu, A.; Li, Y.; Liu, S. Magnetic molecular imprinted polymers-based nanozyme for specific colorimetric detection of protocatechuic acid. *Coatings* **2023**, *13*, 1374. [CrossRef]
49. Quirós-Sauceda, A.E.; Chen, C.Y.O.; Blumberg, J.B.; Astiazaran-Garcia, H.; Wall-Medrano, A.; González-Aguilar, G.A. Processing 'ataulfo' mango into juice preserves the bioavailability and antioxidant capacity of its phenolic compounds. *Nutrients* **2017**, *9*, 1082. [CrossRef]
50. Cruz-Cansino, N.D.; Cariño-Cortés, R.; Neria-de la Cruz, R.; Sandoval-Gallegos, E.M.; Sumaya-Martínez, M.T.; Ramírez-Moreno, E.; Fernández-Martínez, E. Ultrasound with controlled temperature as an emerging technology for extraction of antioxidant compounds from by-products of mango (*Mangifera indica* L. varAtaulfo) juice. *Emir. J. Food Agric.* **2023**, *35*, 162–169. [CrossRef]
51. Dars, A.G.; Hu, K.; Liu, Q.D.; Abbas, A.; Xie, B.J.; Sun, Z.D. Effect of thermo-sonication and ultra-high pressure on the quality and phenolic profile of mango juice. *Foods* **2019**, *8*, 298. [CrossRef]

molecules

MDPI

Article

Rapid Tryptophan Assay as a Screening Procedure for Quality Protein Maize

Gabi Drochioiu [1,*], Elena Mihalcea [1], Jeanclaude Lagobo [1,2] and Catalina-Ionica Ciobanu [3]

1 Faculty of Chemistry, Alexandru Ioan Cuza University of Iasi, 11 Carol I, 700506 Iasi, Romania; lenuta_mihalcea@yahoo.com (E.M.); lagobojeanclaude@gmail.com (J.L.)
2 Environmental Science Laboratory (LSE), Nangui-Abrogoua University, Abidjan 02 BP 801, Côte d'Ivoire
3 Integrated Centre of Environmental Science Studies in the North East Region—CERNESIM Centre, Institute of Interdisciplinary Research, Alexandru Ioan Cuza University of Iasi, 11 Carol I, 700506 Iasi, Romania; catalina.ciobanu@uaic.ro
* Correspondence: gabidr@uaic.ro

Abstract: Tryptophan is an essential amino acid deficient in cereals, especially maize. However, maize (*Zea mays* L.) is the main source of protein in some developing countries in Africa and Latin America. In general, the nutritional profile of cereals is poor, because they are deficient in essential amino acids such as tryptophan and lysine due to a relatively higher proportion of alcohol-soluble proteins. Quality protein maize (QPM) has been developed through genetic manipulation for the nutritional enrichment of maize to address these problems. Nevertheless, methods for protein, lysine and tryptophan are time-consuming and require relatively large amounts of samples. Therefore, we have advanced here a simple, cheap, fast, reliable and robust procedure for the determination of protein and tryptophan in the same biuret supernatant, which can also be used for chemical characterization of other cereals. Samples of 50 mg maize ground to pass through a 0.1 mm screen were sonicated for 5 min. in eppendorf vials with 1.5 mL of a biuret reagent each. After centrifugation and protein determination by biuret, 0.2 mL of supernatant was treated with 0.8 mL of a tryptophan reagent. Both total protein and tryptophan can be determined in microplates at 560 nm to speed up the measurements. The main advantage of the new micro-method is the rapid estimation of the nutrient quality of maize samples by a single weighing of a small amount of valuable plant materials.

Keywords: tryptophan determination; *Zea mays* L.; quality protein maize; protein determination; simultaneous analysis

1. Introduction

The poor protein quality of normal maize seeds is caused by zein proteins, which lack tryptophan and lysine residues [1–3]. Consequently, zein determination has been previously studied to determine the nutritional quality of opaque-2 and common maize [4–7]. However, nutrient-rich maize genotypes by incorporating crtRB1 and o2 genes associated with increased levels of β-carotene, lysine, and tryptophan have been studied and developed [8]. Then, several naturally occurring mutants have been identified that confer high levels of lysine and tryptophan to maize, namely opaque-2, opaque-6, opaque-7, floury-2, and floury-3 [9,10]. For instance, opaque-2 (o2) mutants undergo a decrease in γ-zein and an increase in lysine-rich non-zein proteins through a phenomenon known as proteome rebalancing [9,11,12]. Thus, quality protein maize (QPM) was developed in the late 1960s [13,14] to produce 70% to 100% more lysine and tryptophan than common maize varieties [15–17]. In addition, QPM lines have both high protein and tryptophan levels and can be used as parents for highly nutritious hybrids [18]. Nutritional scoring of QPM in nutrition trials reveals its nutritional superiority over non-QPM varieties for human and animal consumption [13]. Some authors have estimated that the nutritional quality of QPM proteins approaches that of cow's milk proteins [14]. The potential of QPM to help reduce protein deficiency in Africa has also been highlighted [19,20].

In order to determine the nutritional quality of cereals, including maize, it is necessary to determine both the crude or pure protein content and the content of essential amino acids required by the human or monogastric animal body [3,21,22]. Tryptophan is an essential amino acid required by the body, which can only be obtained through daily food intake. The other essential amino acid, lysine, which is also found in a small proportion in maize and other cereals, is actually in a 4:1 ratio with tryptophan [23–25]. For this reason, we consider that the determination of tryptophan by its reaction with glyoxylic acid [26] would be preferable for the estimation of cereal nutritional quality in crop breeding and the selection of QPM maize lines. In addition, the methods of analysis proposed until recently are based on the separate determination of proteins from the determination of tryptophan or lysine [18,27,28].

Protein hydrolysis performed in the presence of $Ba(OH)_2$ was found to yield the highest and least dispersed values for the tryptophan content of plant materials [29]. An electrochemical sensor based on a glassy carbon electrode for tryptophan determination via differential pulse voltammetry was proposed [30]. A glassy carbon electrode modified with nanocomposites of green and chemically synthesized copper oxide-polyaniline was thus applied to determine tryptophan in pineapple [31]. Accurate determination of tryptophan in food was also done by fluorescence analytical technique [32,33]. Many other methods have been employed to detect tryptophan, such as ultra-high performance liquid chromatography with electrospray ionization, cryoelectron microscopy, and spectroscopic techniques like fluorescence microscopy-mass spectroscopy with a triple quadrupole [34,35]. However, the methods for quantification of protein-bound amino acids of plant materials often involve a complicated hydrolysis set-up and several hydrolysate processing steps [36]. Therefore, particularly in plant-breeding work, such as the production of QPM hybrids, rapid analytical methods are required. They should also be high-yielding, low in plant material and sensitive to relevant levels of proteins and essential amino acids of interest.

In this work, we proposed the simultaneous determination of the protein in maize samples and the tryptophan contained in it. This eliminates time-consuming double weighing of plant material samples. To further increase the speed of determination of these two components, we used a microplate spectrophotometer. Indeed, a microplate reader with 96 wells, together with instruments capable of recording measurements from them, allows a large number of samples to be determined rapidly [37]. Here, the method for the determination of tryptophan using a residual biuret solution from a protein assay was tested in comparison with the colorimetric method described in the literature [27]. We also discussed our results in tryptophan determination for their potential utility in maize-breeding programs. In addition, the essential differences between conventional spectrophotometric measurements and those using microplate readers were discussed.

2. Results

2.1. Tryptophan Reaction

To study the reaction of tryptophan with a reagent containing sulfuric acid, acetic acid, and acetic anhydride, a standard solution of 0.1 mM tryptophan (20.4 μg/mL) was prepared [38]. Thus, 20.4 mg of tryptophan was dissolved in 10 mL of an alkaline alcoholic solution. Then, 1 mL of a stock solution was diluted with the same alkaline alcoholic solution to 100 mL. From the resulting 0.1 mM tryptophan alkaline solution, a volume of 1.5 mL was sonicated with 50 mg of insoluble copper phosphate for 15 min. Eppendorf vials were subsequently centrifuged at 15,000 rpm for 5 min, and 1 mL of the supernatant was treated with 4 mL of a tryptophan reagent in tubes with vigorous shaking. The absorbance spectra were plotted against the control made with reagents only in the range of 300–700 nm in 1 cm quartz cuvettes (Figure 1). The molar absorptivity of tryptophan (molecular weight: 204.23) at 558 nm was 4770 L mol^{-1} cm^{-1}. In addition, the reagent components (the blank) did not interfere with the determination of tryptophan at the maximum wavelength (558–560 nm) or even in the range from 450 to 700 nm.

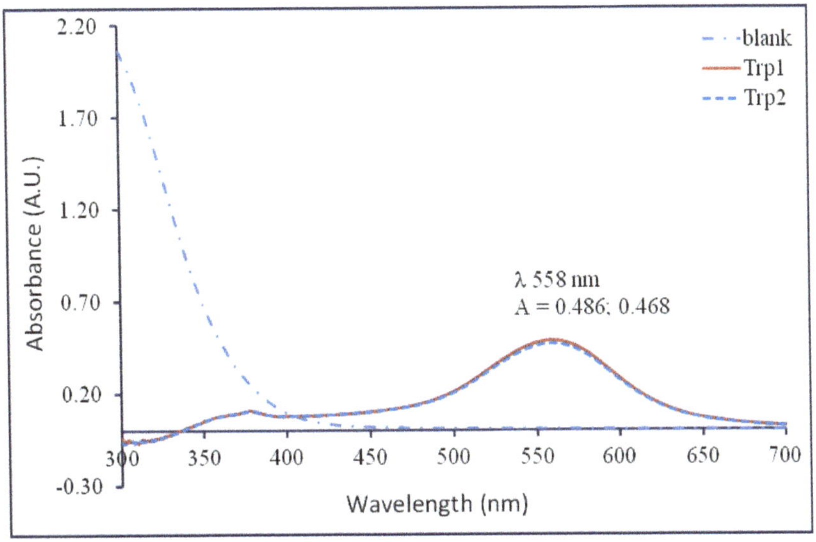

Figure 1. Typical UV-vis spectra of tryptophan (1 mL; 0.1 mM) derivative formed in the reaction with glyoxylic acid in a tryptophan reagent (4 mL).

We also tested the determination of tryptophan in a microplate reader at 600 nm (Table 1). When pipetting six different 0.25 mL samples of 100 μM tryptophan in a biuret solution, the mean absorbance measured in a microplate reader was 0.359 ± 0.0098 AU (blank: 0.060–0.065; samples: 0.409–0.438). The limit of detection, LOD = $3.3{\cdot}s_x/S$ and the limit of quantification, LOQ = $10{\cdot}s_x/S$, respectively, were LOD = 0.016 and LOQ = 0.049 μM, respectively. Here S was the slope of the regression line (S = 280.48). The following regression equation was calculated, T = 280.48·TA − 0.4339 (R^2 = 0.998), where T = tryptophan concentration as μM, and TA, tryptophan absorbance). These results demonstrated the high sensitivity of the proposed micro-method.

Table 1. Tryptophan absorbance measured with a microplate reader at 560 nm.

Volume [1]	Blank1	Blank2	Trp1	Trp2	Trp1−Blank1	Trp1−Blank2	Mean
100 μL	0.072	0.072	0.224	0.217	0.151	0.145	0.148
100 μL	0.073	0.074	0.221	0.215	0.149	0.141	0.145
250 μL	0.064	0.065	0.421	0.438	0.358	0.373	0.365
250 μL	0.062	0.060	0.411	0.409	0.349	0.349	0.349

[1] $C_{Tryptophan}$: 100 mM (Trp1 & Trp2).

We determined the selectivity, linearity, accuracy, precision, robustness, detection limit, and quantitation limit of the proposed method for tryptophan determination. We also calculated RMSE values (0.0109) as well as CV = 0.0489 (RSD = 1.36). A calibration curve was plotted against a reagent blank in the concentration range of 0–100 μM (Figure 2). The regression model is a linear one in which the relationship between tryptophan concentration (C) and the absorbance at 560 nm (A_{560}) is a straight line, A_{560} = 0.0036C, where 0.0036 is the slope of the line. Linearity of the calibration curve was expressed through the coefficient of correlation, r = 0.995, and the coefficient of determination, R^2 = 0.99, although r is not an appropriate measure for the linearity. Therefore, using residual plots is a simple way to check linearity [39]. We thus observed that the A560 values were normally distributed for a linear model, and no curves were observed to suggest a lack of fit due to a nonlinear effect.

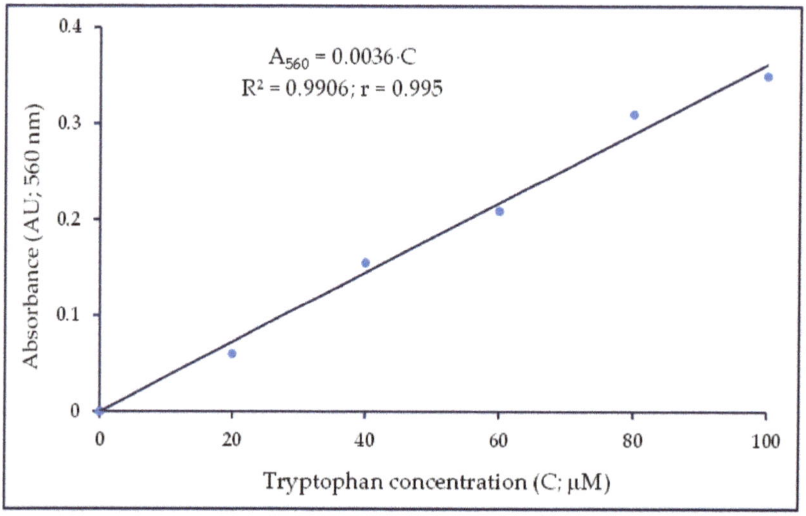

Figure 2. Standard curve of tryptophan using a microplate reader (0–100 µM; volumes of 250 µL in duplicate).

In this work, the assessment of nutritive quality of opaque-2 (QPM) and normal maize samples was performed by simultaneous determination of crude protein and tryptophan as described below, compared with separate determination of crude protein by the micro-Kjeldahl method and tryptophan by a method proposed by Nurit et al. [27]. For this purpose, we selected 25 maize samples, 12 normal maize samples, and 13 QPM-type samples, with tryptophan content determined after 16 h of papain hydrolysis in the range of 0.37–0.82% in crude protein (Table S1). A strong relationship was observed between tryptophan values determined from biuret supernatants and those from enzyme hydrolysates (R^2 = 0.983; r = 0.991). The linear regression equation was y = 1.1086x − 0.0561, where y is %tryptophan in crude protein determined in the biuret supernatants and x is the values of the enzyme-based method. However, the two types of maize samples displayed tryptophan values in two distinct groups with inflection points. For this reason, the Bland–Altman analysis was conducted to compare the obtained results and present quantified measures to decide whether the new method is approvable or not [40]. Figure S1 shows the agreement between two tryptophan measurements, according to the limits of agreement quantified for the two datasets. A one-sample *t*-test was performed to calculate the mean bias and its SD. For our dataset, the mean difference was found as 0.0108 with an SD of 0.0298. Therefore, in our data set, the upper limit can be calculated using mean + 1.71 × SD (0.0108 + 1.71 × 0.0298 = 0.0618), and the lower limit can be calculated using mean − 1.71 × SD (0.0108 + 1.71 × 0.0298 = −0.04027). The appropriate statement used in the manuscript can be as follows: the Bland–Altman plot showed that the mean ± SD between the first and the second tryptophan levels was 0.0108 ± 0.0298% tryptophan, and the limits of agreement were −0.04027 and 0.0618 (Figure S1).

Therefore, the two methods can be employed interchangeably, as the limits vary within 0.1% tryptophan.

In addition, intraday and interday precisions of this analytical method were found to be reliable based on RSD% (<2%). The values of 1.36% (n = 6) and 1.64% (n = 6) were identified for intraday and interday precisions, respectively, confirming that the method is sufficiently precise.

2.2. Protein Determination

The new variant of the biuret micro-method, which was described in Section 4.4.1, total protein determination, was applied for the determination of proteins. The crude protein was thus determined on the basis of a calibration curve using the values of biuret absorbance versus crude protein content of maize samples analyzed by the micro-Kjeldahl method (Figure 3). Indeed, the total protein determined by the biuret method was similar to the crude protein content measured by the micro-Kjeldahl method (Figure 3). Ultrasonic stirring enhanced the biuret reaction in the presence of alkaline alcoholic solutions and insoluble copper phosphate. Ethanol prevented the opalescence generated by the starch of the maize samples, while phosphate powder was the source of copper ions that formed a biuret complex with the peptide bonds of proteins under alkaline conditions. The regression equation was $B = 1.0664 \cdot K - 0.7313$ ($R^2 = 0.9755$; $r = 0.987$), where B is the percent of crude protein determined by the biuret method, and K, the micro-Kjeldahl values.

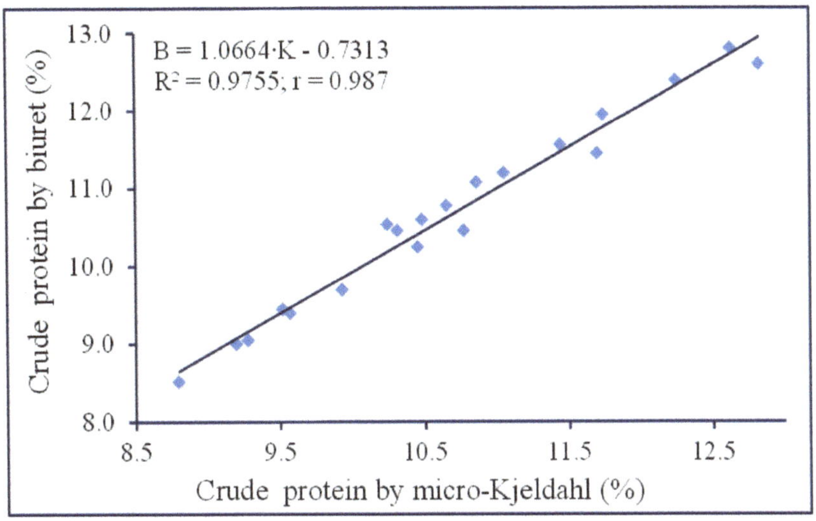

Figure 3. Relationships between the crude protein contents of 23 different maize samples determined by the micro-Kjeldahl method and the protein values determined from the corresponding absorbance values of the biuret supernatant.

The new variant of the biuret method, using insoluble copper phosphate instead of copper sulfate and sodium and potassium tartrate, has been advanced for the determination of proteins in maize samples and for obtaining biuret supernatants for tryptophan quantification. The spectra of the biuret absorbance and the standard curves at various wavelengths of 1–5 mg mL^{-1} bovine serum albumin (BSA) are shown in Figure S2.

The results showed that it is feasible to determine proteins both at 545 nm, where biuret absorbance is maximal, and at 560 nm in a microplate reader. However, if BSA is used as a standard to plot the calibration curve to determine protein in real samples, then the procedure should be the same. In our study, 1 mL of 1–5 mg mL^{-1} BSA was treated with 0.5 mL of an alkaline alcoholic solution, whereas proteins from real samples were extracted in only 1 mL of an alkaline alcoholic solution. In daily determinations, we preferred to analyze 50–100 samples by the biuret method described here, to which we add 4–5 samples with crude protein content determined by the micro-Kjeldahl method and covering a wide range of values (e.g., 8–15% crude protein). These 4–5 samples were sufficient to plot calibration curves each day. The crude protein content was reported on dry matter, determined in an oven at 105 °C for 24 h.

2.3. Characterization of Real Samples

Crude protein contents of 25 maize samples were determined by the micro-Kjeldahl (Table S2) and were found in the concentration range from 7.72% to 15.97%. Increased precision and repeatability were also found in the determination of proteins in maize using the proposed biuret method. A highly significant correlation was observed between the biuret absorbance (BA) of the samples at 545 nm in 1 cm quartz cuvettes and the crude protein content determined by the micro-Kjeldahl method, r = 0.987 with the regression equation, BA = 0.0567·CP + 0.0307, where CP is the percentage of crude protein in the dry matter.

Table S2 also shows that normal maize might be high in protein, but its crude protein contained high amounts of zein, lysine and tryptophan-deficient protein. For example, sample HT-17/81 had 15.97% crude protein and 6.68% zein. Nevertheless, the hybrid HT Sv 17/81 was retained in the breeding program for use in polygastric animal feed. Zein values were particularly low for QPM-type maize, ranging from 1.10% to 2.42%, while normal maize had significantly higher values of this parameter (2.22–6.68%).

A strong relationship was noticed only between zein and crude protein in normal maize and not in QPM. In addition, the QPM samples contained low amounts of zein. Moreover, a negative correlation was found between lysine and zein in maize samples studied (Figure S4). Although, the correlation coefficient was r = −0.823, it is however difficult to differentiate the QPM samples from the normal ones, when compared with tryptophan determination (Figure S5).

Although QPM varieties showed on average relatively lower crude protein values, some opaque-2 maize varieties nevertheless achieved particularly high values for this parameter (O-26/86: 13.94%), suggesting the possibility of increasing the protein content of opaque maize while maintaining high values for lysine and tryptophan. Thus, by applying both the classical methods of lysine and tryptophan determination and that proposed in this paper, QPM samples can be easily differentiated from normal maize (Figure S5). Indeed, the lysine content was higher in opaque maize samples than in normal maize (lysine in total protein: 3.10–3.77% and 1.88–2.43%, respectively). The lysine content of normal maize still showed a decrease with the increasing protein level of the analyzed maize plants (Table S2). However, tryptophan levels followed the same pattern, so that the more rapid and simpler determination of tryptophan can replace more complicated and time-consuming lysine determination. Basically, the sensitive increase in the protein content of the plants did not represent a sensitive change in the lysine content, so that the increase in the protein index was realized only at the expense of the increase in the zein content of the samples, which was independent of lysine content.

In this work, we used a residual biuret supernatant after biuret reaction and treated one volume of the supernatant with four volumes of tryptophan reagents. In fact, the biuret solution can be recovered from protein quantification by the biuret method and used for the determination of tryptophan and lysine. In addition, tryptophan absorbance divided by the biuret absorbance values could be a parameter of choice for assessing the nutritional quality of maize samples.

The strong correlation between the values of the percentage of tryptophan in protein determined by papain hydrolysis and those obtained by the proposed micro-method (r = 0.982), as well as the slope of the regression line (1.108) indicated that essentially all tryptophan was available for the reaction, as shown in Table S1. The statistical analysis revealed that there was no significant difference between the proposed method and the methods proposed by Nurit et al. [27] and by Hernández and Bates [38] for the determination of tryptophan in maize samples ($p > 0.05$).

Tryptophan content in normal maize seeds commonly ranges from 0.2% to 0.5% of the total protein content, while it typically varies from 0.5 to 1.1 in opaque-2 maize [27]. In our study, tryptophan values ranged from 0.37% to 0.82% in crude protein and 35.09–100.36 mg% in dry matter (Table S2). Indeed, normal maize samples showed less than 0.5% tryptophan in DM.

A strong correlation was found between tryptophan and lysine values (r = 0.979***), demonstrating once again the possibility of assessing the quality of maize by determining its tryptophan content (Figure S5). As expected, a negative correlation was determined between the zein content of the analyzed samples and the tryptophan content of their dry matter (r = −0.491), suggesting that zein content is less dependent on the nutritive quality of maize samples. The following regression equation was generated: $y = -6.9527x + 83.755$ ($R^2 = 0.2419$), where y is tryptophan level expressed as mg% in DM and x is the level of zein expressed as % in DM.

3. Discussion

The novelty and originality of these studies consists in the use of a rapid assessment of the nutritional quality of maize protein by the determination of tryptophan and crude protein in the same biuret supernatant. In addition, this simultaneous determination of maize components could be of great interest both for breeding work and for the normal characterization of maize or other cereals.

Lysine, tryptophan, and threonine are limiting amino acids in humans and non-ruminants. Tryptophan and lysine could be determined quickly, simply and accurately from the residual biuret solution from protein determination by the biuret method. However, only tryptophan and crude protein were determined here, given the strong correlation between the two essential amino acids, which allows the nutritional quality of maize to be assessed by knowing the content of only one of them. Although we have advanced this method based on simultaneous determination of protein and tryptophan in the biuret supernatant of maize samples, the proposed procedure can be applied not only for the identification of QPM hybrids, but also for the measurement of protein and tryptophan in normal maize. Indeed, opaque-2 mutants of maize induce increased levels of lysine and tryptophan in maize endosperm of QPM [9]. The development of QPM hybrids through advanced breeding approach like molecular marker-assisted breeding was adopted. It could solve the issue related to protein deficiency in developing countries.

Existing methods for estimating the nutrient quality of QPM and normal maize samples are time-consuming and plant-material-intensive. On the contrary, our innovation eliminates these drawbacks and allows for the rapid and reliable determination of both tryptophan and proteins in biuret supernatants obtained from a single weighing. In addition, the Bradford method is significantly affected by the presence of salts, while the biuret assay remains stable [41]. The biuret reaction is mostly applied to determine the amount of soluble protein in a solution [42]. However, we have used insoluble copper phosphate which released small amounts of soluble copper ions capable of extracting proteins from maize meal. These form the biuret complex with the proteins in alkaline medium, the absorbance of which is proportional to the protein content of the maize samples.

Protein quality was previously expressed as the content of tryptophan in endosperm proteins [43]. According to classical methods, samples were defatted; then protein and tryptophan were determined separately. However, we show in this paper that it is possible to identify QPM varieties by analyzing the whole seeds and not only the endosperms. In addition, corn oil in the maize samples does not interfere with the measurements according to our proposed method. Moreover, only one weighing of a small amount of plant material to be analyzed is required, as this micro-method has been specifically dedicated to breeding work as well as to the use of microplate readers.

Only common, inexpensive reagents and glassware are needed. Besides, no special training is necessary, except the usual reading of colored solutions using a spectrometer or microplate reader. Moreover, the biuret method is quicker and simpler than that of the Kjeldahl method and can be recommended as a replacement for it in breeding work.

If needed, the pass-length of any aqueous solution in a microplate well can be determined by comparing the absorbance of that solution at a known pass-length (e.g., 1 cm) with the absorbance in the microplate well. Nevertheless, correction for the passage length in the microplate wells is not necessary, because the calibration curves are drawn under the same conditions as the samples. In addition, results are obtained more quickly and can be easily transferred to an Excel spreadsheet to be automatically calculated as protein or tryptophan concentrations.

Previously, we also used the simplified methods with good results to differentiate high protein and tryptophan samples from normal maize.

Future research should test the micro-method on a large number of samples of maize and other breeding cereals and to improv it.

4. Materials and Methods

4.1. Reagents

The reagents and solvents used here were of analytic purity, and the aqueous solutions were prepared with pure water (18.2 MΩ·cm) obtained from a Millipore water purification system (Millipore, Bedford, MA, USA).

Corn zein protein and bovine serum albumin (BSA) were purchased from Sigma-Aldrich (Saint Louis, MO, USA) and were used for calibration curves and other analytical studies. A 4 mg mL^{-1} stock solution of zein in 70% ethanol and a 5 mg mL^{-1} stock solution of BSA were prepared for drawing the calibration curves by appropriate dilution. Solvents for degreasing flour samples and protein extraction, such as ethanol, acetone, and petroleum ether, were purchased from Merck (Darmstadt, Germany) and were utilized without further purification. Petroleum ether was the option of choice to defat maize flours by Soxhlet Traditional Solvent Extraction (TSE).

An alkaline-alcoholic solution (AA or biuret reagent) was prepared by dissolving 20 g of KOH in about 200 mL of milliQ water, followed by adding 500 mL of 95% (v/v) ethanol and making up to 1 L with water.

Papain solution. Papain technical powder was dissolved in a 0.1 acetate buffer solution with pH 7 and a final concentration of 4 mg mL^{-1} papain.

Tryptophan reagent. Equal volumes of acetic solution and 30 N sulfuric acid were mixed one hour before use. The acetic solution consisted of 27 mg $FeCl_3 \cdot 6H_2O$ dissolved in 0.5 mL H_2O over which 100 mL of glacial acetic acid containing 4 mL acetic anhydride was added. The 30 N sulfuric acid was prepared by adding dropwise 81.5 mL of concentrated sulphuric acid (98%) over 19 mL of water, under stirring and cooling with tap water, followed by making up to 100 mL with water.

4.2. Plant Material to Be Analyzed

The research samples consisted of normal and opaque-2 maize kernels with a moisture content of about 12% from several maize hybrids and inbred lines taken from the Plant Genetic Resources Bank of Suceava, Romania. When classical analytical methods were applied for comparative studies, the kernels were ground to a fine powder and the resulting flour was defatted using petroleum ether as solvent in a Soxhlet extractor. Dry matter (DM) was determined in a stove at 105 °C for 24 h.

4.3. Instrumentation

Maize kernels were milled to a fine powder using a laboratory electric cereal mill (SAMAP Mod F100, Andolsheim, France) with adjustable millstones. The resulted flours were screened with an analytical vibratory sieve shaker (Retsch Gmbh, Haan, Germany) to obtain flour fully passing a certain sieve (for example, 100 μm mesh sieve).

An ultrasonic bath (J.P. Selecta Ultrasons system, 40 kHz; Barcelona, Spain) was used for both ultrasonic extraction and biuret reaction. Centrifugation at 15,000–18,000 rpm of the solvent mixtures with the working samples was carried out using a Hettich Mikro 22R centrifuge (Tuttlingen, Germany). The UV-vis absorption spectra were taken on a Biochrom Libra S35 PC UV–visible spectrophotometer (Cambridge, England) in quartz cuvettes of 10 mm (1 mL volume) in the range from 200 to 900 nm. In addition, a Modulus™ microplate reader (Turner Biosystems, Sunnyvale, CA, USA) was used. All measurements were carried out at room temperature.

4.4. Procedures

4.4.1. Total Protein Determination

Each 50 mg duplicate of maize flour was treated under shaking with 1 mL of alkaline-alcoholic solution (AA) in a plastic eppendorf vial. Approximately 50 mg of copper phosphate powder was then added and the mixture was sonicated for 30 min. If necessary, other ratios between the amount of flour and the volume of the alkaline-alcoholic solution can be used. The eppendorf vials were centrifuged at 15,000 rpm, and the absorbance spectra were read in the range from 200 to 900 nm; however, the 545 nm absorbance (or the 560 nm absorbance in the microplate reader) of the supernatant was of interest for our experiments, as it was proportional to the protein content of the samples. Biuret absorbance values were compared with those obtained by the micro-Kjeldahl method. In addition, a calibration curve with micro-Kjeldahl values was made for some representative maize samples (high, medium, and low protein contents or biuret absorbances).

4.4.2. Tryptophan Determination in the Biuret Supernatant

After protein determination by the biuret method, 200 μL of biuret supernatant was taken and placed in an eppendorf vial, after which 800 μL of tryptophan reagent was pipetted. The mixture is shaken vigorously and incubated at 70 °C for 15–30 min to perform the color reaction. Vials were cooled to room temperature, and the absorbance of the purple-violet solution is read at 558 nm in 1 cm glass cuvettes against a control using reagents only.

4.4.3. Using the Microchip Reader

After incubation and centrifugation, 200 μL of biuret supernatant was transferred into a 96-well plate, and the absorption signal proportional to the protein content was detected at 560 nm with a Modulus™ microplate reader. The results were expressed as mean relative light units (RLUs) and compared with those obtained by the normal UV–vis spectrophotometry.

Similarly, 250 μL of each violet-violet solution following the reaction of tryptophan with glyoxylic acid was pipetted into a 96-well plate, and the absorbance value proportional to the tryptophan content was also detected at 560 nm. The results were expressed as mean relative light units (RLUs) and compared with those obtained by the normal UV–vis spectrophotometry. Calibration curves were plotted with 0–5 mg mL^{-1} BSA for protein and 0–50 μg mL^{-1} tryptophan.

4.4.4. Other Analytical Methods

The crude protein content of defatted maize meal was determined by the micro-Kjeldahl method [44], and the values obtained are given in Table S2.

The tryptophan content was also quantified by the papain hydrolysis method indicated by Hernández and Bates in 1969 [38], as well as by Nurit in 2009 [27]. However, it was not expressed as percentage of tryptophan in endosperm protein, but in crude protein of the whole maize kernel (Table S1). The sample of 100 mg defatted finely powdered maize flour was mixed with 4 mL of papain solution, incubated for 16 h at 63 ± 2 °C, cooled and centrifuged at 5000 rpm for 5 min. Then, 1 mL of hydrolyzate was transferred to a solution of ferric chloride and sulfuric acid, incubated at the 65 °C for 15 min, allowed to cool, and read at 560 nm to calculate the amount of tryptophan relative to the protein on a standard curve 0–50 μg mL^{-1}.

Lysine was evaluated by the method of Tsai et al. (1975), which was modified by Villegas (1984) [45,46]. From the samples of maize hydrolyzed with papain at 65 ± 2 °C for 16 h, after cooling to room temperature, 1 mL was placed in a copper carbonate solution, vortexed for 5 min and centrifuged at 5000 rpm for 5 min. Afterwards, 1 mL of the supernatant was mixed with 0.1 mL of a 2-chloro-3,5-dinitropyridine solution.

The mixture was then stirred vigorously and allowed to stand for 2 h and shaken every 30 min; hydrochloric acid was added, and the mixture was again stirred until homogenized. Subsequently, an ethyl acetate extraction solution was added, stirred by inversion 10 times, and the upper phase was extracted, repeating three times and read at 390 nm against a blank. The lysine content was calculated from a standard curve and plotted against protein.

In our experiments, zein in the defatted maize samples was extracted with a 70% aqueous ethanol solution [47–49]. Duplicates of 100 mg of maize meal were placed in test tubes over which 2 mL of 70% ethanol was pipetted and ultrasonic stirred for 30 min. 1 mL of zein extract was treated with 0.5 mL of a biuret reagent and 50 mg of insoluble copper phosphate, and after sonication for 5 min and centrifugation at 15,000 rpm, the supernatant was read at 545 nm in a spectrophotometer. The calibration curve was plotted with alpha zein as a standard.

4.4.5. Simplified Methods

For measurements in the field or for poorly equipped laboratories, we provide simpler or naked eye methods. Thus, 100 mg of maize flour is mixed with 3 mL of a biuret reagent in 180/18 mm test tubes fitted with stoppers, and the mixture is shaken. Add about 50 mg of insoluble copper phosphate to the mixture, and shake the mixture from time to time. After approximately one hour, the contents of the tubes are filtered through Whatman No 1 filter paper, and the absorbance is measured at 545 nm using a spectrophotometer or with a colorimeter at 540 nm in 1 cm glass or plastic cuvettes. Plot a calibration curve with representative values obtained with the micro-Kjeldahl method or bovine serum albumin. From the biuret filtrate, take 1 mL which is pipetted into another test tube over which 4 mL of tryptophan reagent is pipetted. Shake the tubes and maintain them at 65 °C for 15 min. The absorbance of the violet- purple solutions is measured at 550–560 nm against a control made with reagents alone. The calibration curve is plotted with tryptophan in the concentration range of 0–50 μg mL^{-1}.

Alternatively, a naked eye method may be used, which involves using a series of 10 test tubes for the calibration curves and visually comparing the sample to be analysed with them.

4.5. Statistics

The coefficient of variation (CV), the coefficient of determination (R^2), the relative standard deviation (RSD%), the root mean square error (RMSE) [50], the limit of detection (LOD), and the limit of quantification (LOQ), as well as the correlation coefficient (r), were calculated to compare the results of the proposed method with those of conventional methods. The Bland–Altman analysis [40] was also applied to investigate the agreement between the proposed micromethod and the reference one based on enzymatic hydrolysis [27] in the determination of tryptophan levels in different maize samples.

5. Conclusions

The method for estimating the nutritional quality of maize samples based on the rapid determination of protein and tryptophan content in a biuret solution is characterized by simplicity, speed, and increased productivity. The method has the advantage to eliminate time-consuming double weighing of plant material samples.

Crude protein from maize samples was ultrasonic-assisted extracted in an alkaline-alcoholic solution in the presence of insoluble copper phosphate, centrifuged and determined in a supernatant at 560 nm as a biuret complex in a microplate reader. The values obtained by the proposed micro-biuret test were in full agreement with those of the micro-Kjeldahl method. In addition, the modified biuret micro-method proved to be simple, productive and reliable.

The tryptophan content determined by the proposed micro-method was also found to be in good agreement with that obtained by reference methods in the literature. Data were achieved with high degree of precision and sensitivity. Relative standard deviation was 1.36% (n = 6).

Analyzing real samples of normal and QPM maize, we observed a strong correlation only between the crude protein contents of normal maize and zein. Therefore, it is difficult to differentiate the two types of maize solely by their zein content values. On the other hand, tryptophan values that are strongly correlated with those of lysine allowed for good differentiation of QPM samples from normal ones, which is an important advantage of the method proposed here.

Application of the proposed method could considerably increase productivity in maize breeding and reduce the requirement of valuable plant materials.

6. Patents

Part of this study is based on our own earlier patent [51]. Thus, we improved the nutritional quality of some normal corn by eliminating a large part of zein, as evidenced by the analysis of the resulting flours by chemical reference methods.

Supplementary Materials: The following supporting information can be downloaded at: https://www.mdpi.com/article/10.3390/molecules29184341/s1, Figure S1: Agreement between two tryptophan measurements (Bland-Altman plot); Figure S2: Standard curves in protein determination using bovine serum albumin (BSA). Here, (a) absorbance spectra of different concentrations of BSA (1–5 mg mL^{-1}); (b) calibration curve with ab-sorbance values at 552 nm; (c) calibration curve with absorbance values at 560 nm; (d) calibration curve with absorbance values at 545 nm, respectively; Figure S3: The relationship between zein and crude protein in some real samples. Here, (a) the strong relationship between zein and crude protein in normal maize; (b) the QPM samples contain low amounts of zein; Figure S4: Relationship between lysine and zein in maize samples studied; Figure S5: Relationship between lysine and tryptophan content of 30 different maize samples; Table S1: Tryptophan levels of representative maize samples determined both in enzyme hydrolyzed solutions and in the biuret supernatants; Table S2: Chemical composition of 25 representative maize samples illustrating their nutritional quality.

Author Contributions: Conceptualization, G.D. and C.-I.C.; methodology, G.D. and C.-I.C.; software, G.D.; validation, G.D., J.L. and C.-I.C.; formal analysis, J.L.; investigation, E.M.; resources, G.D.; data curation, G.D.; writing—original draft preparation, E.M.; writing—review and editing, G.D.; visualization, G.D.; supervision, G.D.; project administration, G.D. All authors have read and agreed to the published version of the manuscript.

Funding: This research received no external funding.

Institutional Review Board Statement: Not applicable.

Informed Consent Statement: Not applicable.

Data Availability Statement: All data in this paper are available from the corresponding author's (GD) Database.

Acknowledgments: We thank Silvia Strajeru and her colleagues for providing plant materials from the Suceava Genebank. In addition, we greatly appreciate the suggestions and pertinent criticisms of the anonymous reviewers, which greatly improved the quality of this article.

Conflicts of Interest: The authors declare no conflicts of interest.

References

1. Hurst, J.P.; Sato, S.; Ferris, T.; Yobi, A.; Zhou, Y.; Angelovici, R.; Clemente, T.E.; Holding, D.R. Editing the 19 kDa alpha-zein gene family generates non-opaque2-based quality protein maize. *Plant Biotechnol. J.* **2024**, *22*, 946–959. [CrossRef] [PubMed]
2. Yau, J.; Bockholt, A.; Smith, J.; Rooney, L.; Waniska, R. Maize endosperm proteins that contribute to endosperm lysine content. *Cereal Chem.* **1999**, *76*, 668–672. [CrossRef]
3. Friedman, M. Analysis, nutrition, and health benefits of tryptophan. *Int. J. Tryptophan Res.* **2018**, *11*, 1178646918802282. [CrossRef] [PubMed]
4. Tufchi, M.; Rashmi, R.; Kumar, A.; Chaudhary, D.P.; Singh, N.K.; Jat, S.L. Breeding quality protein maize (*Zea mays*): Genetic and analytical perspective. *Indian J. Agric. Sci.* **2021**, *91*, 495–502. [CrossRef]

5. Kodrzycki, R.; Boston, R.S.; Larkins, B.A. The opaque-2 mutation of maize differentially reduces zein gene transcription. *Plant Cell* **1989**, *1*, 105–114. [PubMed]
6. Olakojo, S.A.; Omueti, O.; Ajomale, K.; Ogunbodede, B.A. Development of quality protein maize: Biochemical and agronomic evaluation. *Trop. Subtrop. Agroecosys.* **2007**, *7*, 97–104.
7. Duvnjak, M.; Kljak, K.; Grbeša, D. Nitrogen storage in crops: Case study of zeins in maize. In *Nitrogen in Agriculture–Physiological, Agricultural and Ecological Aspect*; IntechOpen: London, UK, 2021; pp. 1–15.
8. Chandrasekharan, N.; Ramanathan, N.; Pukalenthy, B.; Chandran, S.; Manickam, D.; Adhimoolam, K.; Nalliappan, G.K.; Manickam, S.; Rajasekaran, R.; Sampathrajan, V.; et al. Development of β-carotene, lysine, and tryptophan-rich maize (*Zea mays*) inbreds through marker-assisted gene pyramiding. *Sci. Rep.* **2022**, *12*, 8551. [CrossRef]
9. Motukuri, S.R.K. Quality protein maize: An alternative food to mitigate protein deficiency in developing countries. In *Maize-Production and Use*; Hossain, A., Ed.; IntechOpen: London, UK, 2019; pp. 33–43.
10. Vasal, S.K. The quality protein maize story. *Food Nutr. Bull.* **2000**, *21*, 445–450. [CrossRef]
11. Larkins, B.A.; Wu, Y.; Song, R.; Messing, J. Maize Seed Storage Proteins. In *Maize Kernel Development*; Larkins, B., Ed.; CABI International: Wallingford, UK, 2017; Volume 14, pp. 175–189.
12. Woo, Y.M.; Hu, D.W.; Larkins, B.A.; Jung, R. Genomics analysis of genes expressed in maize endosperm identifies novel seed proteins and clarifies patterns of zein gene expression. *Plant Cell* **2001**, *13*, 2297–2317. [CrossRef]
13. Tandzi, L.N.; Mutengwa, C.S.; Ngonkeu, E.L.M.; Woïn, N.; Gracen, V. Breeding for quality protein maize (QPM) varieties: A review. *Agronomy* **2017**, *7*, 80. [CrossRef]
14. Prasanna, B.M.; Vasal, S.K.; Kassahun, B.; Singh, N.N. Quality protein maize. *Curr. Sci.* **2001**, *81*, 1308–1319.
15. Bjarnason, M.; Vasal, S.K. Breeding for quality protein maize. *Plant Breed. Rev.* **1992**, *9*, 181–216.
16. So, P.A.; Wani, S.A.; Rather, A.G.; Wani, S.H. Review article: Quality protein maize (QPM): Genetic manipulation for the nutritional fortification of maize. *J. Plant Breed. Crop Sci.* **2009**, *1*, 244–253.
17. Maqbool, M.A.; Beshir Issa, A.; Khokhar, E.S. Quality protein maize (QPM): Importance, genetics, timeline of different events, breeding strategies and varietal adoption. *Plant Breed.* **2021**, *140*, 375–399. [CrossRef]
18. Amegbor, I.; Van Biljon, A.; Shargie, N.; Tarekegne, A.; Labuschagne, M. Identifying quality protein maize inbred lines for improved nutritional value of maize in Southern Africa. *Foods* **2022**, *11*, 898. [CrossRef]
19. Nuss, E.T.; Tanumihardjo, S.A. Quality protein maize for Africa: Closing the protein inadequacy gap in vulnerable populations. *Adv. Nutr.* **2011**, *2*, 217–224. [CrossRef]
20. Mebratu, A.; Wegary, D.; Teklewold, A.; Tarekegne, A. Testcross performance and combining ability of early-medium maturing quality protein maize inbred lines in Eastern and Southern Africa. *Sci. Rep.* **2024**, *14*, 9151. [CrossRef]
21. D'Amico, S.; Schoenlechner, R.; Tömösközi, S.; Langó, B. Proteins and amino acids of kernels. *Pseudocereals Chem. Technol.* **2017**, 94–118.
22. Kałużna-Czaplińska, J.; Gątarek, P.; Chirumbolo, S.; Chartrand, M.S.; Bjørklund, G. How important is tryptophan in human health? *Crit. Rev. Food Sci.* **2019**, *59*, 72–88. [CrossRef]
23. Tomičić, Z.; Spasevski, N.; Lazarević, J.; Čabarkapa, I.; Tomičić, R. Diversity of amino acids composition in cereals. *Food Feed Res.* **2022**, *49*, 11–22. [CrossRef]
24. Biel, W.; Kazimierska, K.; Bashutska, U. Nutritional value of wheat, triticale, barley and oat grains. *Acta Sci. Pol. Zootech.* **2020**, *19*, 19–28. [CrossRef]
25. Rutherfurd, S.M.; Bains, K.; Moughan, P.J. Available lysine and digestible amino acid contents of proteinaceous foods of India. *Brit. J. Nutr.* **2012**, *108*, S59–S68. [CrossRef] [PubMed]
26. Abdulwahed, M.; Mamoly, L.; Bosnali, W. A simple spectrophotometric method for determination of glyoxylic acid in its synthesis mixture. *Int. J. Anal. Chem.* **2020**, *2020*, 5417549. [CrossRef] [PubMed]
27. Nurit, E.; Tiessen, A.; Pixley, K.V.; Palacios-Rojas, N. Reliable and inexpensive colorimetric method for determining protein-bound tryptophan in maize kernels. *J. Agric. Food Chem.* **2009**, *57*, 7233–7238. [CrossRef] [PubMed]
28. García Rodríguez, J.G.; Cervantes Ortiz, F.; Ramírez Pimentel, J.G.; Aguirre Mancilla, C.; Rodríguez Perez, G.; Ochoa, F.; Mendoza Elos, M. Determination of lysine, tryptophan and protein in germinated of creole and QPM maize. *Rev. Mex. Cienc. Agric.* **2017**, *8*, 877–890.
29. Delhaye, S.; Landry, J. Determination of tryptophan in pure proteins and plant material by three methods. *Analyst* **1992**, *117*, 1875–1877. [CrossRef]
30. Bourigua, S.; Boussema, F.; Bouaazi, D.; Mzoughi, Z.; Barhoumi, H.; Majdoub, H.; Maaref, A.; Jaffrezic-Renault, N. Sensitive electrochemical detection of l-tryptophan using a glassy carbon electrode modified with pectin extracted from Arthrocnemum indicum leaves. *J. Electroanal. Chem.* **2024**, *953*, 117998. [CrossRef]
31. Mokole, S.J.; Aliyu, A.; Fayemi, O.E. Electrochemical detection of tryptophan in pineapple fruit using a green and chemically synthesized PANI/CuO nanocomposite modified electrode. *Sci. Afr.* **2024**, *24*, e02233. [CrossRef]
32. Xiao, X.; Hong, D.; Cao, J.; Ma, A.; Jia, Y. Fluorescence analytical techniques for sensing tryptophan in food and biological samples: Mechanisms, applications, and challenges. *Food Rev. Int.* **2024**, 1–30. [CrossRef]
33. Pandit, S.K.; Das, G. Fluorescent detection of Tryptophan in Physiological media: Potential Applications in Environmental Samples and Living Systems. *Sens. Actuat. B-Chem.* **2024**, *417*, 136131. [CrossRef]

34. Prongmanee, W.; Alam, I.; Asanithi, P. Hydroxyapatite/graphene oxide composite for electrochemical detection of L-tryptophan. *J. Taiwan Inst. Chem. Eng.* **2019**, *102*, 415–423. [CrossRef]
35. Asakaviciute, R.; Maknickiene, Z. Influence of the storage duration on the health promoting tyrosine, tryptophan, and total phenolics in potato tubers. *Czech J. Food Sci.* **2024**, *42*, 93–99. [CrossRef]
36. Teo, M.; Khoo, L.W.; Chew, W. A simplified small-scale workflow for determination of complete protein-bound amino acids using pre-column derivatization HPLC method. *J. Food Compos. Anal.* **2024**, *135*, 106571. [CrossRef]
37. Held, P.G. Nucleic acid and protein quantitation in the microplate format. *Chimia* **2001**, *55*, 40. [CrossRef]
38. Hernández, H.H.; Bates, L.S. A modified method for rapid tryptophan analysis of maize. *CIMMYT Res. Bull.* **1969**, *13*, 1–7.
39. Moosavi, S.M.; Ghassabian, S. Linearity of calibration curves for analytical methods: A review of criteria for assessment of method reliability. In *Calibration and Validation of Analytical Methods—A Sampling of Current Approaches*; Stauffer, M.T., Ed.; InTech: London, UK, 2018; pp. 109–127.
40. Doğan, N.Ö. Bland-Altman analysis: A paradigm to understand correlation and agreement. *Turk. J. Emerg. Med.* **2018**, *18*, 139–141. [CrossRef]
41. Miranda, M.P. Comparison of the effect of sodium chloride concentration on protein determination: Bradford and biuret methods. *Anal. Biochem.* **2024**, *687*, 115450. [CrossRef]
42. Chang, S.K.; Zhang, Y. Protein analysis. In *Food Analysis*; Springer: Berlin/Heidelberg, Germany, 2017; pp. 315–331.
43. Chaudhary, D.P.; Singh, A.; Sekhar, J.C.; Kaul, J.; Yadav, S.; Tufchi, M.; Sethi, M.; Devi, V.; Kumar, R.; Rakshit, S. Analysis of maize populations for developing quality protein maize. *Maize J.* **2022**, *11*, 1–9.
44. AOAC. *Official Methods of Analysis*; Association of Official Agricultural Chemists: Washington, DC, USA, 1975.
45. Tsai, C.Y.; Dalby, A.; Jones, R.A. Lysine and tryptophan increases during germination of maize seed. *Cereal Chem.* **1975**, *52*, 356–360.
46. Villegas, E.; Ortega, E.; Bauer, R. *Chemical Methods Used at CIMMYT for Determining Protein Quality in Cereal Grains*; Folleto Técnico; CIMMYT: Mexico City, Mexico, 1984; p. 35.
47. Drochioiu, G.; Ciobanu, C.I.; Bancila, S.; Ion, L.; Petre, B.A.; Andries, C.; Gradinaru, R.V.; Murariu, M. Ultrasound-based protein determination in maize seeds. *Ultrason. Sonochem.* **2016**, *29*, 93–103. [CrossRef]
48. Darie-Ion, L.; Jayathirtha, M.; Hitruc, G.E.; Zaharia, M.M.; Gradinaru, R.V.; Darie, C.C.; Pui, A.; Petre, B.A. A proteomic approach to identify zein proteins upon eco-friendly ultrasound-based extraction. *Biomolecules* **2021**, *11*, 1838. [CrossRef] [PubMed]
49. Bancila, S.; Ciobanu, C.I.; Murariu, M.; Drochioiu, G. Ultrasound-assisted zein extraction and determination in some patented maize flours. *Rev. Roum. Chim.* **2016**, *61*, 725–731.
50. Hyndman, R.J.; Koehler, A.B. Another look at measures of forecast accuracy. *Int. J. Forecast.* **2006**, *22*, 679–688. [CrossRef]
51. Drochioiu, G. Process for Separating Floury Fractions of High Biological or Industrial Quality. Romania Patent RO128468, 2013; International Patent Classification: A21D-002/14, BOPI 12, 2015, 30 April 2014.

foods

MDPI

Perspective

Soil Microbial Communities and Wine Terroir: Research Gaps and Data Needs

Gabriela Crystal Franco [1], Jasmine Leiva [1], Sanjiev Nand [1], Danica Marvi Lee [1], Michael Hajkowski [1], Katherine Dick [1], Brennan Withers [1], LuzMaria Soto [1], Benjamin-Rafael Mingoa [1], Michael Acholonu [1], Amari Hutchins [2], Lucy Neely [3] and Archana Anand [1,*]

1 Department of Biology, College of Science and Engineering, San Francisco State University, 1600 Holloway Avenue, San Francisco, CA 94132, USA; gfranco3@mail.sfsu.edu (G.C.F.); jleiva2@sfsu.edu (J.L.); snand1@mail.sfsu.edu (S.N.); dlee41@sfsu.edu (D.M.L.); mhajkowski@mail.sfsu.edu (M.H.); kdick@mail.sfsu.edu (K.D.); bwithers@sfsu.edu (B.W.); lsoto@sfsu.edu (L.S.); bmingoa@sfsu.edu (B.-R.M.); macholonu@sfsu.edu (M.A.)

2 Department of Biology, Howard University, 2400 6th St NW, Washington, DC 20059, USA; amari.hutchins@bison.howard.edu

3 Neely Winery, Spring Ridge Vineyard, 555 Portola Road, Portola Valley, CA 94028, USA; lucy@neelywine.com

* Correspondence: archanand@sfsu.edu

Abstract: Microbes found in soil can have a significant impact on the taste and quality of wine, also referred to as wine terroir. To date, wine terroir has been thought to be associated with the physical and chemical characteristics of the soil. However, there is a fragmented understanding of the contribution of vineyard soil microbes to wine terroir. Additionally, vineyards can play an important role in carbon sequestration since the promotion of healthy soil and microbial communities directly impacts greenhouse gas emissions in the atmosphere. We review 24 studies that explore the role of soil microbial communities in vineyards and their influence on grapevine health, grape composition, and wine quality. Studies spanning 2015 to 2018 laid a foundation by exploring soil microbial biogeography in vineyards, vineyard management effects, and the reservoir function of soil microbes for grape-associated microbiota. On the other hand, studies spanning 2019 to 2023 appear to have a more specific and targeted approach, delving into the relationships between soil microbes and grape metabolites, the microbial distribution at different soil depths, and microbial influences on wine flavor and composition. Next, we identify research gaps and make recommendations for future work. Specifically, most of the studies utilize targeted sequencing (16S, 26S, ITS), which only reveals community composition. Utilizing high-throughput omics approaches such as shotgun sequencing (to infer function) and transcriptomics (for actual function) is vital to determining the specific mechanisms by which soil microbes influence grape chemistry. Going forward, understanding the long-term effects of vineyard management practices and climate change on soil microbiology, grapevine trunk diseases, and the role of bacteriophages in vineyard soil and wine-making would be a fruitful investigation. Overall, the studies presented shed light on the importance of soil microbiomes and their interactions with grapevines in shaping wine production. However, there are still many aspects of this complex ecosystem that require further exploration and understanding to support sustainable viticulture and enhance wine quality.

Keywords: wine; vineyard; terroir; soil; microbial composition; function

1. Introduction

The concept of terroir in wine refers to the unique combination of environmental factors, including soil, climate, topography, and human practices, that influence the characteristics of grapes and, ultimately, the flavor and quality of wine [1,2]. While terroir is traditionally associated with macro-level factors such as climate, topography, physical and chemical soil characteristics, recent research has highlighted the role of microbiota, specifically grapevine-associated microbial communities, in potentially shaping the terroir effect [3].

Grapevines host a diverse array of microorganisms, including bacteria, yeasts, and fungi, on the surface of the grapes, within the grapevine itself, and the bulk soil [4,5]. These microbial communities can vary significantly between vineyards, regions, and even individual vines [6]. They play a crucial role in vineyard ecology, interacting with the plant and influencing its growth [7], health, and the development of grapes. For example, the grape-associated yeast community is a vital component of the vine–wine system contributing to terroir [8]. Additionally, the diversity and proportion of yeast species change with the grape's maturation stage. As grapes begin to ripen, they are predominantly inhabited by basidiomycetous yeasts [9]. As maturation continues, these initial colonizers are replaced by ascomycetous species that exhibit oxidative or weak fermentative properties, including *Hanseniaspora*, *Metschnikowia*, *Pichia*, and *Candida* [9]. Notably, *Saccharomyces cerevisiae*, the primary yeast responsible for wine fermentation, is infrequently observed. In contrast, overmatured, damaged, or botrytized grapes favor the growth of yeasts with robust fermentative characteristics and others like *Pichia*, *Zygoascus hellenicus*, *Zygosaccharomyces*, and *Torulaspora* [9,10].

The microbiota associated with grapevines can influence terroir in several ways, based on which they can be classified into three categories:

Soil microbiota: Soil is an essential component of terroir, and the microbial communities within the soil can likely impact vine health and grape characteristics. Microbes in the soil interact with the vine's root system, affecting nutrient availability, water uptake, and overall vine physiology. Links to the production of metabolites that influence grapevine metabolism and flavor compounds in the grapes are suggested in some literature but not conclusive [11–13]. Additionally, knowledge concerning variability within and between vineyards and regions and their contribution to wine terroir is still fragmented.

Epiphytic microbiota: Microorganisms present on the grape, leaf, and bark surfaces are the epiphytic microbiota. To date, the literature suggests that the composition of epiphytic microbiota can be influenced by vineyard management practices, such as the use of pesticides or fungicides [1,11,14]. However, the extent to which epiphytic microbiota can affect the fermentation process and contribute to the sensory attributes of the resulting wine is largely unknown. Recent research has begun to indicate that yeasts and bacteria on grape skins can influence the initiation and progression of fermentation, leading to different flavor profiles [15].

Endophytic microbiota: Endophytes are microorganisms that live within the tissues of grapevines. These microbes can have various effects on the vine, including enhancing nutrient uptake, modulating the plant's immune system, and producing bioactive compounds. The presence and diversity of endophytic microbiota can vary between grape varieties and vineyard sites, contributing to the unique terroir expression as suggested by Compant et al. [16], Pacifico et al. [17], and Hamaoka et al. [18], to name a few.

Among the studies to date on the influence of microbial communities on wine terroir, the contribution of the soil microbiome (in comparison to epiphytic and endophytic microbiota) remains inconclusive and least scientifically explored. To address this gap, we review the current state of knowledge of soil microbiota contribution to terroir expression. We present research gaps and highlight future areas of research that warrant attention. For the purpose of this paper, we focus on literature that specifically looks at microbial communities in the soil.

Although understanding the influence of soil microbial communities on wine production is crucial, it is also important to explore how microbial biogeography and activity might respond to climate changes. By studying the complex interactions between soil microorganisms and the environment, we can gain valuable insights into the role of soil microbiota in shaping terroir. This knowledge allows us to comprehend how soil microbial communities contribute to the unique characteristics and flavors found in wines, ultimately helping us manage and manipulate these communities to enhance desired terroir traits or preserve the distinctiveness of specific terroirs. Therefore, in addition to reviewing the literature on soil microbiota in vineyards, we also discuss vine management practices, including disease and topics of interest such as bacteriophages and their likely relationship with soil microbiota and wine terroir.

2. Current State of Knowledge on the Microbiota Contribution to Terroir Expression

We conducted a PubMed literature search using the keywords "soil", "microbial communities", and "wine terroir", resulting in 24 studies published between 2015 and 2023 (Table 1) and five review papers [3,19–22]. Studies focusing on grapevine microbiomes in plant parts, rather than bulk soil, were excluded from this study.

Table 1. List of studies on soil microbial communities in vineyards.

Year	Source	Methodology
2015	Burns et al., [23]	16S rRNA
	Morrison-Whittle and Goddard [24]	26S rDNA
	Zarraonaindia et al., [13]	16S rRNA and shotgun metagenomics
2016	Burns et al., [25]	16S rRNA
2017	Casteñada and Barbosa [26]	shotgun metagenomics
	Mezzasalma et al., [27]	16S rRNA and ITS
2018	Hendgen et al., [28]	16S rRNA and ITS
	Chou et al., [29]	16S rRNA
	Wei et al., [30]	16S rRNA and ITS
	Morrison-Whittle and Goddard [31]	26S rDNA
2019	Gupta et al., [32]	16S rRNA and ITS
	Liang et al., [33]	16S rRNA
2020	Ramirez et al., [34]	16S rDNA
	Liu et al., [35]	16S rRNA and ITS
	Oyuela Aguilar et al., [36]	16S rRNA, ITS1, and ITS2
2021	Teixeira et al., [37]	DNA-based assays to detect single nucleotide polymorphisms (SNPs) on three genes of the anthocyanin pathway (UFGT, F3H, and LDOX)
	Rivas et al., [38]	16S rRNA
	Torres et al., [39]	16S rRNA and ITS1
2022	Yan et al., [40]	ITS1
	Geiger et al., [41]	ITS2, ITS4
	Gobbi et al., [42]	16S rRNA and ITS
	Regecová et al., [43]	ITS
2023	Larsen et al., [44]	16S rRNA and ITS1
	Nanetti et al., [45]	16S rRNA

2.1. *The "Core Microbiome" Concept*

Early studies from 2015 to 2018 [13,23–31] laid the foundation by exploring soil microbial biogeography in vineyards, vineyard management effects, and the reservoir function of soil microbes for grape-associated microbiota. For instance, Zarronaindia et al. [13] mention that belowground bacterial communities differed significantly from those aboveground, and yet the communities associated with leaves, flowers, and grapes shared a greater proportion of taxa with soil communities than with each other, suggesting that soil may serve as a bacterial reservoir. Additionally, the authors define "core" microbiota found below-ground that included those associated with *Bradyrhizobium*, *Steroidobacter*, and *Acidobacteria* species. Mezzasalma et al. [27] shared that the grape microbiome could be influenced by a "common microbiome" influenced by farming practices and climate conditions. This was deduced by observing microbes present at harvest and prior to fermentation that were commonly observed in soil bacteria *Pasteurellales* and *Bacteroidales* as well as *Rhodospirillales* and *Lactobacillales*. The studies point out the consistent presence of common soil bacteria as a source of microorganisms in grapefruit; however, we speculate that this could be driven by rain splash-, wind-, and insect-driven microbial migration [11].

Although there is limited information specifically about bulk soil microbes and their connection to wine terroir or grapefruit microbiomes, later studies (2019 onwards) appear to have a more specific and targeted approach, delving into the relationships between soil microbes and grape metabolites, the microbial distribution at different soil depths, and microbial influences on wine flavor and composition. Darriaut et al. [46] discussed microbes found in vineyard soils that could potentially influence terroir (*Actinobacteria* was prevalent in vineyard soils and linked to the production of secondary metabolites that may influence grape and wine characteristics; *Proteobacteria* was commonly found in vineyard soils and associated with nutrient cycling and plant growth promotion, potentially affecting grape quality and wine terroir. The same study mentioned that *Enterobacter* and *Paenibacillus* were observed to promote plant growth and cause metabolic changes when added to the soil. While not specifically linked to wine terroir, these changes could potentially influence grape characteristics. Beule et al. [47] discuss the abundance, diversity, and function of soil microorganisms in temperate agroforestry systems, which could be relevant to vineyard soil microbiology. Liu et al. [35] and Oyuela Aguilar et al. [36] examined both bacterial and fungal communities at different soil depths. However, specific findings from these studies in connection with wine terroir are not provided in the search results. Regecová et al. [43] observed a dominating presence of lactic acid bacteria during fermentation, specifically the *Lactiplantibacillus*, *Leuconostoc*, *Oenococcus*, and *Pediococcus* genera. These genera contain species associated with malolactic fermentation, which results in acidity reduction while enriching the wine with diacetyl and esters compounds. Regecová et al. [43] also noted a close relationship between population densities of lactic acid bacteria and non-*Saccharomyces* yeasts and the concentrations of metabolites such as biogenic amines, including histamine and tyramine. Notably, there were overlapping yeast species found in soil, leaf, and grape berries (*Metschnikowia pulcherrima*), also validating the "core microbiota" theory that came up in the literature from earlier years.

2.2. Dominant Groups in Vineyard Soil Microbial Communities

Concerning bacterial populations (Figure 1), twelve studies found *Proteobacteria* to be the most dominant phylum present [13,23,26,28,30,32,33,35,36,38,39,42], majority of which also found *Actinobacteria* to be present in high abundance. Ten studies noted *Acidobacteria* to be present in high abundance [13,23,26,28,32,33,35,36,38,42], while eight studies identified the presence of *Bacteroidetes* in high abundance [13,23,26,28,30,35,36,42]. *Gemmatimonadetes* was noted in six studies, though not as dominant as the aforementioned phyla [26,28,32,35,39,42], eight studies identified *Firmicutes* as one of the present phyla [13,26,28,30,33,35,38,42], and ten studies identified *Planctomycetes* in a mix of high, medium, and low abundance [13,23,26, 28,30,32,36,38,39,42].

Commonly observed fungal phyla were *Ascomycota*, *Basidiomycota*, *Chytridiomycota*, *Mucoromycota*, and *Clomeromycota*. Ten studies found *Ascomycota* in high abundance and, in some cases, the most abundant [26,28,30–33,35,36,39,40]. *Basidiomycota* was observed in ten studies, although in high, medium, and low abundance [26,28,30–33,35,36,39,40]. *Chytridiomycota* and *Mucoromycota* were observed but not as dominant as the other phyla—Basidiomycota and Ascomycota [28,30,35,36,40]. *Glomeromycota* was observed but in low abundance [32,35,36].

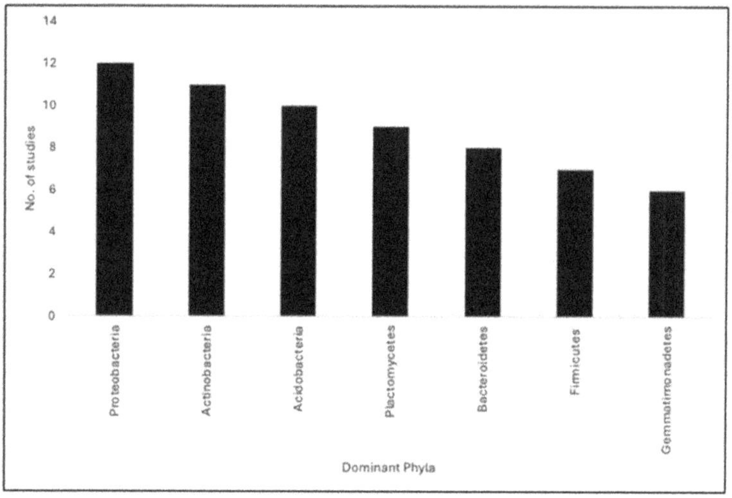

Figure 1. Dominant bacterial phyla in vineyard soil microbial communities.

2.3. Contributing Factors That Link Soil-Associated Microbiota to Wine Terroir

While there is growing evidence supporting the significance of soil microbial influence on grapevines and their associated microbiota, aspects of this interaction remain subject to debate and further investigation. For example, Zarraonaindia et al. [13] focused on the spatial and temporal dynamics of bacterial communities associated with grapevine organs (leaves, flowers, grapes, and roots) and soils. The study explored factors like vine cultivar, edaphic parameters, vine developmental stage, and vineyard that influence the microbial communities, but it did not directly address the influence of soil on grape microbiota. On the other hand, Chou et al. [29] investigated the impact of under-vine soil management practices (herbicide application, soil cultivation, and natural vegetation) on the microbiomes of soil and grapes in a Riesling vineyard. The study showed that soil management practices influenced the soil microbiome but did not have corresponding changes in the grape-associated microbiome, suggesting that other vineyard management practices or environmental factors may be more influential in shaping the grape microbiota.

The PubMed search also resulted in 5 reviews. We discuss them briefly in this section: Belda et al. [3] highlight the underestimated role of the soil microbiome in wine production. The study reveals that the soil-associated microbiota is likely to influence soil chemistry, grapevine health, and the final sensory properties of wines, calling for a deeper understanding of these crucial interactions for precision enology practices. Liu et al. [35] emphasize the role of microbial biogeography, shaped by geographical, climatic, and viticultural factors, as a new perspective to enhance regional characteristics and optimize wine production by managing the present microbes. Relevant to the current study, a review [20] highlights that the role of region-specific microbial communities (microbial terroir) in defining wine characteristics is still debated, requiring further research for a clearer understanding. Cobos et al. [21] discuss how the grapevine microbiome offers potential sources for new and promising biocontrol agents that could serve as effective tools in controlling grapevine trunk diseases. Lastly, Wei et al. [22] discuss the benefits of mimicking natural ecological cultivation to enhance microbial diversity and sustainability in large-scale natural wine practices. Together, these studies highlight the complex interactions between soil microbiomes, grapevines, and the production of high-quality wine. However, there are still many aspects of this complex ecosystem that require further exploration and understanding to support sustainable viticulture and enhance wine quality. Therefore, in the following sections, we discuss contributing factors that link soil-associated microbiota to wine terroirs, such as vineyard management, diseases, phages, and next-generation sequencing as topics that are important drivers of our knowledge of the changing landscape of soil microbes in vineyards.

2.3.1. Climate Change Impacts

Investigating the effects of climate change on microbial community dynamics, their functional roles, and their implications for wine production can help prepare the wine industry for potential challenges [38]. Notably, one of the few papers that discusses soil diversity impacts through climatic condition changes was Rivas et al. [38]. The study indicates a consistent set of microorganisms in both soil and wine (*Proteobacteria, Actinobacteria*) that remain steady over multiple vintage years from Argentina. Consistently, these two groups were also documented (Table 1) as the most dominant phyla in the other short-term studies from 2015–2023 that were included in this perspective.

2.3.2. Vineyard Management Practices

Vineyard management practices have profound impacts on both the immediate and long-term health of vineyards, influencing grapevine productivity and overall ecosystem sustainability, with soil health being a key concern [48]. The vineyard management techniques discussed in the literature (Table 1) are irrigation management [34,36], canopy management [34], pest and disease control [27,43], nutrient management [23,30], precision viticulture [37], and soil tilling [28]. Liu et al. [35] compares organic and conventional vineyard management practices and their effects on soil health and microbial diversity. Vineyard managers use techniques like cover cropping, composting, and organic fertilizers to maintain soil health, which not only ensures grapevines' viability but also contributes to the sustainability of the vineyard ecosystem. Understanding these effects is crucial for developing environmentally friendly and economically viable viticultural methods. Although the studies focus on the effects of some vineyard management practices, the emphasis is not on providing in-depth descriptions of the practices and mechanisms of influence. To fully grasp the implications of various agricultural approaches on soil biodiversity and vineyard microbiomes, further research is warranted.

2.3.3. Comparative Arguments on Soil Tilling

In addition to investigating vineyard management effects locally, comparative studies spanning different viticultural regions worldwide can offer valuable insights [49]. Specifically, soil tilling in viticulture is likely to have significant implications on nutrient and soil organic carbon content with correlations to the microbial community composition and associated function [50]. Yet, the impact of this practice on soil microbial abundance and richness, as well as its link to wine terroir, remains fragmented and contradictory. Andrade et al. [51], Buckley & Schmidt [52], and Jangid et al. [53] observed that no-till fields tended to have higher microbial diversity compared to tilled fields. However, the studies report highly variable results on the number and type of unique operational taxonomic units (OTUs) in tilled vs. no-till fields. *Pseudomonas, Bacillus, Streptomyces, Rhizobium*, and *Actinobacteria* in Andrade et al. [51], vs. *Acidobacteria, Proteobacteria, Firmicutes*, and *Bacteroidetes* in Buckley & Schmidt [52] and *Nitrosospira, Sphingomonas, Flavobacterium, Burkholderia*, and *Mycobacterium* in Jangid et al. [53]. Such cross-regional analyses have the potential to reveal both common patterns and unique characteristics associated with specific wine-producing areas.

2.3.4. Precision Viticulture

Advancements in precision viticulture offer significant potential for targeted microbial management in vineyards [54]. However, among the 24 studies listed, only Teixeira et al. [37] directly discuss precision viticulture by employing advanced molecular techniques to enhance vineyard management. Their study employed DNA-based assays to detect single nucleotide polymorphisms (SNPs) in three genes of the anthocyanin pathway (UFGT, F3H, and LDOX), demonstrating how advanced molecular techniques can be integrated into vineyard management strategies. It is worth pointing out that the study appears to be more focused on the genetics of the grapevine itself rather than soil microbes. Other studies, while not explicitly labeled as precision viticulture, may involve practices and methodologies that align with the principles of precision viticulture, such as detailed monitoring and management of soil microbial communities in Hendgen et al. [28], Oyuela Aguilar et al. [36], and Gobbi et al. [42].

2.4. Future Directions for Soil Microbial Communities and Wine Terroir Research

Despite the growing body of research implying connections between soil microbes and wine terroir, the precise mechanisms underlying these interactions remain elusive. Next-generation sequencing (NGS) has emerged as a promising tool to unravel these complexities, offering detailed insights into microbial diversity and function. However, additional approaches such as precision viticulture, bacteriophages, and grapevine trunk disease management also hold the potential to enhance our understanding of these interactions. This section explores these future research areas that could provide a deeper understanding of soil microbial communities and their impact on wine terroir.

2.4.1. Next-Generation Sequencing

To identify and quantify the microorganisms present in vineyard soil (Table 1), several studies have opted to use next-generation sequencing. Among the studies we reviewed, 16 focused on amplicon sequencing of the 16S rRNA gene, which is widely used in microbiology for the identification and classification of microorganisms, particularly bacteria and archaea. Twelve papers focused on sequencing the ITS regions and their subregions, and two papers used 26S rDNA. ITS and 26S rDNA are used widely to characterize eukaryotic organisms. Relevant to wine making, yeast communities were of interest to these papers. Only two studies utilized shotgun sequencing that analyzed entire genomes and complex microbial communities without the need for prior knowledge of specific DNA regions.

Although targeted sequencing (16S rRNA, ITS, and 26S rDNA) can reveal insights into the microbial community composition, shotgun sequencing offers a broader picture of the entire genome, making it suitable for functional information as well. However, the information obtained from shotgun sequencing (also sometimes referred to as whole genome sequencing or metagenomics, derived from DNA) can only lead to inferences about function. On the other hand, transcriptomics (derived from RNA rather than DNA) provides insights into the active metabolic pathways and biological processes occurring in the soil, giving a more dynamic view of microbial activity than just analyzing the microbial composition. To our knowledge, no studies have utilized transcriptomics to investigate the actual function of soil microbial communities in vineyard settings.

Metagenomics and transcriptomics of soil microbial communities in vineyards offer a powerful tool for gaining functional insights into soil microbiomes, supporting sustainable vineyard management, and contributing to the production of high-quality wines with a distinct terroir. Firstly, it helps predict microorganisms' roles in nutrient cycling, organic matter decomposition, and other essential processes for vineyard health [55]. Secondly, transcriptomics helps in monitoring how soil microbial communities respond to changes in environmental factors, such as climate, soil management practices, and agricultural inputs [56]. Thirdly, this type of analysis can help identify specific microbial species or groups that play essential roles in promoting soil health, enhancing nutrient availability, and protecting grapevines from diseases. Such beneficial microbes can be targeted for potential use as biofertilizers or biocontrol agents. Lastly, such methods help in understanding how soil microbes contribute to the regional identity of wines (terroir), which is essential for promoting authenticity and quality [57].

2.4.2. Expanding Precision Viticulture

The integration of cutting-edge technologies into viticulture can optimize soil microbial communities, bolster grapevine health, and elevate overall wine quality. Delving into the potential of precision viticulture to influence the dynamics of vineyard microbiomes can lead to innovative practices that maximize wine production efficiency while maintaining environmental sustainability. For example, integrating multispectral, hyperspectral, and thermal sensing (which are among the most widely used sensors for vineyard monitoring over the last two decades) with soil microbial community characterization using next-generation sequencing while also utilizing vineyard canopy images [58] would be vital in predicting an impending poor harvest following severe environmental stress such as drought or heavy precipitation events.

2.4.3. Grapevine Trunk Diseases

Grapevine trunk diseases remain a significant challenge to the wine industry [59–62]. The most common microorganisms that grapevines tend to be most susceptible to are *Plasmopara viticola* (downy mildew), *Elsinoe ampelina* (anthracnose), *Guignardia bidwellii* (black rot), *Erysiphe neator* (powdery mildew). Interestingly, recent research [63] has demonstrated associations between belowground microbiota *Fusarium* spp. and exacerbating progression of grapevine trunk disease. However, confusion remains concerning the cause and progression of the disease, many of which result in serious infections, loss of yield, or quality. In-depth studies are needed to decipher the interactions between the host grapevine and the diverse fungal communities, some of which may act as opportunistic pathogens under specific conditions. Several studies from the review indicate a growing interest in understanding soil microbial impacts on grapevine health. For example, Liu et al. [35] highlight the importance of microbial diversity in organic versus conventional vineyard management, which can influence disease resistance. Additionally, Mezzasalma et al. [27] and Ramirez et al. [34] discuss pest and disease control practices, emphasizing the need for integrated management strategies that consider microbial interactions. Understanding these interactions could lead to more effective prevention and treatment methods for trunk diseases. Such insights can aid in devising targeted strategies for disease management and prevention. Another particularly intriguing area of research is the role of both asymptomatic and symptomatic grapevines in harboring pathogenic fungi. Understanding the differences in microbial communities between these two states can shed light on the mechanisms underlying the progression of trunk diseases [64].

2.4.4. Bacteriophages for Vineyard Soil Management

Bacteriophages, viruses that infect and replicate within bacterial cells, are abundant in vineyard soils and play a crucial role in regulating bacterial communities. Their ability to selectively target specific bacteria can alter soil microbial composition and dynamics, influencing nutrient cycling, plant health, and likely also grape quality [65]. The field of bacteriophage research in viticulture is still in its infancy [66], and further studies are needed to understand their interactions with bacterial communities and their impact on wine characteristics. For instance, Morrison-Whittle & Goddard [24,31] focused on 26S rDNA to explore yeast diversity, indirectly hinting at the microbial dynamics that bacteriophages might influence. Investigating how phage pressure changes soil bacterial community composition and functions [67] could reveal new strategies for enhancing soil health and grapevine resilience.

3. Conclusions

The influence of soil microbial communities on grapevine-associated microbiota is an area of active research and ongoing exploration. Precision viticulture, disease management, bacteriophage research, and next-generation sequencing represent promising areas for future investigation. By delving deeper into these areas, researchers can develop innovative strategies to enhance vineyard health, productivity, and wine quality, contributing to a more sustainable and nuanced understanding of terroir.

Author Contributions: Conceptualization, A.A.; methodology, G.C.F., J.L. and D.M.L.; formal analysis, G.C.F., J.L., S.N., B.W., K.D., L.S., B.-R.M., M.H., M.A. and A.H.; resources, L.N.; writing—A.A., G.C.F., J.L., D.M.L., S.N., B.W., K.D., L.S., B.-R.M., M.H., M.A. and A.H.; review and editing—A.A., G.C.F., J.L., D.M.L., S.N., B.W., K.D., L.S., B.-R.M., M.H., M.A. and A.H.; visualization, A.A. and G.C.F.; supervision, A.A.; project administration, A.A.; funding acquisition, A.A. All authors have read and agreed to the published version of the manuscript.

Funding: The College of Science and Engineering, Department of Biology, San Francisco State University—Anand Lab Seed Fund.

Institutional Review Board Statement: Not applicable.

Informed Consent Statement: Not applicable.

Data Availability Statement: No new data were created or analyzed in this study. Data sharing is not applicable to this article.

Acknowledgments: The authors would like to thank the Department of Biology and College of Science and Engineering at the San Francisco State University and the team at the Neely Vineyard in Portola Valley, CA, USA, for their support.

Conflicts of Interest: Author Lucy Neely is employed by the company Neely Winery, Spring Ridge Vineyard, 555 Portola Road, Portola Valley, CA 94028, USA. The remaining authors declare that the research was conducted in the absence of any commercial or financial relationships that could be construed as a potential conflict of interest.

References

1. Guzzon, R.; Bertoldi, D.; Roman, T.; Zanzotti, R.; Franciosi, E. Spatial and Seasonal Structure of Bacterial Communities within Alpine Vineyards: Trentino as a Case Study. *Microb. Ecol.* **2023**, *85*, 108–120. [CrossRef] [PubMed]
2. Van Leeuwen, C.; Friant, P.; Chone, X.; Tregoat, O. Influence of climate, soil, and cultivar on terroir. *Am. J. Enol. Vitic.* **2004**, *55*, 207–217. [CrossRef]
3. Belda, I.; Zarraonaindia, I.; Perisin, M.; Palacios, A.; Acedo, A. From vineyard soil to wine fermentation: Microbiome approximations to explain the "terroir" concept. *Front. Microbiol.* **2017**, *8*, 821. [CrossRef] [PubMed]
4. Gilbert, J.A.; Van Der Lelie, D.; Zarraonaindia, I. Microbial terroir for wine grapes. *Proc. Natl. Acad. Sci. USA* **2014**, *111*, 5–6. [CrossRef] [PubMed]
5. Bokulich, N.A.; Thorngate, J.H.; Richardson, P.M.; Mills, D.A. Microbial biogeography of wine grapes is conditioned by cultivar, vintage, and climate. *Proc. Natl. Acad. Sci. USA* **2014**, *111*, E139–E148. [CrossRef] [PubMed]
6. Griggs, R.G.; Steenwerth, K.L.; Mills, D.A.; Cantu, D.; Bokulich, N.A. Sources and Assembly of Microbial Communities in Vineyards as a Functional Component of Winegrowing. *Front. Microbiol.* **2021**, *12*, 673810. [CrossRef] [PubMed] [PubMed Central]
7. Wagner, M.R.; Lundberg, D.S.; Coleman-Derr, D.; Tringe, S.G.; Dangl, J.L.; Mitchell-Olds, T. Natural soil microbes alter flowering phenology and the intensity of selection on flowering time in a wild *Arabidopsis* relative. *Ecol. Lett.* **2014**, *17*, 717–726. [CrossRef] [PubMed]
8. Chalvantzi, I.; Banilas, G.; Tassou, C.; Nisiotou, A. Biogeographical Regionalization of Wine Yeast Communities in Greece and Environmental Drivers of Species Distribution at a Local Scale. *Front. Microbiol.* **2021**, *12*, 705001. [CrossRef] [PubMed] [PubMed Central]
9. Barata, A.; Malfeito-Ferreira, M.; Loureiro, V. The microbial ecology of wine grape berries. *Int. J. Food Microbiol.* **2012**, *153*, 243–259. [CrossRef]
10. Nisiotou, A.A.; Nychas, G.J.E. Yeast populations residing on healthy or botrytis-infected grapes from a vineyard in Greece. *Appl. Environ. Microbiol.* **2007**, *73*, 2765–2768. [CrossRef]
11. Martins, G.; Lauga, B.; Miot-Sertier, C.; Mercier, A.; Lonvaud, A.; Soulas, M.L.; Soulas, G.; Masneuf-Pomarède, I. Characterization of Epiphytic Bacterial Communities from Grapes, Leaves, Bark and Soil of Grapevine Plants Grown, and Their Relations. *PLoS ONE* **2013**, *8*, e73013. [CrossRef] [PubMed]
12. Zhou, J.; Cavagnaro, T.R.; De Bei, R.; Nelson, T.M.; Stephen, J.R.; Metcalfe, A.; Gilliham, M.; Breen, J.; Collins, C.; López, C.M.R. Wine Terroir and the Soil Bacteria: An Amplicon Sequencing-Based Assessment of the Barossa Valley and Its Sub-Regions. *Front. Microbiol.* **2021**, *11*, 597944. [CrossRef] [PubMed] [PubMed Central]
13. Zarraonaindia, I.; Owens, S.M.; Weisenhorn, P.; West, K.; Hampton-Marcell, J.; Lax, S.; Bokulich, N.A.; Mills, D.A.; Martin, G.; Taghavi, S.; et al. The Soil Microbiome Influences Grapevine-Associated Microbiota. *mBio* **2015**, *6*, e02527-14. [CrossRef] [PubMed]

14. Bokulich, N.A.; Collins, T.S.; Masarweh, C.; Allen, G.; Heymann, H.; Ebeler, S.E.; Mills, D.A. Associations among Wine Grape Microbiome, Metabolome, and Fermentation Behavior Suggest Microbial Contribution to Regional Wine Characteristics. *mBio* **2016**, *7*, e00631-16. [CrossRef] [PubMed] [PubMed Central]
15. Sancho-Galan, P.; Amores-Arrocha, A.; Jimenez-Cantizano, A.; Palacios, V. Influence of the presence of grape skins during white wine alcoholic fermentation. *Agronomy* **2021**, *11*, 452. [CrossRef]
16. Compant, S.; Mitter, B.; Colli-Mull, J.G.; Gangl, H.; Sessitsch, A. Endophytes of grapevine flowers, berries, and seeds: Identification of cultivable bacteria, comparison with other plant parts, and visualization of niches of colonization. *Microb. Ecol.* **2011**, *62*, 188–197. [CrossRef] [PubMed]
17. Pacifico, D.; Squartini, A.; Crucitti, D.; Barizza, E.; Lo Schiavo, F.; Muresu, R.; Carimi, F.; Zottini, M. The Role of the Endophytic Microbiome in the Grapevine Response to Environmental Triggers. *Front. Plant Sci.* **2019**, *10*, 1256. [CrossRef] [PubMed] [PubMed Central]
18. Hamaoka, K.; Aoki, Y.; Takahashi, S.; Enoki, S.; Yamamoto, K.; Tanaka, K.; Suzuki, S. Diversity of endophytic bacterial microbiota in grapevine shoot xylems varies depending on wine grape-growing region, cultivar, and shoot growth stage. *Sci. Rep.* **2022**, *12*, 15772. [CrossRef]
19. Liu, D.; Zhang, P.; Chen, D.; Howell, K. From the Vineyard to the Winery: How Microbial Ecology Drives Regional Distinctiveness of Wine. *Front. Microbiol.* **2019**, *10*, 2679. [CrossRef]
20. Alexandre, H. Wine Yeast Terroir: Separating the Wheat from the Chaff-for an Open Debate. *Microorganisms* **2020**, *8*, 787. [CrossRef] [PubMed] [PubMed Central]
21. Cobos, R.; Ibañez, A.; Diez-Galán, A.; Calvo-Peña, C.; Ghoreshizadeh, S.; Coque, J.J.R. The Grapevine Microbiome to the Rescue: Implications for the Biocontrol of Trunk Diseases. *Plants* **2022**, *11*, 840. [CrossRef] [PubMed] [PubMed Central]
22. Wei, R.; Wang, L.; Ding, Y.; Zhang, L.; Gao, F.; Chen, N.; Song, Y.; Li, H.; Wang, H. Natural and sustainable wine: A review. *Crit. Rev. Food Sci. Nutr.* **2023**, *63*, 8249–8260. [CrossRef] [PubMed]
23. Burns, K.L.; Kluepfel, D.A.; Strauss, S.L.; Bokulich, N.A.; Cantu, D.; Steenwerth, K.L. Vineyard soil bacterial diversity and composition revealed by 16S rRNA genes: Differentiation by geographic features. *Soil Biol. Biochem.* **2015**, *91*, 232–247. [CrossRef]
24. Morrison-Whittle, P.; Goddard, M. Quantifying the relative roles of selective and neutral processes in defining eukaryotic microbial communities. *ISME J.* **2015**, *9*, 2003–2011. [CrossRef] [PubMed]
25. Burns, K.L.; Bokulich, N.A.; Cantu, D.; Greenhut, R.F.; Kluepfel, D.A.; O'Geen, A.T.; Strauss, S.L.; Steenwerth, K.L. Vineyard soil bacterial diversity and composition revealed by 16S rRNA genes: Differentiation by vineyard management. *Soil Biol. Biochem.* **2016**, *103*, 337–348. [CrossRef]
26. Castañeda, L.E.; Barbosa, O. Metagenomic analysis exploring taxonomic and functional diversity of soil microbial communities in Chilean vineyards and surrounding native forests. *PeerJ* **2017**, *5*, e3098. [CrossRef] [PubMed] [PubMed Central]
27. Mezzasalma, V.; Sandionigi, A.; Bruni, I.; Bruno, A.; Lovicu, G.; Casiraghi, M.; Labra, M. Grape microbiome as a reliable and persistent signature of field origin and environmental conditions in Cannonau wine production. *PLoS ONE* **2017**, *12*, e0184615. [CrossRef] [PubMed]
28. Hendgen, M.; Hoppe, B.; Döring, J.; Friedel, M.; Kauer, R.; Frisch, M.; Dahl, A.; Kellner, H. Effects of different management regimes on microbial biodiversity in vineyard soils. *Sci. Rep.* **2018**, *8*, 9393. [CrossRef]
29. Chou, M.Y.; Vanden Heuvel, J.; Bell, T.H.; Panke-Buisse, K.; Kao-Kniffin, J. Vineyard under-vine floor management alters soil microbial composition, while the fruit microbiome shows no corresponding shifts. *Sci. Rep.* **2018**, *8*, 11039. [CrossRef]
30. Wei, Y.-J.; Wu, Y.; Yan, Y.-Z.; Zou, W.; Xue, J.; Ma, W.-R.; Wang, W.; Tian, G.; Wang, L.-Y. High-throughput sequencing of microbial community diversity in soil, grapes, leaves, grape juice and wine of grapevine from China. *PLoS ONE* **2018**, *13*, e0193097. [CrossRef]
31. Morrison-Whittle, P.; Goddard, M. From vineyard to winery: A source map of microbial diversity driving wine fermentation. *Environ. Microbiol.* **2017**, *20*, 75–84. [CrossRef] [PubMed]
32. Gupta, V.V.S.R.; Bramley, R.G.V.; Greenfield, P.; Yu, J.; Herderich, M.J. Vineyard Soil Microbiome Composition Related to Rotundone Concentration in Australian Cool Climate 'Peppery' Shiraz Grapes. *Front. Microbiol.* **2019**, *10*, 1607. [CrossRef] [PubMed]
33. Liang, H.; Wang, X.; Yan, J.; Luo, L. Characterizing the Intra-Vineyard Variation of Soil Bacterial and Fungal Communities. *Front. Microbiol.* **2019**, *10*, 1239. [CrossRef]
34. Ramirez, R.; Lopez-Pinero, A.; Velazquez, R.; Munoz, A.; Regodon, J.A. Analysing the vineyard soil as a natural reservoir for wine yeasts. *Food Res. Int.* **2020**, *129*, 108845. [CrossRef]
35. Liu, D.; Chen, Q.; Zhang, P.; Chen, D.; Howell, K.S. The Fungal Microbiome Is an Important Component of Vineyard Ecosystems and Correlates with Regional Distinctiveness of Wine. *mSphere* **2020**, *5*, e00534-20. [CrossRef] [PubMed] [PubMed Central]
36. Oyuela Aguilar, M.; Gobbi, A.; Browne, P.D.; Ellegaard-Jensen, L.; Hansen, L.H.; Semorile, L.; Pistorio, M. Influence of vintage, geographic location and cultivar on the structure of microbial communities associated with the grapevine rhizosphere in vineyards of San Juan Province, Argentina. *PLoS ONE* **2020**, *15*, e0243848. [CrossRef]
37. Teixeira, R.J.S.; Gomes, S.; Malheiro, V.; Pereira, L.; Fernandes, J.R.; Mendes-Ferreira, A.; Gomes, M.E.P.; Martins-Lopes, P. A Multidisciplinary Fingerprinting Approach for Authenticity and Geographical Traceability of Portuguese Wines. *Foods* **2021**, *10*, 1044. [CrossRef]

38. Rivas, G.A.; Guillade, A.C.; Semorile, L.C.; Delfederico, L. Influence of Climate on Soil and Wine Bacterial Diversity on a Vineyard in a Non-Traditional Wine Region in Argentina. *Front. Microbiol.* **2021**, *12*, 726384. [CrossRef] [PubMed] [PubMed Central]
39. Torres, N.; Yu, R.; Kurtural, S.K. Inoculation with Mycorrhizal Fungi and Irrigation Management Shape the Bacterial and Fungal Communities and Networks in Vineyard Soils. *Microorganisms* **2021**, *9*, 1273. [CrossRef]
40. Yan, H.; Ge, C.; Zhou, J.; Li, J. Diversity of soil fungi in the vineyards of Changli region in China. *Can. J. Microbiol.* **2022**, *68*, 341–352. [CrossRef]
41. Geiger, A.; Karacsony, Z.; Golen, R.; Vaczy, K.Z.; Gemi, J. The Compositional Turnover of Grapevine-Associated Plant Pathogenic Fungal Communities Is Greater among Intraindividual Microhabitats and Terroirs than among Healthy and Esca-Diseased Plants. *Phytopathology* **2022**, *112*, 1029–1035. [CrossRef] [PubMed]
42. Gobbi, A.; Acedo, A.; Imam, N.; Santini, R.G.; Ortiz-Álvarez, R.; Ellegaard-Jensen, L.; Belda, I.; Hansen, L.H. A global microbiome survey of vineyard soils highlights the microbial dimension of viticultural terroirs. *Commun. Biol.* **2022**, *5*, 241. [CrossRef] [PubMed]
43. Regecová, I.; Semjon, B.; Jevinová, P.; Očenáš, P.; Výrostková, J.; Šuľáková, L.; Nosková, E.; Marcinčák, S.; Bartkovský, M. Detection of Microbiota during the Fermentation Process of Wine in Relation to the Biogenic Amine Content. *Foods* **2022**, *11*, 3061. [CrossRef] [PubMed] [PubMed Central]
44. Larsen, S.; Albanese, D.; Stegen, J.; Franceschi, P.; Coller, E.; Zanzotti, R.; Ioriatti, C.; Stefani, E.; Pindo, M.; Cestaro, A.; et al. Distinct and Temporally Stable Assembly Mechanisms Shape Bacterial and Fungal Communities in Vineyard Soils. *Microb. Ecol.* **2023**, *86*, 337–349. [CrossRef] [PubMed]
45. Nanetti, E.; Palladino, G.; Scicchitano, D.; Trapella, G.; Cinti, N.; Fabbrini, M.; Cozzi, A.; Accetta, G.; Tassini, C.; Iannaccone, L.; et al. Composition and biodiversity of soil and root-associated microbiome in *Vitis vinifera* cultivar Lambrusco distinguish the microbial terroir of the Lambrusco DOC protected designation of origin area on a local scale. *Front. Microbiol.* **2023**, *14*, 1108036. [CrossRef] [PubMed] [PubMed Central]
46. Darriaut, R.; Tran, J.; Martins, G.; Ollat, N.; Masneuf-Pomarede, I.; Lauvergeat, V. In grapevine decline, microbiomes are affected differently in symptomatic and asymptomatic soils. *Appl. Soil Ecol.* **2023**, *183*, 104767. [CrossRef]
47. Beule, L.; Vaupel, A.; Moran-Rodas, V.E. Abundance, Diversity, and Function of Soil Microorganisms in Temperate Alley-Cropping Agroforestry Systems: A Review. *Microorganisms* **2022**, *10*, 616. [CrossRef] [PubMed]
48. Giffard, B.; Winter, S.; Guidoni, S.; Nicolai, A.; Castaldini, M.; Cluzeau, D.; Coll, P.; Cortet, J.; Le Cadre, E.; d'Errico, G.; et al. Vineyard Management and Its Impacts on Soil Biodiversity, Functions, and Ecosystem Services. *Front. Ecol. Evol.* **2022**, *10*, 850272. [CrossRef]
49. Tofalo, R. Microbial Dynamics in Wine Production. *Microorganisms* **2021**, *9*, 700. [CrossRef] [PubMed] [PubMed Central]
50. Smith, C.R.; Blair, P.L.; Boyd, C.; Cody, B.; Hazel, A.; Hedrick, A.; Kathuria, H.; Khurana, P.; Kramer, B.; Muterspaw, K.; et al. Microbial community responses to soil tillage and crop rotation in a corn/soybean agroecosystem. *Ecol. Evol.* **2016**, *6*, 8075–8084. [CrossRef] [PubMed] [PubMed Central]
51. Andrade, D.S.; Colozzi-Filho, A.; Giller, K.E. The soil microbial community and soil tillage. In *Soil Tillage in Agroecosystems*; El Titi, A., Ed.; CRC Press: Boca Raton, FL, USA, 2002; p. 51.
52. Buckley, D.H.; Schmidt, T.M. Diversity and dynamics of microbial communities in soils from agro-ecosystems. *Environ. Microbiol.* **2003**, *5*, 441–452. [CrossRef] [PubMed]
53. Jangid, K.; Williams, M.A.; Franzluebbers, A.J.; Schmidt, T.M.; Coleman, D.C.; Whitman, W.B. Land-use history has a stronger impact on soil microbial community composition than aboveground vegetation and soil properties. *Soil Biol. Biochem.* **2011**, *43*, 2184–2193. [CrossRef]
54. Ferro, M.V.; Catania, P. Technologies and Innovative Methods for Precision Viticulture: A Comprehensive Review. *Horticulturae* **2023**, *9*, 399. [CrossRef]
55. Zarik, L.; Meddich, A.; Hijri, M.; Hafidi, M.; Ouhammou, A.; Ouahmane, L.; Duponnois, R.; Boumezzough, A. Use of arbuscular mycorrhizal fungi to improve the drought tolerance of *Cupressus atlantica* G. *C. R. Biol.* **2016**, *339*, 185–196. [CrossRef] [PubMed]
56. Stewart, F.J.; Sharma, A.K.; Bryant, J.A.; Eppley, J.M.; DeLong, E.F. Community transcriptomics reveals universal patterns of protein sequence conservation in natural microbial communities. *Genome Biol.* **2011**, *12*, R26. [CrossRef] [PubMed]
57. Nwachukwu, B.C.; Babalola, O.O. Metagenomics: A tool for exploring key microbiome with the potentials for improving sustainable agriculture. *Front. Sustain. Food Syst.* **2022**, *6*, 886987. [CrossRef]
58. Loggenberg, K.; Strever, A.; Greyling, B.; Poona, N. Modelling Water Stress in a Shiraz Vineyard Using Hyperspectral Imaging and Machine Learning. *Remote Sens.* **2018**, *10*, 202. [CrossRef]
59. Kenfaoui, J.; Radouane, N.; Mennani, M.; Tahiri, A.; El Ghadraoui, L.; Belabess, Z.; Fontaine, F.; El Hamss, H.; Amiri, S.; Lahlali, R.; et al. A Panoramic View on Grapevine Trunk Diseases Threats: Case of Eutypa Dieback, Botryosphaeria Dieback, and Esca Disease. *J. Fungi* **2022**, *8*, 595. [CrossRef] [PubMed] [PubMed Central]
60. Gramaje, D.; Armengol, J. Fungal trunk pathogens in the grapevine propagation process: Potential inoculum sources, detection, identification, and management strategies. *Plant Dis.* **2011**, *95*, 1040–1055. [CrossRef]
61. Gramaje, D.; Baumgartner, K.; Halleen, F.; Mostert, L.; Sosnowski, M.R.; Úrbez-Torres, J.R.; Armengol, J. Fungal trunk diseases: A problem beyond grapevines? *Plant Pathol.* **2016**, *65*, 355–356. [CrossRef]
62. Gramaje, D.; Urbez-Torres, J.R.; Sosnowski, M.R. Managing grapevine trunk diseases with respect to etiology and epidemiology: Current strategies and future prospects. *Plant Disease* **2018**, *102*, 12–39. [CrossRef] [PubMed]

63. Li, Y.; Li, X.; Zhang, W.; Zhang, J.; Wang, H.; Peng, J.; Wang, X.; Yan, J. Belowground microbiota analysis indicates that *Fusarium* spp. exacerbate grapevine trunk disease. *Environ. Microbiome* **2023**, *18*, 29. [CrossRef] [PubMed] [PubMed Central]
64. Fall, M.L.; Poursalavati, A.; Sidibe, A.; Xu, D.; Lemoyne, P.; Martins, G.S.; Javaran, V.J.; Moffett, P.; Odile, C. From asymptomatic to symptomatic: Multiomics profiling of the temporal response of grapevine viral-mixed infection. *bioRxiv* **2023**. [CrossRef]
65. Ye, M.; Sun, M.; Huang, D.; Zhang, Z.; Zhang, H.; Zhang, S.; Hu, F.; Jiang, X.; Jiao, W. A review of bacteriophage therapy for pathogenic bacteria inactivation in the soil environment. *Environ. Int.* **2019**, *129*, 488–496. [CrossRef] [PubMed]
66. Chaib, A.; Philippe, C.; Joamanjaka, F.; Barchi, Y.; Oviedo-Hernandez, F.; Claisse, O.; Le Marrec, C. Phage-host interactions as a driver of population dynamics during wine fermentation: Betting on underdogs. *Int. J. Food Microbiol.* **2022**, *383*, 109936. [CrossRef]
67. Braga, L.P.P.; Spor, A.; Kot, W.; Breuil, M.-C.; Hansen, L.H.; Setubal, J.C.; Philippot, L. Impact of phages on soil bacterial communities and nitrogen availability under different assembly scenarios. *Microbiome* **2020**, *8*, 52. [CrossRef]

foods

Article

Evaluation of the Impact of the Ripening Stage on the Composition and Antioxidant Properties of Fruits from Organically Grown Tomato (*Solanum lycopersicum* L.) Spanish Varieties

Ivan Cruz-Chamorro [1,2,†], Guillermo Santos-Sánchez [3,†], Franz Martín [2], María-Soledad Fernández-Pachón [2,*], Dámaso Hornero-Méndez [4] and Isabel Cerrillo [2]

1 Departamento de Biología y Geología, Escuela Superior de Ingeniería, Universidad de Almería, 04120 Almería, Spain; icc967@ual.es
2 Área de Nutrición y Bromatología, Departamento de Biología Molecular e Ingeniería Bioquímica, Universidad Pablo de Olavide, Ctra. Utrera Km 1, 41013 Seville, Spain; fmarber@upo.es (F.M.); icergar@upo.es (I.C.)
3 Instituto de Investigación en Ciencias de la Alimentación, CIAL (CSIC-UAM), 28049 Madrid, Spain; g.santos@csic.es
4 Departamento de Fitoquímica de los Alimentos, Instituto de la Grasa-CSIC, Campus Universitario Pablo de Olavide, Ctra. Utrera Km 1, 41013 Seville, Spain; hornero@ig.csic.es
* Correspondence: msferpac@upo.es
† These authors contributed equally to this work.

Abstract: Tomato (*Solanum lycopersicum* L.) is a widely cultivated horticultural crop. It belongs to the Solanaceae family and is known for its high concentration of essential nutrients, including vitamins, minerals, and bioactive compounds with antioxidant properties. The Mediterranean countries, including Italy, Spain, and Greece, have a diverse range of tomato landraces. Assessing the nutritional and bioactive composition of different tomato varieties and their ripening stages is crucial to determine their suitability for the market. Therefore, the aim of this study was to investigate the effect of ripening on nutritional composition (including carotenoids and polyphenols content) and antioxidant activities of fruits of three specific tomato varieties grown in Spain: *Josefina* and *Karelya*, which are cherry-like tomatoes, and *Muchamiel*, a type of salad tomato. In addition to evaluating their characteristics and composition (including carotenoids and polyphenol content), the antioxidant activities of these varieties at three different ripening stages were quantified. As expected, the results reveal that, as the tomatoes matured, their antioxidant capacity increased along with higher levels of carotenoids and polyphenols. Interestingly, cherry-like tomatoes showed a higher antioxidant activity than the salad tomatoes. This investigation emphasizes the role of fruit ripening in increasing carotenoid levels, which contribute to the antioxidant activity of three tomato varieties.

Keywords: cherry-like tomato; salad tomato; carotenoid; polyphenol; antioxidant activity

1. Introduction

Tomato is one of the most popular and widely consumed vegetables in the Mediterranean diet and has therefore been extensively characterized in terms of its nutritional profile and bioactive compounds, being an excellent source of ascorbic acid, carotenoids, and flavonoids [1]. In order to maximize its sensory quality, nutritional composition, and bioactivity, several studies have investigated how different factors, such as the ripening state or tomato variety, influence these aspects [2]. Among the 17 goals of the 2030 Agenda for Sustainable Development, adopted by all United Nations member states in 2015, are (i) responsible production and consumption, (ii) climate action, and (iii) the life of terrestrial ecosystems, with a strong emphasis on reducing greenhouse gas emissions, land use, energy consumption and, to a lesser extent, water use in the cultivation and production of food [3]. Furthermore, organically grown vegetables are a good sustainable alternative, due to the exclusion of the use of synthetic chemicals and respect for the environment [4]. The manner in which tomatoes are grown, as well as the variety of tomato that is grown, can influence the quality of the fruit and may be of significant importance in the production of the crop from a profitable standpoint (obtaining a tomato with optimal quality and long commercial shelf life), reducing costs at the industrial level, and health (obtaining a tomato with a high nutritional density, rich in bioactive compounds, and high antioxidant activity) [5]. Furthermore, organic farming compared to conventional cultivation in terms of quality, nutritional intake, and bioactivity is better [6]. Currently, there are few data on the characterization of organic tomato quality, and the information available to the farmer about the factors that influence it is scarce.

Noncommunicable diseases (NCDs), such as cardiovascular diseases, diabetes, and cancer, killed 41 million people (74% of global deaths) in 2022 [7]. Research supports that oxidative stress, known as an imbalance between reactive oxygen species (ROS) and antioxidant defenses, is part of the triggering and development of the pathophysiological changes associated with noncommunicable diseases [8,9]. In this sense, numerous epidemiological and clinical studies have shown how tomato consumption reduces the risk of contracting numerous diseases, such as cardiovascular diseases or cancer [10]. This beneficial role of tomato consumption has been attributed to its content of polyphenols and carotenoids (mainly lycopene and β-carotene) since these compounds inhibit reactions mediated by ROS. According to numerous studies, the content of these antioxidant compounds in tomato fruits depends on genetic and environmental factors, as well as on the ripening stage of the fruit [11]. In addition, functional foods are gaining attention in the field of nutrition because of their beneficial effects on health and the prevention and treatment of many diseases [12]. In this sense, the extraction of tomato antioxidant compounds, mainly the carotenoid lycopene, for the subsequent generation of functional foods, is gaining attention within the food and pharmaceutical industries [13,14]. Therefore, it is of a great interest to know the variations in the content of antioxidant compounds during the ripening process in order to obtain fruits with the highest bioactivity potential. In light of this well-known evidence presented above, the objective of this study was to evaluate, for the first time, the quality, nutritional, and antioxidant (composition and activity) variations throughout the ripening process of fruits from three Spanish tomato varieties (*Josefina*, *Karelya*, and *Muchamiel*) grown under organic farming practices. The aim was to determine the optimal ripening stage to maximize the nutritional and antioxidant potential of the fruits.

2. Materials and Methods

2.1. Reagents

Folin and Ciocalteu phenol reagent, gallic acid, Trolox, 2,2-azino-bis-(3-ethylbenzothia zoline-6-sulfonic acid) (ABTS), azo2,20-azobis(2-methylpropionamidine) dihydrochloride (AAPH), sodium fluorescein, 2,2-diphenyl-1-picrylhydrazyl (DPPH), and 2,4,6-Tris(2-pyridyl)-s-triazine (TPTZ) were purchased from Sigma-Aldrich (St. Louis, MO, USA).

2.2. Experimental Design and Sample Collection

Bioalverde S.L. (Seville, Spain) supplied the organic tomato samples. Specifically, 30 plants of each variety, *Josefina* and *Karelya*, which are cherry-like tomatoes, and *Muchamiel*, a type of salad tomato, were planted in the greenhouse, divided into 2 different plots (with 15 plants of each species/area).

Three consecutive ripening stages (RS) were selected according to the external color as follows: RS1—changing color fruit, in which green-colored areas are more prevalent than red ones; RS2—changing color fruit, in which red-colored areas are more prevalent than green ones; and RS3—fully mature fruit with an intense red color. A minimum of 40 tomato fruits were collected for each ripening stage and for each variety. Nondestructive determinations (such as fruit size, weight, and color) were carried out in 20 fruits (size and weight) or 10 fruits (color). The same fruits used for the measurement of the color (*n* = 10) were used for the determination of the texture (firmness). For the rest of the measurements, homogenates of tomatoes were used. Specifically, determinations of moisture (water content), pH, and SST were performed immediately after the preparation of the homogenate, while determinations of antioxidant activity, total phenols, and titratable acidity were carried out in frozen aliquots of the homogenate supernatant. The determinations of the carotenoid and nutritional composition were carried out on lyophilized aliquots of the homogenate.

2.3. *Physicochemical Characterization and Quality Parameters*

2.3.1. Fruit Weight and Size

Fruits of each variety were weighed using an analytical balance (model AX224, Sartorius AG, Goettingen, Germany). Fruit sizes were determined as the equatorial diameter and measured with a digital caliper (VWR).

2.3.2. Color

The evaluation of the fruit color was carried out using the colorimetry technique. The color was also expressed in terms of the CIE L* (luminosity, whiteness or brightness/darkness), a* (redness/greenness), and b* (yellowness/blueness) coordinates [15]. Colorimetric measurements on the tomatoes were performed using a BYK-Gadner, model 9000, Color-view™ spectrophotometer (Silver Spring, MD, USA). Results are expressed as a/b.

2.3.3. Total Soluble Solid (TSS) Content

The TSS content was measured in a small sample (2 drops) of the fruit homogenate supernatant using a portable handheld refractometer (model 0-32° BRIX ATC, Bellingham + Stanley, Kent, UK) with an accuracy of ±0.1° Brix [16].

2.3.4. pH Determination

The pH value was determined using a pH meter (model pH 1000L, VWR International Eurolab S.L., Barcelona, Spain) using the homogenate prepared from the collected fruits.

2.3.5. Texture

Texture (firmness) was measured with an Instron Universal Testing Machine (Canton, MA, USA) fitted with a Kramer shear compression cell. The cross-head speed was 200 mm/min. The firmness (shear compression force) of the tomato fruits was expressed as N/100 g, and the value was the mean of ten measurements, each of which was performed on one fruit for cherry-like varieties and with one quarter of the fruit for the *Muchamiel* cultivar.

2.3.6. Proximate Composition

The proximate composition of the fruits was evaluated for each cultivar at the selected ripening stages (RS1, RS2, and RS3). Dry matter, moisture, ash, protein, fat, and carbohydrate determinations were made using AOAC methods [17], and the results are expressed in relative units (%). The dry matter content and moisture were determined by gravimetry using a moisture balance (model MB35, OAHUS, Barcelona, Spain). The mineral content (ash) was determined by the calcination method. The total protein content was estimated using a LECO Elemental Analyzer (model CHNS-932, Leco Corporation, St. Joseph, MI, USA) for nitrogen determination. The protein content was calculated from the nitrogen content (Dumas method) as follows: Protein (%) = Nitrogen (%) × 6.25 [18]. The total lipid (fat) content was determined using the Soxhlet method [19]. The determination of total carbohydrates (including fiber) was carried out using the differential method by subtracting the sum of percentages of the rest of nutrients (total proteins, total lipids, minerals, and moisture) from 100 [17]. Each parameter was analyzed in triplicate.

2.3.7. Carotenoid Analysis

Carotenoid pigments were extracted and analyzed using HPLC following the procedure described in de los Santos et al. [20]. Briefly, an aliquot (0.045–0.050 g) of the lyophilized and homogenized sample was placed in a 2 mL micro-centrifuge tube and extracted with 1 mL of acetone-THF (4:1; containing 0.1% BHT) by sonication for 1 min, followed by a vigorous shaking in a mixer for 10 min. The extraction procedure was repeated twice, without the addition of new solvent. The resulting extract was centrifuged at 12,000× *g* for 10 min and 4 °C, and the upper layer stored at −30 °C until analyzed by HPLC. The quantitative analysis of carotenoids was carried out by HPLC according to the method of Mínguez-Mosquera and Hornero-Méndez [21] with minor modifications. The HPLC system consisted of a Waters e2695 Alliance chromatograph equipped with a Waters 2998 photodiode array detector and controlled with Empower2 software Build 2154 (Waters Cromatografía, SA, Barcelona, Spain). A reversed phase C18 column (200 mm × 4.6 mm, 3 μm, Mediterranea SEA18; Teknokroma, Barcelona, Spain), fitted with a guard column of the same material (10 mm × 4.6 mm), was used. Separation was achieved by a binary gradient elution with an initial composition of 75% acetone and 25% deionized water, linearly increased to 95% acetone in 10 min, then held for 7 min, raised to 100% in 3 min, and held constant for 10 min. The initial conditions were reached in 5 min. The column temperature was maintained at 25 °C and the sample compartment was cooled to 15 °C. An injection volume of 10 μL and a flow rate of 1 mL/min were used. Detection was carried out at 450 nm, and online spectra were acquired in the wavelength range of 330 to 650 nm with a resolution of 1.2 nm. Quantification was carried out using external standard calibration curves prepared with previously isolated lycopene and β-carotene standards and purified in our laboratory. Calibration curves were prepared in the range of 0.5–50.0 μg/mL and were constructed by plotting the peak area at 450 nm versus the pigment concentration. All operations were performed under dim light to prevent isomerization and degradation of carotenoids. Each sample was analyzed in triplicate, and a chromatographic analysis was carried out on the same day as the preparation of the extracts. The results are reported as mg/kg dry weight.

2.3.8. Total Polyphenol Determination

The total polyphenols were quantified using the Folin and Ciocalteu method. A 20 μL homogenate sample (diluted 1:2 with distilled water) was mixed with 100 μL Folin and Ciocalteu phenol reagent and with 80 μL 7.5% Na_2CO_3 solution. After incubation for 2 h at room temperature and in the dark, the absorbance was read at 765 nm with a Synergy™ HT-multimode microplate reader (Biotek Instruments, Winooski, VT, USA). A calibration curve using different concentrations of gallic acid (25–250 mg/L) was used to calculate the results as gallic acid equivalents (mg/kg dry weight).

2.4. *Antioxidant Capacity*

All assays were performed using the homogenate of the collected fruits at each stage of ripening.

2.4.1. Ferric Reducing Antioxidant Power (FRAP) Assay

The ferric reducing ability was estimated according to the procedure of Delgado-Andrade et al. [22]. A 20 μL homogenate sample (diluted 1:20 with water) was mixed with 280 μL FRAP solution (0.83 mM TPTZ and 1.66 mM $FeCl_3 \times 6H_2O$ in 0.25 M acetate buffer, pH 3.6). After 30 min of incubation at 37 °C, the absorbance at 595 nm was measured with the microplate reader. The values were extrapolated by a Trolox standard curve.

2.4.2. Trolox Equivalent Antioxidant Capacity (TEAC) Assay

The TEAC assay was performed following the procedure described by Delgado-Andrade et al. [22]. To perform the assay, the ABTS radical solution was used. A total of 280 μL of this solution was mixed with 20 μL of homogenate sample (diluted 1:30 with water). After 30 min of incubation at 30 °C, the ABTS radical content was quantified by a microplate reader at 730 nm. The TEAC values were extrapolated using a Trolox standard curve.

2.4.3. Oxygen Radical Absorbance Capacity (ORAC) Assay

The ORAC assay was performed according to Ou et al. [22], with some minor modifications. A 50 μL homogenate sample (diluted 1:500 with phosphate buffer) was added to 100 μL of sodium fluorescein (2.93 μg/μL) and incubated for 15 min at 37 °C. Then, 50 μL AAPH (60.84 mM) was added to generate peroxyl radicals. Therefore, every 5 min, over 2 h, the decay of fluorescein at its maximum emission of 528 nm was measured using the microplate reader. The area under the curve (AUC) was calculated using a Trolox calibration curve, and the data are expressed as millimole Trolox equivalent/L.

2.4.4. 2,2-Diphenyl-1-picrylhydrazyl (DPPH) Radical Scavenging Activity

The DPPH assay was carried out according to a previous study [22]. A 40 μL homogenate sample (diluted 1:10 with distilled water) were added to 200 μL methanol and mixed with 60 μL DPPH radical solution (0.23 mg/mL). After 1 h of incubation at 30 °C, the DPPH radical was measured at 520 nm with a microplate reader. The results were calculated as a percentage of DPPH radical scavenging activity by the following equation:

$$\%\text{DPPH radical scavenging activity} = \frac{\text{Abs C} - \text{Abs S}}{\text{Abs C}} \times 100$$

where C is the control group (H_2O + methanol + DPPH radical solution) and S is the sample.

2.5. Statistical Analysis

The results are shown as the mean ± standard deviation (SD) and a two-way ANOVA test was applied, followed by multiple comparisons and Tukey's test correction. Correlations between the antioxidant activity assayed by the DPPH radical and the different carotenoids and polyphenol content were analyzed by Pearson's correlation. Differences with a p-value ≤ 0.05 were considered statistically significant. Data were graphed and analyzed with GraphPad Prism 8 (GraphPad Software, San Diego, CA, USA).

3. Results

3.1. Quality Parameters of Tomato Varieties during the Ripening Process

The fruit quality parameters obtained for the three studied cultivars throughout the ripening process are summarized in Table 1. The cherry-like tomato fruits showed a mean weight of 17 g, which contrasted with the salad tomato, presenting a mean fruit weight of up to 244 g. This observation was also in accordance with the size of the fruit (determined as the equatorial diameter) of 31 mm and 83 mm for the cherry-like and salad tomato fruits, respectively. Interestingly, there was a marked difference in firmness between the two types of tomatoes: the average firmness in cherry-like fruits was double compared to that of the salad tomatoes, 21.4 N/g and 10.6 N/g respectively. Finally, although the two types of tomatoes possessed a mean pH of 4, the TSS of the cherry-like tomatoes was around 6.9 °Brix, while that of the salad tomatoes was 3.8 °Brix. In terms of variations during the ripening process, the weight and size of the fruits varied over time, depending on the variety. Although the *Josefina* variety showed an increase in weight and size during the ripening process, the *Muchamiel* variety reduced its weight and size in RS3. The *Karelya* variety, however, did not show changes in weight and size throughout the process. The pH value also evolved differently depending on the tomato variety. The two cherry varieties showed an increase in pH levels throughout the ripening process, while the pH values in the *Muchamiel* variety did not change.

Table 1. Changes in the quality parameters of tomato fruits cultivars *Josefina*, *Karelya*, and *Muchamiel*, at three different ripening stages.

	Josefina			*Karelya*			*Muchamiel*		
	RS1	**RS2**	**RS3**	**RS1**	**RS2**	**RS3**	**RS1**	**RS2**	**RS3**
Weight (g)	15.1 ± 2.81 ^{a}	16.5 ± 3.21 ^{b}	20.6 ± 4.68 ^{c}	17.3 ± 2.49 ^{a}	17.9 ± 1.94 ^{a}	18.2 ± 2.80 ^{a}	254 ± 74.10 ^{a,b}	258 ± 88.00 ^{a}	225 ± 79.50 ^{b}
Size (equatorial diameter, mm)	30.2 ± 2.10 ^{a}	29.3 ± 1.92 ^{b}	33.0 ± 2.78 ^{c}	31.4 ± 2.16 ^{a}	31.6 ± 1.62 ^{a}	31.3 ± 1.68 ^{a}	84.2 ± 10.20 ^{a}	84.6 ± 12.40 ^{a}	80.9 ± 12.10 ^{b}
Firmness (N/g)	27.4 ± 2.32 ^{a}	24.9 ± 4.11 ^{a}	14.5 ± 6.38 ^{b}	25.3 ± 3.75 ^{a}	19.2 ± 2.91 ^{b}	17.0 ± 1.64 ^{b}	15.0 ± 3.88 ^{a}	10.0 ± 1.75 ^{b}	6.74 ± 1.57 ^{c}
pH	3.88 ± 0.01 ^{a}	3.92 ± 0.01 ^{b}	4.07 ± 0.01 ^{c}	3.88 ± 0.09 ^{a}	3.91 ± 0.03 ^{a}	4.09 ± 0.08 ^{b}	4.23 ± 0.06 ^{a}	4.21 ± 0.03 ^{a}	4.23 ± 0.05 ^{a}
TSS (°Brix)	5.97 ± 0.10 ^{a}	6.70 ± 0.06 ^{b}	7.37 ± 0.08 ^{c}	6.25 ± 0.05 ^{a}	6.75 ± 0.05 ^{b}	8.10 ± 0.09 ^{c}	3.49 ± 0.14 ^{a}	3.80 ± 0.12 ^{b}	4.03 ± 0.09 ^{c}
Color (a/b)	0.07 ± 0.13 ^{a}	0.32 ± 0.13 ^{b}	0.75 ± 0.10 ^{c}	0.16 ± 0.10 ^{a}	0.45 ± 0.10 ^{b}	0.79 ± 0.16 ^{c}	0.28 ± 0.15 ^{a}	0.59 ± 0.14 ^{b}	0.81 ± 0.13 ^{c}

RS, ripening stages. For each parameter and each cultivar, different letters indicate a statistically significant difference ($p \leq 0.05$) among ripening stages.

The three tomato varieties showed a significant reduction in the firmness values throughout the ripening process, while the TSS and color increased significantly, reaching the highest levels in RS3. The color changes along the ripening process are shown in Figure 1.

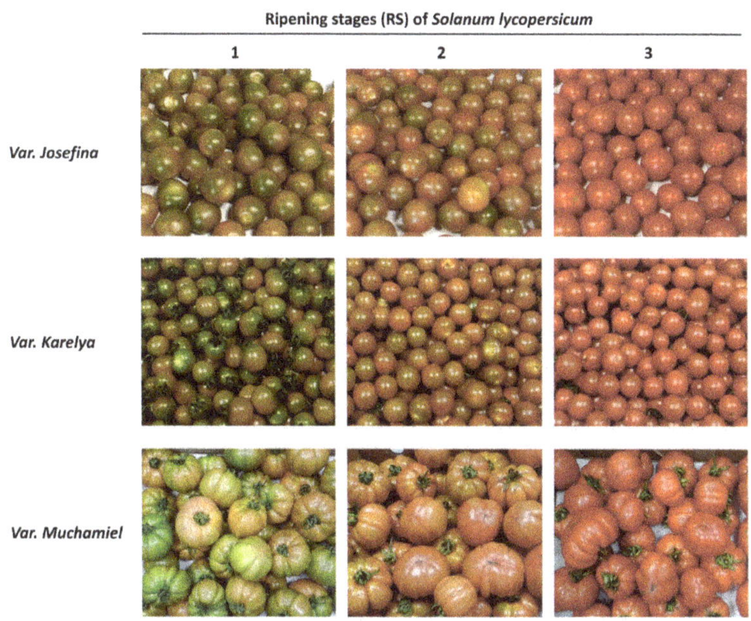

Figure 1. Selected ripening stages (RS), based on fruit color, for the studied cultivars (*Josefina*, *Karelya*, and *Muchamiel*).

3.2. Nutritional Composition of Tomato Fruits during the Ripening Process

As shown in Table 2, water was the main constituent of the fruits of the three varieties of tomatoes, with *Muchamiel* being the variety that presented the highest moisture content (95.50 ± 0.84%, value in RS1), which was significantly different from the moisture content of the *Josefina* (RS1: 92.9 ± 0.55%, $p < 0.0001$) and *Karelya* (RS1: 91.9 ± 0.33%, $p < 0.0001$) varieties. Furthermore, carbohydrates were the most representative macronutrients in tomatoes, followed by proteins and lipids. The *Karelya* variety showed significantly higher levels of proteins (0.27 ± 0.02%) and carbohydrates (9.11 ± 0.03%) in RS3 compared to *Josefina* (0.20 ± 0.01%, $p = 0.029$; 7.67 ± 0.19%, $p < 0.0001$, respectively) and *Muchamiel* (0.22 ± 0.03%, $p = 0.044$; 3.91 ± 0.94%, $p = 0.05$, respectively). *Muchamiel* was the variety with the highest levels of lipids (0.19 ± 0.02%). No significant differences in lipid content were observed between *Muchamiel* and *Josefina* ($p = 0.166$) or *Karelya* ($p = 0.258$). Regarding minerals, the *Karelya* variety had the highest levels in RS3 (0.46 ± 0.02), which was significantly higher than the values of *Muchamiel* (0.34 ± 0.02%, $p < 0.0001$) and *Josefina* (0.24 ± 0.03%, $p < 0.0001$).

Table 2. Proximate composition of tomato fruits (cultivars *Josefina*, *Karelya*, and *Muchamiel*) at three different ripening stages.

	Josefina			*Karelya*			*Muchamiel*		
	RS1	**RS2**	**RS3**	**RS1**	**RS2**	**RS3**	**RS1**	**RS2**	**RS3**
Moisture (%)	92.90 ± 0.55 a	92.40 ± 0.51 a,b	91.10 ± 1.68 b	91.90 ± 0.33 a	91.50 ± 0.12 b	90.00 ± 0.21 c	95.50 ± 0.84 a	95.70 ± 0.36 a	95.30 ± 0.77 a
Proteins (%)	0.19 ± 0.02 a	0.18 ± 0.02 a	0.20 ± 0.01 a	0.20 ± 0.01 a	0.23 ± 0.01 a	0.27 ± 0.02 b	0.20 ± 0.02 a	0.19 ± 0.02 a	0.22 ± 0.03 a
Lipids (%)	0.10 ± 0.02 a	0.13 ± 0.02 a	0.16 ± 0.02 b	0.11 ± 0.02 a	0.12 ± 0.01 a	0.17 ± 0.03 b	0.15 ± 0.04 a	0.14 ± 0.02 a	0.19 ± 0.02 b
Carbohydrates (%)	6.54 ± 0.10 a	7.29 ± 0.15 b	7.67 ± 0.19 b	7.40 ± 0.17 a	7.89 ± 0.09 b	9.11 ± 0.01 c	3.70 ± 1.00 a	3.56 ± 0.47 a	3.91 ± 0.94 a
Minerals (%)	0.23 ± 0.02 a	0.22 ± 0.03 a	0.24 ± 0.03 a	0.38 ± 0.02 a	0.40 ± 0.03 a	0.46 ± 0.02 b	0.35 ± 0.08 a	0.33 ± 0.02 a	0.34 ± 0.02 a

RS, ripening stages. For each parameter and each cultivar, different letters indicate a statistically significant difference ($p \leq 0.05$) among ripening stages.

Similarly, the nutritional composition of the fruits of each tomato variety varied throughout the ripening process (Table 2). The water content showed a statistically significant reduction in RS3 for both cherry varieties compared to RS1 (*Josefina*: $p = 0.027$; *Karelya*: $p < 0.0001$), while the *Muchamiel* variety did not show changes in the value of the moisture content (RS1 compared to RS3 $p = 0.789$). In terms of macronutrients, the three varieties showed a statistically significant increase ($p < 0.05$) in the lipid content during the ripening process, while the carbohydrate content increased only in the cherry varieties (*Josefina*: $p = 0.0002$; *Karelya*: $p < 0.0001$) but not in the *Muchamiel* variety ($p > 0.05$). Furthermore, *Karelya* was the only variety to show a statistically significant increase ($p = 0.003$) in protein levels during the ripening process. However, the mineral content increased significantly in *Karelya* ($p = 0.016$), without any changes in the other two varieties ($p > 0.05$).

3.3. Evolution of Polyphenol Content in Tomato Varieties during the Ripening Process

As shown in Table 3, the three tomato varieties showed an increase in the total polyphenol levels throughout the ripening process. In particular, *Josefina* and *Karelya* varieties showed a significant increase in total polyphenols already in RS2 of the ripening process (*Josefina*: $p = 0.022$; *Karelya*: $p = 0.003$, with respect to RS1), without significant differences between RS2 and RS3. However, the *Muchamiel* variety did not show a significant increase in the levels of the total polyphenol content in RS2 ($p = 0.725$ compared to the RS1 stage), but it increased its polyphenol content in the advanced RS3 stage (compared to RS1, $p = 0.0001$; and RS2, $p = 0.003$).

Table 3. Polyphenol and carotenoid content of the *Josefina*, *Karelya*, and *Muchamiel* tomato varieties at the different ripening status.

	Josefina			*Karelya*			*Muchamiel*		
	RS1	**RS2**	**RS3**	**RS1**	**RS2**	**RS3**	**RS1**	**RS2**	**RS3**
Total polyphenols (mg/kg dry weight)	106.75 ± 12.38 a	118.38 ± 4.71 b	120.04 ± 7.50 b	84.86 ± 5.05 a	101.59 ± 4.07 b	109.56 ± 9.58 b	60.10 ± 3.37 a	60.99 ± 1.87 a	65.24 ± 2.98 b
Total carotenoids (mg/kg dry weight)	97.80 ± 9.10 a	178.00 ± 9.11 b	310.00 ± 23.90 c	82.70 ± 3.78 a	135.00 ± 8.52 b	294.00 ± 26.00 c	133.00 ± 9.62 a	271.00 ± 23.20 b	451.00 ± 28.90 c
Lycopene (mg/kg dry weight)	42.30 ± 3.77 a	103.00 ± 5.62 b	236.00 ± 18.60 c	41.40 ± 2.27 a	88.00 ± 5.43 b	245.00 ± 22.30 c	90.00 ± 7.69 a	220.00 ± 19.10 b	392.00 ± 28.60 c
β-Carotene (mg/kg dry weight)	55.50 ± 5.56 a	74.30 ± 3.84 b	73.80 ± 5.38 b	41.40 ± 1.61 a	46.90 ± 3.32 b	48.30 ± 3.75 b	42.90 ± 2.87 a	51.10 ± 9.19 b	59.50 ± 12.20 c

For each parameter and each cultivar, different letters indicate a statistically significant difference ($p \leq 0.05$) among ripening stages (RS).

3.4. Changes in the Carotenoid Levels in Tomato Fruits during the Ripening Process

Variations in total and individual carotenoid contents (total carotenoid, lycopene, and β-carotene) were evaluated (Table 3). Therefore, the fruits of the three tomato varieties showed a significant increase ($p < 0.0001$) in total carotenoids during the ripening process, reaching the highest levels in RS3 (Table 3). In more detail, the ripening process allowed for a significant increase ($p < 0.0001$) in lycopene levels in all three varieties, reaching the highest levels in the RS3 stage (236 ± 18.60 mg/kg dry weight, 245 ± 22.30 mg/kg dry weight, and 392 ± 28.60 mg/kg dry weight, respectively). On the other hand, β-carotene levels increased differently. While the *Muchamiel* variety reached its highest β-carotene levels in RS3 (59.50 ± 12.20) ($p < 0.0001$), the *Josefina* and *Karelya* varieties reached the highest levels in RS2 and RS3, respectively (*Josefina*: 74.30 ± 3.84; $p < 0.0001$; *Karelya*: 48.30 ± 3.75 $p = 0.0003$), without observed differences between stages RS2 and RS3 (*Josefina*: $p = 0.968$; *Karelya*: $p = 0.507$).

3.5. Effect of Ripening Process on Antioxidant Capacity of Tomato Varieties

To find out if the antioxidant capacity of tomatoes increased during the ripening process, the FRAP, ORAC, TEAC, and DPPH antioxidant tests were carried out at each ripening stage.

As observed in Figure 2A, the FRAP value increased in the RS2 and RS3 stages, reaching the maximum value in RS3 in the three tomato varieties studied (RS3: *Josefina*: 2.70 ± 0.08 mmol/L; *Karelya*: 4.07 ± 0.22 mmol/L; and *Muchamiel*: 1.56 ± 0.05 mmol/L, $p < 0.0001$ with respect to RS1). As shown in Figure 2B, the capacity to scavenge the ABTS radical (TEAC assay) increased throughout the ripening process for the *Josefina* (RS1: 3.72 ± 0.19 mmol/L; RS2: 4.17 ± 0.10 mmol/L; RS3: 4.72 ± 0.13 mmol/L) and *Muchamiel* (RS1: 1.52 ± 0.07 mmol/L; RS2: 1.81 ± 0.06 mmol/L; RS3: 2.14 ± 0.09 mmol/L) varieties. However, the ABTS scavenging capacity decreased significantly in the *Karelya* variety (RS1: 5.74 ± 0.44 mmol/L; RS2: 4.52 ± 0.21 mmol/L; RS3: 5.26 ± 0.19 mmol/L). The results of the ORAC assay show how the oxygen radical absorption capacity increased in RS2 (*Josefina*: $p = 0.002$; *Karelya*: $p = 0.005$) and RS3 (*Josefina*: $p = 0.001$; *Karelya*: $p = 0.0002$) for the cherry varieties, without significant differences (*Josefina*: $p = 0.934$; *Karelya*: $p = 0.194$) between these stages (Figure 2C). However, the *Muchamiel* variety did not show significant differences in ORAC at any ripening stage (Figure 2C).

Regarding DPPH radical scavenging activity, the varieties *Josefina* and *Karelya* increased their activity in stages RS2 (*Josefina*: 54.8 ± 1.22%, $p = 0.006$; *Karelya*: 66.4 ± 2.93%, $p = 0.03$ with respect to RS1) and RS3 (*Josefina*: 55.4 ± 2.15%, $p = 0.005$; *Karelya*: 69.7 ± 5.69%, $p = 0.001$ with respect to RS1), without significant differences were found between them (Figure 2D). However, the *Muchamiel* variety significantly ($p = 0.0009$) increased the DPPH radical scavenging capacity in RS3 (35.2 ± 4.35%), with no significant differences being found between stages RS1 (26.8 ± 1.90%) and RS2 (29.3 ± 6.72) (Figure 2D).

Regarding the differences in the antioxidant capacity between varieties, both cherry tomatoes stood out for their higher antioxidant activity compared to the *Muchamiel* salad tomato. Among the cherry varieties, *Karelya* showed the highest antioxidant activity in all assays, except for ORAC.

3.6. Correlation between Bioactive Compounds and Antioxidant Activity

To identify a possible relationship between the carotenoid or polyphenol content and the antioxidant capacity, the data were analyzed using Pearson's correlation. As shown in Table 4, in the *Josefina* variety, significant and positive correlations were observed between the levels of total carotenoids, β-carotene, and lycopene with the values of FRAP, TEAC, and ORAC, as well as negative correlations between these and DPPH.

In contrast, the total polyphenols showed positive correlations with FRAP and TEAC values, but no statistically significant correlation was observed with ORAC and DPPH. In the *Karelya* variety (Table 5), the total carotenoids, β-carotene, and lycopene showed positive and significant correlations with FRAP and ORAC values and negative correlations with DPPH values. However, the carotenoid levels did not show a significant correlation with TEAC.

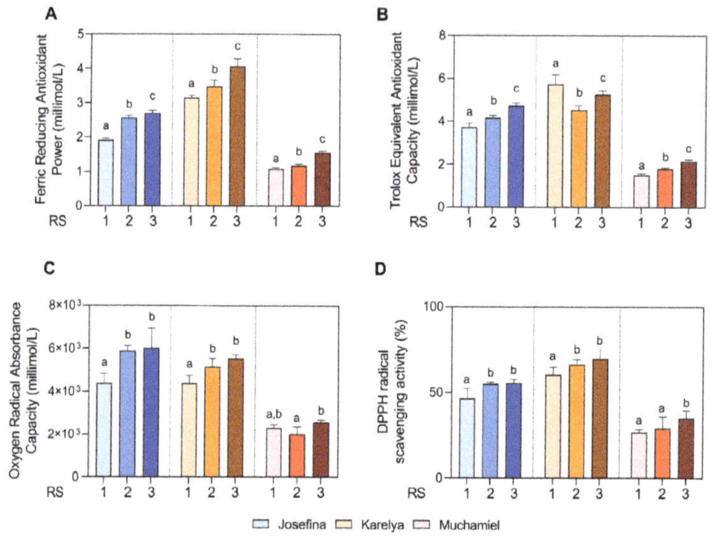

Figure 2. Antioxidant activity in the three tomato varieties according to ripening stage using different assays: (**A**) Ferric Reducing Antioxidant Power. (**B**) Trolox Equivalent Antioxidant Capacity. (**C**) Oxygen Radical Absorbance Capacity. (**D**) DPPH (2,2-diphenyl-1-picrylhydrazyl). RS, ripening stage. Different letters represent a statistical difference ($p \leq 0.05$) within each variety.

Table 4. Pearson correlations between antioxidant activity and carotenoid and polyphenol contents in tomato fruits from the *Josefina* variety.

	Total Polyphenols		Total Carotenoids		β-Carotene		Lycopene	
	r²	*p*-Value	r²	*p*-Value	r²	*p*-Value	r²	*p*-Value
vs. FRAP	0.541	**0.020**	0.865	**<0.0001**	0.899	**<0.0001**	0.828	**<0.0001**
vs. TEAC	0.469	**0.049**	0.959	**<0.0001**	0.718	**0.0008**	0.952	**<0.0001**
vs. ORAC	0.271	0.277	0.640	**0.004**	0.633	**0.005**	0.616	**0.006**
vs. DPPH radical	−0.048	0.855	−0.521	**0.026**	−0.528	**0.029**	−0.515	**0.029**

DPPH, 2,2-Diphenyl-1-picrylhydrazyl radical; FRAP, ferric reducing antioxidant power; ORAC, oxygen radical absorbance capacity; r², Pearson correlation coefficient; TEAC, Trolox equivalent antioxidant capacity. The bold number indicates a statistically significant difference.

Table 5. Pearson correlations between antioxidant activity and carotenoid and polyphenol contents in tomato fruits from the *Karelya* variety.

	Total Polyphenols		Total Carotenoids		β-Carotene		Lycopene	
	r²	*p*-Value	r²	*p*-Value	r²	*p*-Value	r²	*p*-Value
vs. FRAP	0.711	**0.0009**	0.921	**<0.0001**	0.585	**0.017**	0.922	**<0.0001**
vs. TEAC	−0.431	0.074	−0.112	0.659	−0.47	0.064	−0.085	0.736
vs. ORAC	0.718	**0.002**	0.782	**0.0003**	0.669	**0.005**	0.769	**0.0005**
vs. DPPH radical	−0.377	0.123	−0.576	**0.012**	−0.518	**0.028**	−0.560	**0.016**

DPPH, 2,2-Diphenyl-1-picrylhydrazyl radical; FRAP, ferric reducing antioxidant power; ORAC, oxygen radical absorbance capacity; r², Pearson correlation coefficient; TEAC, Trolox equivalent antioxidant capacity. The bold number indicates a statistically significant difference.

In the case of polyphenols, positive and significant correlations were observed with FRAP and ORAC, but not with DPPH and ABTS. For the *Muchamiel* variety (Table 6), positive and significant correlations were observed between total polyphenols, total carotenoids, β-carotene, and lycopene and FRAP and TEAC. Regarding the other assays, a positive correlation was observed between β-carotene and ORAC values, while negative correlations were found between total carotenoid and lycopene levels and DPPH radical values.

Table 6. Pearson correlations between antioxidant activity and carotenoid and polyphenol contents in tomato fruits from the *Muchamiel* variety.

	Total Polyphenols		Total Carotenoids		β-Carotene		Lycopene	
	r^2	*p*-Value	r^2	*p*-Value	r^2	*p*-Value	r^2	*p*-Value
vs. FRAP	0.674	**0.002**	0.950	**<0.0001**	0.935	**<0.0001**	0.939	**<0.0001**
vs. TEAC	0.595	**0.019**	0.955	**<0.0001**	0.819	**0.0002**	0.957	**<0.0001**
vs. ORAC	0.495	0.072	0.410	0.145	0.598	**0.024**	0.384	0.175
vs. DPPH radical	0.225	0.186	−0.387	**0.032**	−0.119	0.489	−0.391	**0.030**

DPPH, 2,2-Diphenyl-1-picrylhydrazyl radical; FRAP, ferric reducing antioxidant power; ORAC, oxygen radical absorbance capacity; r^2, Pearson correlation coefficient; TEAC, Trolox equivalent antioxidant capacity. The bold number indicates a statistically significant difference.

4. Discussion

This study evaluated the quality, nutritional, and antioxidant variations during the ripening process of three Spanish tomato varieties (including two cherry-like and one salad tomatoes) were evaluated. Tomatoes were grown under organic farming conditions and evaluated at three different stages of ripening, according to the color of the fruit. The ripening process was shown to affect each tomato variety differently in terms of quality parameters, nutritional properties, and antioxidant activities.

First, the quality of each tomato fruit was assessed by analyzing the pH, TSS, and moisture content. The pH values, which determine the acidity of the fruit, were similar to those observed in other studies [23,24]. The TSS is the index that most influences the production performance of fruits for industrial processing. This depends on the variety and the agronomic conditions, including the irrigation and ripening period. In this sense, tomato fruits grown under organic conditions have been observed to have higher TSS values (3.90 °Brix) compared to those fruits grown under conventional conditions (3.21 °Brix) [6]. The tomatoes in our study showed an increase in TSS throughout the ripening process. However, the observed levels were comparable to those previously documented for other tomato varieties [23,25,26], regardless of the cultivation method employed (organic or conventional). Finally, it was found that the content of dry matter, which is essential for the quality of the fruit and has a positive effect on the organoleptic characteristics of the product, is within the normal range from 5 to 7.5% [27]. The rest of the quality parameters, including weight, size, color, and firmness, showed values similar to those previously described in the same or others tomato varieties [28,29], except the weight of the *Karelya* and *Muchamiel* varieties, which were higher than those established by other authors [30,31].

In terms of nutritional analysis, the levels of carbohydrates, lipids, and minerals were similar to those of other previously characterized tomato cultivars [32–34]. However, the protein fraction was slightly below average levels, probably due to the tomato varieties themselves and/or the effect of cultivation under organic conditions. Additionally, the ripening process did not affect the nutritional composition of the *Muchamiel* variety, which only showed a slight increase in lipid fraction in RS3. However, cherry varieties, in addition to showing a decrease in moisture in RS2, showed an increase in protein, carbohydrate, and mineral contents, in some cases (such as carbohydrates) already in the RS2 stage.

Polyphenol and carotenoids levels increased significantly during the ripening process. This is of great interest as the functional value of tomatoes is mainly determined by the content of carotenoids and polyphenols, which have attracted the interest of many researchers in the field because of their biological and physicochemical properties. Lycopene is the main carotenoid in tomatoes and is responsible for their characteristic red color. This bioactive component, along with other carotenoids, has been shown to inhibit cancer cell proliferation, as well as exert a powerful antioxidant effect, of great importance in counteracting the development of numerous diseases, such as cardiovascular disease and those related to age [35–38]. In this sense, the carotenoid content has been shown to depend on numerous factors such as the stage of ripening, the variety of tomato, or the cultivation conditions. With regard to this last factor, the lycopene content of tomatoes grown under organic conditions has been shown to be higher than those grown under conventional practices [6]. In our study, the levels of lycopene, β-carotene, and total carotenoids in the ripening stage RS3 were in agreement with those of previous studies [1,15,24,26,39,40].

Furthermore, total polyphenol levels were comparable to those observed in previous studies [40–42] using conventional cultivation conditions. Our results show that tomato ripening significantly increases the levels of bioactive compounds, thus enhancing the functional value of tomatoes. Our results can also indicate the optimal harvesting time of tomatoes according to the bioactive compound of interest to optimize its content (Table 7).

Table 7. Schematic summary of the ripening stages of the three varieties in which the tomatoes could be harvested in order to obtain a specific characteristic (more antioxidant, with more polyphenols, etc.).

Bioactive Compounds	Tomato Varieties		
	Cherry-like		Salad
	Josefina	*Karelya*	*Muchamiel*
Total polyphenol index	RS3	RS3	RS3
Total carotenoids	RS3	RS3	RS3
Lycopene	RS3	RS3	RS3
β-Carotene	RS2	RS3	RS3
Antioxidant capacity			
Total antioxidant activity (FRAP)	RS3	RS3	RS3
Radical scavenging capacity (TEAC, ORAC, DPPH)	RS2	RS2	RS3

DPPH, 2,2-Diphenyl-1-picrylhydrazyl radical; FRAP, ferric reducing antioxidant power; ORAC, oxygen radical absorbance capacity; RS, ripening status, TEAC, Trolox equivalent antioxidant capacity.

To evaluate the antioxidant activity of tomatoes during the ripening process, four antioxidant assays were used: FRAP, to measure the total antioxidant status, and ORAC, TEAC, and DPPH to evaluate the radical scavenging capacity. The correlation analysis showed how the exerted antioxidant effect depends on each bioactive compound and the different types of antioxidant assays used. In fact, due to the chemical nature of the method used to measure the antioxidant activity, it is important to use several methods to evaluate a possible effect mediated by different mechanisms of action. While the TEAC and DPPH assays measure the antioxidant capacity exerted by water-soluble (polyphenols) and lipid-soluble (carotenoids) compounds, the ORAC and FRAP assays can only measure those that are water-soluble. Our results show that the global antioxidant status increased significantly during the ripening process, reaching the highest levels in RS3. Although other authors have studied the antioxidant capacity of several tomato varieties [15,24,33,40,43], this is the first time that the antioxidant activity was evaluated in the *Karelya*, *Josefina*, and *Muchamiel* varieties at three different stages of ripening and grown under organic conditions. The total antioxidant capacity has been shown to be higher in tomatoes that have been grown under organic conditions compared to those grown following conventional practices [34,44]. This is consistent with the findings of Vinha et al. [6], who reported DPPH inhibition values in tomatoes grown under organic practices (62.1 $\pm$ 1.3%) similar to those observed in the present study. Both values were higher than those observed for tomatoes grown using conventional cultivation methods (58.4 $\pm$ 0.7%).

Considering the optimal time to harvest the fruit based on its maximum antioxidant capacity, harvesting should be conducted when the tomatoes are fully ripe (stage RS3). However, in some cases, harvesting could be brought forward depending on whether the final goal is to obtain the greatest scavenging activity. In this sense, cherry tomatoes reached their highest scavenging activity in RS2, while salad tomatoes had their highest values of

DPPH, ORAC, and TEAC in RS3. Acquiring this knowledge would lead to optimizing the antioxidant activity and scavenging capacity of tomatoes according to the variety.

5. Conclusions

This study examines the various effects of ripening on different tomato varieties, including their impact on quality, nutritional content, and antioxidant levels. As expected, ripening significantly improves the levels of carotenoids and polyphenols in the varieties studied, thus improving the functional value of tomatoes and their antioxidant activity, especially in fully ripe tomatoes. Harvesting tomatoes during ripening stage 3 maximizes the bioactive content and antioxidant capacity in salad tomatoes, but not in cherry tomato varieties. In particular, the TEAC, ORAC, and DPPH assays showed a higher antioxidant capacity in RS2 in the *Josefina* and *Karelya* varieties. The results obtained in this study provide useful information on the optimal ripening times for the tomato varieties studied, which can be used to guarantee their quality even before harvest.

Author Contributions: Conceptualization, I.C., M.-S.F.-P. and D.H.-M.; design of experiments, D.H.-M., M.-S.F.-P. and I.C.; methodology, D.H.-M. and I.C.; investigation, D.H.-M., I.C.-C. and G.S.-S., data analysis, I.C.-C., G.S.-S., M.-S.F.-P. and I.C.; writing—original draft preparation, I.C.-C. and G.S.-S.; writing—review and editing, D.H.-M., M.-S.F.-P. and I.C.; funding acquisition, D.H.-M., F.M., M.-S.F.-P. and I.C. All authors have read and agreed to the published version of the manuscript.

Funding: The research was funded by the Centro para el Desarrollo Tecnológico Industrial-CDTI (2019/00064/001) and the PAIDI Program from the Andalusian Government (BIO311). G.S.-S. was supported by Ministerio de Ciencia e Innovación, Gobierno de España (JDC2022-048411-I). I.C.-C. was supported by the grant I-006-2024 (Universidad de Almería, Spain).

Institutional Review Board Statement: Not applicable.

Informed Consent Statement: Not applicable.

Data Availability Statement: The original contributions presented in the study are included in the article, further inquiries can be directed to the corresponding author.

Acknowledgments: The authors thank the company Bioalverde S.L. for providing the samples.

Conflicts of Interest: The authors declare no conflicts of interest.

References

1. Fattore, M.; Montesano, D.; Pagano, E.; Teta, R.; Borrelli, F.; Mangoni, A.; Seccia, S.; Albrizio, S. Carotenoid and flavonoid profile and antioxidant activity in "Pomodorino Vesuviano" tomatoes. *J. Food Compos. Anal.* **2016**, *53*, 61–68. [CrossRef]
2. Bertin, N.; Génard, M. Tomato quality as influenced by preharvest factors. *Sci. Hortic.* **2018**, *233*, 264–276. [CrossRef]
3. Arora, N.K.; Mishra, I. United Nations Sustainable Development Goals 2030 and environmental sustainability: Race against time. *Environ. Sustain.* **2019**, *2*, 339–342. [CrossRef]
4. Ali, S.; Ullah, M.I.; Sajjad, A.; Shakeel, Q.; Hussain, A. Environmental and health effects of pesticide residues. In *Sustainable Agriculture Reviews 48: Pesticide Occurrence, Analysis and Remediation Vol. 2 Analysis*; Springer: Cham, Switzerland, 2021; pp. 311–336.
5. Collins, E.; Bowyer, C.; Tsouza, A.; Chopra, M. Tomatoes: An Extensive Review of the Associated Health Impacts of Tomatoes and Factors That Can Affect Their Cultivation. *Biology* **2022**, *11*, 239. [CrossRef] [PubMed]
6. Vinha, A.F.; Barreira, S.V.; Costa, A.S.; Alves, R.C.; Oliveira, M.B.P. Organic versus conventional tomatoes: Influence on physicochemical parameters, bioactive compounds and sensorial attributes. *Food Chem. Toxicol.* **2014**, *67*, 139–144. [CrossRef]
7. Khaltaev, N.; Axelrod, S. Cardiovascular disease mortality and air pollution in countries with different socioeconomic status. *Chronic Dis. Transl. Med.* **2024**, *10*, 247–255. [CrossRef]
8. Sharifi-Rad, M.; Anil Kumar, N.V.; Zucca, P.; Varoni, E.M.; Dini, L.; Panzarini, E.; Rajkovic, J.; Tsouh Fokou, P.V.; Azzini, E.; Peluso, I. Lifestyle, oxidative stress, and antioxidants: Back and forth in the pathophysiology of chronic diseases. *Front. Physiol.* **2020**, *11*, 694. [CrossRef]
9. Seyedsadjadi, N.; Grant, R. The potential benefit of monitoring oxidative stress and inflammation in the prevention of non-communicable diseases (NCDs). *Antioxidants* **2020**, *10*, 15. [CrossRef]
10. Kumar, K.S.; Paswan, S.; Srivastava, S. Tomato-a natural medicine and its health benefits. *J. Pharmacogn. Phytochem.* **2012**, *1*, 33–43.
11. Dumas, Y.; Dadomo, M.; Di Lucca, G.; Grolier, P. Effects of environmental factors and agricultural techniques on antioxidantcontent of tomatoes. *J. Sci. Food Agric.* **2003**, *83*, 369–382. [CrossRef]

12. Granato, D.; Barba, F.J.; Bursać Kovačević, D.; Lorenzo, J.M.; Cruz, A.G.; Putnik, P. Functional foods: Product development, technological trends, efficacy testing, and safety. *Annu. Rev. Food Sci. Technol.* **2020**, *11*, 93–118. [CrossRef]
13. Madia, V.N.; De Vita, D.; Ialongo, D.; Tudino, V.; De Leo, A.; Scipione, L.; Di Santo, R.; Costi, R.; Messore, A. Recent advances in recovery of lycopene from tomato waste: A potent antioxidant with endless benefits. *Molecules* **2021**, *26*, 4495. [CrossRef]
14. Ashraf, W.; Latif, A.; Lianfu, Z.; Jian, Z.; Chenqiang, W.; Rehman, A.; Hussain, A.; Siddiquy, M.; Karim, A. Technological advancement in the processing of lycopene: A review. *Food Rev. Int.* **2022**, *38*, 857–883. [CrossRef]
15. Kaur, C.; Walia, S.; Nagal, S.; Walia, S.; Singh, J.; Singh, B.B.; Saha, S.; Singh, B.; Kalia, P.; Jaggi, S. Functional quality and antioxidant composition of selected tomato (*Solanum lycopersicum* L) cultivars grown in Northern India. *LWT-Food Sci. Technol.* **2013**, *50*, 139–145. [CrossRef]
16. AOAC International. *Official Methods of Analysis of the AOAC International*, 18th ed.; The Association: Arlington County, VA, USA, 2005.
17. AOAC International. *Official Methods of Analysis of the AOAC International*, 17th ed.; The Association: Arlington County, VA, USA, 2000.
18. Bradstreet, R.B. Kjeldahl method for organic nitrogen. *Anal. Chem.* **1954**, *26*, 185–187. [CrossRef]
19. De Castro, M.L.; Garcıa-Ayuso, L. Soxhlet extraction of solid materials: An outdated technique with a promising innovative future. *Anal. Chim. Acta* **1998**, *369*, 1–10. [CrossRef]
20. de los Santos, B.; García-Serrano, P.; Romero, C.; Aguado, A.; García-García, P.; Hornero-Méndez, D.; Brenes, M. Effect of fertilisation with black table olive wastewater solutions on production and quality of tomatoes cultivated under open field conditions. *Sci. Total Environ.* **2021**, *790*, 148053. [CrossRef]
21. Mínguez-Mosquera, M.I.; Hornero-Méndez, D. Separation and quantification of the carotenoid pigments in red peppers (*Capsicum annuum* L.), paprika, and oleoresin by reversed-phase HPLC. *J. Agric. Food Chem.* **1993**, *41*, 1616–1620. [CrossRef]
22. Ou, B.; Hampsch-Woodill, M.; Prior, R.L. Development and validation of an improved oxygen radical absorbance capacity assay using fluorescein as the fluorescent probe. *J. Agric. Food Chem.* **2001**, *49*, 4619–4626. [CrossRef] [PubMed]
23. Klunklin, W.; Savage, G. Effect on quality characteristics of tomatoes grown under well-watered and drought stress conditions. *Foods* **2017**, *6*, 56. [CrossRef] [PubMed]
24. Vinha, A.F.; Alves, R.C.; Barreira, S.V.; Castro, A.; Costa, A.S.; Oliveira, M.B.P. Effect of peel and seed removal on the nutritional value and antioxidant activity of tomato (*Lycopersicon esculentum* L.) fruits. *LWT-Food Sci. Technol.* **2014**, *55*, 197–202. [CrossRef]
25. Flores, P.; Sánchez, E.; Fenoll, J.; Hellín, P. Genotypic variability of carotenoids in traditional tomato cultivars. *Food Res. Int.* **2017**, *100*, 510–516. [CrossRef]
26. Lahoz García, I.; Leiva-Brondo, M.; Martí, R.; Macua, J.I.; Campillo, C.; Roselló, S.; Cebolla-Cornejo, J. Influence of high lycopene varieties and organic farming on the production and quality of processing tomato. *Sci. Hortic.* **2016**, *204*, 128–137. [CrossRef]
27. Stoyanova, A.; Veleva, P.; Valkova, E.; Pevicharova, G.; Georgiev, M.; Valchev, N. Dry matter content and organic acids in Tomatoes, Greenhouse grown under different Manuring and Irrigation Modes. In Proceedings of the 2nd International Conference on Food and Agricultural Economics, Alanya, Turkey, 27–28 April 2018; pp. 257–265.
28. Quamruzzaman, A.; Islam, F.; Akter, L.; Mallick, S.R. Effect of Maturity Indices on Growth and Quality of High Value Vegetables. *Am. J. Plant Sci.* **2022**, *13*, 1042–1062. [CrossRef]
29. Skolik, P.; Morais, C.L.; Martin, F.L.; McAinsh, M.R. Determination of developmental and ripening stages of whole tomato fruit using portable infrared spectroscopy and Chemometrics. *BMC Plant Biol* **2019**, *19*, 236. [CrossRef]
30. Alfosea-Simón, M.; Simón-Grao, S.; Zavala-Gonzalez, E.A.; Navarro-Morillo, I.; Martínez-Nicolás, J.J.; Alfosea-Simón, F.J.; Simon, I.; García-Sánchez, F. Ionomic, metabolic and hormonal characterization of the phenological phases of different tomato genotypes using omics tools. *Sci. Hortic.* **2022**, *293*, 110697. [CrossRef]
31. Pérez-Marín, J.; Issa-Issa, H.; Clemente-Villalba, J.; García-Garví, J.M.; Hernández, F.; Carbonell-Barrachina, Á.A.; Calín-Sánchez, Á.; Noguera-Artiaga, L. Physicochemical, volatile, and sensory characterization of promising cherry tomato (*Solanum lycopersicum* L.) cultivars: Fresh market aptitudes of pear and round fruits. *Agronomy* **2021**, *11*, 618. [CrossRef]
32. Guil-Guerrero, J.; Rebolloso-Fuentes, M. Nutrient composition and antioxidant activity of eight tomato (*Lycopersicon esculentum*) varieties. *J. Food Compos. Anal.* **2009**, *22*, 123–129. [CrossRef]
33. Pinela, J.; Barros, L.; Carvalho, A.M.; Ferreira, I.C. Nutritional composition and antioxidant activity of four tomato (*Lycopersicon esculentum* L.) farmer'varieties in Northeastern Portugal homegardens. *Food Chem. Toxicol.* **2012**, *50*, 829–834. [CrossRef]
34. de Lima, D.P.; Júnior, E.d.S.P.; de Menezes, A.V.; de Souza, D.A.; de São José, V.P.B.; da Silva, B.P.; de Almeida, A.Q.; de Carvalho, I.M.M. Chemical composition, minerals concentration, total phenolic compounds, flavonoids content and antioxidant capacity in organic and conventional vegetables. *Food Res. Int.* **2024**, *175*, 113684. [CrossRef] [PubMed]
35. Przybylska, S. Lycopene–a bioactive carotenoid offering multiple health benefits: A review. *Int. J. Food Sci. Technol.* **2020**, *55*, 11–32. [CrossRef]
36. Przybylska, S.; Tokarczyk, G. Lycopene in the prevention of cardiovascular diseases. *Int. J. Mol. Sci.* **2022**, *23*, 1957. [CrossRef] [PubMed]
37. Jiménez Bolaño, D.C.; Insuasty, D.; Rodríguez Macías, J.D.; Grande-Tovar, C.D. Potential Use of Tomato Peel, a Rich Source of Lycopene, for Cancer Treatment. *Molecules* **2024**, *29*, 3079. [CrossRef] [PubMed]
38. Carvalho, G.C.; de Camargo, B.A.F.; de Araújo, J.T.C.; Chorilli, M. Lycopene: From tomato to its nutraceutical use and its association with nanotechnology. *Trends Food Sci. Technol.* **2021**, *118*, 447–458. [CrossRef]
39. Erba, D.; Casiraghi, M.C.; Ribas-Agustí, A.; Cáceres, R.; Marfà, O.; Castellari, M. Nutritional value of tomatoes (*Solanum lycopersicum* L.) grown in greenhouse by different agronomic techniques. *J. Food Compos. Anal.* **2013**, *31*, 245–251. [CrossRef]

40. Ilahy, R.; Hdider, C.; Lenucci, M.S.; Tlili, I.; Dalessandro, G. Phytochemical composition and antioxidant activity of high-lycopene tomato (*Solanum lycopersicum* L.) cultivars grown in Southern Italy. *Sci. Hortic.* **2011**, *127*, 255–261. [CrossRef]
41. Riahi, A.; Hdider, C. Bioactive compounds and antioxidant activity of organically grown tomato (*Solanum lycopersicum* L.) cultivars as affected by fertilization. *Sci. Hortic.* **2013**, *151*, 90–96. [CrossRef]
42. Vallverdú-Queralt, A.; Medina-Remón, A.; Casals-Ribes, I.; Lamuela-Raventos, R.M. Is there any difference between the phenolic content of organic and conventional tomato juices? *Food Chem.* **2012**, *130*, 222–227. [CrossRef]
43. Choi, H.-G.; Park, K.-S. Ripening Process of Tomato Fruits Postharvest: Impact of Environmental Conditions on Quality and Chlorophyll a Fluorescence Characteristics. *Horticulturae* **2023**, *9*, 812. [CrossRef]
44. Peralta, M.I.; Fuentes, K.N.; Canalis, A.M.; Soria, E.A. Effect of cultivation method and processing on total polyphenols content and antioxidant capacity of tomatoes (*Solanum lycopersicum*). *Clin. Nutr. Hosp. Diet.* **2020**, *40*, 126–131.

foods MDPI

Article

Analyzing the Effects of Rapid and Natural Cooling Techniques on the Quality of Hand-Shaken Green Tea Beverages

Yuan-Ke Chen [1], Tuzz-Ying Song [2], Chi-Yu Chang [2], Shiann-Cherng Sheu [3] and Chih-Wei Chen [3,*]

1 Ph.D. Program of Biotechnology and Bioindustry, College of Biotechnology and Bioresources, Da-Yeh University, Changhua 515006, Taiwan; fsancohor@yahoo.com.tw
2 Department of Medicinal Botanicals and Foods on Health Applications, Da-Yeh University, Changhua 515006, Taiwan; song77@mail.dyu.edu.tw (T.-Y.S.); feccy2@yahoo.com.tw (C.-Y.C.)
3 Bachelor Degree Program in Food Safety/Hygiene and Laboratory Science, Chang Jung Christian University, Tainan 711301, Taiwan; scschem@mail.cjcu.edu.tw
* Correspondence: ccwlly@mail.cjcu.edu.tw; Tel.: +886-6-2785-123 (ext. 7607)

Abstract: This study compared the quality of hand-shaken green tea prepared through rapid and natural cooling methods. Cooling is crucial in preserving green tea's flavor, aroma, and nutritional components. In the rapid cooling method, green tea is freshly brewed at an initial temperature of 95 °C for 25 min, and then rapidly cooled to 18 °C for 25 min. Conversely, the natural cooling method involves brewing tea at the same initial temperature and time, but allowing it to cool gradually to 30 °C over approximately 4–5 h at room temperature. This study's findings indicate that the rapid cooling method produced green tea with a more vibrant color and improved clarity versus the natural cooling method. Sensory analysis revealed that the taste and aroma of the hand-shaken green tea prepared using rapid cooling were perceived to be more refreshing and invigorating. However, the natural cooling method preserved a higher level of chemical components, including individual catechin caffeine, total polyphenol, soluble solids, reducing sugar, and total tannins. The essential amino acid content of the rapidly and naturally cooled green tea infusions was 6.85 and 13.55 μg/mL, respectively. The γ-Aminobutyric acid (GABA) content was 439.82 and 457.31 μg/mL, respectively. This study's findings suggest that rapid cooling during the preparation of hand-shaken green tea enhances its overall quality. The vibrant color, improved clarity, refreshing taste, and invigorating aroma make it a preferable choice for tea enthusiasts who seek an enhanced sensory experience and excellent quality.

Keywords: hand-shaken green tea; rapid cooling; natural cooling; quality

1. Introduction

Following water, tea is the most widely consumed beverage worldwide. Its health benefits and medicinal use have been extensively investigated since early Chinese civilizations [1,2]. Three main types of tea have been classified based on their manufacturing process: green, black, and oolong tea. Green tea is a highly consumed beverage globally, particularly in Asia, due to its unique sensory characteristics and cultural significance [3].

Studies indicate that green tea provides numerous health benefits, including inhibiting the growth of harmful microbes and viruses, reducing the risk of cardiovascular diseases, and preventing the development of diabetes, obesity, cancer, and other ailments that involve oxidative stress [3–6]. Recent research indicates that green tea can help protect against coronaviruses, including COVID-19, by reducing inflammation-related complications [7,8]. The array of bioactive components in tea infusions, including polyphenols, caffeine, theanine, catechin, and free amino acids, provides such protection [3–5,9].

Two essential qualities of green tea infusion are its astringency and bitterness. Studies suggest that bitter substances often produce astringency, leaving a dry, rough sensation in the mouth [6,10,11]. Bitterness is typically a warning signal to avoid harmful substances; too much bitterness can lead to nausea [12]. However, some bitter and astringent components can enhance the taste of foods and beverages, such as green tea [11,13]. Moderate bitterness and astringency can enrich the flavor of green tea, and a sweet aftertaste often follows its bitterness or astringency [6,11]. Additionally, some bitter and astringent substances in tea have health benefits for humans [14].

The biological components of tea infusion impact human health and affect its quality. Certain compounds, such as tannins, catechins, and other phenolic compounds, can cause bitterness and astringency during tea infusion [11]. Alkaloids such as caffeine, theobromine, and theophylline are the primary bitter and astringent substances, with caffeine being the prime flavor-enhancing substance [8,10,11]. Green tea infusion typically comprises bitterness, umami, and a sweet aftertaste [6,15]. Astringency, a feeling on the tongue caused by the interaction of polyphenols (such as catechins, quercetin, and gallic acid) and salivary proteins, is a significant factor in the taste quality of green tea [6,15–17].

However, when controlling the concentration of bitter and astringent substances in green tea infusion, it is challenging to balance the flavor and health benefits while reducing discomfort from excessive bitterness and astringency [11].

The content of bioactive components in tea infusion depends primarily on the time and temperature at which the tea is brewed. Therefore, it is essential to pay attention to these factors. When brewing green tea, the brewing process plays a crucial role in the final quality of the infusion process. For traditional green tea, 3 g of tea should be steeped in 150 mL of boiling water for five minutes before it is ready to drink [18]. However, hand-shaken green tea requires 250 g of tea leaves and 18 L of boiling water. After steeping for 25 min, the tea leaves and the tea infusion should be separated and left to cool for approximately 4–5 h until the temperature reaches 27–30 °C. Therefore, the quality of hand-shaken green tea is determined by the speed at which it cools.

This study investigated the effects of two cooling methods, rapid and natural, on hand-shaken green tea infusion. We examined the levels of catechins, caffeine, total polyphenols, soluble solids, reducing sugar, total tannins, free amino acids, and GABA, and how they contribute to the bitterness and astringency of tea.

2. Materials and Methods

2.1. Chemicals

This study's reagents and chromatographic grade standards, such as individual catechin (catechin (C), catechin gallate (CG), gallocatechin (GC), epicatechin (EC), epicatechin gallate (ECG), epigallocatechin (EGC), gallocatechin gallate (GCG), epigallocatechin gallate (EGCG)), caffeine, gallic acid, tannin, amino acid, and γ-Aminobutyric acid (GABA), were purchased from Sigma-Aldrich Chemical Co. (St. Louis, MO, USA). All chemical standards had a minimum purity of 98%. The green teas were obtained from the Tea Research and Extension Station (TRES) in Nantou, Taiwan.

2.2. Preparing the Green Tea Infusions

For this experiment, Taiwan Tea Experiment Station No. 12 (TTES No. 12, Jinxuan, Nantou, Taiwan) green tea, purchased from the Ruicheng Tea Factory in Nantou, Taiwan, was used. The tea infusions were separated into two groups for sensory evaluation and physicochemical analysis based on the use of either the natural or rapid cooling method.

To naturally cool the tea, we brewed it with boiling water and allowed it to cool. We poured 18,000 mL of hot water at 95 °C into a tea barrel, then added 250 g of green tea leaves and stirred them well. We let the leaves brew for approximately 25 min before filtering them. The resulting tea infusion (Figure 1A) was cooled in a cold room to 30 °C, which took about 4–5 h, before packaging.

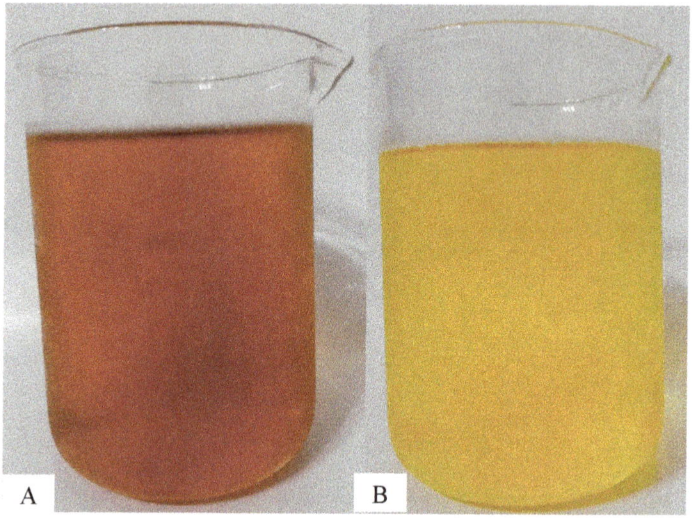

Figure 1. The green tea infusions prepared using two different cooling methods: (A) natural cooling (95 °C to 30 °C in 4–5 h, brewing time: 25 min) and (B) rapid cooling (95 °C to 18 °C in 25 min, brewing time: 25 min).

A NE-15 cooling machine (Chenguang Energy Technology Co., Ltd., Nantou, Taiwan) was used for rapid cooling. We followed the same process of boiling and pouring 18,000 mL of hot water at 95 °C into the tea barrel, adding 250 g of green tea leaves, and stirring well. It took about 25 min to brew and cool to 18 °C. We filtered out the tea infusion (see Figure 1B) with a filter and proceeded with subsequent packaging. The detailed temperature changes were as follows:

Natural Cooling Process:

- Initial temperature: 95 °C
- Brewing time: 25 min
- Final cooled temperature: 30 °C
- Cooling duration: Approximately 4–5 h

Rapid Cooling Process:

- Initial temperature: 95 °C
- Brewing time: 25 min
- Final cooled temperature: 18 °C
- Cooling duration: Approximately 25 min

2.3. Determining the Total Polyphenol Content

The measurements and modifications in the present study were based on Chiang et al. [5]. For each 0.5 mL of tea infusion, we added 7 mL of water, followed by 0.5 mL of Folin–Ciocalteu reagent, which was mixed for three minutes. Next, 2 mL of 20% Na_2CO_3 was added, and the samples were left in the dark at room temperature for two hours. The absorbance was read at 750 nm, and the calibration curve was determined using gallic acid solutions (R^2 = 0.999). The phenolic content was expressed as mg gallic acid equivalents per gram of the extract (mg gallic acid/g).

2.4. Determining the Reducing Sugars and Soluble Solid Content

The measurements and modifications were performed following the method described by Chiang et al. [4]. Specifically, 2 mL of DNS reagent was added to each 1 mL of tea infusion and thoroughly mixed. The samples were then subjected to a five-minute boiling water bath and cooled to room temperature. We measured the absorbance at 540 nm and constructed a calibration curve using glucose concentrations ranging from 0.2 to 1.6 mg/mL (R^2 = 0.995). The soluble solid content was measured using the AOAC method 22.024 [19]. The reducing sugars and soluble solids content were expressed in μg/mL.

2.5. Determining the Amino Acid and GABA Content

The green tea infusions were hydrolyzed using 10 mL of 6 M hydrogen chloride at 110 °C for 22 h under a nitrogen atmosphere to obtain the total amino acids. We used the Hitachi L-8900 Amino Acid Analyzer (Hitachi Co., Ltd., Tokyo, Japan) for the amino acid composition analysis. The amino acid and GABA content were expressed as μg/mL.

2.6. Determining the Total Tannin Content

Total tannin content was determined using the Folin–Ciocalteu method with modifications as per Farahmandfar et al. [20]. A 1 g sample was sonicated in 20 mL water for 30 min, followed by centrifugation. We mixed the supernatant (100 μL) with 1 mL of diluted Folin–Ciocalteu reagent and left it for five minutes. Next, 2 mL 7.5% sodium carbonate was added and incubated at 40 °C for 30 min. Absorbance was measured at 765 nm. A gallic acid standard curve (0–100 mg/L, R^2 = 0.997) was used to calculate tannin content, expressed as mg gallic acid equivalents per gram (mg gallic acid/g) of the sample.

2.7. Determining the Individual Catechins and Caffeine Content

We analyzed the individual catechins and caffeine content using reversed-phase high-performance liquid chromatography (HPLC) with a UV-VIS spectrometric detector (Shimadzu RID-6A, Tokyo, Japan), as described by Chiang et al. [4]. A 5 μm Diamonsil C18 column (4.6 mm × 250 mm) was used, and the mobile phase comprised a mixture of water (0.1 mol/L H_3PO_4, 1.0×10^{-4} mol/L EDTA), acetonitrile, and ethyl acetate in a ratio of 85:12:3 (v/v). Elution was performed at a 1.0 mL/min solvent flow rate, and chromatograms were recorded at 280 nm. The individual catechins and caffeine content were expressed in μg/mL.

2.8. pH Value and Color Measurement

The pH value of the tea infusion was determined using a pH meter (model number: MP220, Swiss Merchant Mettler-Toledo Co., Ltd., Taiwan Branch (Kaohsiung, Taiwan). The colorimeter (model number: NE4000, Nippon Denshoku, Tokyo, Japan) was used according to the description of Lee et al. [21] to measure Hunter L, a, and b values to determine the colors of the two kinds of tea infusion. The comprehensive color difference was represented by ΔE using the following formula:

$$\Delta E = [(\Delta L)^2 + (\Delta a)^2 + (\Delta b)^2]^{1/2}$$

2.9. Sensory Analysis-Consumer Hedonic Tests

Concerning the sensory evaluation section of this study, we consulted the Research Ethics Committee (DYU-REC, 25 November 2023) at the outset. As a result, since the study did not involve sensitive data or risks to the participants, the experiment was not formally subjected to a research ethics review. To analyze the samples, this study recruited a panel of 60 semi-trained members (30 males and 30 females) aged 22 to 50 from the Department of Medicinal Botanicals and Foods on Health Applications at Da-Yeh University, Changhua, Taiwan. We conducted sensory evaluations using a 9-point hedonic scale (1 point: Dislike Extremely; 2 points: Dislike Very Much; 3 points: Dislike Moderately; 4 points: Dislike Slightly; 5 points: Neither Like nor Dislike; 6 points: Like Slightly; 7 points: Like Moderately; 8 points: Like Very Much; and 9 points: Like Extremely). The evaluations were conducted at a controlled temperature of 27 ± 2 °C. The overall acceptability of the samples was determined by calculating the average scores given for the attributes evaluated.

2.10. Statistical Analysis

The reported data were obtained through triplicate measurements of the samples. We performed statistical analysis using one-way analysis of variance (ANOVA) and t-test multiple comparisons in SAS 9.0 (SAS Institute, Cary, NC, USA). A significance level of $p < 0.05$ was considered statistically significant.

3. Results and Discussion

3.1. Effects of Different Brewing Cooling Methods on the pH Value, Soluble Solids, and Reducing Sugar Content of Tea Infusions

The pH values of the rapidly and naturally cooled green tea infusions were 6.36 and 7.44, respectively (Table 1).

Table 1. The pH value, soluble solids, reducing sugar, total tannins, total polyphenol, and caffeine contents of rapid cooling and natural cooling green tea infusions.

Items	Green Tea Infusion	
	Rapid Cooling Green Tea	Natural Cooling Green Tea
pH value	6.36 ± 0.15 ^{B,*}	7.44 ± 0.11 ^{A}
Soluble solids (μg/mL)	2714.58 ± 143.74 ^{B}	3476.94 ± 188.39 ^{A}
Reducing sugar (μg/mL)	609.93 ± 73.41 ^{B}	799.84 ± 46.83 ^{A}
Total tannins (mg gallic acid/g)	785.47 ± 36.72 ^{B}	1157.65 ± 61.88 ^{A}
Total polyphenol (mg gallic acid/g)	972.85 ± 53.82 ^{B}	1449.21 ± 84.51 ^{A}
Caffeine (μg/mL)	263.41 ± 26.75 ^{B}	398.49 ± 19.63 ^{A}

* Results from three separate experiments are expressed as mean ± SD. A,B: Data with identical letters in the same row are not significantly different ($p < 0.05$).

The observed difference in pH between the two cooling methods can be attributed to several factors. During the rapid cooling process, the green tea infusion temperature drops quickly, potentially leading to increased proton release from certain compounds, thus lowering the pH. In contrast, the natural cooling process allows for a slower temperature decline, which may result in reduced proton release and a higher pH value.

The pH of green tea plays a significant role in determining its taste, flavor, and overall quality. Research indicates that variations in pH can influence the solubility and stability of tea components, such as catechins and amino acids, which are critical contributors to its sensory attributes and health benefits [22]. Moreover, pH affects the sensory perception of green tea, with lower pH values associated with a more pronounced and vibrant taste. Comparatively, higher pH values may result in a smoother, milder taste [23].

The difference in pH between the two cooling methods can have significant implications for green tea quality and consumer acceptance. Careful control of the cooling process to achieve the desired pH level is crucial for optimizing green tea production and meeting consumer preferences.

The cooling method used when preparing green tea infusions affects the levels of soluble solids and reducing sugars. Table 1 shows that this study's green tea infusion cooled rapidly had a soluble solids content of 2714.58 μg/mL and a reducing sugar content of 609.93 mg/mL. In contrast, the natural cooling tea infusion had a soluble solids content of 3476.94 μg/mL and a reducing sugar content of 799.84 mg/mL. The cooling rate and duration caused these differences. When cooled rapidly, the soluble solids may be better preserved due to the sudden temperature drop, inhibiting the breakdown of certain compounds. In contrast, a slower cooling rate during natural cooling may allow for more chemical reactions, leading to a higher release of soluble solids and reducing sugars in the tea infusion. Soluble solids content is a vital indicator of the dissolved compounds in green tea that contribute to its taste, flavor, and overall quality. These compounds include catechins, amino acids, and other flavor components [22]. Similarly, reducing sugar enhances a tea's sweetness and mouthfeel.

The difference in soluble solids and reducing sugar contents between the two cooling methods can affect green tea's perceived taste and flavor profile. Consumers may have varying preferences for sweetness levels and mouthfeel, making the control of cooling conditions crucial in meeting these preferences.

The soluble solid and reducing sugar contents are crucial in green tea infusions' sensory qualities and overall quality. Soluble solid content affects a tea's body and mouthfeel, while the reducing sugars add to the sweetness and flavor profile. Higher soluble solid content produces a fuller-bodied tea with a richer taste, while increased reducing sugar content enhances a tea's sweetness and overall sensory perception. These attributes benefit those seeking a more flavorful and enjoyable tea experience.

Furthermore, green tea's soluble solids and reducing sugars interact with other components, such as catechins and amino acids, affecting their stability and bioavailability. These interactions influence the sensory attributes, antioxidant activity, and potential health benefits of green tea consumption [4,5].

3.2. Effects of Different Brewing Cooling Methods on the Total Tannins, Total Polyphenol, and Caffeine Content of Tea Infusions

The cooling method used in preparing green tea infusions significantly impacts the total tannin, total polyphenol, and caffeine content. The naturally cooled green tea infusion in the present study had higher levels of total tannins (1157.65 mg gallic acid/g), total polyphenol (1449.21 mg gallic acid/g), and caffeine (398.49 μg/mL) compared to the rapidly cooled infusion (785.47 mg gallic acid/g, 972.85 mg gallic acid/g, and 263.41 μg/mL, respectively) (Table 1). The longer steeping time during natural cooling resulted in a more efficient extraction of tannins, polyphenols, and caffeine from the tea leaves, leading to higher concentrations in the infusion. The differences in total tannin and polyphenol content can be attributed to the cooling process and its effect on the extraction efficiency of these compounds from the tea leaves. Natural cooling allows for a longer steeping time, which promotes greater extraction of tannins and polyphenols. Tannins and polyphenols contribute to the astringency and bitterness of tea, and longer steeping times result in higher levels of these compounds. In a previous study [4], the caffeine and polyphenol content was shown to increase significantly with longer steeping times in green tea infusions.

The higher caffeine content in the natural cooling green tea infusion may be attributed to the prolonged exposure to caffeine during natural cooling. Caffeine is a naturally occurring compound in tea extracted during the steeping process. The longer steeping time with natural cooling allows for more efficient extraction of caffeine from the tea leaves, resulting in higher levels in the infusion.

The total tannins, total polyphenols, and caffeine are essential parameters that can impact the sensory qualities and overall quality of green tea infusions. Tannins and polyphenols contribute to a tea's astringency, bitterness, and antioxidant activity. Higher levels of these compounds may result in a more pronounced astringent and bitter taste. Caffeine content contributes to tea's stimulant effects and can influence the overall perception of its sensory attributes. Moreover, tannins, polyphenols, and caffeine are bioactive compounds in green tea associated with various health benefits. These compounds possess antioxidative, anti-inflammatory, and potential anticancer properties. Higher tannins, polyphenols, and caffeine in a naturally cooled green tea infusion may enhance its potential health benefits. Research suggests that the polyphenol content and antioxidant activity of green tea infusions are positively correlated [4,5].

The present study's ratio of polyphenols to amino acids in green tea infusions showed a significant difference between the rapidly and naturally cooled methods (Table 1). The rapidly cooled green tea infusion had a ratio of polyphenols to amino acids of 12.17, while the naturally cooled green tea infusion had a higher ratio of 20.02.

The ratio of polyphenols to amino acids is an essential indicator of the balance between these two groups of compounds in green tea. Polyphenols, such as catechins, are abundant in green tea and contribute to its antioxidative, health-promoting properties. On the other hand, amino acids play a vital role in the taste and flavor of tea, particularly in providing its umami and sweet notes [4,5,24]. The ratio of polyphenols to amino acids in green tea infusions significantly impacts their taste, flavor, and overall quality. A higher ratio may lead to a more robust and astringent taste due to the increased presence of polyphenols. A lower ratio may result in a milder flavor with more pronounced umami and sweetness from the amino acids [15].

3.3. Effect of Different Brewing Cooling Methods on Individual Catechin Content of Tea Infusions

The amounts of individual catechins in green tea infusions made using the rapid and natural cooling methods are shown in Table 2. The individual catechin contents in the green tea infusion made using the rapid cooling process were as follows: gallocatechin (GC) 33.62 μg/mL, epigallocatechin (EGC) 126.01 μg/mL, catechin (C) 254.64 μg/mL, epicatechin (EC) 23.98 μg/mL, epigallocatechin gallate (EGCG) 5.28 μg/mL, gallocatechin gallate (GCG) 83.61 μg/mL, epicatechin gallate (ECG) 46.54 μg/mL, and catechin gallate (CG) 52.15 μg/mL. The individual catechin contents in the naturally cooled green tea infusion were as follows: GC 85.65 μg/mL, EGC 91.16 μg/mL, C 246.32 μg/mL, EC 127.36 μg/mL, EGCG 25.72 μg/mL, GCG 100.53 μg/mL, ECG 55.50 μg/mL, and CG 74.21 μg/mL.

Table 2. Individual catechin content of rapid cooling and natural cooling green tea infusions.

Type of Individual Catechins (μg/mL)	Rapid Cooling Green Tea	Natural Cooling Green Tea
Gallocatechin (GC)	$33.62 \pm 1.54^{B,*}$	85.65 ± 1.39^{A}
Epigallocatechin (EGC)	126.01 ± 5.54^{A}	91.16 ± 1.46^{B}
Catechin (C)	254.64 ± 12.71^{A}	246.32 ± 2.78^{A}
Epicatechin (EC)	23.98 ± 1.62^{B}	127.36 ± 1.86^{A}
Epigallocatechin gallate (EGCG)	5.28 ± 0.74^{B}	25.72 ± 1.01^{A}
Gallocatechin gallate (GCG)	83.61 ± 3.43^{B}	100.53 ± 0.24^{A}
Epicatechin gallate (ECG)	46.54 ± 2.69^{B}	55.50 ± 0.83^{A}
Catechin gallate (CG)	52.15 ± 1.98^{B}	74.21 ± 0.70^{A}

* Results from three separate experiments are expressed as mean ± SD. A,B: Data with identical letters in the same row are not significantly different ($p < 0.05$).

The differences in individual catechin contents can be attributed to the cooling process and its impact on extraction and transformation. Natural cooling, with a longer steeping time, allows for more efficient extraction of catechins from the tea leaves. Additionally, catechins can undergo oxidation and conversion into other forms during natural cooling. Studies indicate that the longer steeping time and exposure to oxygen during natural cooling can lead to the conversion of catechins into more complex, polymerized forms, such as theaflavins (TF) and thearubigins (TA) [5,6,25].

Different catechin profiles can significantly influence the sensory qualities and overall quality of green tea infusions. For example, EGCG is one of green tea's most abundant, biologically active catechins, known for its antioxidative and potential health-promoting effects [26]. Higher levels of EGCG, as observed in the natural cooling infusion, may contribute to increased antioxidant activity and potential health benefits.

Furthermore, individual catechins contribute to green tea's flavor profile and astringency. Catechins such as EGC, C, and EC are associated with bitterness and astringency [5,11,27]. Higher levels of these catechins, as seen in the natural cooling infusion, may produce a more pronounced bitter and astringent taste.

3.4. Effect of Different Brewing Cooling Methods on Amino Acids, EAA, and GABA Contents of Tea Infusions

The amino acid content of the green tea infusions in the present study showed significant differences between the rapid and natural cooling methods (Table 3). The rapid cooling green tea infusion had a total amino acid content of 79.96 μg/mL, while the natural cooling green tea infusion had a slightly lower content of 72.40 μg/mL. Among the essential amino acids (EAA), the naturally cooled green tea infusion exhibited higher levels of glycine (Gly), alanine (Ala), arginine (Arg), threonine (Thr), valine (Val), isoleucine (Ile), leucine (Leu), and phenylalanine (Phe) than its rapidly cooled counterpart. In contrast, the rapidly cooled green tea infusion had higher glutamic acid (Glu), aspartic acid (Asp), tyrosine (Tyr), tryptophan (Trp), and lysine (Lys) levels. However, the cysteine (Cys), serine (Ser), proline (Pro), and methionine (Met) levels did not differ significantly between the two cooling methods. Additionally, no significant difference was observed in GABA content, with values of 439.82 and 457.31 μg/mL for the rapidly and naturally cooled infusions, respectively.

Table 3. Amino acid and GABA content of the rapid and natural cooling green tea infusions.

Free Amino Acid (μg/mL)	Green Tea Infusion	
	Rapid Cooling	**Natural Cooling**
Glu	21.68 ± 0.27 ^{A,*}	17.12 ± 0.35 ^{B}
Asp	24.0 ± 0.52 ^{A}	20.05 ± 0.28 ^{B}
Cys	4.10 ± 0.13 ^{A}	4.35 ± 0.04 ^{A}
Ser	2.13 ± 0.11 ^{A}	2.47 ± 0.03 ^{A}
Gly	0.65 ± 0.01 ^{B}	2.12 ± 0.02 ^{A}
His	ND **	ND
Ala	1.18 ± 0.02 ^{B}	1.93 ± 0.05 ^{A}
Arg	1.37 ± 0.06 ^{B}	2.65 ± 0.05 ^{A}
Thr	0.25 ± 0.00 ^{B}	0.78 ± 0.02 ^{A}
Pro	0.18 ± 0.01 ^{A}	0.17 ± 0.00 ^{A}
Tyr	11.82 ± 0.13 ^{A}	7.99 ± 0.06 ^{B}
Val	0.46 ± 0.01 ^{B}	1.68 ± 0.04 ^{A}
Met	3.43 ± 0.07 ^{A}	3.61 ± 0.02 ^{A}
Ile	0.32 ± 0.01 ^{B}	0.55 ± 0.02 ^{A}
Lue	0.51 ± 0.01 ^{B}	4.33 ± 0.05 ^{A}
Phe	0.26 ± 0.00 ^{B}	1.46 ± 0.03 ^{A}
Trp	1.45 ± 0.05 ^{A}	0.98 ± 0.01 ^{B}
Lys	0.26 ± 0.01 ^{A}	0.16 ± 0.01 ^{B}
EAA ***	6.85 ± 0.13 ^{B}	13.55 ± 0.19 ^{A}
Total	79.96 ± 0.38 ^{A}	72.40 ± 0.34 ^{B}
GABA (μg/mL)	439.82 ± 15.93 ^{A}	457.31 ± 11.58 ^{A}

* Results from three separate experiments are expressed as mean ± SD. A,B: Data with identical letters in the same row are not significantly different ($p < 0.05$). ** ND: not detected. *** EAA: essential amino acid.

Variations in amino acid content can significantly impact the taste, flavor, and overall quality of green tea infusions. Amino acids contribute to a tea's umami, sweetness, bitterness, and astringency [28,29]. Therefore, the balance and levels of different amino acids play a crucial role in determining the sensory attributes of tea. Interestingly, the two cooling methods showed no significant difference in this study's GABA (gamma-aminobutyric acid) content. GABA is known for its calming and relaxing effects and contributes to the unique sensory profile of green tea [29].

Notably, the amino acids Cys, Ser, Pro, and Met showed no significant differences between the two cooling methods. These amino acids appear to have stable levels in green tea infusions and are less affected by the cooling process.

3.5. Effect of Different Brewing Cooling Methods on the Color of Tea Infusions

Two different cooling methods for green tea infusion are shown in Figure 1. The color of the green tea infusions exhibited significant differences between the rapid and natural cooling methods we tested during this study (Figure 2). The L-values, representing lightness, were higher in the rapid cooling green tea infusion (L = 54.13) than in the natural cooling green tea infusion (L = 20.68). The a-values, indicating redness-greenness, were lower in the rapid cooling green tea infusion (a = −10.71) than in the natural cooling green tea infusion (a = 18.73). The b-values, representing yellowness-blueness, were higher in the rapid cooling green tea infusion (b = 27.19) than in the natural cooling infusion (b = 4.53). Moreover, the ΔE-values, indicating the overall color difference, were more significant in the rapid cooling infusion (ΔE = 61.51) than in the natural cooling infusion (ΔE = 28.25).

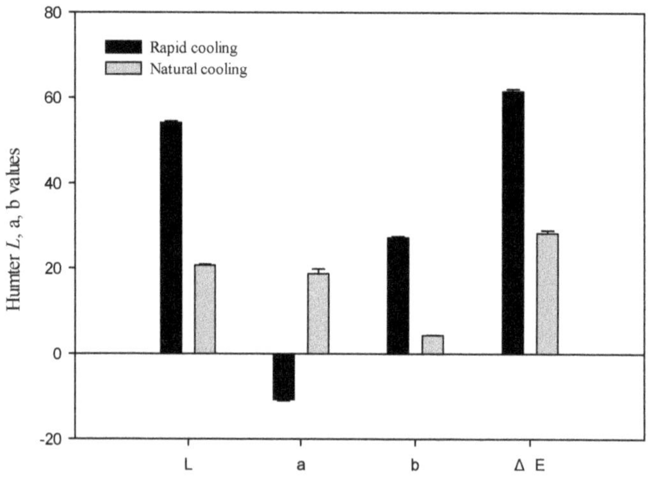

Figure 2. Chromatic aberration analysis of rapid and natural cooling green tea infusion. Results from three separate experiments are expressed as mean ± SD.

The color of green tea is influenced by various factors, including the content of chlorophyll, flavonoids, and other pigments, as well as by oxidation and degradation processes. Steeping conditions, such as temperature and duration, can affect the extraction and transformation of these pigments, resulting in color variations.

Therefore, the color of green tea is a critical quality attribute that can affect consumer perception, as it is often associated with freshness, taste, and overall quality. Lighter and yellow-green colors are generally preferred, indicating higher quality, better appearance, and a tea with improved mouthfeel [15].

3.6. *Effect of Brewing Temperature and Duration on the Sensory Analysis-Consumer Hedonic Tests*

According to the consumer hedonic tests in Figure 3, the sensory analysis indicates that the green tea infusion with rapid cooling had a better aroma, flavor, taste, aftertaste, appearance, and higher overall acceptability than the natural cooling green tea infusion. The differences between the two cooling methods were particularly significant in terms of taste, appearance, and preference.

Green tea's taste and appearance are crucial factors influencing consumer preferences. This study's naturally cooled green tea had significantly higher bitter and astringent tastes than the rapidly cooled green tea. This finding could be attributed to the longer steeping time during the natural cooling process, which increased the extraction of bitter-tasting catechins and astringent compounds from the tea leaves.

Rapid cooling preserves green tea's delicate and fresh characteristics better than natural cooling. A faster cooling rate helps retain volatile aroma compounds essential for the overall sensory experience. The higher aroma, flavor, and aftertaste scores for the rapid cooling infusion might be due to the better preservation of these volatile compounds. Figure 3 indicates that consumers preferred the rapidly cooled green tea infusions regarding color, aroma, and taste.

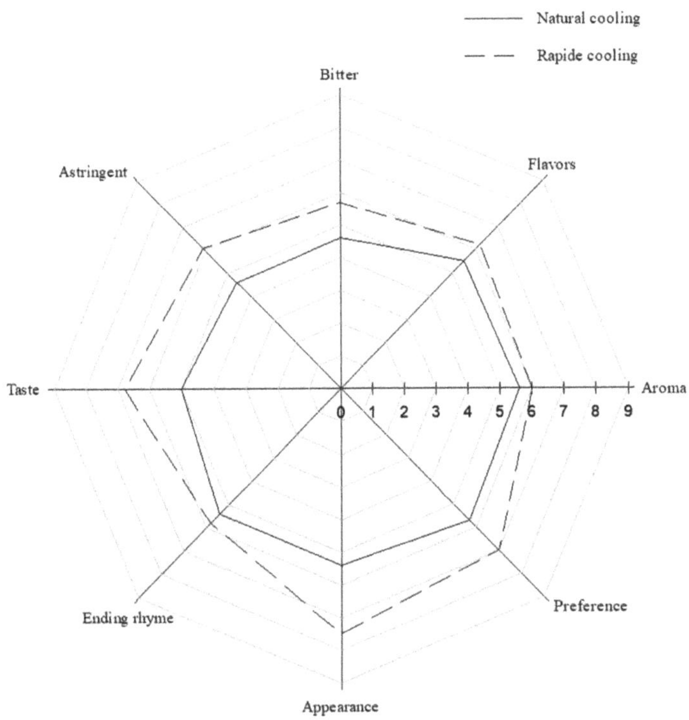

Figure 3. Radar chart of consumers' acceptance of tea sensory evaluation under rapid and natural cooling.

4. Conclusions

This study comparing rapid and natural cooling methods for green tea revealed significant differences in pH, soluble solids, reducing sugars, total tannins, polyphenols, caffeine, catechins, amino acids, and color between the two methods. These differences are crucial for determining the flavor, aroma, and overall sensory experience of tea. Rapid cooling resulted in a lower pH (6.36) compared to natural cooling (7.44), likely due to different rates of chemical reactions. Additionally, rapid cooling yielded lower soluble solids and reducing sugars, indicating decreased solubility and sugar extraction. In contrast, natural cooling produced higher amounts of total tannins, polyphenols, and caffeine, enhancing the extraction of these compounds.

The catechin profile varied between the two methods, influencing the flavor and aroma of the produced teas. Natural cooling increased essential amino acids and those associated with pleasant tastes and aromas. Color differences were also noted, with rapid cooling resulting in a lighter, greener, and yellower tea, while natural cooling produced a darker, redder infusion. Sensory analysis favored the rapid cooling method, with higher scores in aroma, flavor, taste, aftertaste, and appearance, while natural cooling led to more bitter and astringent tastes, potentially affecting consumer perception.

These findings underscore the importance of the cooling method in green tea production. It impacts consumer acceptability and perception of tea quality. Tea producers and researchers should consider the effects of the cooling process in order to optimize tea quality and meet consumer preferences and market demands.

Author Contributions: Methodology, Y.-K.C. and T.-Y.S.; software, T.-Y.S.; formal analysis, C.-Y.C. and S.-C.S.; investigation, C.-W.C.; data curation, C.-W.C.; writing—original draft preparation,

C.-W.C.; writing—review and editing, C.-W.C. All authors have read and agreed to the published version of the manuscript.

Funding: This research received no external funding.

Institutional Review Board Statement: Ethical review and approval were waived for this study due to Reason. Minimal Risk to Participants: Justification: The study poses no more than minimal risk to participants. Minimal risk means that the probability and magnitude of harm or discomfort anticipated in the research are not greater than those ordinarily encountered in daily life or during routine physical or psychological examinations. Quality Improvement or Evaluation Projects: Justification: The project is a quality improvement or evaluation project, not intended to contribute to generalizable knowledge but to improve the quality of services within an organization. Such projects are often exempt from ethical review unless they involve significant risks or burdens to participants.

Informed Consent Statement: Informed consent was obtained from all subjects involved in the study.

Data Availability Statement: The original contributions presented in the study are included in the article, further inquiries can be directed to the corresponding author.

Conflicts of Interest: The authors declare no conflicts of interest.

References

1. Mckay, D.L.; Blumberg, J.B. The role of tea in human health: An update. *J. Am. Coll. Nutr.* **2002**, *21*, 1–3. [CrossRef] [PubMed]
2. Venditti, E.; Bacchetti, T.; Tiano, L.; Carloni, P.; Greci, L.; Damiani, E. Hot vs. cold water. steeping of different teas: Do they affect antioxidant activity? *Food Chem.* **2010**, *119*, 1597–1604. [CrossRef]
3. Pérez Burillo, S.; Giménez, R.; Rufián-Henares, J.A.; Pastoriza, S. Effect of brewing time and temperature on antioxidant capacity and phenols of white tea: Relationship with sensory properties. *Food Chem.* **2018**, *248*, 111–118. [CrossRef] [PubMed]
4. Chiang, S.H.; Tsou, M.F.; Chang, C.Y.; Chen, C.W. Physicochemical characteristics, sensory quality, and antioxidant properties of Paochung tea infusion brewed in cold water. *Int. J. Food Prop.* **2020**, *23*, 1611–1623. [CrossRef]
5. Chiang, S.H.; Yang, K.M.; Wang, S.Y.; Chen, C.W. Enzymatic treatment in black tea manufacturing processing: Impact on bioactive compounds, quality, and bioactivities of black tea. *LWT* **2021**, *163*, 113560. [CrossRef]
6. Xu, Y.Q.; Zhang, Y.N.; Chen, J.X.; Wang, F.; Du, Q.Z.; Yin, J.F. Quantitative analyses of the bitterness and astringency of catechins from green tea. *Food Chem.* **2018**, *258*, 16–24. [CrossRef]
7. Park, R.; Jang, M.; Park, Y.I.; Park, Y.; Jung, W.; Park, J.; Park, J. Epigallocatechin gallate (EGCG), a green tea polyphenol, reduces coronavirus replication in a mouse model. *Viruses* **2021**, *13*, 2533. [CrossRef] [PubMed]
8. Zhang, Z.; Hao, M.; Zhang, X.; He, Y.; Chen, X.; Taylor, E.W.; Zhang, J. Potential of green tea EGCG in neutralizing SARS-CoV-2 Omicron variant with greater tropism toward the upper respiratory tract. *Trends Food Sci. Technol.* **2023**, *132*, 40–53. [CrossRef]
9. Vuong, Q.V. Epidemiological evidence linking tea consumption to human. health: A review. *Crit. Rev. Food Sci. Nutr.* **2014**, *54*, 523–536. [CrossRef]
10. Dai, W.; Qi, D.; Yang, T.; Lv, H.; Guo, L.; Zhang, Y.; Zhu, Y.; Peng, Q.; Xie, D.; Tan, J.; et al. Non-targeted analysis using ultraperformance liquid chromatography-quadrupole-time of flight mass spectrometry uncovers the effects of harvest season on the metabolites and taste quality of tea (*Camellia sinensis* L.). *J. Agric. Food Chem.* **2015**, *63*, 9869–9878. [CrossRef]
11. Deng, S.; Zhang, G.; Aluko, O.O.; Mo, Z.; Mao, J.; Zhang, H.; Liu, X.; Ma, M.; Wang, Q.; Liu, H. Bitter and astringent substances in green tea: Composition, human perception mechanisms, evaluation methods and factors influencing their formation. *Food Res. Int.* **2022**, *157*, 111262. [CrossRef] [PubMed]
12. Mura, E.; Yagi, M.; Yokota, K.; Seto, E.; Matsumiya, K.; Matsumura, Y.; Hayashi, Y. Tolerance of bitter stimuli and attenuation/accumulation of their bitterness in humans. *Biosci. Biotechnol. Biochem.* **2018**, *82*, 1539–1549. [CrossRef]
13. Rozin, P.; Vollmecke, T.A. Food likes and dislikes. *Annu. Rev. Nutr.* **1986**, *6*, 433–456. [CrossRef]
14. Fang, J.; Sureda, A.; Silva, A.S.; Khan, F.; Xu, S.W.; Nabavi, S.M. Trends of tea in cardiovascular health and disease: A critical review. *Trends Food Sci. Technol.* **2019**, *88*, 385–396. [CrossRef]
15. Li, Y.; Ran, W.; He, C.; Zhou, J.; Chen, Y.; Yu, Z.; Ni, D. Effects of different tea tree varieties on the color, aroma, and taste of Chinese Enshi green tea. *Food Chem. X* **2022**, *14*, 100289. [CrossRef]
16. Yin, J.; Zhang, Y.N.; Du, Q.Z.; Chen, J.X.; Yuan, H.B.; Xu, Y.Q. Effect of Ca^{2+} concentration on the tastes from the main chemicals in green tea infusions. *Food Res. Int.* **2014**, *62*, 941–946. [CrossRef]
17. Yu, P.G.; Yeo, A.S.L.; Low, M.Y.; Zhou, W.B. Identifying key non-volatile. compounds in ready-to-drink green tea and their impact on taste profile. *Food Chem.* **2014**, *155*, 9–16. [CrossRef]
18. Yan, J.; Zhao, J.; Kim, E.M.; Kim, K.H.; Kang, S.; Lee, H.; Lee, J. Comprehensive Investigation of the Effects of Brewing Conditions in Sample Preparation of Green Tea Infusions. *Molecules* **2019**, *24*, 1735. [CrossRef] [PubMed]
19. AOAC. Official Methods of Analysis. In *Association of Official Analytical Chemists*, 18th ed.; AOAC international: Gaithersburg, MD, USA, 2006.

20. Farahmandfar, R.; Kenari, R.E.; Asnaashari, M.; Shahrampour, D.; Bakhshandeh, T. Bioactive compounds, antioxidant and antimicrobial activities of *Arum maculatum* leaves extracts as affected by various solvents and extraction methods. *Food Sci. Nutr.* **2019**, *7*, 465–475. [CrossRef]
21. Lee, S.M.; Chen, Z.Y.; Sheu, S.C.; Chen, C.W. Optimizing brewing beer production using *Aspergillus oryzae* solid-state fermentation of sorghum koji as an adjunct. *Int. J. Food Prop.* **2023**, *26*, 3065–3081. [CrossRef]
22. Cabrera, C.; Giménez, R.; López, M.C. Determination of tea components with antioxidant activity. *J. Agric. Food Chem.* **2003**, *51*, 4427–4435. [CrossRef] [PubMed]
23. Amarowicz, R.; Dykes, G.A.; Pegg, R.B. Antibacterial activity of tannin. constituents from Phaseolus vulgaris, Fagoypyrum esculentum, Corylus avellana and Juglans nigra. *Fitoterapia* **2008**, *79*, 217–219. [CrossRef]
24. Damiani, E.; Bacchetetti, T.; Padella, L.; Tiano, L.; Carloni, P. Antioxidant activity of different white teas: Comparison of hot and cold tea infusions. *J. Food Compost. Anal.* **2014**, *33*, 59–66. [CrossRef]
25. Liu, P.P.; Deng, Y.L.; Yin, J.F.; Zhang, Y.N.; Chen, G.S.; Wang, F.; Chen, J.X.; Yuan, H.B.; Xu, Y.Q. Quantitative analysis of the taste and its correlation research of chemical constituents of green tea (in Chinese). *J. Chin. Inst. Food Sci. Technol.* **2014**, *14*, 173–181.
26. Kumar, V.; Joshi, V.K.; Vyas, G.; Thakur, N.S.; Sharma, N. Process Optimization for the Preparation of Apple Tea Wine with Analysis of Its Sensory and Physico-chemical Characteristics and Antimicrobial Activity against Food-borne Pathogens. *Nutrafoods* **2016**, *15*, 111–121. [CrossRef]
27. Xu, Y.-Q.; Zou, C.; Gao, Y.; Chen, J.-X.; Wang, F.; Chen, G.-S.; Yin, J.-F. Effect of the type of brewing water on the chemical composition, sensory quality and antioxidant capacity of Chinese teas. *Food Chem.* **2017**, *263*, 142–151. [CrossRef] [PubMed]
28. Zhang, L.; Cao, Q.Q.; Granato, D.; Xu, Y.Q.; Ho, C.T. Association between chemistry and taste of tea: A review. *Trends Food Sci. Technol.* **2020**, *101*, 139–149. [CrossRef]
29. Alcázar, A.; Ballesteros, O.; Jurado, J.M.; Pablos, F.; Martín, M.J.; Vilches, J.L.; Navalón, A. Differentiation of Green, White, Black, Oolong, and Pu-erh Teas According to Their Free Amino Acids Content. *J. Agric. Food Chem.* **2007**, *55*, 5960–5965. [CrossRef]

nutrients

MDPI

Article

Assessment of Water Intake among Chinese Toddlers: The Report of a Survey

Yiding Zhuang [1,†], Zhencheng Xie [1,†], Minghan Fu [1], Hongliang Luo [2], Yitong Li [2], Ye Ding [1,*] and Zhixu Wang [1,3,*]

1 Department of Maternal, Child and Adolescent Health, School of Public Health, Nanjing Medical University, Nanjing 211166, China; zhuangyiding@stu.njmu.edu.cn (Y.Z.); zhenchengxie@njmu.edu.cn (Z.X.)
2 Danone Open Science Research Center for Life-Transforming Nutrition, Shanghai 201204, China
3 The Institute of Nutrition and Food Science, Nanjing Medical University, Nanjing 211166, China
* Correspondence: dingye@njmu.edu.cn (Y.D.); zhixu_wang@163.com (Z.W.); Tel.: +86-025-8686-8459 (Y.D.); +86-025-8686-8458 (Z.W.)
† These authors contributed equally to this work.

Abstract: Toddlerhood (aged 13~36 months) is a period of dietary transition, with water intake being significantly influenced by parental feeding patterns, cultural traditions, and the availability of beverages and food. Nevertheless, given the lack of applicable data, it is challenging to guide and evaluate the water intake of toddlers in China. In this study, our objectives were to assess the daily total water intake (TWI), evaluate the consumption patterns of various beverages and food sources contributing to the TWI, determine the conformity of participants to the adequate intake (AI) recommendation of water released by the Chinese Nutrition Society, and analyze the various contributors to the daily total energy intake (TEI). The data for the assessment of water and dietary intake were obtained from the cross-sectional dietary intake survey of infants and young children (DSIYC, 2018–2019). A total of 1360 eligible toddlers were recruited in the analysis. The differences in related variables between two age groups were compared by Mann–Whitney U test and Chi-Square test. The potential correlation between water and energy intake was examined utilizing age-adjusted partial correlation. Toddlers consumed a median daily TWI of 1079 mL, with 670 mL (62.3%, $r = 0.752$) derived from beverages and 393 mL (37.7%, $r = 0.716$) from foods. Plain water was the primary beverage source, contributing 300 mL (52.2%, $r = 0.823$), followed by milk and milk derivatives (MMDs) at 291 mL (45.6%, $r = 0.595$). Notably, only 28.4% of toddlers managed to reach the recommended AI value. Among these, toddlers obtain more water from beverages than from foods. The median daily TEI of toddlers was 762 kcal, including 272 kcal from beverages (36.4%, $r = 0.534$) and 492 kcal from foods (63.6%, $r = 0.894$). Among these, the median daily energy intake from MMDs was 260 kcal, making up 94.6% of the energy intake from beverages ($r = 0.959$). As the pioneer survey on TWI of toddlers in China based on nationally representative data, attention to the quality and quantity of water intake and actions to better guide parents by both individuals and authorities are eagerly anticipated. Additionally, the revision of the reference value of TWI for Chinese toddlers is urgently required.

Keywords: China; toddlers; total water intake; water from beverages; water from foods; adequate intake

1. Introduction

Toddlerhood (aged 13~36 months) is a crucial period during which toddlers gradually shift from a milk-based diet into a more diversified eating pattern. This transition is imperative for ensuring the intake of energy, water, and other nutrients, which are crucial for supporting the growth and development of toddlers. Furthermore, this transition contributes to the formation of healthy eating behaviors, which have a long-lasting and profound impact on their overall well-being [1,2].

Water makes up a significant portion of human tissue, accounting for 60~65% of adult body weight and 70~80% in toddlers [3]. Water also plays a crucial role in a variety of physiological processes, metabolic functions, and temperature balance [4]. Healthy adults have efficient mechanisms for regulating water balance in their bodies, whereas toddlers lack these mechanisms. Toddlerhood is typically one of the periods where children engage in increased physical activities, accompanied by more active sweat glands and higher water loss. However, their renal function remains immature, and their ability to perceive and express thirst is relatively weak. Furthermore, toddlers tend to be more susceptible to diarrhea, vomiting, and fever [5–7]. Consequently, all these factors put them at risk of dehydration. The total water intake (TWI) comprises water obtained from both beverages and foods. Understanding the TWI of a certain population and its sources is crucial to ensure adequate water intake for that population. In recent years, many countries have developed recommended values for the TWI of residents by utilizing population water intake data. For toddlers, however, the deficiency of basic data on their water intake poses a challenge to the development of adequate intake (AI) value. For instance, for toddlers aged 13~23 months, the European Food Safety Authority established an AI value of 1100 to 1200 mL/day for TWI, determined through interpolation due to the unavailability of intake data [8]. The AI value of TWI for toddlers, as proposed by the Chinese Nutrition Society (CNS), is mathematically calculated from the sum of water provided by breast milk and complementary food. The CNS reported that the average water intake from breast milk was 480 mL/day, while water from complementary food was unknown. Thus, in accordance with the WHO recommendation of 550 kcal/d of energy for complementary foods and in consideration of recommended dietary allowance raised in the United Sates, which indicated that infants and children required 1.5 mL of water for every 1 kcal of energy consumed, the water intake from complementary food is extrapolated at 825 mL/d. Therefore, the AI value of TWI for toddlers in China is set as 1300 mL/d (480 mL from breast milk and 825 mL from complementary foods) [9].

As mentioned above, toddlerhood is a period of transformation in dietary patterns, so toddlers' TWI is highly influenced by their parents' feeding patterns, cultural customs, and accessibility to beverages and food, which vary between China and other countries. These differences among various countries may result in toddlers having different levels of TWI, with varying contributions from beverages and food to their TWI. Quite a few researchers have voiced their concerns about whether these recommendations, not grounded on actual survey data, were suitable in China [10,11]. Therefore, initiating a nationwide survey on the TWI of toddlers (aged 13~36 months) in China will serve as a pivotal step in developing recommended values for water intake for this population. Currently, as far as we know, there have been only a handful of surveys available on the water intake of children in China, and none of these surveys specifically targeted toddlers or considered the concept of TWI [11–14].

Therefore, based on the nationally representative sample in China, our primary objective was to assess the daily TWI and total energy intake (TEI) among toddlers of different ages, as well as the beverage-food consumption patterns and contributions to their TWI and TEI. The secondary aim was to conduct a comparative analysis between the TWI of toddlers and the AI value established by the CNS, and to assess the extent to which beverages and food sources contribute to the TWI in scenarios based on their compliance with the AI value [9]. The ultimate goal was to examine the correlation between the water and energy intake from different sources, the TWI and TEI, as well as water or energy derived from beverages and foods. We further hypothesized that the participants' TWI did not meet the recommended AI value, and there was a strong correlation between TWI and the intake of plain water.

2. Materials and Methods

2.1. Study Sample

Our study utilized data from the 2018–2019 cross-sectional dietary intake survey on infants and young children in China (DSIYC). In view of the regional differences in geographic location, urban-rural disparities, income inequality, and annual birth rate, the provinces and municipalities covered were as follows: Beijing, Shanghai, Anhui, Fujian, Guangdong, Hebei, Henan, Hubei, Jiangsu, Liaoning, Sichuan, Yunnan, and Zhejiang, representing 13 out of 34 provincial administrative regions in China. Each region was subdivided into urban and rural areas, with a city established in each region to facilitate the survey. Subsequently, all the study samples were randomly recruited guided by the Maternal and Child Health Centre.

The inclusion criteria for children were: healthy children aged 13 to 36 months who were full-term singletons born at 37 to 42 weeks' gestation, weighing 2500 to 4000 g. Children who were born preterm (gestational age < 37 weeks), post-term (gestational age ≥ 42 weeks), low birth weight (birth weight < 2500 g), or macrosomia (birth weight > 4000 g) were excluded. The parents/guardians of these children must have sufficient understanding, cooperation, and cognitive ability as well as question-and-answer capacity.

To determine the required sample size for the study, we used the formula $N = \frac{(Z_\alpha \times S)^2}{d^2}$ with a 95% confidence level ($Z_\alpha = 1.96$), an estimated standard deviation of 1524 mL for TWI intake (S), and the maximum permissible error of 100 mL (d). This calculation suggested we needed at least 887 participants. Finally, water and dietary data of 1360 participants aged 13~36 months were collected, including 682 aged 13~24 months, and 678 aged 25~36 months [10,15].

2.2. Data Collection and Analysis

Well-trained investigators utilized an online diary, compatible with Phone/Tablet/iPad devices to assess the dietary intake of children. The online diary included a comprehensive food list with a food atlas to accurately estimate the water and dietary intake. Additional details have been presented in our previous report [16]. For children who were being breastfed, the frequency and duration of their mother's daily breastfeeding were recorded, providing an estimate of their breast milk intake. Due to practical considerations, the online diaries were recorded two days (one weekday and one weekend) for breastfed children and four days (two weekdays and two weekends) for non-breastfed children.

The study was conducted in collaboration with Danone Open Science Research Centre for Life-Transforming Nutrition (Shanghai, China). Data collection was outsourced to a third-party organization, Taylor Nelson Sofres (Shanghai, China). Prior to the formal survey, a pilot survey was conducted to testify the feasibility and scientific validity of dietary data collection. All investigators must undergo training in survey techniques prior to conducting the study. In order to ensure authenticity of the data submitted by the participants and to prevent duplication, investigators were available to guide participants to upload their dietary data, thus maintaining a high level of participation. The raw data retrieved from the original recordings were promptly reviewed and scrutinized for the completeness and validity of the results.

Beverages are categorized into the following five main groups: plain water, milk and milk derivatives (MMDs), fruit and vegetables juices (FVJs), sugar-sweetened drinks (SSDs), and plant-based protein drinks (PBDs) (Table 1). Foods are categorized into the following five main groups: staple food, dishes, soup, porridge, and snacks (Table 2). Furthermore, TEI was also determined. The water content and energy for each food were calculated according to the Chinese Food Composition Table (6th edition) [17]. The primary reference level for TWI was derived from the AI values established by CNS for different age groups [9].

Table 1. Classification of the beverage types.

Classification	Detailed Beverage Types
Plain water	Tap water, Bottled water (mineral water, purified water).
Milk and milk derivatives	Liquid milk (low fat milk, full fat milk), Powdered milk (whole or skimmed powdered milk), Yogurt, etc.
Fruit and vegetables juices	100% Fruit juice (bottled or freshly squeezed watermelon juice, pear juice, orange juice, etc.), 100% Fruit and vegetable juice (bottled or freshly squeezed cucumber and pineapple mixed juice, orange and carrot mixed juice, apple, and celery mixed juice, etc.).
Sugar-sweetened drinks	Carbonated beverages (Cola, Sprite, Fanta, etc.), Tea flavored beverage (bottled/boxed ice red tea, ice green tea, jasmine tea, etc.), Fruit flavored beverage (peach flavored, lemon flavored, grapefruit flavored, etc.), Milk flavored beverage (chocolate milk, banana milk, coconut milk, etc.), Functional beverages (probiotics, dietary fiber, vitamins, etc.), Milky tea (bubble tea, coconut jelly milk tea, fruit milk tea, etc.), etc.
Plant-based protein drinks	Soybean milk, almond juice, walnut juice, etc.

Table 2. Classification of food types.

Classification	Detailed Food Types
Staple food	
Cereals and cereals products	Rice, Steamed buns, Bread, Steamed twisted rolls, Fried dough sticks, Noodles, etc.
Compound processed products	Dumplings, Wonton, Hamburger, Sandwich, etc.
Others	Sweet potato, Starch vermicelli, etc.
Dishes	
Vegetables	Spinach, Cabbage, Skunk cabbage, Rape, Leaf lettuce, Amaranth, Chinese leek, Chrysanthemum, Garlic bolt, Cauliflower, Broccoli flower, Rice stem, Chinese cabbage, Carrot, Potato, Chinese yam, Taro, Bamboo shoot, Onion, Garlic, Lotus root, Cucumber, Summer squash, Towel gourd, Pumpkin, White gourd, Bitter gourd, Eggplant, Cowpea, Snap bean, Snow pea, Green soybean, Pea, Soybean sprout, Green gram sprouts, Mushroom, Laver, Sea frozen vegetables, Kelp seaweed, etc.
Soybeans and their products	Tofu, Shredded tofu, Dried bean curd, Tofu skin, etc.
Seafoods	River white fish, Grass carp, Catfish, Loach, Carp, Yellow croaker, Perch, Sea shrimp, River shrimp, Lobster, Sea crab, River crab, Oysters, Clams, Scallops, Cuttlefish, Octopus, etc.
Livestock meat	Pork, Pork (heart/liver/kidney/ear/blood/intestine), Trotters, Bacon, Spam, Beef, Beef tongue, Tripe, Mutton, Lamb liver, Lamb belly, etc.
Poultry	Chicken, Chicken (wings/legs/claw/liver, heart/gizzards/blood), Duck, Duck (wings/claw/tongue/intestine/liver/gizzard/blood), Goose, Goose (wings/foot/liver/gizzard/blood), Quail, Pigeon, etc.
Eggs	Hen's egg, Duck egg, Goose egg, Quail egg, etc.
Porridge	Rice porridge, Millet porridge, Corn porridge, Black rice, and mixed bean porridge, etc.
Soup	Tomato egg soup, Bone soup, etc.
Snacks	
Fruits	Apple, Pear, Hawthorn, Peach, Apricot, Plum, Papaya, Cherry, Grape, Persimmon, Tangerines, Grapefruit, Watermelon, Honeydew melon, Strawberry, Kiwifruit, Mulberry, Banana, Mango, Litchi, Pitaya, Durian, etc.
Nuts	Walnut, Chestnut, Pine nut, Peanut, Sunflower seed, Pumpkin seed, watermelon seed, etc.
Others	Cake, Cookies, Chocolate, Ice cream, etc.

2.3. Statistical Analysis

Dietary data from the two- or four-day surveys were inputted into Excel sheets. Continuous variables with a normal distribution were expressed by their mean and standard deviation (SD). Those with a non-normal distribution were denoted by their median and interquartile range (IQR). Categorical variables were denoted by their frequency (n) and percentage (%). Mann–Whitney U test and Chi-Square test were employed to compare the differences in related variables between age groups. Moreover, age-adjusted partial correlation was utilized to assess the potential correlation. All data were calculated and analyzed via the SPSS software package version 26.0 (IBM, New York, NY, USA). A two-sided p value of <0.05 was considered as statistical significance.

3. Results

3.1. Age of the Study Participants

In summary, a total of 1360 eligible participants aged from 13 to 36 months (mean age: 24.4 ± 6.9 months) were included in the analysis. Among them, there were 670 boys (mean age: 24.4 ± 6.9 months) and 690 girls (mean age: 24.4 ± 6.9 months), accounting for 49.3% and 50.7%, respectively.

3.2. The Contribution of Water Intake from Different Sources

The data in Table 3 showed that the median daily TWI of the study population aged 13 to 36 months was 1079 mL. Of this, 670 mL (62.3%) came from beverages and 393 mL (37.7%) came from foods. In the water intake from beverages, plain water served as the main source, accounting for 300 mL (52.2%), followed by MMDs at 291 mL (45.6%), while the water intake from FVJs, SSDs, and PBDs was almost negligible, with a median of 0 mL. In terms of water intake from foods, staple food accounted for the highest contribution at 27.3%, followed by porridge (24.7%), dishes (22.4%), snacks (13.0%), and soup (12.6%).

Table 3. Daily water intake (mL/day) and contribution of water intake from different sources (%) among children aged 13~36 months.

Variables	Total (n = 1360)		13~24 Months (n = 682)		25~36 Months (n = 678)		Z	P_1	Z	P_2
	P50 (P25, P75)	%	P50 (P25, P75)	%	P50 (P25, P75)	%				
Total water intake	1079 (886, 1336)	/	1015 (812, 1218)	/	1180 (949, 1428)	/	8.400	<0.001	/	/
Water from beverages	670 (516, 860)	62.3	651 (508, 834)	64.6	696 (527, 898)	60.1	2.927	0.003	−6.698	<0.001
Plain water	300 (200, 500)	52.2	300 (200, 400)	46.9	400 (250, 550)	57.6	7.768	<0.001	9.874	<0.001
MMDs	291 (200, 401)	45.6	336 (223, 430)	51.8	263 (183, 364)	39.6	−7.018	<0.001	−10.868	<0.001
FVJs	0 (0, 0)	0.6	0 (0, 0)	0.5	0 (0, 0)	0.6	0.999	0.318	0.935	0.350
SSDs	0 (0, 0)	0.8	0 (0, 0)	0.5	0 (0, 0)	1.1	5.651	<0.001	5.593	<0.001
PBDs	0 (0, 0)	0.8	0 (0, 0)	0.3	0 (0, 0)	1.1	6.471	<0.001	6.444	<0.001
Water from foods	393 (270, 529)	37.7	336 (229, 463)	35.4	449 (319, 589)	39.9	10.306	<0.001	6.698	<0.001
Staple food	84 (52, 138)	27.3	78 (44, 133)	29.3	91 (60, 144)	25.4	4.561	<0.001	−2.805	0.005
Dishes	77 (41, 129)	22.4	65 (32, 110)	22.0	90 (52, 138)	22.8	7.321	<0.001	1.751	0.080
Soup	38 (1, 80)	12.6	25 (0, 65)	11.6	50 (15, 95)	13.5	7.321	<0.001	4.395	<0.001
Porridge	76 (22, 160)	24.7	59 (16, 120)	23.6	95 (26, 192)	25.8	5.647	<0.001	1.848	0.065
Snacks	39 (18, 74)	13.0	36 (15, 95)	13.5	42 (20, 81)	12.5	3.569	<0.001	−1.480	0.139

MMDs: Milk and milk derivatives; FVJs: fruit and vegetable juices; SSDs: sugar-sweetened drinks; PBDs: plant-based protein drinks. P_1 was for the differences in the daily water intake (mL/day) between two age groups through statistical comparisons. P_2 was for the differences in the contribution of water intake from different sources to TWI, water from beverages and water from foods (%) between two age groups through statistical comparisons.

Upon further analysis by age group, the older group exhibited a significant increase in TWI, the water intake coming from beverages, plain water, SSDs, PBDs, food, staple foods, dishes, soup, porridge, and snacks ($p < 0.05$), compared to the younger group; however, their water intake from MMDs significantly decreased from 336 mL to 263 mL ($p < 0.05$). The results with regard to the contribution of water intake from various sources revealed that, in the older group, the contribution of water from beverage to TWI, the contribution of water from MMDs to the water from beverages, and the contribution of water from staple food to the water from foods were significantly lower than those in the younger group ($p < 0.05$). Conversely, the water intake from beverages derived from plain water,

SSDs, or PBDs, food-derived water, and soup-derived water all contributed considerably more water to TWI than did the younger group ($p < 0.05$).

3.3. Adherence to the AI Value: Comparisons with the TWI

As shown in Figure 1, based on their TWI, the participants were divided into four groups compared with the AI value, <50%, 50~75%, 75~100%, and ≥100%, which corresponded to <650 mL, 650~975 mL, 975~1300 mL, and ≥1300 mL, respectively. Among the total population, only 28.4% of children had a TWI equal to or exceeding 1300 mL. After being categorized by age, the results indicated that the proportion of children with a TWI of 1300 mL or more and within the range of 975–1300 mL was notably higher in the older group compared to the younger group (36.3% vs. 20.5%, 35.3% vs. 33.6%), which were statistically significant ($p < 0.05$).

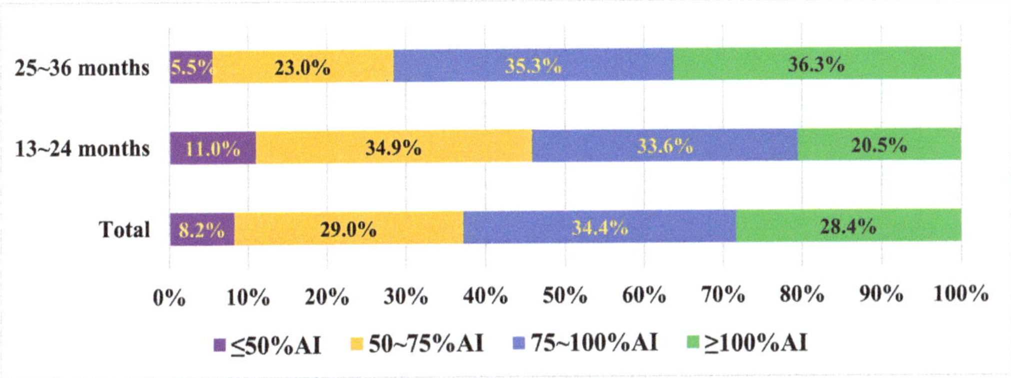

Figure 1. Percentages (%) of children aged 13~36 months according to compliance with AI value of TWI set by the Chinese Nutrition Society by age, segmented based on 50%AI, 75%AI, and 100%AI; AI: adequate intake, The chi-square test (χ^2) was used to analyze the differences, yielding a chi-square value of 59.270 ($p < 0.05$), indicating a statistically significant difference between the two groups.

As depicted in Table 4, based on whether their actual TWI reached the AI of 1300 mL, children aged 13~24 months and 25~36 months were divided into two groups. Their daily TWI and contributions of water intake from different sources were also analyzed.

The median daily TWI among children aged 13~24 months whose daily TWI met the AI value (n = 140, group 1) was 1511 mL. This intake was primarily derived from beverages (989 mL, 63.9%) compared to foods (530 mL, 36.1%). For those who failed to meet the AI value (n = 542, group 2), the median daily TWI was much lower (926 mL) compared to group 1. The water intake from beverages (601 mL) and foods (306 mL) was significantly lower in group 2 compared to group 1, even though the water contributions to TWI from beverages (64.8%) and foods (35.2%) were similar between the two groups. When analyzing water from beverages, it was shown that the water intake from plain water and MMDs was significantly higher in group 1 compared to group 2 (500 mL vs. 265 mL and 408 mL vs. 307 mL, respectively) ($p < 0.05$). The water intake from staple foods, dishes, soup, porridge, and snacks in group 1 significantly surpassed those in group 2 in terms of water intake from foods ($p < 0.01$).

Table 4. Daily water intake (mL/day) and contribution of water intake from different sources (%) among children aged 13~36 months categorized by AI values of TWI set by the Chinese Nutrition society.

Variables	13~24 Months (n = 682)								25~36 Months (n = 678)							
	≥AI (n = 140), Group 1		<AI (n = 542), Group 2		Z	P_1	Z	P_2	≥AI (n = 246), Group 3		<AI (n = 432), Group 4		Z	P_1	Z	P_2
	P50 (P25, P75)	%	P50 (P25, P75)	%					P50 (P25, P75)	%	P50 (P25, P75)	%				
Total water intake	1511 (1366, 1689)	/	926 (759, 1073)	/	−18.256	<0.001	/	/	1517 (1401, 1712)	/	1008 (835, 1164)	/	−21.669	<0.001	/	/
Water from beverages	989 (831, 1186)	63.9	601 (475, 714)	64.8	−14.486	<0.001	0.839	0.401	955 (774, 1136)	61.0	578 (459, 724)	59.5	−16.968	<0.001	−1.154	0.248
Plain water	500 (400, 800)	53.9	265 (150, 400)	45.1	−12.155	<0.001	−5.387	<0.001	600 (500, 800)	62.4	300 (200, 500)	54.8	−14.716	<0.001	−5.192	<0.001
MMDs	408 (340, 526)	44.4	307 (210, 405)	53.6	−7.635	<0.001	5.400	<0.001	301 (221, 421)	34.7	241 (172, 337)	42.3	−5.932	<0.001	5.315	<0.001
FVJs	0 (0, 0)	0.7	0 (0, 0)	0.5	−2.661	0.008	−2.391	0.017	0 (0, 0)	0.7	0 (0, 0)	0.6	−3.037	0.002	−2.626	0.009
SSDs	0 (0, 0)	0.5	0 (0, 0)	0.5	0.014	0.989	0.122	0.903	0 (0, 0)	1.2	0 (0, 0)	1.0	−2.659	0.008	−2.107	0.035
PBDs	0 (0, 0)	0.5	0 (0, 0)	0.3	−0.586	0.558	−0.522	0.602	0 (0, 0)	1.0	0 (0, 0)	1.3	−0.121	0.904	0.341	0.733
Water from foods	530 (400, 669)	36.1	306 (213, 404)	35.2	−11.649	<0.001	−0.839	0.401	581 (470, 741)	39.0	386 (276, 490)	40.5	−12.496	<0.001	1.154	0.248
Staple food	121 (75, 231)	30.5	69 (39, 113)	29.1	−7.716	<0.001	−0.443	0.658	112 (70, 178)	23.9	83 (55, 126)	26.3	−5.497	<0.001	2.963	0.003
Dishes	94 (56, 156)	21.3	56 (28, 101)	22.3	−6.123	<0.001	0.673	0.501	118 (63, 186)	21.4	79 (49, 125)	23.7	−5.709	<0.001	2.326	0.020
Soup	58 (0, 100)	13.4	20 (0, 58)	11.0	−4.886	<0.001	−1.948	0.051	75 (31, 138)	14.4	38 (10, 77)	13.1	−6.143	<0.001	−1.957	0.050
Porridge	85 (27, 179)	21.3	56 (16, 114)	24.1	−3.283	0.001	1.231	0.218	160 (53, 289)	27.8	80 (17, 152)	24.4	−6.103	<0.001	−2.037	0.042
Snacks	57 (28, 95)	13.5	31 (14, 59)	13.5	−5.613	<0.001	0.176	0.861	59 (27, 103)	12.5	35 (19, 67)	12.5	−5.730	<0.001	0.107	0.915

MMDs: Milk and milk derivatives; FVJs: fruit and vegetable juices; SSDs: sugar-sweetened drinks; PBDs: plant-based protein drinks. P_1 was for the differences in the daily water intake (mL/day) between two age groups through statistical comparisons. P_2 was for the differences in the contribution of water intake from different sources to TWI, water from beverages and water from foods (%) between two age groups through statistical comparisons.

The median daily TWI of children aged 25–36 months was 1517 mL for those whose daily TWI achieved the AI value ($n = 246$, group 3). The contributions from food and beverages were 581 mL (39.0%) and 955 mL (61.0%), respectively. For those whose daily TWI did not reach the AI value ($n = 432$, group 4), the median daily TWI (1008 mL) was significantly lower than that of group 3. Although the contribution of water in beverages (59.5%) and food (40.5%) in group 4 was similar to that in group 3, the water intake from beverages (578 mL) and foods (386 mL) was significantly lower than those in group 3 ($p < 0.05$). Further analysis of the contribution of water intake from different sources revealed a trend similar to that of children aged 13~24 months.

3.4. Daily TEI from Different Sources

As detailed in Table 5, study participants had a median daily TEI of 762 kcal, with 272 kcal from beverages and 492 kcal from foods, representing 36.4% and 63.6% of TEI, respectively. In the energy intake from beverages, the median daily energy intake from MMDs was 260 kcal, making up 94.6% of the energy intake from beverages, while the energy intake from FVJs, SSDs, and PBDs was minimal, with a median of 0 kcal/day. In terms of daily energy intake from foods, staple food contributed the most calories with a total of 201 kcal (43.0%), followed by dishes (143 kcal, 33.4%), snacks (74 kcal, 18.3%), and porridge (8 kcal, 5.3%).

Table 5. Daily energy intake (kcal/day) and contribution of energy intake from different sources (%) among children aged 13~36 months.

Variables	Total (n = 1360)		13~24 Months (n = 682)		25~36 Months (n = 678)		Z	P_1	Z	P_2
	P50 (P25, P75)	%	P50 (P25, P75)	%	P50 (P25, P75)	%				
Total energy intake	762 (589, 990)	/	719 (559, 923)	/	811 (622, 1035)	/	5.250	<0.001	/	/
Energy from beverages	272 (185, 367)	36.4	285 (207, 373)	40.8	248 (164, 356)	32.0	−4.694	<0.001	−9.390	<0.001
MMDs	260 (174, 349)	94.6	279 (195, 364)	96.6	229 (154, 336)	92.6	−5.285	<0.001	−5.218	<0.001
FVJs	0 (0, 0)	1.7	0 (0, 0)	1.4	0 (0, 0)	2.0	0.840	0.401	1.076	0.282
SSDs	0 (0, 0)	2.0	0 (0, 0)	1.2	0 (0, 0)	2.8	6.127	<0.001	6.302	<0.001
PBDs	0 (0, 0)	1.7	0 (0, 0)	0.8	0 (0, 0)	2.6	6.442	<0.001	6.612	<0.001
Energy from foods	492 (331, 684)	63.6	422 (280, 611)	59.2	535 (388, 740)	68.0	8.526	<0.001	9.390	<0.001
Staple food	201 (115, 311)	43.0	179 (100, 284)	43.6	223 (144, 336)	42.4	6.199	<0.001	−1.171	0.242
Dishes	143 (94, 216)	33.4	119 (70, 187)	31.9	167 (121, 244)	34.8	10.680	<0.001	4.488	<0.001
Porridge	8 (2, 30)	5.3	6 (2, 25)	5.6	11 (3, 32)	5.1	3.422	0.001	1.135	0.256
Snacks	74 (37, 135)	18.3	67 (32, 127)	18.9	81 (41, 142)	17.7	3.364	0.001	−1.194	0.232

MMDs: Milk and milk derivatives; FVJs: fruit and vegetable juices; SSDs: sugar-sweetened drinks; PBDs: plant-based protein drinks. P_1 was for the differences via statistical comparisons between age groups in the daily energy intake (kcal/day) of different sources; P_2 was for the differences via statistical comparisons between age groups in the contribution of energy intake from different sources to total energy intake/energy from beverages/energy from foods (%).

Further analysis by age group found that, compared with the younger group, the older group showed a significant increase in TEI, energy intake from food, staple foods, dishes, porridge, and snacks ($p < 0.05$) while experiencing a significant decrease in the energy intake from beverages and MMDs ($p < 0.05$). As for the contribution of energy intake, the results revealed that the older group had significantly lower contributions of energy from beverages to TEI and from MMDs to beverage energy than those in the younger group ($p < 0.05$), whereas the contribution of energy from SSDs or PBDs to the energy from beverages, energy from food to TEI, and the energy from dishes to energy from foods were significantly greater than those in the younger group ($p < 0.05$).

3.5. Partial Correlation Analysis

A partial correlation analysis was performed after controlling for age to investigate the correlations between various variables, including TWI, water and energy intake from beverages and different categories, water and energy intake from food and different categories, and TEI (Table 6). In terms of water intake, the TWI demonstrated a moderate correlation with water from beverages (r = 0.752), plain water (r = 0.634), and foods (r = 0.716). The water intake from beverages exhibited a strong correlation with plain water (r = 0.823), while the water intake from foods was moderately correlated with that from staple food, dishes, and porridge (r = 0.671, 0.694, and 0.654, respectively). In terms of energy intake, there was a robust correlation between TEI and energy from foods (r = 0.894), whereas TEI and energy from staple food and dishes had a moderate correlation (r = 0.621 and 0.708, respectively). We observed a robust correlation between energy from beverages and energy from MMDs (r = 0.959), while the energy from foods was moderately correlated with energy from staple food and dishes (r = 0.713 and 0.775, respectively). Regarding the correlation between energy and water intake, the TEI showed moderate correlations with water from foods and dishes (r = 0.657 and 0.644, respectively), while energy from foods showed moderate correlations with water from foods (r = 0.750) and dishes (r = 0.718).

Table 6. Partial correlation through various variables, including TWI, water and energy intake from beverages and their different sources, water and energy intake from foods and their different sources, and TEI (adjusted for age).

Variables	Total Water Intake	Water from Beverages	Water from Foods	Total Energy Intake	Energy from Beverages	Energy from Foods
Total water intake	1.000	0.752 ***	0.716 ***	0.562 ***	0.325 ***	0.490 ***
Water from beverages	0.752 ***	1.000	0.077 **	0.182 ***	0.421 ***	−0.009
Water from plain water	0.634 ***	0.823 ***	0.087 **	0.011	0.024	0.000
Water from MMDs	0.382 ***	0.595 ***	−0.052	0.249 ***	0.681 ***	−0.073 **
Water from FVJs	0.308 ***	0.114 ***	0.345 ***	0.295 ***	0.073 **	0.309 ***
Water from SSDs	0.066 *	0.067 *	0.029	0.116 ***	0.133 ***	0.067 *
Water from PBDs	0.249 ***	0.044	0.331 ***	0.170 ***	−0.068 *	0.236 ***
Water from foods	0.716 ***	0.077 **	1.000	0.657 ***	0.045	0.750 ***
Water from staple food	0.513 ***	0.099 ***	0.671 ***	0.510 ***	0.046	0.576 ***
Water from dishes	0.472 ***	0.019	0.694 ***	0.644 ***	0.075 **	0.718 ***
Water from soup	0.295 ***	0.119 ***	0.320 ***	0.129 ***	0.086 **	0.106 ***
Water from porridge	0.421 ***	−0.017	0.654 ***	0.227 ***	−0.076 **	0.307 ***
Water from snacks	0.377 ***	0.048	0.519 ***	0.508 ***	0.100 ***	0.545 ***
Total energy intake	0.562 ***	0.182 ***	0.657 ***	1.000	0.534 ***	0.894 ***
Energy from beverages	0.325 ***	0.421 ***	0.045	0.534 ***	1.000	0.099 ***
Energy from MMDs	0.313 ***	0.438 ***	0.009	0.461 ***	0.959 ***	0.036
Energy from FVJs	0.217 ***	0.101 ***	0.221 ***	0.233 ***	0.087 **	0.228 ***
Energy from SSDs	0.040	0.048	0.010	0.124 ***	0.216 ***	0.032
Energy from PBDs	0.243 ***	0.037	0.328 ***	0.173 ***	−0.066 *	0.239 ***
Energy from foods	0.490 ***	−0.009	0.750 ***	0.894 ***	0.099 ***	1.000
Energy from staple food	0.367 ***	0.057 *	0.494 ***	0.621 ***	0.034	0.713 ***
Energy from dishes	0.353 ***	−0.062 *	0.599 ***	0.708 ***	0.112 ***	0.775 ***
Energy from porridge	0.354 ***	−0.018	0.554 ***	0.293 ***	−0.056*	0.374 ***
Energy from snacks	0.256 ***	0.05	0.009	0.378 ***	0.587 ***	0.109 ***

MMDs: Milk and milk derivatives; FVJs: fruit and vegetable juices; SSDs: sugar-sweetened drinks; PBDs: plant-based protein drinks. * $p < 0.05$, ** $p < 0.01$, *** $p < 0.001$.

4. Discussion

As far as current research is concerned, this represents the first-ever and an innovative attempt to look at the daily TWI and TEI as well as consumption patterns of toddlers (13–36 months) in China using a nationally representative sample. Encouragingly, the study revealed positive consumption trends, with plain water and MMDs being the primary contributors to water from beverages, and minimal consumption of SSDs. Nevertheless, a tremendous gap existed, as the water intake of most participants failed to meet the recommended AI value of water set by the CNS.

Before the study was conducted, the TWI of Chinese children aged 13~36 months remained unknown [9]. Until now, there have been only a few studies on the water intake of Chinese children, each with its own limitations. For instance, in Wang XJ et al.'s survey, despite having a large study sample of 5914, it only included children from four cities with higher economic levels, which was not representative of the general population [18]. Moreover, some studies focused only on measuring total fluid intake (TFI) rather than TWI, and have mistakenly equated these two [11,14,19]. During the toddler stage, both liquid and solid foods that provide energy, water, and other nutrients are crucial for meeting the child's energy and nutrient needs [20,21]. As the study's results showed, water intake from food in this age group did contribute to variations in TWI, and water intake from food should be calculated instead of being neglected. Therefore, our present survey could provide some valuable insights and strategies for subsequent investigations into TWI of children.

Currently, reports on toddlers' water intake have been limited to only three countries. Of these studies, the ones conducted in France and Mexico were particularly noteworthy as they considered both water intake from food and beverages, whereas the study in the United States solely focused on beverage consumption [15,22,23]. Our research indicated that the median daily TWI of toddlers aged 13~24 months in China was 1015 mL, with 651 mL derived from beverages and 336 mL from foods, while that of toddlers aged 25~36 months was 1180 mL, with 696 mL coming from beverages and 449 mL from foods. This intake was slightly higher than the median daily TWI reported in France for different age groups: 999 mL (12~17 months), 985 mL (18~23 months), 991 mL (24~29 months), and 991 mL (30~35 months) [22], and lower than the mean daily TWI of toddlers aged 1–3 years reported in Mexico, which was 1283 mL, with 743 mL from beverages and 540 mL from foods [15]. These differences may be attributed to racial diversity, geographical location, cultural customs, economic level, dietary habits, feeding patterns, accessibility to beverages and food, and survey methodologies.

Plain water contributed differently to TWI, as we discovered when we looked more closely at the contributions of other sources. In our study, plain water contributed 30.3% to TWI for 13~24-month-olds, compared to 13% (12~17 months) and 19% (18~23 months) in France. For 25~36-month-olds, plain water contributed 34.6% in our study, versus 23% (24~29 months) and 25% (30~35 months) in France [22]. Additionally, our study determined that the median daily plain water intake among toddlers aged 13 to 36 months was 300 mL, whereas the average plain water intake in Mexico was 227 mL [15]. The intake of plain water and its contribution to TWI among toddlers in China are significantly higher compared to those in other regions. This difference could be explained by the widespread preference among Chinese individuals for drinking boiled water. This preference is not only a convention in China that is believed to aid digestion, improve circulation, and maintain internal balance, but also a common practice to eliminate pathogens from tap water, thereby enhancing its safety for drinking [24,25]. Despite the notably higher intake of plain water among Chinese toddlers, it did not translate into a significantly elevated water intake from beverages compared to toddlers in Mexico. This might be due to the higher intake of coffee, soda, and other beverages among Mexican toddlers, potentially influenced by the Mexican culture that often includes these beverages as part of their daily routine [15]. In our previous survey, we found that the daily consumption of soft drinks among preschoolers in China was 40 g/d [26]. However, in this study, we were pleased to see that Chinese toddlers consumed significantly fewer soft drinks, especially SSBs. The rising popularity of SSBs poses risks of obesity, dental issues, and other health concerns [26–29]. Therefore, it is advisable to avoid their consumption in early childhood.

In our study, Chinese toddlers showed poor compliance with the AI value (1300 mL/d) for TWI set by the CNS, with 71.6% of all ages failing to reach this value (79.5% for aged 13~24 months and 63.7% for aged 25~36 months). This is similar to the situation of 13~36-month-old toddlers in Mexico, with 62% of them not reaching the AI value (1300 mL/d) set by the Institute of Medicine in TWI [15]. However, in the survey conducted in France, nearly half of the toddlers aged 12~23 months had a TWI that reached the AI value (49% for those aged 12~17 months and 50% for those aged 18~23 months), whereas the proportion of toddlers aged 24~29 months and 30~35 months whose TWI did not reach the AI value was 88% and 84%, respectively. This may be due to the different AI recommendations set by EFSA (1100 mL/d for aged 12~23 months and 1300 mL/d for aged 24~35 months) [8,22]. The discrepancy in TWI recommendations among regions worldwide underscores the limitations of relying solely on TWI to assess hydration status. Introducing a gold-standard biomarker or other physical methods is deemed necessary, although some studies suggested that such approaches were still in their early stages [30].

Based on whether the TWI of toddlers reached the AI value set by the CNS, two age groups in our study showed similar trends in the intake of plain water and MMDs, which were the main contributors to water from beverages. For those with TWI reaching AI levels, their intake of plain water and MMDs was higher compared to those without TWI reaching AI levels. While parents often focus on offering complementary foods, maintaining a regular and adequate intake of water is important because toddlers have limited ability to perceive and express thirst, and they often encounter difficulties in self-feeding [30,31]. In addition to hydration, MMDs are considered to be good sources of calcium, phosphorus, magnesium, vitamin A, B vitamins, and vitamin D, making them ideal dietary components for improving the nutritional status of toddlers [26,32]. However, our research findings indicate that as children age, the consumption of MMDs significantly decreases. Especially for toddlers whose TWI did not reach the AI value, the median intake of MMDs in both age groups was lower than the recommended level of 350 g/d by the CNS [31]. These findings prompt us to reconsider strategies for enhancing dietary nutrition during early life stages and emphasize the importance of cultivating the habit of drinking MMDs and adequate water in toddlers under practical guidance.

There are some limitations to be acknowledged. First, the water demand of the human body is closely influenced by climate, temperature, and humidity [33]. Despite using a nationwide population sample, the collection of short-term water and dietary data cannot fully represent the annual food and water intake status of toddlers. Second, although a food atlas was used in this study, it should be noted that toddlers have relatively low levels of food and beverage intake each time, which may introduce bias into their estimation. Even with well-trained researchers conducting surveys, the amount of food or beverage provided by guardians to toddlers may differ from their actual intake, as adding supplementary food and learning to eat independently can lead to wastage or extra intake [30]. Third, toddlers are naturally active, and their physical activity levels have potential impacts on the water intake. However, the level of physical activity was not considered in this study. Finally, as a cross-sectional study, the relationship between water intake and children's growth and development cannot be explored.

5. Conclusions

In summary, the results of this study suggest the need to pay attention to the quality of water intake among Chinese toddlers, which can be achieved by strengthening the education of guardians and implementing more scientific drinking water plans. In the future, broader coverage and more scientific data collection methods can be carried out to further verify the results of this study, and longitudinal follow-up can be conducted to explore the impact of water intake on children's growth and development.

Author Contributions: Design and concept of the work, Y.D. and Z.W.; analysis and interpretation of the work, Y.D., Y.Z., Z.X., M.F. and H.L.; writing of the manuscript, Y.Z., Z.X. and Y.D.; review and editing of the manuscript, Z.W., M.F. and Y.L.; supervision, Y.D. and Z.W.; funding acquisition, Y.D. All authors have read and agreed to the published version of the manuscript.

Funding: This research was provided by the Undergraduate Training Programs for Innovation and Entrepreneurship of Jiangsu province (No. 202210312048Z), as well as a project supported by the Priority Academic Program Development of Jiangsu Higher Education Institutions (Public Health and Preventive Medicine).

Institutional Review Board Statement: The study was conducted in accordance with the Declaration of Helsinki, and approved by the Ethics Committee of Nanjing Medical University on 21 August 2018. The ethical approval project identification code is 2018-1123.

Informed Consent Statement: Informed consent was obtained from all subjects involved in the study.

Data Availability Statement: The data utilized in this study are available upon request from the corresponding author.

Acknowledgments: We extend our heartfelt gratitude to all staffs conducting the DSIYC from 2018 to 2019, as well as all the participants, for their collaboration and support throughout the duration of the study.

Conflicts of Interest: The authors declare that they have no competing interests. Authors H.L. and Y.L., who are employees of Danone, were involved in this study. H.L. participated in the analysis and interpretation of data, and Y.L. contributed to the review and editing of the manuscript. We affirm that any potential conflicts of interest were managed appropriately to ensure the objectivity and integrity of the research. The study design, data collection, analysis, interpretation, manuscript writing, and decision to publish were all conducted independently, without any commercial influence from Danone or any other entity.

Abbreviations

TWI	Total water intake
AI	Adequate intake
TEI	Total energy intake
MMDs	Milk and milk derivatives
FVJs	Fruit and vegetables juices
SSDs	Sugar sweetened drinks
PBDs	Plant-based protein drinks
SD	Standard deviation
IQR	Median and interquartile range
CNS	China Nutrition Society
DSIYC	The cross-sectional dietary intake survey of infants and young children

References

1. Pang, W.W.; McCrickerd, K. The impact of feeding experiences during infancy on later child eating behaviours. *Curr. Opin. Clin. Nutr. Metab. Care* **2021**, *24*, 246–251. [CrossRef] [PubMed]
2. Riley, L.K.; Rupert, J.; Boucher, O. Nutrition in Toddlers. *Am. Fam. Physician* **2018**, *98*, 227–233. [PubMed]
3. Jéquier, E.; Constant, F. Water as an essential nutrient: The physiological basis of hydration. *Eur. J. Clin. Nutr.* **2010**, *64*, 115–123. [CrossRef] [PubMed]
4. Popkin, B.M.; D'Anci, K.E.; Rosenberg, I.H. Water, hydration, and health. *Nutr. Rev.* **2010**, *68*, 439–458. [CrossRef] [PubMed]
5. Chang, H.; Guo, J.; Wei, Z.; Huang, Z.; Wang, C.; Qiu, Y.; Xu, X.; Zeng, M. Aetiology of acute diarrhoea in children in Shanghai, 2015–2018. *PLoS ONE* **2021**, *16*, e0249888. [CrossRef]
6. Shields, T.M.; Lightdale, J.R. Vomiting in Children. *Pediatr. Rev.* **2018**, *39*, 342–358. [CrossRef] [PubMed]
7. Vega, R.M.; Avva, U. Pediatric Dehydration. In *StatPearls*; StatPearls Publishing LLC.: Treasure Island, FL, USA, 2023.
8. EFSA Panel on Dietetic Products, Nutrition, and Allergies (NDA). Scientific Opinion on Dietary Reference Values for water. *EFSA J.* **2010**, *8*, 1459.
9. Chinese Nutrition Society. *Chinese Dietary Reference Intakes for China 2023*; People's Health Publishing House: Beijing, China, 2023.
10. Ding, Y.; Xie, Z.; Lu, X.; Luo, H.; Pan, H.; Lin, X.; Wu, J.; Wang, Z. Water Intake in Pregnant Women in China, 2018: The Report of a Survey. *Nutrients* **2021**, *13*, 2219. [CrossRef]
11. Zhang, N.; Morin, C.; Guelinckx, I.; Moreno, L.A.; Kavouras, S.A.; Gandy, J.; Martinez, H.; Salas-Salvadó, J.; Ma, G. Fluid intake in urban China: Results of the 2016 *Liq.In*[7] national cross-sectional surveys. *Eur. J. Nutr.* **2018**, *57*, 77–88. [CrossRef]
12. Du, S.M.; Pan, H.; Hu, X.Q.; Zhang, Q.; Wang, X.J.; Lu, L.X.; Zou, S.R.; Fan, J.W.; Li, X.H.; Ma, G.S. Water intake of primary and middle school students in four cities of China. *Zhonghua Yu Fang Yi Xue Za Zhi* **2013**, *47*, 210–213.
13. Morin, C.; Gandy, J.; Brazeilles, R.; Moreno, L.A.; Kavouras, S.A.; Martinez, H.; Salas-Salvadó, J.; Bottin, J.; Guelinckx, I. Fluid intake patterns of children and adolescents: Results of six *Liq.In*[7] national cross-sectional surveys. *Eur. J. Nutr.* **2018**, *57*, 113–123. [CrossRef] [PubMed]
14. Du, S.M.; Hu, X.Q.; Zhang, Q.; Wang, X.J.; Pan, H.; Gao, J.M.; Song, J.; Gao, C.L.; He, Z.F.; Ma, G.S. Daily intake of plain water and beverages of primary and middle school students in four cities of China. *Zhonghua Yu Fang Yi Xue Za Zhi* **2013**, *47*, 202–205. [PubMed]
15. Piernas, C.; Barquera, S.; Popkin, B.M. Current patterns of water and beverage consumption among Mexican children and adolescents aged 1–18 years: Analysis of the Mexican National Health and Nutrition Survey 2012. *Public Health Nutr.* **2014**, *17*, 2166–2175. [CrossRef] [PubMed]

16. Ding, Y.; Han, F.; Xie, Z.; Li, G.; Zhuang, Y.; Yin, J.; Fu, M.; You, J.; Wang, Z. Dairy fortification as a good option for dietary nutrition status improvement of 676 preschool children in China: A simulation study based on a cross-sectional diet survey (2018–2019). *Front. Nutr.* **2022**, *9*, 1081459. [CrossRef] [PubMed]
17. Yang, Y. *China Food Composition Tables*, 6th ed.; Peking University Medical Press: Beijing, China, 2018.
18. Wang, X.J.; Hu, X.Q.; Du, S.M.; Pan, H.; Zhang, Q.; Yan, L.; Wang, Z.Y.; Nie, S.P.; Yang, J.B.; Ma, G.S. The relationship between within-day drinking occasion and intake amount of water of primary and middle school students in four cities of China. *Zhonghua Yu Fang Yi Xue Za Zhi* **2013**, *47*, 206–209. [PubMed]
19. Martinez, H.; Guelinckx, I.; Salas-Salvadó, J.; Gandy, J.; Kavouras, S.A.; Moreno, L.A. Harmonized Cross-Sectional Surveys Focused on Fluid Intake in Children, Adolescents and Adults: The Liq.In^7 Initiative. *Ann. Nutr. Metab.* **2016**, *68* (Suppl. S2), 12–18. [CrossRef] [PubMed]
20. WHO Guidelines Approved by the Guidelines Review Committee. *Infant and Young Child Feeding: Model Chapter for Textbooks for Medical Students and Allied Health Professionals*; World Health Organization: Geneva, Switzerland, 2009.
21. Kuo, A.A.; Inkelas, M.; Slusser, W.M.; Maidenberg, M.; Halfon, N. Introduction of Solid Food to Young Infants. *Matern. Child Health J.* **2011**, *15*, 1185–1194. [CrossRef] [PubMed]
22. Chouraqui, J.P.; Thornton, S.N.; Seconda, L.; Kavouras, S.A. Total water intake and its contributors in infants and young children. *Br. J. Nutr.* **2022**, *128*, 531–541. [CrossRef] [PubMed]
23. Grimes, C.A.; Szymlek-Gay, E.A.; Nicklas, T.A. Beverage Consumption among U.S. Children Aged 0–24 Months: National Health and Nutrition Examination Survey (NHANES). *Nutrients* **2017**, *9*, 264. [CrossRef]
24. Yang, H.; Wright, J.A.; Gundry, S.W. Household water treatment in china. *Am. J. Trop. Med. Hyg.* **2012**, *86*, 554–555. [CrossRef]
25. World Health Organization. *WHO Technical Breif: Boiled Water*; World Health Organization: Geneva, Switzerland, 2015.
26. Zhuang, Y.; Yin, J.; Han, F.; You, J.; Ding, Y.; Wang, Z. Impact of Replacing Soft Drinks with Dairy Products on Micronutrient Intakes of Chinese Preschool Children: A Simulation Study. *Nutrients* **2023**, *15*, 4071. [CrossRef] [PubMed]
27. Chi, D.; Scott, J.M. Added Sugar and Dental Caries in Children: A Scientific Update and Future Steps. *Dent. Clin. N. Am.* **2019**, *63*, 17–33. [CrossRef] [PubMed]
28. Gui, Z.; Zhu, Y.; Cai, L.; Sun, F.; Ma, Y.; Jing, J.; Chen, Y. Sugar-Sweetened Beverage Consumption and Risks of Obesity and Hypertension in Chinese Children and Adolescents: A National Cross-Sectional Analysis. *Nutrients* **2017**, *9*, 1302. [CrossRef] [PubMed]
29. Luger, M.; Lafontan, M.; Bes-Rastrollo, M.; Winzer, E.; Yumuk, V.; Farpour-Lambert, N. Sugar-Sweetened Beverages and Weight Gain in Children and Adults: A Systematic Review from 2013 to 2015 and a Comparison with Previous Studies. *Obes. Facts* **2017**, *10*, 674–693. [CrossRef] [PubMed]
30. Warren, J.; Guelinckx, I.; Livingstone, B.; Potischman, N.; Nelson, M.; Foster, E.; Holmes, B. Challenges in the assessment of total fluid intake in children and adolescents: A discussion paper. *Eur. J. Nutr.* **2018**, *57*, 43–51. [CrossRef]
31. Chinese Nutrition Society. *Dietary Guidelines for Chinese*; People's Medical Publishing House: Beijing, China, 2016.
32. Scholz-Ahrens, K.; Ahrens, F.; Barth, C. Nutritional and health attributes of milk and milk imitations. *Eur. J. Nutr.* **2020**, *59*, 19–34. [CrossRef]
33. Manz, F. Hydration in Children. *J. Am. Coll. Nutr.* **2007**, *26*, 562S–569S. [CrossRef]

beverages

MDPI

Article

Emphasizing the Potential of Attenuated Total Reflectance–Fourier Transform Infrared (ATR-FTIR) Spectroscopy Combined with Chemometrics, for Classification of Greek Grape Marc Spirits

Evangelia Anastasia Tsapou [1], Vassilia J. Sinanoglou [2], George Ntourtoglou [1] and Elisabeth Koussissi [1,*]

[1] Department of Wine, Vine and Beverage Sciences, University of West Attica, Agiou Spyridonos, 12243 Egaleo, Greece; etsapou@uniwa.gr (E.A.T.); gntourtoglou@uniwa.gr (G.N.)
[2] Laboratory of Chemistry, Analysis & Design of Food Processes, Department of Food Science and Technology, University of West Attica, Agiou Spyridonos, 12243 Egaleo, Greece; vsina@uniwa.gr
* Correspondence: ekoussissi@uniwa.gr

Abstract: Grape marc spirits, such as the Greek tsipouro/tsikoudia, reflect the cultural heritage of winemaking traditions worldwide. This study explored the application of Attenuated Total Reflectance–Fourier Transform Infrared (ATR-FTIR) spectroscopy combined with chemometrics for its potential as a fast classification methodology for spirit characterization. ATR-FTIR spectra from thirty-nine products revealed distinctive bands corresponding to various chemical constituents, such as alcohols, organic acids, water, carbohydrates, and phenols. Principal Component Analysis (PCA) was performed on all acquired ATR-FTIR data and 78.50% of the total variance in the data was explained. Also, partial least squares–discriminant analysis (PLS-DA), used for the classification of products based on their major geographic origin, gave a correct classification of 89.5% for the north and 83.3% for the south of Greece. Classification of the type of distillations used was with 74.36% accuracy. Significant markers were identified through analysis, such as those associated with the O-H bending vibrations of phenols or alcohols, contributing to the discrimination of grape marc spirits from Crete when compared with the other four main geographical origin designations. By combining ATR-FTIR spectroscopy with chemometrics, this research gave insights into the origins and compositional variations of the spirits, providing an opportunity for a quality control assessment tool.

Keywords: attenuated total reflectance–fourier transform infrared (ATR-FTIR); classification; grape marc spirits; tsipouro/tsikoudia; chemometrics

1. Introduction

The historical significance of grape marc spirits is deeply rooted in the cultural tapestry of winemaking traditions, offering a journey of technology and tradition through the ages. Each variant, whether the Iberian Peninsula's Orujo, Italy's Grappa, Cyprus's Zivania, or the diverse expressions of grape marc worldwide, encapsulates the essence of its region and the craftsmanship of its makers [1]. These spirits emerge as liquid ambassadors, reflecting the historical evolution of viticulture and the utilization of grape pomace—a byproduct of winemaking that, rather than being discarded, undergoes meticulous distillation to yield beverages of distinct character [2]. The global diversity of grape marc spirits is a testament to the terroir-driven approach, where geographical and climatic conditions shape the unique qualities of the pomace. EU Regulation 1576/89 established overarching production protocols and set standard analytical composition limits for traditional alcoholic beverages. Additionally, it delineated specific production methods and stricter analytical parameters tailored to particular geographical denominations [3]. Further guidance on procedural matters such as labeling and geographical indications for European grape marc distillates was outlined in EU Regulation 2019/78.

Greece, in accordance with Annex III of Regulation (EC) No. 110/2008, designated four viticultural regions for the production of protected designation of origin (PDO) grape marc spirits. Those regions include tsipouro of Thessaly, tsipouro of Macedonia, tsikoudia of Crete, and tsipouro of Tyrnavos [1]. The first three derive their names from their respective geographical areas, while Tyrnavos, situated in the Thessaly region, serves as a significant viticultural hub. Before 1988 in Greece, production licenses for tsipouro were restricted to certain days of the year and typically granted only to farmers, often grape growers residing in economically disadvantaged areas. However, the landscape has evolved significantly since then, with tsipouro gaining recognition and popularity among consumers worldwide, including wine enthusiasts. Homemade tsipouro, now available commercially, is typically consumed in its undiluted form, boasting an ethanol content usually exceeding 50% v/v. Conversely, commercially produced tsipouro undergoes dilution with water before bottling, resulting in an alcohol content ranging between 37.5% and 50% vol. [3,4].

The surge in the popularity of grape marc spirits has, unfortunately, given rise to authenticity challenges [5–7]. Quality control of alcoholic beverages can be conducted through two distinct approaches: sensory and instrumental quality control [5,8,9]. Although sensory information will always be important and relevant, instrumental quality control methods might offer several advantages in terms of reproducibility speed and not being susceptible to adaptation or fatigue [8–10]. Nevertheless, for practical application in the alcoholic beverage industry, those methods must be swift, cost-effective, and capable of delivering reproducible results through continuous operation [11].

Techniques such as UV, Vis, near-infrared (NIR), mid-infrared (MIR), Raman, Fluorescence, nuclear magnetic resonance spectroscopy, and voltammetric techniques have emerged as promising solutions to address the challenges previously outlined [12–17]. Adulteration and counterfeiting pose threats to the distinctive characteristics of these spirits, underscoring the need for advanced analytical methods. Traditional methods like gas chromatography are time-consuming and have a lengthy, complex, and labor-intensive sample preparation process with increased analysis times and capital equipment requirements. In contrast, Fourier Transform Infrared (FT-IR) and MIR spectroscopy appear simpler, faster, and more available with reduced analysis times and instrument costs for comprehensive analysis. Specifically, MIR spectroscopy offers rapid and high-resolution analysis of grape marc spirits, with the fingerprint region (1800–1200 cm^{-1}) revealing unique signatures. Operating between 4000 and 600 cm^{-1}, mid-infrared spectroscopy explores overtone and combination bands, offering nuanced insights into molecular composition. As the demand for authentic and unadulterated beverages grows, those advanced analytical techniques serve as invaluable tools in uncovering the true essence of grape marc spirits. Attenuated Total Reflectance–Fourier Transform Infrared (ATR-FTIR) was chosen due to its widespread use as an analytical instrument, offering the advantage of rapidly generating data with minimal or no sample preparation. ATR-FTIR has been a method of choice in food characterization previously, due to its widespread use as an analytical instrument, offering the advantage of rapidly generating data with minimal or no sample preparation. Additionally, it has the capability to provide key information about the chemicals present in the sample. Previous applications of FTIR data, in conjunction with Principal Component Analysis (PCA), have successfully classified various substances such as vegetable oils, fruit juices, and coffee [18–21], while it was recently suggested as a valid tool for wine authenticity and traceability evaluation [22].

Previously, we utilized novel sensory methodologies [2] and Gas Chromatography–Mass Spectrometry (GC-MS) analyses combined with chemometrics [1] to understand product differentiation within that complex category of spirits. GC-MS analyses were also applied in other studies of grape marc spirits from Greece [3] and also recently together with chemometrics [23]. Schiavone and co-workers (2020) studied the authentication of Grappa spirits by mid- and near-infrared spectroscopies coupled with chemometrics. Nevertheless, to our knowledge there has not been any other published study on FTIR analyses of grape marc spirits from Greece, previously.

The main objective of this work was to study thirty-nine (39) samples of Greek grape marc spirits from diverse distillation processes [24] and specifically and/or broadly defined origin areas [1,23], employing ATR-FTIR spectroscopy band transformation. Additionally, the study aimed to assess the capability of FTIR spectroscopy combined with chemometric tools such as PCA and partial least squares–discriminant analysis (PLS-DA) for classifying Greek grape marc spirits through meaningful insights into the quality information of a representative set of samples.

2. Material and Methods

2.1. Materials

Thirty-nine (39) grape marc spirits were collected from all over Greece, representing all major Greek PDOs (Table 1). The samples were either provided by the producers themselves or purchased in retail markets. Information regarding the grape variety used and production techniques was mostly directly obtained in parallel, from the producers themselves. In the case of samples bought from retailers, information was taken from product labels. Among those products, 29 were produced by local distilleries of the regions, and the remaining 10 were homemade (traditional grape marc spirits), while 20 tsipouro/tsikoudia spirits were made from a single grape variety and 19 were obtained by the distillation of the grape marc from multiple grape varieties. Last but not least, 10 products came out of single distillation, 20 out of double, and 10 out of smaller-scale home distillation (Table 1).

Table 1. Grape marc spirits used in the study. Coding, specific and major regions of origin, and distillation processes used in their making.

Spirit Code	Specific Origin	Major Region	Type (Tsipouro/Tsikoudia)	Distillation Process	Code	Varieties
T1	Tyrnavos	North	Tsipouro	double distillation	CAD2	Moschato Mavro
T2	Tyrnavos	North	Tsipouro	double distillation	CAD2	Moschato Mavro
T3	Tyrnavos	North	Tsipouro	double distillation	CAD2	Moschato Mavro
T4	Tyrnavos	North	Tsipouro	double distillation	CAD2	Moschato Mavro
T5	Tyrnavos	North	Tsipouro	double distillation	CAD2	Malagouzia
T6	Tyrnavos	North	Tsipouro	double distillation	CAD2	Roditis
T7	Thessaly	North	Tsipouro	double distillation	CAD2	Muscat, Roditis, Savatiano
T8	Thessaly	North	Tsipouro	double distillation	CAD2	Xinomavro, Cabernet sauvignon, Chardonnay, Muscat
T9	Thessaly	North	Tsipouro	double distillation	CAD2	Savatiano, Roditis, Muscat
T10	Peloponnese	South	Tsipouro	single distillation	CAD1	Roditis
T11	Peloponnese	South	Tsipouro	double distillation	CAD2	Malagouzia, Chardonnay, Merlot
T12	Peloponnese	South	Tsipouro	single distillation	CAD1	Agiorgitiko, Moschofilero, Muscat
T13	Peloponnese	South	Tsipouro	double distillation	CAD2	Moschofilero
T14	Peloponnese	South	Tsipouro	single distillation	CAD1	Moschofilero, Agiorgitiko
T15	Peloponnese	South	Tsipouro	double distillation	CAD2	Agiorgitiko
T16	Thrace	North	Tsipouro	double distillation	CAD2	Roditis, Moschato Mavro
T17	Thrace	North	Tsipouro	single distillation	CAD1	Muscat, Cabernet Sauvihnon, Syrah, Merlot
T18	Thrace	North	Tsipouro	single distillation	CAD1	Muscat, Merlot, Syrah

Table 1. *Cont.*

Spirit Code	Specific Origin	Major Region	Type (Tsipouro/Tsikoudia)	Distillation Process	Code	Varieties
T19	Epirus	North	Tsipouro	double distillation	CAD2	Debina
T20	Macedonia	North	Tsipouro	double distillation	CAD2	Malagousia
T21	Macedonia	North	Tsipouro	single distillation	CAD1	Muscat, Asyrtiko, Cabernet Sauvignon
T22	Central Greece	North	Tsipouro	single distillation	CAD1	Sabbatiano
T23	Central Greece	North	Tsipouro	single distillation	CAD1	Savatiano, Muscat, Chardonnay
T24	Crete	South	Tsikoudia	double distillation	CAD2	Moschato Mavro
T25	Crete	South	Tsikoudia	double distillation	CAD2	Moschato Mavro
T26	Crete	South	Tsikoudia	double distillation	CAD2	Moschato Mavro
T27	Crete	South	Tsikoudia	single distillation	CAD1	Moschato Mavro
T28	Cyclades	Island	Tsipouro	double distillation	CAD2	Asyrtiko
T29	Cyclades	Island	Tsipouro	double distillation	CAD2	Asyrtiko, Mavrotragano
T30	Tyrnavos	North	Tsipouro	home distillation	HD	Moschato Mavro
T31	Tyrnavos	North	Tsipouro	home distillation	HD	Moschato Mavro
T32	Tyrnavos	North	Tsipouro	home distillation	HD	Moschato Mavro
T33	Tyrnavos	North	Tsipouro	home distillation	HD	Moschato Mavro
T34	Tyrnavos	North	Tsipouro	home distillation	HD	Moschato Mavro
T35	Tyrnavos	North	Tsipouro	home distillation	HD	Moschato Mavro
T36	Crete	South	Tsikoudia	home distillation	HD	Romeiko
T37	Crete	South	Tsikoudia	home distillation	HD	Romeiko
T38	Central Greece	North	Tsipouro	home distillation	HD	Malagouzia, Muscat, Roditis
T39	Central Greece	North	Tsipouro	home distillation	HD	Savatiano, Roditis

2.2. Fourier Transform Infrared Spectroscopy with Attenuated Total Reflectance (ATR-FTIR)

Fourier Transform Infrared Spectroscopy with Attenuated Total Reflectance (ATR-FTIR) was obtained using an IRAffinity-1S FTIR spectrometer (Shimadzu, Kyoto, Japan). The IRE used was diamond. The samples and the background spectra were screened from 4000 to 499 cm^{-1} and the average of 20 scans at 4 cm^{-1} resolution was recorded, while the ATR reference was adjusted to 3284.77 cm^{-1}. The FTIR spectra of tsipouro/tsikoudia spirit samples were subjected to ATR correction, normalization, smoothing, and peak picking treatment. Data processing and analysis was performed by LabSolutions IR software (version 2.21) [25].

2.3. Statistical Analysis

The statistical analysis of Principal Component Analysis (PCA) and partial least squares (PLS) was performed using XLSTAT Addinsoft statistical and data analysis solutions (New York, NY, USA). PCA was initially employed to map all thirty-nine products, based on their total ATR FTIR data. Outlier detection and removal were conducted according to PCA analysis and specifically according to the values on the PC1 axis. Subsequently, the ATR-FTIR spectral data from all spirits were subjected to Variable Importance in Projection (VIP) in conjunction with partial least squares analyses to determine which of the FTIR variables were the most relevant to utilize in further classification analyses, according to different production parameters. Namely, the parameters of raw materials (grapes), specific and major regions of origin, as well as type of distillation method used were used in the PLS-DA. The VIP approach is mainly used in a PLS-based context, where the influence of each instrumental variable in defining the variable space is determined. When a variable has a VIP index higher than 1, it is considered relevant for the classification [1,26].

3. Results and Discussion

3.1. ATR-FTIR Spectral Interpretation of Spirits

ATR-FTIR spectra (Table 2, Figure 1) were interpretated in order to ascertain the chemical profile of the tsipouro/tsikoudia products. Table 2 showcases the FTIR spectral bands spanning from 4000 to 499 cm^{-1}. In this spectral data table, each row represents a different spirit, while the columns denote specific wavelength ranges. The values in each cell correspond to the intensity of spectral response for the respective spirit at the given wavelength range. This dataset provides insights into the chemical compositions and structural properties of the analyzed spirits. The bands ranging from 3500 to 3200 and 1642 to 1648 cm^{-1} signify the stretching and bending vibrations of hydroxyl groups found in water and organic acids [6,27]. Furthermore, the bands at 2986 and 2901 cm^{-1} are corresponded to the C-H stretching vibrations of methyl and methylene groups in alcohols and organic acids, especially ethanol and tartaric acid [28]. The intensity ratio of 2986:2901 cm^{-1} ranges from 4.53 to 6.33, for all tsipouro/tsikoudia samples examined. The band at 1420 cm^{-1} denotes the C-H rocking vibration [27] and at 1450–2460 cm^{-1} is associated with O-H bending deformation [6], while the band at 1385 cm^{-1} is attributed to the C-O stretching vibrations of tartaric acid [28].

Table 2. The spectral absorbance bands (intensities) of thirty-nine (39) grape marc spirits (tsipouro/tsikoudia) after a duplicate analysis.

Spirit Code	590–700	877–879	1043–1045	1083–1085	1275	1350–1310	1384–1385	1419–1420	1450–1460	1643–1646	2901–2904	2980–2986	3500–3200
T1	0.001	0.053	0.223	0.061	0.008	0.003	0.010	0.003	0.019	0.158	0.022	0.124	0.006
T2	0.005	0.049	0.211	0.057	0.008	0.003	0.010	0.003	0.019	0.005	0.019	0.119	0.004
T3	0.003	0.046	0.204	0.056	0.007	0.002	0.011	0.003	0.018	0.003	0.018	0.114	0.003
T4	0.000	0.043	0.195	0.053	0.007	0.003	0.009	0.008	0.016	0.003	0.018	0.109	0.006
T5	0.003	0.050	0.222	0.060	0.008	0.011	0.012	0.009	0.018	0.166	0.021	0.124	0.004
T6	0.002	0.053	0.226	0.062	0.008	0.011	0.012	0.009	0.018	0.169	0.022	0.126	0.006
T7	0.000	0.045	0.202	0.054	0.007	0.003	0.011	0.008	0.017	0.004	0.019	0.110	0.003
T8	0.000	0.051	0.222	0.060	0.008	0.010	0.012	0.003	0.019	0.159	0.021	0.122	0.006
T9	0.001	0.050	0.215	0.058	0.008	0.010	0.009	0.008	0.018	0.004	0.023	0.121	0.014
T10	0.002	0.053	0.226	0.062	0.009	0.011	0.012	0.009	0.018	0.003	0.021	0.128	0.014
T11	0.001	0.050	0.222	0.060	0.008	0.009	0.012	0.009	0.018	0.009	0.020	0.123	0.019
T12	0.000	0.049	0.216	0.059	0.008	0.003	0.012	0.003	0.019	0.157	0.019	0.120	0.003
T13	0.002	0.056	0.237	0.064	0.009	0.011	0.012	0.010	0.019	0.010	0.026	0.132	0.013
T14	0.004	0.047	0.205	0.056	0.008	0.000	0.008	0.003	0.018	0.159	0.019	0.115	0.006
T15	0.000	0.048	0.213	0.058	0.008	0.010	0.012	0.008	0.017	0.170	0.019	0.120	0.005
T16	0.002	0.049	0.215	0.059	0.008	0.010	0.011	0.008	0.017	0.170	0.023	0.120	0.007
T17	0.003	0.052	0.216	0.056	0.009	0.011	0.010	0.009	0.019	0.008	0.021	0.119	0.013
T18	0.004	0.044	0.202	0.056	0.007	0.002	0.008	0.007	0.017	0.004	0.020	0.114	0.003
T19	0.002	0.052	0.222	0.062	0.008	0.002	0.010	0.009	0.018	0.004	0.028	0.127	0.003
T20	0.003	0.049	0.208	0.056	0.007	0.003	0.009	0.008	0.017	0.004	0.020	0.118	0.007
T21	0.002	0.049	0.214	0.058	0.007	0.010	0.010	0.008	0.018	0.003	0.023	0.121	0.006
T22	0.004	0.051	0.222	0.061	0.008	0.010	0.010	0.009	0.019	0.166	0.022	0.124	0.004
T23	0.003	0.048	0.209	0.057	0.008	0.003	0.012	0.002	0.018	0.003	0.021	0.116	0.012
T24	0.000	0.050	0.220	0.060	0.009	0.010	0.012	0.009	0.017	0.169	0.021	0.121	0.002
T25	0.001	0.051	0.223	0.061	0.008	0.009	0.011	0.009	0.018	0.008	0.021	0.125	0.007
T26	0.002	0.048	0.211	0.058	0.008	0.011	0.011	0.009	0.017	0.174	0.022	0.119	0.006
T27	0.004	0.051	0.223	0.060	0.008	0.010	0.011	0.009	0.019	0.175	0.022	0.123	0.013
T28	0.002	0.047	0.210	0.057	0.008	0.002	0.011	0.002	0.018	0.003	0.019	0.115	0.013
T29	0.003	0.045	0.198	0.053	0.007	0.003	0.010	0.007	0.016	0.003	0.022	0.110	0.005
T30	0.002	0.079	0.312	0.087	0.012	0.015	0.019	0.010	0.023	0.164	0.034	0.178	0.007
T31	0.002	0.077	0.304	0.085	0.012	0.014	0.018	0.011	0.022	0.008	0.032	0.173	0.013
T32	0.003	0.081	0.314	0.087	0.012	0.014	0.019	0.011	0.023	0.009	0.032	0.180	0.020
T33	0.002	0.108	0.394	0.114	0.014	0.019	0.024	0.012	0.027	0.009	0.044	0.231	0.008
T34	0.002	0.100	0.375	0.107	0.014	0.017	0.023	0.012	0.024	0.008	0.040	0.219	0.007
T35	0.003	0.069	0.274	0.077	0.011	0.004	0.016	0.010	0.023	0.158	0.028	0.155	0.004
T36	0.000	0.050	0.217	0.059	0.008	0.010	0.012	0.008	0.018	0.164	0.020	0.119	0.013
T37	0.002	0.051	0.221	0.060	0.008	0.009	0.012	0.008	0.017	0.003	0.020	0.123	0.008
T38	0.001	0.045	0.199	0.054	0.007	0.002	0.009	0.008	0.017	0.003	0.019	0.109	0.006
T39	0.003	0.043	0.192	0.052	0.007	0.002	0.008	0.008	0.016	0.004	0.017	0.105	0.013

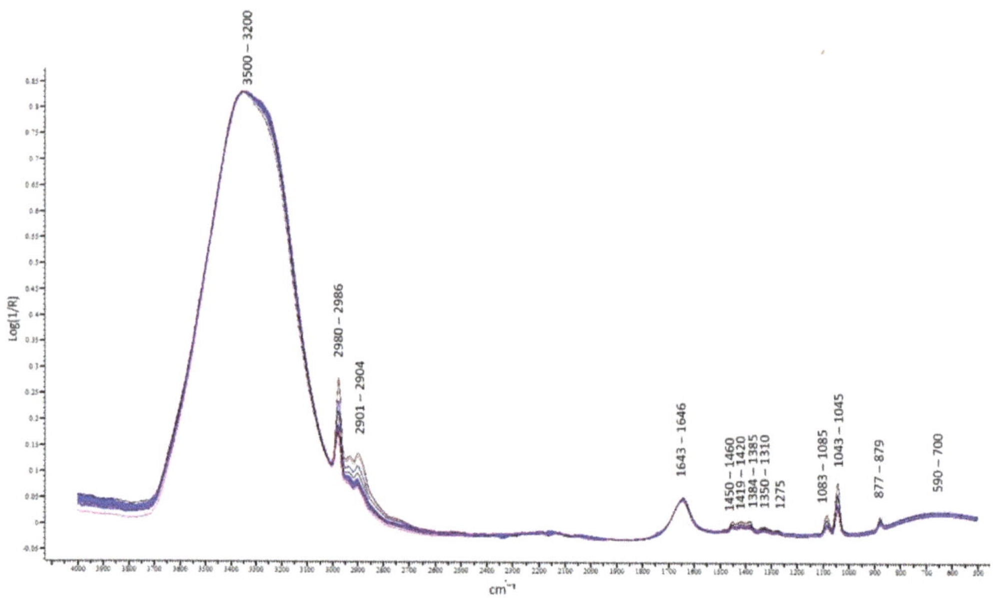

Figure 1. Overlay MIR spectra of thirty-nine (39) samples of tsipouro.

Moreover, the band from 1350 to 1310 cm^{-1} signifies the O-H bending vibration (out of plane) of the phenols or/and alcohols [27], and the band at 1275 cm^{-1} is assigned to the C-O stretching vibrations of organic acids [29]. Additionally, the bands at 1083–1085 cm^{-1} and 1043–1045 cm^{-1} are corresponded to the C-O stretching vibrations of primary alcohols and are very important regions for ethanol and methanol quantification [6,27]. Coldea et al. (2013) [30] noted that ethanol exhibits distinct peaks at 1047 and 1087 cm^{-1}, with the former being the major signal and the latter the minor signal. On the other hand, methanol displays characteristic vibrations at 1020 and 1112 cm^{-1}, with the former being the major signal and the latter the minor signal. Interestingly, regarding ethanol, the intensity ratio of major/minor signals (1043–1045:1083–1085) ranges from 3.46 to 3.85, for all tsipouro samples examined. Moreover, Yadav and Sharma (2019) [11] reported that the bands at 1047 and 1087 cm^{-1} could be attributed to the C-O stretching vibration in fermentation products such as carbohydrates, phenolics, esters, and lactones. Furthermore, the band at 877–879 cm^{-1} is related to the C-H out of plane bending vibrations of aromatic compounds [11,27,29], while the region from 590 to 700 cm^{-1} represents the O-H out of plane bending vibrations of various alcohols [27]. The presence of outliers in the spectral data highlights the diversity and complexity of tsipouro/tsikoudia compositions. According to Table 2 the samples that show the greatest variation in terms of their values in almost all bands are those made from the homemade distillation process [1].

3.2. Classification of the Spirits through Chemometrics

Initially, Principal Component Analysis (PCA) was performed on the averaged values of the duplicate measurements from all acquired ATR-FTIR data directly and 78.50% of the total variance in the data was explained in two principal components (PC1: 66.76%, PC2: 10.74%) with a p-value (two-tailed) < 0.0001 (Figure 1).

Graphical representation of the results is shown in Figure 1, where PC1 is plotted against PC2. In that representation, the main separation of products is driven by some of the distillates from Tyrnavos being clearly differentiated from all other samples with very high scores in PC1. Those were the samples T30–32 and T34, with the spirit T32 showing the highest PC1 score and clearly separated from all the rest. Products T33–35 and T30 were all traditional homemade tsipouro samples from the PDO of Tyrnavos, but so were products T31 and T32, which were largely clustered in first and fourth quadrant all with positive PC1 values but at the same time with variation in their PC1 scores. These results are in agreement with the results from the HS-SPME GC/MS analyses of the same samples [1]. Samples T36–T39 were also traditional homemade spirits from different geographical areas and were all clustered in the second quadrant (Figure 2). The grape varieties used for making the samples did not appear significant in the classification of the spirits from ATR-FTIR data.

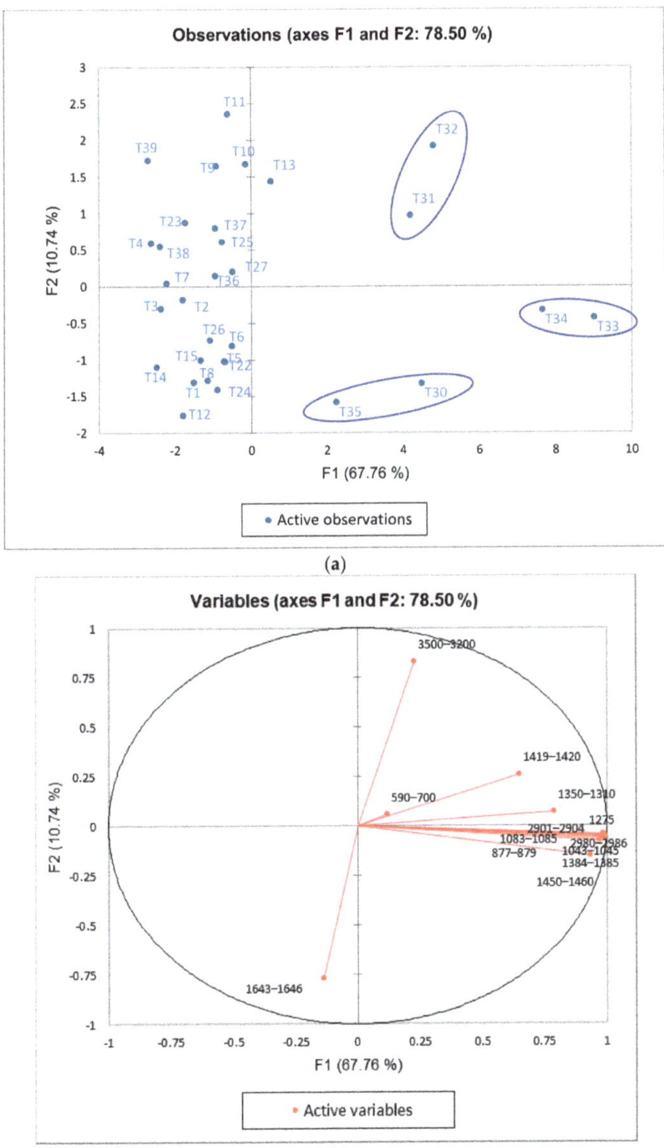

(a)

(b)

Figure 2. Principal Component Analysis (PCA) run on the FT-IR data of 39 Greek grape marc spirit samples: (**a**) PCA biplot and (**b**) the correlation circle of the initial variables used.

3.2.1. Type of Distillation Effect

The clear clusters of homemade spirits that appear in Figure 2 led to the need for further analysis to investigate the factor of the distillation method. Partial least squares–discriminant analysis (PLS-DA) is one of the most widely used classification techniques in chemometrics [1,31,32]. Therefore, a partial least squares–discriminant analysis (PLS-DA) was used to run the same data matrix in order to evaluate the effect of distillation process on the acquired spectral data (Figure 3). Distillation process could be classified in three techniques of copper alempic and fractional distillation among the spirits, namely that of: one distillation (CAD1), two distillations (CAD2), and homemade distillation (HD). The outcome of the PLS-DA approach was a product space with clear clusters according to the distillation technique used to make the spirits at 74.36% correct classification (accuracy) (Figure 3). The sensitivity and selectivity of classification of the PLS-DA method were calculated at 95.40% and 71.43%, respectively. Specifically, the initial classification values were 44.44%, 95.00% and 60.00% for the single distillation (CAD1), double distillation (CAD2), and homemade products (HD), respectively (Figure 2).

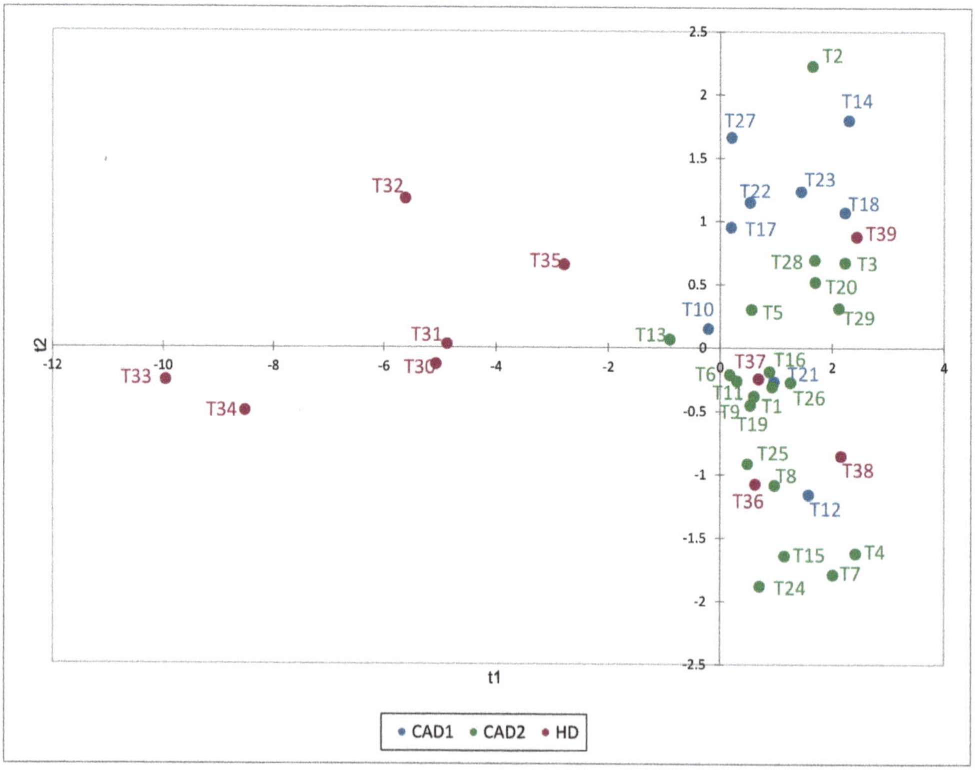

Figure 3. Partial least squares–discriminant analysis (PLS-DA) regression of ATR-FTIR data of the spirits when coded according to the distillation technique used for production. CAD1: spirits made by single distillation; CAD2: spirits made by double distillation; and HD: homemade traditional spirits.

The most important marker for the classification of double-distilled products (CAD2), differentiating them from products of the other distillation techniques, was in the range of 590–700 cm^{-1}, which corresponds to the O-H out of plane bend of alcohols with a variable importance in projection (VIP) value of 1805. The other significant bends in the classification with VIP > 1 were at 1384–1385 cm^{-1}, 1275 cm^{-1}, 877–879 cm^{-1}, 1083–1085 cm^{-1}, 1043–1045 cm^{-1}, and 2980–2986 cm^{-1}. Those markers indicate the C-O and C-H stretching vibrations of alcohols and organic acids (Table 3). In the research of Kokoti et al. (2023), a similar discrimination of distillation process was carried out in tsipouro samples from fractional distillation, copper alembic, and homemade distillation using the volatile profile of the products, measured with SPME-GC/MS analysis. In the main findings of the aforementioned research, the authors demonstrated that the most significant variables in product classification were ethanol, higher alcohols, esters, and terpenes. These results show great relevance to those of our study [24]. Extensive research has underscored the pivotal role of distillation techniques in shaping the distinct qualities of grape marc spirits. Investigation into this domain has unveiled a fundamental divergence in alcohol concentration between traditionally homemade Orujo spirits and their industrially manufactured counterparts in the research of Cortes et al. (2005) [33], who also analyzed their volatile profile, indicating the importance of fraction collection through the distillation process. In spirit production, irrespective of the raw materials utilized, increased distillation iterations are commonly associated with elevated quality attributes of the resulting spirit. Additionally, in the grape marc spirit market, a spirit made with double distillation is considered of higher quality and is priced more expensively. ATR-FTIR analysis, and specifically the aforementioned marker, could play a significant role in the quality control of the category in general, and specifically in the detection of the distillation process used.

Table 3. Variable Importance in Projection in the classification of the spirits based on: (1) number of distillations, (2) major region classification (north–south), and (3) the differentiation of spirits from Crete from those of Thessaly, Tyrnavos, Central Greece, and Peloponnese, respectively.

Classification		Variable	VIP
Number of distillations		590–700	1.805
		877–879	1.238
		1275	1.230
		1083–1085	1.227
		1043–1045	1.226
		1384–1385	1.213
		2980–2986	1.202
North–South		1384–1385	1.409
		1083–1085	1.219
		1043–1045	1.213
		1275	1.255
Crete	Thessaly	1350–1310	1.987
		1419–1420	1.746
	Tyrnavos	1350–1310	1.648
		1419–1420	1.399
		1384–1385	1.302
	Central Greece	1350–1310	1.454
		1384–1385	1.202
	Peloponnese	1350–1310	1.846
		1419–1420	1.823
		1450–1460	1.606

3.2.2. Geographical Origin Effect

Further in the investigation of the important variables from the ATR-FTIR analysis, a differentiation was observed in terms of the geographical compartment of the spirit origin. In particular, samples were divided into two main groups, namely, the northern geographical divisions (Thrace, Macedonia, Epirus, Thessaly, Tyrnavos, and Central Greece) and the southern geographical divisions (Peloponnesus and Crete) (Figure 4). Outlier samples of the initial PCA were not included in the above classification. Those two groups consisted of nineteen and twelve samples for the north and south of Greece, respectively. PLS-DA was again used, and the results are displayed in Figure 4 with a total correct classification of the method of 87.10% (sensitivity and selectivity) and specifically 89.5% correct classification for the northern and 83.3% for the southern-origin spirits (Figure 4). Markers with VIP > 1, which were more important for that classification, were at 1275 cm^{-1} for the C-O stretch of organic acids, at 1384–1385 cm^{-1} for the C-O stretching vibration of tartaric acid, and at 1043–1045 cm^{-1} and 1083–1085 cm^{-1} for ethanol (Table 3).

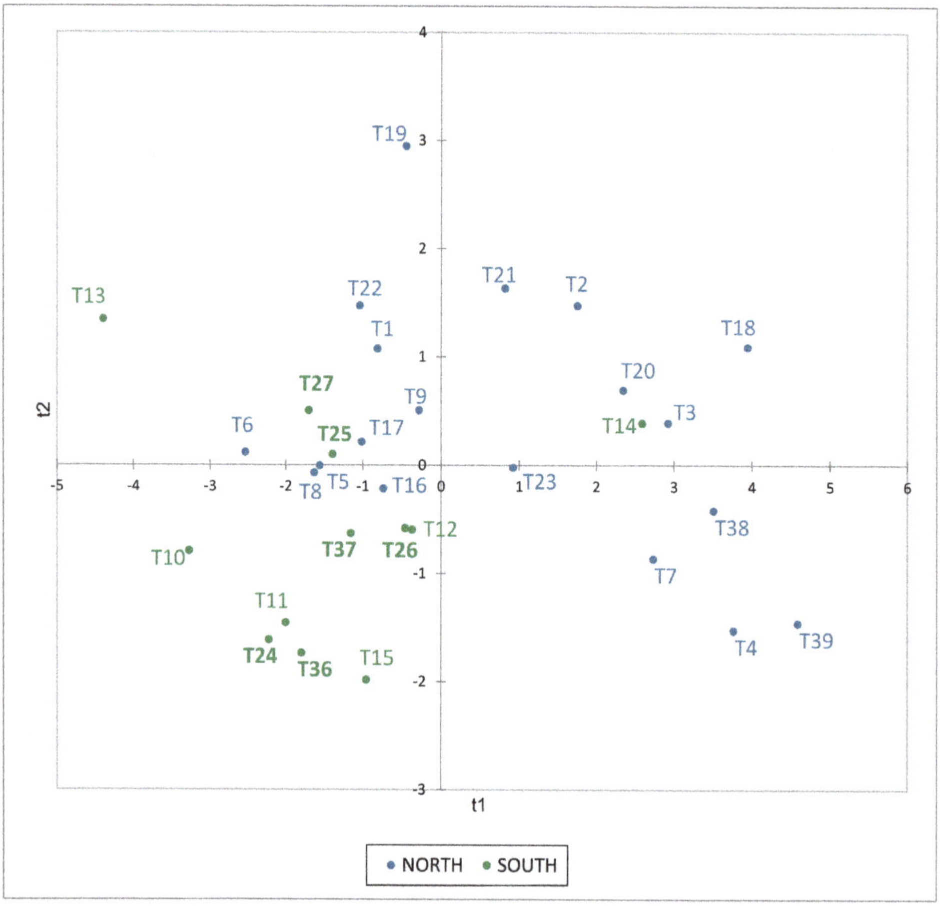

Figure 4. Partial least squares–discriminant analysis (PLS-DA) regression of ATR-FTIR data of the spirits when coded according to major geographical division. Samples coded for north represent samples originating from Thrace, Macedonia, Epirus, Thessaly, Tyrnavos, and Central Greece. Samples coded for south represent samples originating from Peloponnese and Crete.

ATR-FTIR is a method that has been previously applied to other spirits with the aim of categorizing them based on their geographical origin. Specifically, Yadav and Sharma in 2019 used ATR-FTIR in their research to classify illicit liquor from seven different regions of Punjab state of Northern India and achieved an overall rate of 93% of correct discrimination. Also, ATR-FTIR analysis has been previously employed for the classification of unifloral honeys sourced from Sardinia, Italy [34]. This analytical technique yielded promising results in discerning the unique chemical fingerprints of each honey variety, facilitating the identification of their respective origins. Furthermore, in the review research of Basalekou et al. (2020) [22], FT-IR spectroscopy is presented as a successful tool in wine authentication and traceability and can provide classifications based on a large number of parameters such as origin, grape variety, maturation, vintage, and adulteration. Additionally, in the research of Ioannou et al. (2023) [25], Greek honey samples from different botanical and geographical origins were investigated and the results predicted with high overall accuracy the geographical (93.55%) and botanical (96.77%) origin of the honey samples. Moreover, ATR-FTIR spectroscopy holds considerable promise for the classification of natural resins. Research has demonstrated its efficacy in providing a valuable approach to categorize resins into distinct families, leveraging a set of band positions for precise classification [35].

According to Figure 5, the region with the clearest separation from the rest was Crete, whose samples appear to have much stronger clustering compared to the rest of the geographical divisions and can be viewed in that figure by bold writing. Interestingly, the traditional name given for marc spirits from Crete differs from all other Greek grape marc spirits. Namely, grape marc spirits from Crete are called "tsikoudia", whereas marc spirits from all other areas in Greece are called: "tsipouro" (Table 1). Subsequently, we proceeded to a more detailed classification of the spirits by comparing them in groups of two; we used each one of the four represented geographical regions with the largest number of samples, namely: Thessaly, Tyrnavos, Central Greece, and Peloponnese. All above regions were compared to that of Crete with respect to product representation (Figure 5a–c). The aim of this classification was the detection of the most significant FTIR markers for the separation of the Cretan products from those from the rest of the geographical origins. Results are presented in Figure 5a–d from a PLS-DA analysis and in the two principal components a total of 44.78%, 95.34%, 81.84%, and 44.03% of the total of variance was explained for the comparison with Thessaly, Tyrnavos, Central Greece, and Peloponnese, respectively. When a variable has a VIP index higher than one, it is considered relevant for the classification [26]. The most important marker for the classification of Crete from the other regions was in the band 1350–1310 cm^{-1} at all four plots (Table 3). According to the results of Tsapou et al. (2023) [1], where HS-SPME GC/MS analysis was conducted, the region of Crete was also classified mainly because of phenols. Also, as noted in prior studies, materials enriched with phenolic compounds demonstrate a propensity for superior classification in FTIR. FTIR analysis is effective for detecting phenolic compounds due to their unique chemical structures and corresponding infrared absorption spectra. Phenolic compounds contain one or more hydroxyl groups (-OH) attached to an aromatic ring, which gives rise to characteristic absorption bands in the infrared region [36–38].

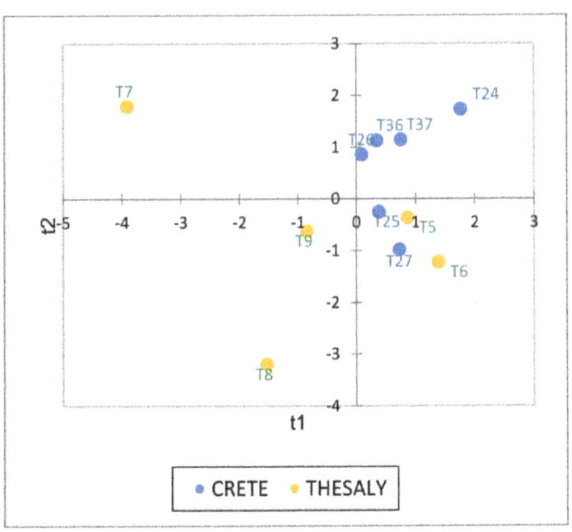

(a)

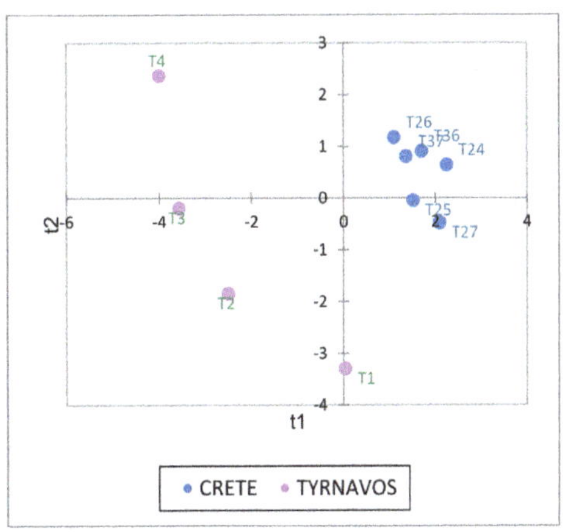

(b)

Figure 5. *Cont.*

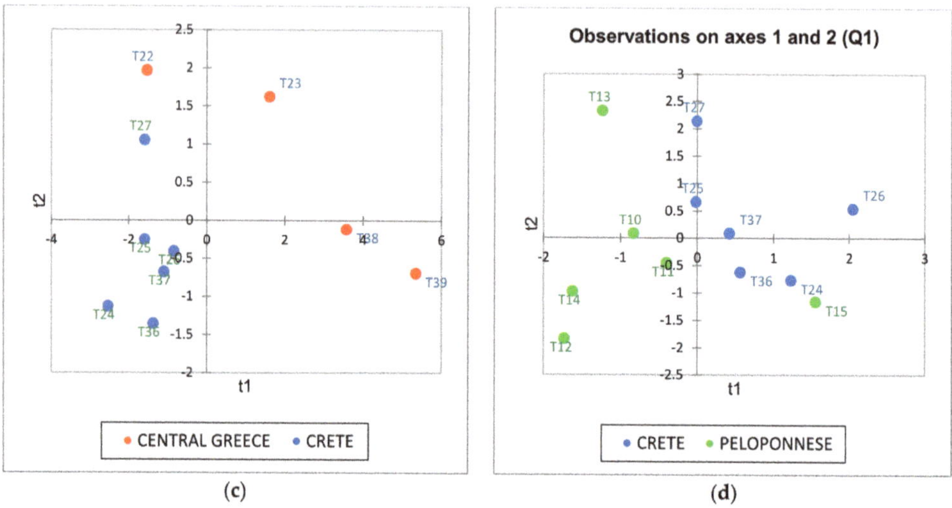

Figure 5. Partial least squares–discriminant analysis (PLS-DA) to compare (**a**) Crete–Thesaly; (**b**) Crete–Tyrnavos; (**c**) Crete–Central Greece; and (**d**) Crete—Peloponnese.

4. Conclusions

Spectroscopic techniques offer comparable accuracy to traditional chemical composition or sensory methods, with the added advantages of being cost-effective and less time-intensive. Although the integration of multivariate data analysis and spectroscopy primarily yields qualitative insights, it is important to note potential inaccuracies stemming from similar environmental conditions, raw materials, preparation techniques, and limited sample sizes for calibration models. The original hypothesis in this study was that we could employ FTIR analysis combined with chemometrics to quickly and accurately classify a relatively large set of grape marc spirits from Greece. Despite the above-mentioned limitations, the combination of ATR–FTIR spectroscopy and chemometric tools proved in our case to be an efficient and accurate approach for classifying the spirit samples. Specifically, according to our findings, the combination of ATR–FTIR with chemometric methods like PCA and PLS-DA led to highly accurate classification of grape marc spirits with 87.1% sensitivity and selectivity of the method based on their geographical origin. From all four designated viticultural regions for the production of PDO grape marc spirits represented in the sample set, the region of Crete was the one most profoundly distinguished. Additionally, when we attempted to classify those spirits based on the distillation technique used, we achieved an accuracy of 74.36% with specific accuracy of 95% for the products resulting from double distillations. Therefore, ATR-FTIR spectroscopy combined with chemometrics could be a useful tool for a fast but efficient classification of complex food products such as the grape marc spirits from Greece.

Author Contributions: Conceptualization E.A.T. and E.K.; methodology and validation E.A.T. and V.J.S. statistical analysis E.A.T. and G.N.; formal analysis V.J.S. and E.A.T., writing—original draft preparation E.A.T. and G.N.; writing—review and editing E.K. and V.J.S.; visualization and supervision E.K. All authors have read and agreed to the published version of the manuscript.

Funding: This research did not receive any specific grant from funding agencies in the public, commercial, or not-for-profit sectors.

Data Availability Statement: The original contributions presented in the study are included in the article, further inquiries can be directed to the corresponding author.

Acknowledgments: Most products used in the study were donated by the producers themselves, a contribution greatly appreciated.

Conflicts of Interest: The authors have no competing interests to declare.

References

1. Tsapou, E.A.; Dourtoglou, V.; Dourtoglou, T.; Sinanoglou, V.; Koussissi, E. Volatile Profile in Greek Grape Marc Spirits with HS-SPME-GC-MS and Chemometrics: Evaluation of Terroir Impact. *ACS Omega* **2023**, *8*, 42803–42814. [CrossRef] [PubMed]
2. Tsapou, E.A.; Tzortzis, P.M.; Koussissi, E. Application of Polarized Projective Mapping combined with Ultra Flash Profiling to a complex– High fatigue product category: The case of Greek grape marc spirits. *Food Qual. Prefer.* **2024**, 105182. [CrossRef]
3. Apostolopoulou, A.A.; Flouros, A.I.; Demertzis, P.G.; Akrida-Demertzi, K. Differences in concentration of principal volatile constituents in traditional Greek distillates. *Food Control* **2005**, *16*, 157–164. [CrossRef]
4. Soufleros, E.H. *Wine and Distillates*; Papageorgiou Pub: Thessaloniki, Greece, 1987.
5. Arslan, M.; Tahir, H.E.; Zareef, M.; Shi, J.; Rakha, A.; Bilal, M.; Huang, X.; Li, Z.; Zou, X. Recent trends in quality control, discrimination and authentication of alcoholic beverages using nondestructive instrumental techniques. *Trends Food Sci. Technol.* **2021**, *107*, 80–113. [CrossRef] [PubMed]
6. Anjos, O.; Santos, A.J.; Estevinho, L.M.; Caldeira, I. FTIR–ATR spectroscopy applied to quality control of grape-derived spirits. *Food Chem.* **2016**, *205*, 28–35. [CrossRef] [PubMed]
7. Schiavone, S.; Marchionni, B.; Bucci, R.; Marini, F.; Biancolillo, A. Authentication of *Grappa* (Italian grape marc spirit) by Mid and Near Infrared spectroscopies coupled with chemometrics. *Vib. Spectrosc.* **2020**, *107*, 103040. [CrossRef]
8. Ross, C.F. Sensory science at the human–machine interface. *Trends Food Sci. Technol.* **2009**, *20*, 63–72. [CrossRef]
9. Sáenz-Navajas, M.P.; Fernández-Zurbano, P.; Ferreira, V. Contribution of nonvolatile composition to wine flavor. *Food Rev. Int.* **2012**, *28*, 389–411. [CrossRef]
10. Tan, J.; Li, R.; Jiang, Z.-T. Chemometric classification of Chinese lager beers according to manufacturer based on data fusion of fluorescence, UV and visible spectroscopies. *Food Chem.* **2015**, *184*, 30–36. [CrossRef]
11. Yadav, P.K.; Sharma, R.M. Classification of illicit liquors based on their geographic origin using Attenuated total reflectance (ATR)–Fourier transform infrared (FT-IR) spectroscopy and chemometrics. *Forensic Sci. Int.* **2019**, *295*, e1–e5. [CrossRef]
12. Arslan, M.; Zou, X.; Tahir, H.E.; Hu, X.; Rakha, A.; Basheer, S.; Hao, Z. Near-infrared spectroscopy coupled chemometric algorithms for prediction of antioxidant activity of black goji berries (*Lycium ruthenicum* Murr.). *J. Food Meas. Charact.* **2018**, *12*, 2366–2376. [CrossRef]
13. Arslan, M.; Zou, X.; Tahir, H.E.; Zareef, M.; Hu, X.; Rakha, A. Total polyphenol quantitation using integrated NIR and MIR spectroscopy: A case study of Chinese dates (*Ziziphus jujuba*). *Phytochem. Anal.* **2019**, *30*, 357–363. [CrossRef]
14. Cozzolino, D.; Kwiatkowski, M.J.; Dambergs, R.G.; Cynkar, W.U.; Janik, L.J.; Skouroumounis, G.; Gishen, M. Analysis of elements in wine using near infrared spectroscopy and partial least squares regression. *Talanta* **2008**, *74*, 711–716. [CrossRef]
15. Tahir, H.E.; Zou, X.; Huang, X.; Shi, J.; Mariod, A.A. Discrimination of honeys using colorimetric sensor arrays, sensory analysis and gas chromatography techniques. *Food Chem.* **2016**, *206*, 37–43. [CrossRef]
16. Tahir, H.E.; Zou, X.; Li, Z.; Shi, J.; Zhai, X.; Wang, S.; Mariod, A.A. Rapid prediction of phenolic compounds and antioxidant activity of Sudanese honey using Raman and Fourier transform infrared (FT-IR) spectroscopy. *Food Chem.* **2017**, *226*, 202–211. [CrossRef]
17. Amorello, D.; Barreca, S.; Gulli, E.; Orecchio, S. Platinum and rhodium in wine samples by using voltammetric techniques. *Microchem. J.* **2017**, *130*, 229–235. [CrossRef]
18. Akbar, Z.; Idroes, R.; Ginting, B.; Karma, T.; Rahimah, S.; Helwani, Z.; Yusuk, M. Identification of Gayo Arabic Coffee Beans and Powder using the FTIR-PCA Combination Method. *IOP Conf. Ser. Mater. Sci. Eng.* **2021**, *1087*, 012059. [CrossRef]
19. Leopold, L.F.; Leopold, N.; Diel, H.; Socaciu, C. Quantification of Carbohydrates in Fruit Juices Using FTIR Spectroscopy and Multivariate Analysis. *Spectroscopy* **2011**, *26*, 93–104. [CrossRef]
20. Rusak, D.A.; Brown, L.M.; Martin, S.D. Classification of vegetable oils by principal component analysis of FTIR spectra. *J. Chem. Educ.* **2003**, *80*, 541. [CrossRef]
21. Zhang, X.-J.; Meng, L.-J.; Lu, Z.-M.; Chai, L.-J.; Wang, S.-T.; Shi, J.-S.; Shen, C.-H.; Xu, Z.-H. Identification of Age-markers Based on Profiling of Baijiu Volatiles over a Two-year Maturation Period: Case Study of Lu-flavor Baijiu. *LWT Food Sci. Technol.* **2021**, *141*, 110913. [CrossRef]
22. Basalekou, M.; Pappas, C.; Tarantilis, P.A.; Kallithraka, S. Wine Authenticity and Traceability with the Use of FT-IR. *Beverages* **2020**, *6*, 30. [CrossRef]
23. Marinaki, M.; Sampsonidis, I.; Nakas, A.; Arapitsas, P.; Assimopoulou, A.N.; Theodoridis, G. Analysis of the Volatile Organic Compound Fingerprint of Greek Grape Marc Spirits of Various Origins and Traditional Production Styles. *Beverages* **2023**, *9*, 65. [CrossRef]
24. Kokoti, K.; Kosma, I.S.; Tataridis, P.; Badeka, A.V.; Kontominas, M.G. Volatile aroma compounds of distilled "tsipouro" spirits: Effect of distillation technique. *Eur. Food Res. Technol.* **2023**, *249*, 1173–1185. [CrossRef]
25. Ioannou, A.G.; Kritsi, E.; Sinanoglou, V.J.; Cavouras, D.; Tsiaka, T.; Houhoula, D.; Zoumpoulakis, P.; Strati, I.F. Highlighting the Potential of Attenuated Total Reflectance—Fourier Transform Infrared (ATR-FTIR) Spectroscopy to Characterize Honey Samples with Principal Component Analysis (PCA). *Anal. Lett.* **2023**, *56*, 789–806. [CrossRef]
26. Nassirpour, S.; Chang, P.; Henning, A. MultiNet PyGRAPPA: Multiple neural networks for reconstructing variable density GRAPPA (a 1H FID MRSI study). *NeuroImage* **2018**, *183*, 336–345. [CrossRef]

27. Nandiyanto, A.B.D.; Oktiani, R.; Ragadhita, R. How to read and interpret FTIR spectroscope of organic material. *Indones. J. Sci. Technol.* **2019**, *4*, 97–118. [CrossRef]
28. Nagarajan, R.; Gupta, A.; Mehrotra, R.; Bajaj, M.M. Quantitative analysis of alcohol, sugar, and tartaric acid in alcoholic beverages using attenuated total reflectance spectroscopy. *J. Anal. Methods Chem.* **2006**, *2006*, 45102. [CrossRef]
29. Anjos, O.; Martínez Comesaña, M.; Caldeira, I.; Pedro, S.I.; Eguía Oller, P.; Canas, S. Application of functional data analysis and FTIR-ATR spectroscopy to discriminate wine spirits ageing technologies. *Mathematics* **2020**, *8*, 896. [CrossRef]
30. Coldea, T.E.; Socaciu, C.; Fetea, F.; Ranga, F.; Pop, R.M.; Florea, M. Rapid quantitative analysis of ethanol and prediction of methanol content in traditional fruit brandies from Romania, using FTIR spectroscopy and chemometrics. *Not. Bot. Horti Agrobot. Cluj-Napoca* **2013**, *41*, 143–149. [CrossRef]
31. Giannetti, V.; Mariani, M.B.; Marini, F.; Torrelli, P.; Biancolillo, A. Flavour fingerprint for the differentiation of Grappa from other Italian distillates by GC-MS and chemometrics. *Food Control* **2019**, *105*, 123–130. [CrossRef]
32. García-Martín, S.; Herrero, C.; Peña, R.M.; Barciela, J. Solid-phase microextraction gas chromatography–mass spectrometry (HS-SPME-GC–MS) determination of volatile compounds in orujo spirits: Multivariate chemometric characterisation. *Food Chem.* **2010**, *118*, 456–461. [CrossRef]
33. Cortés, S.; Gil, M.L.; Fernández, E. Volatile composition of traditional and industrial Orujo spirits. *Food Control* **2005**, *16*, 383–388. [CrossRef]
34. Ciulu, M.; Oertel, E.; Serra, R.; Farre, R.; Spano, N.; Caredda, M.; Malfatti, L.; Sanna, G. Classification of Unifloral Honeys from SARDINIA (Italy) by ATR-FTIR Spectroscopy and Random Forest. *Molecules* **2021**, *26*, 88. [CrossRef]
35. Martín-Ramos, P.; Fernández-Coppel, I.A.; Ruíz-Potosme, N.M.; Martín-Gil, J. Potential of ATR-FTIR Spectroscopy for the Classification of Natural Resins. *Biol. Eng. Med. Sci. Rep.* **2018**, *4*, 3–6. [CrossRef]
36. Oliveira, R.N.; Mancini, M.C.; de Oliveira, F.C.S.; Passos, T.M.; Quilty, B.; da Silva Moreira Thiré, R.M.; McGuinness, G.B. FTIR analysis and quantification of phenols and flavonoids of five commercially available plants extracts used in wound healing. *Matéria* **2016**, *21*, 767–779. [CrossRef]
37. Thummajitsakul, S.; Paensanit, P.; Saeieo, T.; Sirirat, J.; Silprasit, K. FTIR and multivariate analysis of total phenolic content, antioxidant and anti-amylase activities of extracts and milk of *Glycine max* L. and *Phaseolus vulgaris* L. *Electron. J. Biotechnol.* **2023**, *64*, 69–75. [CrossRef]
38. Wongsa, P.; Phatikulrungsun, P.; Prathumthong, S. FT-IR characteristics, phenolic profiles and inhibitory potential against digestive enzymes of 25 herbal infusions. *Sci. Rep.* **2022**, *12*, 6631. [CrossRef]

foods

Review

NIR Sensing Technologies for the Detection of Fraud in Nuts and Nut Products: A Review

Miguel Vega-Castellote [1], María-Teresa Sánchez [1,*], Irina Torres-Rodríguez [2], José-Antonio Entrenas [2] and Dolores Pérez-Marín [2,*]

[1] Department of Bromatology and Food Technology, University of Cordoba, Rabanales Campus, 14071 Córdoba, Spain; g32vecam@uco.es
[2] Department of Animal Production, University of Cordoba, Rabanales Campus, 14071 Córdoba, Spain; g72toroi@uco.es (I.T.-R.); p82enlej@uco.es (J.-A.E.)
* Correspondence: teresa.sanchez@uco.es (M.-T.S.); dcperez@uco.es (D.P.-M.); Tel.: +34-957-212576 (M.-T.S.); Fax: +34-957-212000 (M.-T.S.)

Abstract: Food fraud is a major threat to the integrity of the nut supply chain. Strategies using a wide range of analytical techniques have been developed over the past few years to detect fraud and to assure the quality, safety, and authenticity of nut products. However, most of these techniques present the limitations of being slow and destructive and entailing a high cost per analysis. Nevertheless, near-infrared (NIR) spectroscopy and NIR imaging techniques represent a suitable non-destructive alternative to prevent fraud in the nut industry with the advantages of a high throughput and low cost per analysis. This review collects and includes all major findings of all of the published studies focused on the application of NIR spectroscopy and NIR imaging technologies to detect fraud in the nut supply chain from 2018 onwards. The results suggest that NIR spectroscopy and NIR imaging are suitable technologies to detect the main types of fraud in nuts.

Keywords: food authentication; detection of adulterants; near-infrared spectroscopy; hyperspectral imaging; targeted and non-targeted approaches

1. Introduction

Food fraud is a long-standing issue that has increased in occurrence in recent years given the sophisticated nature of the food supply chain and the growth of dishonest practices of reducing production costs or deceiving costumers regarding product quality to obtain a higher market price. In Regulation (EU) 2017/625 [1], the European Union (EU) established the general rules in official controls to identify risks associated with the use of products, processes, or materials that could have a negative influence on the integrity of foodstuffs and to account for any information regarding the composition, properties, origin, or production method that could mislead consumers. This regulation arose as a consequence of the billions of euros that the fraudulent practices cost the EU every year and the effects that these practices can have on consumers' health, the quality of the commercialized foodstuffs, and on public confidence in the agri-food chain [2]. Although efforts have been made to counter food fraud [see, for example, the FoodIntegrity European Project (ID: 613688), with participants from 18 European countries, which aims to ensure the quality, safety, and authenticity of the European food chain], there is unfortunately still a long way to go to reduce and prevent fraud in the agri-food chain.

The Knowledge Centre for Food Fraud and Quality (KC-FFQ) of the European Commission provides expertise for policy making, creating a network of food fraud experts [3] and publishing the 'Monthly Food Fraud Summary Reports'. In these reports, several cases of food fraud related to nuts have been published in the past few years, such as the reported lack of documentation of 98 tons of almonds sold as organic, valued at EUR 753,000 in Italy in 2020, or the incorrect labelling of pistachio nuts in France and Spain, valued at EUR 6 million, in 2021 [4]. According to the Combined Nomenclature (CN) code provided in Regulation (EU) 2017/1925, the fruits most commonly included in the 'nuts' category are almonds, Brazil nuts, cashew nuts, chestnuts, hazelnuts, macadamia nuts, peanuts, pecan nuts, pine nuts, pistachio nuts, and walnuts [5]. It is key to note that although peanuts are, in botanical terms, strictly legumes, they are commonly included in the nuts group, as indicated by Buthelezi et al. [6]. Nuts are high-added-value products with a total production over 1 million tons in the EU in 2022 [7] and are a valuable source of energy, healthy fats, proteins, fiber, and vitamins. They are commonly sold as whole nuts or as ingredients in other formats, e.g., ground, to be processed by the manufacturing industry. Fraudulent practices in nuts are a major issue that can involve economic fraud and even pose a threat to human health [8,9], and for this reason, there is an urgent need to make reliable and rapid techniques available to assist in fraud detection throughout the food supply chain [10].

A wide range of analytical techniques have been proposed to assure the authenticity of foodstuffs [11,12]. In nuts, the most common ones are those based on chromatographic techniques [13–15], polymerase chain reaction (PCR) methods [16,17], spectroscopic technologies [18–22], and mass spectrometry fingerprinting alternatives [23,24] among others. The chromatographic techniques and PCR methods require the preparation of the sample prior to analysis, which involves destroying the original structure of the product. In addition, these methods entail a high cost per analysis, are time-consuming, require trained staff, and are not suitable for assessing large volumes of product, which, in many cases, is a key factor in identifying fraud in foodstuffs. However, spectroscopy-based applications offer the possibility of assuring the authenticity of nuts in a fast and non-destructive way while analyzing large amounts of product. Among these, the most popular technique for fraud detection in nuts is near-infrared (NIR) spectroscopy [25]. This technique is based on the absorbance that takes place when a molecule bond vibrates at the same specific frequency as the incident NIR light irradiating the product. A wide range of applications can be found using NIR spectroscopy in nuts [6,26], such as the successful quantitative estimation of the lipid oxidation in in-shell and shelled hazelnuts, carried out by Pannico et al. [27], or the qualitative application for the identification of commercial sweet almond batches adulterated with bitter almond kernels, carried out by Torres-Rodríguez et al. [28], which showed an excellent capability for discrimination. Additionally, the detection of adulterated products with other non-compliant ingredients, e.g., different nuts and fragments of nutshells or non-edible elements, could benefit from the combination of conventional spectroscopy and imaging techniques, providing not only the spectral information but also the spatial properties of the commodity analyzed [29]. NIR imaging technology offers the possibility to obtain both spectral and spatial features of an object, integrating the power of spectroscopy and digital imaging, which can be of great utility in specific applications and complex issues. This technology is known as hyperspectral imaging (HSI) if the image is composed of hundreds of contiguous wavebands and as multispectral imaging (MSI) if only a few spectral bands (usually between three and ten bands) are included in the spectral dimension [30,31].

In the scientific literature, we found only one review (Teixeira & Sousa [25], submitted for publication in May 2018) that focuses on NIR applications for origin authentication and the detection of adulteration in nuts. Thus, the aim of this work is to provide a review on the application of NIR-based techniques for the detection of fraud in nuts and nut products over the last 7 years (2018–2024).

2. Bibliographic Search

The literature review was carried out using three scientific databases, Scopus, Web of Science, and Google Scholar. The keywords used (in different combinations) included the range of nuts considered—'almonds', 'Brazil nuts', 'cashew nuts', 'chestnuts', 'hazelnuts', 'macadamia nuts', 'peanuts', 'pecan nuts', 'pine nuts', 'pistachio nuts', 'walnuts', and the term 'nut'—together with the techniques reviewed (in this case, 'NIR spectroscopy' and 'hyperspectral imaging') and words that made any reference to fraud, such as 'adulteration', 'authentication', 'fraud', and 'counterfeit'. Once the list of potential research papers to be included in the review was made, we used the following criteria to accept or reject a study: (1) Assurance of the quality of the documents—only peer-reviewed papers published in journals included in the Journal Citation Reports (JCR) index were considered. The peer-review criterion was also used for books and chapters of books. (2) Nuts were presented in intact, ground, powder, or paste formats, excluding other presentations such as nut oils. (3) Only NIR-based techniques were used, excluding other spectroscopic techniques such as mid-infrared or Raman spectroscopy. (4) Any reference to the authentication of nuts or issues of fraud needed to be reported in the document in question. For example, a document would be accepted if the objective was the classification of pistachio cultivars to detect fraud in their commercialization; however, a paper would not be considered in this review if the objective was the classification of pistachio cultivars per se, with no mention or consideration of food fraud. (5) The earliest date for a document to be included in this review was set at 2018. The number of documents published in the past 7 years regarding fraud detection in nuts using NIR spectroscopy and HSI was significantly greater than the number of documents available in the years before 2018 (Figure 1). Consequently, this review is of great importance to summarize the research information available on this topic from the past seven years.

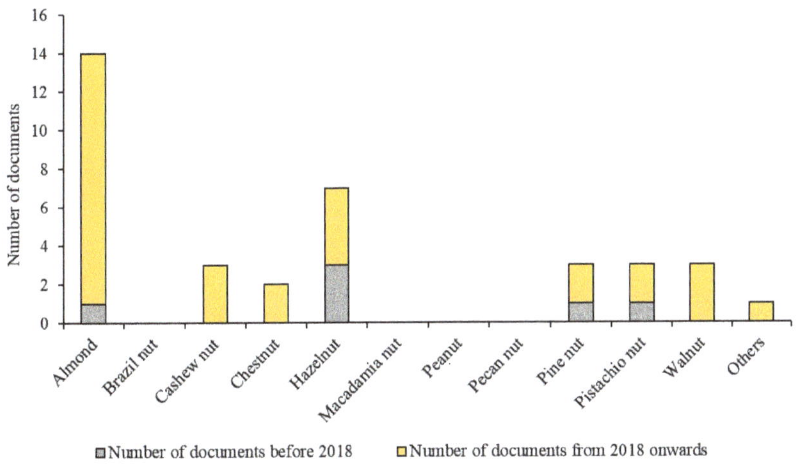

Figure 1. Numbers of studies related to food fraud detection in nuts before 2018 and from 2018 onwards.

3. The Nature of Fraud in Nuts and Types of NIR Applications Reviewed

Nowadays, several approaches to food fraud can be taken due to the great complexity of the food supply chain and the latest trends in which foodstuffs are produced, processed, distributed, and sold [32]. Momtaz et al. [33] indicated the main mechanisms of food adulteration, including the adulteration of food properties by adding artificial coloring agents or preservatives and the substitution of a specific food ingredient by adding another ingredient of inferior quality and value. Moreover, other types of fraud related to authentication issues can be found in the scientific literature, such as the mislabeling of foodstuffs, which takes advantage of the high economic value of products of a superior quality category or those that have been produced in specific regions and associated with a Protected Geographical Indication (PGI), Protected Designation of Origin (PDO), or similar quality standards [34]. In nuts, different type of fraud can be found depending on the format of the commercialized product. One of the most common types identified in the papers we reviewed was the adulteration of ground, powder, or paste nut products by using other foodstuffs of a lower value, including different nuts or fruit, such as the adulteration of almond powder with apricot or peanut powder [35,36]. This can mainly lead to food safety issues (allergies) and consumer dissatisfaction, given the lower quality of the product they purchase and the financial deceit involved. In addition, although it is not so common, given the ease of committing fraud in the previously mentioned nut formats, several cases can be found of authenticity issues with intact nuts, including the year of stock, type of production (for example, organic farming) or origin, or their adulteration, such as the commercialization of non-compliant intact nuts. In this context, of the studies reviewed in this work (Tables 1 and 2), three main targets were identified related to food fraud detection in different formats of nuts using NIR-based techniques: the identification of their origin, the authentication of the cultivar, and the detection of adulterants in the commercialized product.

The origin of nuts is a key issue affecting the characteristics of their products since factors such as the climatologic conditions of the area or the soil properties and the management of the orchards have a significant influence on the chemical composition of the nuts and, thus, on their quality and final market price. Biancolillo et al. [37] and Sammarco et al. [38] developed methods to authenticate different PDOs and PGIs of Italian hazelnuts harvested in specific locations such as the Roma and Viterbo (Lazio, Italy) areas. The authentication of the origins of different nuts was also evaluated in the studies carried out by Firmani et al. [20] for almonds, Amendola et al. [39] for walnuts, and Nardecchia et al. [40] for chestnuts. Furthermore, nut cultivars are also a key factor in the market value of the product since its shape, taste, and/or aroma can differ from one cultivar to another. Hence, different studies have been carried out in nuts such as almonds, chestnuts, and pine nuts [41–44].

Finally, regarding the detection of adulterants in the commercialized product, a wide range of adulterants can be found in the studies reviewed, such as was suggested by the presence of green peas in pistachio nut products, given the lower price of the former [45,46]; the presence of bitter almonds in batches of sweet almonds, which can affect consumer safety [47,48]; the adulteration of cashew nuts with other allergenic nuts [49,50]; and the detection of foreign bodies in Chinese hickory nuts, such as shell fragments that can be harmful for human consumption [51]. To address these issues, different chemometric approaches were tested in the reviewed documents by means of the development of quantitative and qualitative models to detect fraud in nuts. The former aim at the estimation of the quantity of the undesired component in the adulterated product while the latter apply methods designed for the classification of the product according to its membership to a certain class.

Table 1. Applications for fraud detection in nuts using NIR point spectroscopy techniques. N: number of samples; Ref.: reference.

Type of Nut	Aim of the Study	Sample Presentation and Analysis	Technique	Acquisition Mode	Spectral Range (nm)	Chemometric Technique	N	Performance	Ref.
Almonds	Cultivar authentication	Intact shelled. Two spectra per kernel.	NIR spectroscopy	Reflectance	1000–1700	Partial least squares–discriminant analysis (PLS-DA) and quadratic discriminant analysis (QDA)	120	Test set: 94% of almonds correctly classified	[41]
Almonds	Origin authentication (Avola almonds from Syracuse area, southeast of Sicily, Italy)	Intact. Two spectra per sample.	Fourier-Transformed (FT) NIR spectroscopy	Reflectance	1000–2500	PLS-DA and soft independent modeling of class analogies (SIMCA)	227	Test set: 95% of almonds correctly classified	[20]
Almonds	Origin authentication	Intact and ground. Up to six spectra per sample.	FT NIR spectroscopy	Reflectance	866–2532	Support vector machine (SVM)	64	Nested cross-validation: 62.6% of intact almonds and 79.1% of ground almonds correctly classified	[52]
Almonds	Detection of adulterants	Powder. One spectrum per sample.	FT NIR spectroscopy	Reflectance	1000–2500	Data-driven SIMCA (DDSIMCA) and one-class PLS (OCPLS)	260	Test set: 91–100% of adulterated samples correctly classified	[35]
Almonds	Cultivar authentication	Ground, five spectra per sample.	FT NIR spectroscopy	Reflectance	866–2532	SVM	250	Nested cross-validation: 80.3% of almonds correctly classified	[42]
Almonds	Detection of adulterants	Intact shelled almonds.	NIR spectroscopy	Reflectance	908–1676	PLS-DA	216	Cross-validation: 90% samples correctly classified	[28]
Almonds	Detection of adulterants	Intact shelled almonds.	NIR spectroscopy	Reflectance	950–1650	PLS-DA	216	Test set: 98.6% samples correctly classified	[28]
Almonds	Detection of adulterants	Intact in-shell and shelled. Batch spectra in dynamic mode. Four spectra per sample.	NIR spectroscopy	Reflectance	950–1650	PLS-DA	145	Test set: 95 and 100% of in-shell and shelled samples, respectively, correctly classified	[47]

Table 1. *Cont.*

Type of Nut	Aim of the Study	Sample Presentation and Analysis	Technique	Acquisition Mode	Spectral Range (nm)	Chemometric Technique	N	Performance	Ref.
Almonds	Detection of adulterants	Intact in-shell and shelled. Batch spectra in dynamic mode. Four spectra per sample.	NIR spectroscopy	Reflectance	908–1676	PLS-DA	145	Test set: 100% of in-shell and shelled samples correctly classified	[47]
Almonds	Detection of adulterants	Intact in-shell and shelled. Batch spectra in dynamic mode. Four spectra per sample.	NIR spectroscopy	Reflectance	950–1650	Modified PLS (MPLS) regression, LOCAL algorithm	145	Test set MPLS: R^2 = 0.53 for in-shell samples; R^2 = 0.96 for shelled samples. Test set LOCAL: R^2 = 0.98 for shelled samples	[47]
Almonds	Detection of adulterants	Intact in-shell and shelled. Batch spectra in dynamic mode. Four spectra per sample.	NIR spectroscopy	Reflectance	908–1676	MPLS regression, LOCAL algorithm	145	Test set MPLS: R^2 = 0.53 for in-shell samples; R^2 = 0.96 for shelled samples. Test set LOCAL: R^2 = 0.95 for shelled samples	[47]
Almonds	Cultivar authentication	Intact shelled. Batch spectra in dynamic mode. Four spectra per sample.	NIR spectroscopy	Reflectance	950–1650	Principal component analysis (PCA), Shewhart control chart, GH distance	140	Test set: 52–90% of almond batches correctly classified	[53]
Almonds	Cultivar authentication	Intact shelled. Batch spectra in dynamic mode. Two spectra per sample.	FT NIR spectroscopy	Reflectance	834–2502	PCA, Shewhart control chart, GH distance	140	Test set: 57–100% of almond batches correctly classified	[53]
Almonds	Origin authentication	Powder. Five spectra per sample.	FT NIR spectroscopy	Reflectance	866–2532	Linear discriminant analysis (LDA)	72	Nested cross-validation: 92.5–95.0% of almonds correctly classified	[54]
Almonds	Detection of adulterants	Powder. Three spectra per sample.	NIR spectroscopy	Reflectance	900–1700	DDSIMCA, SIMCA and OCPLS	182	Test set: 98.5% of almond samples correctly classified	[55]
Almonds	Detection of adulterants	Powder. Three spectra per sample.	NIR spectroscopy	Reflectance	950–1650	DDSIMCA, SIMCA and OCPLS	182	Test set: 98.5% of almond samples correctly classified	[55]

Table 1. *Cont.*

Type of Nut	Aim of the Study	Sample Presentation and Analysis	Technique	Acquisition Mode	Spectral Range (nm)	Chemometric Technique	N	Performance	Ref.
Almonds	Detection of adulterants	Powder. Three spectra per sample.	FT NIR spectroscopy	Reflectance	1350–2500	DDSIMCA, SIMCA and OCPLS	182	Test set: 96.4% of almond samples correctly classified	[55]
Almonds	Detection of adulterants	Powder. Three spectra per sample.	FT NIR spectroscopy	Reflectance	900–2500	DDSIMCA, SIMCA and OCPLS	182	Test set: 100% of almond samples correctly classified	[55]
Almonds	Detection of adulterants	Ground.	NIR spectroscopy	Reflectance	908–1676	SIMCA, decision tree (DT), logistic regression (LR), naive Bayes (NB), SVM, k-Nearest Neighbor (KNN), Gaussian Process (GP)	120	Test set: 100% of almond samples correctly classified	[36]
Almonds	Detection of adulterants	Ground.	FT NIR spectroscopy	Reflectance	1000–2500	SIMCA, DT, LR, NB, SVM, KNN, GP	120	Test set: 100% of almond samples correctly classified	[36]
Almonds	Detection of adulterants	Ground.	NIR spectroscopy	Reflectance	908–1676	PLS regression	120	Test set: $R^2 = 0.96$ for the determination of adulterant concentration	[36]
Almonds	Detection of adulterants	Ground.	FT NIR spectroscopy	Reflectance	1000–2500	PLS regression	120	Test set: $R^2 = 0.96$ for the determination of adulterant concentration	[36]
Cashew nuts	Detection of adulterants	Ground. One spectrum per sample.	NIR spectroscopy	Reflectance	908–1676	PLS-DA and SIMCA	280	Test set: 55.4–100% reliability rate	[49]
Cashew nuts	Detection of adulterants	Ground. One spectrum per sample.	NIR spectroscopy	Reflectance	908–1676	SIMCA	280	Test set: 95.7–98.7% of adulterated samples correctly classified	[50]
Cashew nuts	Detection of adulterants	Ground. One spectrum per sample.	NIR spectroscopy	Reflectance	908–1676	SIMCA	280	Test set: 95.7–100% of adulterated samples correctly classified	[56]
Chestnuts	Origin authentication ('Vallerano' chestnuts, Italy)	Intact in-shell. Four spectra on the pericarp and two on the hilum per individual fruit.	FT NIR spectroscopy	Reflectance	1000–2500	PLS-DA and SIMCA	441	Test set: 95% of in-shell and 97% of shelled walnuts correctly classified	[40]

Table 1. *Cont.*

Type of Nut	Aim of the Study	Sample Presentation and Analysis	Technique	Acquisition Mode	Spectral Range (nm)	Chemometric Technique	N	Performance	Ref.
Chestnuts	Cultivar authentication	Intact shelled. One spectrum per kernel.	FT NIR spectroscopy	Reflectance	1000–2500	PLS-DA	96	Cross-validation: 98% correctly classified	[43]
Hazelnuts	Origin authentication (Italian PDO "Nocciola Romana")	Intact in-shell. Two spectra per intact nut.	FT NIR spectroscopy	-	1000–2500	PLS-DA and SIMCA	376	Test set: 94% of hazelnuts correctly classified	[37]
Hazelnuts	Origin authentication (different regions of Turkey)	Intact in-shell and shelled. Two spectra per kernel.	FT NIR spectroscopy	Reflectance	1000–2500	PLS-DA and LR	280	Test set: 100% of in-shell and shelled hazelnuts correctly classified	[57]
Hazelnuts	Origin authentication (from five different countries)	Ground. Five spectra per sample.	FT NIR spectroscopy	Reflectance	866–2532	Discriminant classifiers using random subspaces for constructing DT	233	Nested cross-validation: 90.6% of hazelnuts correctly classified	[58]
Hazelnuts	Origin authentication (Italy: PGI from Piedmont and PDO from Campania)	Ground and paste. Fourteen spectra per sample.	NIR spectroscopy	Reflectance	400–2500	PLS-DA	216	Test set: 100% of intact hazelnuts correctly classified and 72% of paste hazelnuts correctly classified	[38]
Pine nuts	Cultivar authentication	Intact shelled. Two spectra per kernel.	NIR spectroscopy	Reflectance	1100–2300	PLS-DA and Interval-PLS-DA (iPLS-DA)	900	Test set: 97% of pine nuts correctly classified	[44]
Pine nuts	Cultivar authentication	Intact in-shell. One spectrum per kernel.	FT NIR spectroscopy	Reflectance	781–2632	DT, random forest (RF), MLP, SVM, NB	210	Test set: 99% of pine nuts correctly classified	[59]
Pistachio	Detection of adulterants	Ground. Three spectra per sample.	FT NIR spectroscopy	Reflectance	1250–2500	SIMCA	60	Test set: 100% of pistachio adulterated samples correctly classified	[46]
Pistachio	Detection of adulterants	Powder. Ten spectra per sample.	NIR spectroscopy	-	908–1695	PCA and PLS regression for adulterant quantification	143	PCA scores plot reported a clear grouping of classes. Test set PLS: $R^2 >$ 0.99	[45]

Table 1. *Cont.*

Type of Nut	Aim of the Study	Sample Presentation and Analysis	Technique	Acquisition Mode	Spectral Range (nm)	Chemometric Technique	N	Performance	Ref.
Walnuts	Origin authentication	Powder. One spectrum per sample.	FT NIR spectroscopy	Reflectance	833–2500	PLS-DA	555	Test set: 98.8–100% of walnuts correctly classified	[60]
Walnuts	Origin authentication (Sorrento area, south of Italy)	Intact in-shell and shelled. Two spectra per individual kernel.	NIR spectroscopy	-	-	PLS-DA	237	Test set: 95% of in-shell and 97% of shelled walnuts correctly classified	[39]
Walnuts	Origin authentication	Ground. Three spectra per kernel.	FT NIR spectroscopy	Reflectance	866–2532	LDA	212	Nested cross-validation: 77% of walnuts correctly classified	[61]

Table 2. Applications for fraud detection in nuts using hyperspectral imaging techniques. N, number of samples; Ref., Reference.

Type of Nut	Aim of the Study	Sample Presentation and Analysis	Technique	Acquisition Mode	Spectral Range (nm)	Chemometric Technique	N	Performance	Ref.
Almonds	Detection of adulterants	Intact shelled almond kernels	NIR HSI system	Reflectance	900–1700	PLS-DA	448	Test set: 97.1% of samples correctly classified	[62]
Almonds	Detection of adulterants	Intact shelled almond kernels	NIR HSI system	Reflectance	946–1648	PLS-DA	158	Test set: 75% of samples correctly classified	[48]
Chinese hickory nuts	Detection of adulterants	Intact shelled Chinese hickory nut kernels	Visible (VIS)–NIR HSI system	Reflectance	400–1000	PCA-KNN, and SVM	213	Test set: 99% of samples correctly classified	[51]

4. NIR-Based Instrumentation and Analysis Mode for the Detection of Fraud in Different Formats of Nuts

The wide range of applications for the detection of fraud in nuts using NIR-based technologies have been made possible due to the great development in terms of instrumentation over the last few years, enabling this technology to be used throughout the different steps of the food supply chain to meet the needs of producers, industry, retailers, and consumers [63–65]. In this work, the research studies reviewed were classified into two groups, according to the type of instrumentation used: conventional NIR spectroscopy instruments or multispectral/hyperspectral camera systems.

4.1. NIR Spectroscopy

The versatility of NIR spectroscopy enables us to analyze nuts in the different formats available in the market such as intact, ground, powder, or paste, which can be considered as a major advantage for fraud detection in these products. This technology can be applied in various modes, namely reflectance, interactance, and transmission, using three main types of instruments, namely laboratory devices, instruments located in the sorting and grading industrial lines, and portable devices [66,67]. It permits the analysis of food products in the different steps of the supply chain, i.e., in the field (in-situ), in the laboratory (off-line), on the production line (at-line, on-line), etc.

The most common instruments used among the studies reviewed (Table 1) were those designed for at-line or off-line applications in different formats of nuts, such as the FT-NIR TANGO spectrometer (Bruker Optics, Bremen, Germany) [58,61] or the FT-NIR Nicolet 6700 spectrophotometer (Thermo Fisher Scientific, Madison, WI, USA) [20,37], which both work in reflectance mode. Although these FT-NIR laboratory devices offer high-spectral-resolution NIR measurements, their main disadvantage is that they can be only incorporated at-line and in the laboratory under controlled conditions and that the amount of product which can be inspected using them is limited compared to that using other types of instruments. In addition to the laboratory systems, other research papers used devices suitable for the on-line NIR analysis of nuts. One of the main advantages of this type of instrument is the possibility to analyze—directly from the processing line—greater amounts of product compared to the laboratory devices and, thus, to collect a greater variability, which is of major importance when dealing with food fraud detection. In this context, Vega-Castellote et al. [53] used the Matrix-F spectrophotometer (Bruker Optik GmbH, Ettlingen, Germany), which is an FT-NIR instrument interfaced to a fiber-optic NIR illumination and a detection head working in reflectance mode in the 834–2502 nm spectral range. The spectral range in which most of the benchtop and on-line instruments used in the studies we reviewed work (approx. 1000–2500 nm) provides key information for the detection of fraud in nuts. For example, the 1000–1100 nm band, at the beginning of the second overtone region, can be useful to discriminate hazelnuts of different geographical origins since that band can be associated with the presence of polyphenols in aromatic compounds [38] while the 1200 nm wavelength can be related to absorbances of the second overtone of C–H bonds of lipids [68]. Furthermore, the 1408–1650 nm region can be related to the N-H stretching (first overtone) of proteins, which can be useful for determining the origins of almonds [52]. In addition, the 1724 nm wavelength, linked to C-H stretching (first overtone), can provide information about the lipid contents of walnuts, which can be of great interest to identify the geographical origin of their products [61].

In addition, apart from these applications using laboratory and on-line NIR instruments, various research papers can be found that aimed to detect food fraud using portable NIR instruments. These devices belong to a new generation of NIR handheld instruments which permit the in-situ analysis of products at different steps throughout the food supply chain. Despite the reduced sizes of these instruments, the technological developments enable one to ensure the satisfactory performance of these devices [69]. The portable instruments used in the papers we reviewed work in reflectance mode in the spectral range of approx. 900–1700 nm so that key information for fraud detection in nuts can be extracted, as mentioned above. It is important to highlight that compared to on-line instruments, the area from which the NIR information is collected when using portable devices is smaller, given the limited window size of these devices. Nevertheless, different strategies to deal with this issue have been reported in the studies we reviewed, such as analyzing the nuts in dynamic mode (moving the portable sensor along the surface of the product) to collect the maximum amount of variability as possible, and to obtain more representative information from a sample [28], or taking more than one spectrum per sample, which, in addition, would enable one to guarantee the spectral repeatability of the data acquired. For example, Lösel et al. [54] acquired five spectra per powdered almond sample, and Arndt et al. [52] took six spectra per each intact almond sample. It is worth noting that even when the purpose is to obtain a representative measurement of a single kernel, it is of paramount importance to collect more than one spectrum per sample since, as reported by Nardecchia et al. [40], there is a difference in the spectral signatures of the measurements taken from different parts of a chestnut (pericarp versus hilum), which highlights the wide variability within a single kernel. In addition, these authors reported that taking the spectral information from the pericarp or the hilum could have an influence on the model's performance when classifying PDO and non-PDO Italian 'Vallerano' chestnuts.

4.2. NIR Imaging Systems

The studies available for fraud detection in nuts with NIR imaging devices all used hyperspectral systems (Table 2). These systems are generally composed of charged-coupled device (CCD) cameras, which include sensors with small photodiodes (pixels) made of materials sensitive to light such as silicon (Si) or indium gallium arsenide (InGaAs) [70], along with point, line, or area-scan spectrographs. The line-scan methods are well suited for taking information from moving samples by acquiring a slit of spatial and spectral information for each point in the linear field of view using conveyor belts or electric displacement platforms, which is of particular interest for analyzing products in food industry processing lines. These methods can be considered as extensions of the point-scan approach and are both known as spatial-scan methods [70]. For the line-scan systems, the speed of the displacement platforms is dependent on the experiment settings and may need to be adapted to industrial operation conditions when developing an application for routine analysis. Qin et al. [70] stated that HSI applications are the first step to determining the optimal wavebands with the aim of reducing the total volume of data to be used later as an MSI solution for a particular application. Faqeerzada et al. [62] used a commercial Pika NIR-640 (Resonon, Bozeman, MT) line-scan push-broom hyperspectral camera to acquire images in the 900–1700 nm range in reflectance mode in order to detect apricot kernels in almonds—the former is a cheaper product with similar color, texture, or odor compared to almonds—and reported excellent classification accuracy (~97%). Feng et al. [51] worked with an HSI system to identify shell fragments in Chinese hickory nuts. The authors reported that the color similarities between kernels and shell fragments can make it difficult to differentiate small but harmful endogenous foreign bodies in these nuts. To discriminate between these types of products, a line-scan HSI system consisting of a CCD camera working in reflectance mode in the VIS–NIR range (400–1000 nm) was used, which enabled the researchers to separate kernels from shell fragments with a high degree of accuracy (99%). Furthermore, Torres-Rodríguez et al. [48] also used a line-scan HSI system to detect bitter almonds in batches of sweet almonds by preparing mixtures in which the proportion of bitter kernels was increased from 5 to 20% in 5% steps. The system also featured a CCD camera working in the NIR range between 946 and 1648 nm, acquiring data in reflectance mode. Based on the results obtained in their study, the authors concluded that it would be possible to classify almonds by bitterness although further studies would be needed to be able to identify each individual bitter almond present in a mixture.

5. Strategies for the Treatment of the NIR Data and Chemometric Approaches

The complexity of the food matrices and the large number of data obtained when working with NIR-based techniques require using multivariate statistical tools for data interpretation. One of the first steps prior to the development of NIR models is to pretreat the spectral data in order to remove unwanted signals [71]. In addition, different chemometric approaches can be found to develop quantitative or qualitative applications to detect fraud in nuts. These approaches, in turn, can be targeted or not targeted, as explained below.

5.1. Spectral Preprocessing of Nuts

In nuts, it is important to study the influence that the physical characteristics of the sample, like the shape of the nut or particle size (i.e., kernels analyzed in intact versus ground format), can have on the captured spectrum, given the absorbance displacement and the slope shift related to the scatter effect of the light when it interacts with a sample. In addition to the correction of the scatter effects—using, for example, standard normal variate (SNV) or multiplicative scatter correction (MSC) specifically suggested to eliminate drifts (non-linear baseline deviations) caused by scatter in reflectance measurements—other preprocessing methods should also be tested, such as the filtering and/or the scaling/centering signal pretreatments, since the latter, for example, often enables one to obtain a more simple and interpretable regression model [72]. Corona et al. [43] tested different pretreatments on the spectra of intact chestnuts such as SNV, MSC, and Savitzky–Golay (SG) derivatives to remove the unwanted spectral variance that could be associated with the variability in the refractive index and the morphology or density of the sample. These authors reported that the best modeling results were obtained using SNV in combination with derivatives. This pretreatment combination of scatter correction followed by a filtering method was the one most frequently used among the studies reviewed and could be seen in the research carried out by Moscetti et al. [44], Torres et al. [28], Vega-Castellote et al. [53], Shakiba et al. [58], and Netto et al. [55] among others. Likewise, Arndt et al. [61], who tested more than eight different pretreatments applied in different orders to classify walnuts analyzed in ground format based on their geographical origin, also concluded that the most suitable option to remove unwanted variations from the spectral data was the application of a scatter correction method—followed, in this case, by a centering pretreatment. It should be mentioned that while these studies selected the optimal pretreatment combination based on the models' performance, none of them showed significance tests comparing the models' performance when the different pretreatments were applied and, therefore, the fact that most of them used a scatter correction method along with other spectral pretreatments should be viewed with caution. Finally, other types of pretreatments can be found among the papers reviewed, such as the use of binning as a preprocessing technique to average the intensity of a different number of adjacent, strongly correlated wavenumbers to decrease the computational time of the spectral data [54].

5.2. Chemometric and Machine Learning Approaches for Fraud Detection in Nuts

Two types of chemometric approaches can be differentiated to deal with food fraud using NIR techniques. When the aim is the quantification of a specific compound/property of the matrix analyzed or the discrimination of previously defined classes, targeted chemometric methods are used. However, when the objective is to assess product deviations compared with a library of previously recorded compliant samples according to a standard (without focusing on any specific compound, but rather focusing on the whole fingerprint of the product in question), non-targeted methods are applied [73].

The targeted approach is the most traditional method in the NIR community for quality and safety determination in food products. Quantitative and qualitative applications are commonly carried out using regression and classification techniques such as the PLS regression or PLS-DA, respectively. However, non-targeted approaches are a recent, promising trend in NIR applications and have great potential to assure the integrity of food products. Many examples of non-targeted approaches can be found, such as the use of PCA or the OCPLS method [74,75].

5.2.1. Targeted Approaches

The most frequently used approach to detect fraud in nuts is the application of the targeted linear PLS-DA technique (Tables 1 and 2), for example, to detect adulterants in almonds [28] or cashew nuts [49], to authenticate the origin of almonds [20] or hazelnuts [37,38], and to authenticate the cultivars of almonds [41] or pine nuts [44]. One important aspect to take into consideration when using the PLS-DA technique, as mentioned by Vega-Castellote et al. [47], who assessed the presence of bitter almond kernels in commercial batches of sweet almonds, is the number of samples included per class since unbalanced models can bias the PLS-DA prediction boundary towards the smaller class. After the application of PLS-DA, different authors have proposed ways of identifying which variables have the greatest influence on the model using different methods such as Biancolillo et al. [37], who used the variable importance in projection (VIP) score values and found that the most important spectral regions for the origin authentication of hazelnuts were at approximately 1700–1785 nm, related to the presence of lipids, and 2000–2170 nm, associated with combination bands of the peptide bonds. The VIP values were also studied by Torres-Rodríguez et al. [48], who used a PLS-DA model to detect the presence of bitter almonds in batches of sweet almonds using an HSI system. The results obtained with the full spectral range (179 bands) and only the ones selected using the VIP values (25 bands) reported a similar accuracy, which is of special interest in HSI applications to reduce the amount of available data.

In addition to PLS-DA, other chemometric targeted approaches have been used to detect fraud in nuts. Arndt et al. [61] used LDA to determine the geographical origin of walnuts. Prior to the application of this technique, these authors reduced the dimensions of the spectral data by applying PCA. Likewise, Cortés et al. [41] performed QDA to determine the cultivars of almonds in which, prior to the application of this algorithm, PCA was performed and the QDA was applied to the principal component scores obtained. When applying these types of algorithms (LDA or QDA), the dimension of the NIR data usually needs to be reduced—by using PCA or PLS, for example—since the number of variables must be lower than the number of samples [76].

Other approaches based on machine learning techniques have also been used such as the application of discriminant classifiers using random subspaces to construct DTs for the geographical determination of hazelnuts [58], in which the authors carried out a data fusion strategy using NIR and nuclear magnetic resonance (NMR) spectroscopies. Arndt et al. [52] and Arndt et al. [42] developed multiclass models to determine the geographical origin of almonds using SVM, which is a method that can be used to train linear and non-linear classifiers, for classification [77]. Huang et al. [59] also used machine learning methods, such as the RF, NB, or the multilayer perceptron (MLP), to identify the cultivars of pine nuts. Menevseoglu et al. [36] successfully identified ground almonds that had been adulterated with apricot kernels by applying different machine learning algorithms. The results obtained in these studies showed that the machine learning approaches, along with the NIR techniques tested, were effective means of geographical determination and cultivar identification in different nuts.

In addition to these classification techniques, quantitative approaches have also been tested in studies aiming at the detection of adulterants in nuts. Genis et al. [45] applied the PLS regression method to detect the presence of green peas and/or spinach powder in ground samples of pistachio, reporting a high degree of prediction accuracy. Vega-Castellote et al. [47] also used the MPLS regression technique and the LOCAL algorithm to identify adulterants in intact almonds by quantifying the amount of a toxic compound in the samples (Table 1).

5.2.2. Non-Targeted Approaches

There is a current and important discussion in the food authentication literature regarding the 'discriminant' and 'class-modeling' approaches [78]. Non-targeted techniques, which are focused on a single class and can be related to the class-modeling approach, can also be found among the studies reviewed. As can be seen from Table 1, the most frequently used is the SIMCA method. This class-modeling approach, based on PCA, models each of the classes individually, with new unknown samples accepted or rejected by each class model [79]. Firmani et al. [20] tested this method for the authentication of the origin of almonds, indicating the importance of presenting the results of the analysis in terms of sensitivity and specificity—samples correctly identified in their class and correctly rejected samples, respectively—given the fact that when the SIMCA method is used, one sample can be associated with more than one class. Biancolillo et al. [37] also applied the SIMCA method for to authenticate the origin of hazelnuts. In their study, they focused on a specific class (an Italian PDO hazelnut), with all the rest of the samples not belonging to that PDO assigned to a second class. This type of approach is known as asymmetric classification. In addition, other studies using the SIMCA method can be found for authenticating the origin of chestnuts [40], as well as for detecting adulteration in pistachios [46], cashew nuts [49], or almond flour [55] among others. In all cases, satisfactory results have been obtained using the SIMCA method. Nevertheless, it is important to note that in many studies in which PLS-DA and the SIMCA method were tested, the authors highlighted that PLS-DA outperformed the SIMCA method [40,49], given the fact that the SIMCA method models the intra-class variance, which is usually low, as was in the case of the study conducted by Miaw et al. [49], while PLS-DA maximizes the between-group variance [80]. An adaptation of the SIMCA method, the DDSIMCA method, was tested by Faqeerzada et al. [35] and Netto et al. [55] to detect adulterants in almonds in powder format. Both studies also applied the OCPLS algorithm, which is another non-targeted class-modeling method and which models only the target class and sets threshold boundaries during the development of the model. Although, in these studies, high accuracy (~90–100%) was obtained for the detection of adulterants in almonds in powder format, in both cases, it was reported that the DDSIMCA models performed better for classification when compared to OCPLS.

Another example of a non-targeted method was used by Vega-Castellote et al. [53], who developed a system based on PCA, Mahalanobis distances, and the construction of Shewhart control charts to detect non-compliant batches of sweet almonds, i.e., batches of sweet almonds adulterated with bitter kernels. A standard or control population of sweet almonds was characterized by means of PCA, and new adulterated samples (with 5, 10, 15, and 20% of bitter kernels) were projected into that PCA space. The distance (global Mahalanobis distance—GH) of each adulterated sample to the center of the control population was plotted on the Shewhart control charts and GH limits were set in order to detect the batches of adulterated almonds. The results obtained suggested that this approach, which uses solely the NIR spectral information of the product, would enable one to successfully perform conformity tests on batches of sweet almonds processed in the industry.

6. Conclusions and Future Prospects

The adulteration of the physicochemical characteristics of nut products through the addition or substitution of any of their components not only has an economic impact -given the high value of this commodity—but can also have an impact on consumer health. In nuts, the main types of fraud are related to the identification of the origin, authentication of the cultivar, and the detection of adulterants in the commercialized product. The papers included in this review show that NIR technology is a suitable analytical alternative for the rapid detection of fraud in different nuts. The at-line or off-line applications were the most common type and were developed using laboratory devices that can be used in limited scenarios. However, only a few applications were available in the literature that use the new generation of portable and on-line NIR sensors to detect fraud in nuts. The feasibility of portable devices to perform NIR measurements in different steps of the food supply chain and the possibility of on-line sensors analyzing large amounts of produce, along with the need for fast, non-destructive analytical alternatives for routine operations to detect fraud in the food supply chain, mean there are promising perspectives for the application of these types of sensors in the upcoming years. In addition, HSI technology has become a powerful tool that combines characteristics of machine vision and point spectroscopy and is of great interest in the identification of fraud in intact nuts, where spatial information plays a significant role. Future prospects for this technology seek to improve the implementation of HSI systems by reducing the workload of imaging acquisition and processing so that real-time routine operations can be used for fast, non-destructive fraud detection and thus meet the needs of the nut industry. Although considerable advances have been achieved in the past few years in the field of HSI systems in terms of hardware and software, the speed of these systems is still the greatest handicap for their adoption in inspection processes in the food industry.

In the reviewed papers, different chemometric approaches were used to assess the occurrence of fraud in nuts, and these can be divided into targeted and non-targeted methods. Several papers included in this review used non-targeted methods, given that fraudsters often develop sophisticated alternatives to avoid detection through targeted methods. Thus, non-targeted methods are a promising tool for the screening of fraud in nuts, reducing the volume of product analyzed using confirmatory techniques. However, among the key challenges still to be addressed in the following years in order to implement these methods for fraud detection in nuts in routine operations in the industry are the development of legislative guidelines to apply these methods in routine analysis for official control and the validation of the models using independent sample tests on the industrial sorting lines.

Author Contributions: M.V.-C.: investigation, conceptualization, and writing—review and editing. M.-T.S.: investigation, conceptualization, supervision, and writing—review and editing. I.T.-R.: investigation, conceptualization, and writing—review and editing. J.-A.E.: investigation, conceptualization, and writing—review and editing. D.P.-M.: investigation, conceptualization, supervision, and writing—review and editing. All authors have read and agreed to the published version of the manuscript.

Funding: This research received no external funding.

Institutional Review Board Statement: Not applicable.

Informed Consent Statement: Not applicable.

Data Availability Statement: No new data were created or analyzed in this study. Data sharing is not applicable to this article.

Acknowledgments: The authors wish to express their gratitude to the Spanish Ministry of Universities for the support offered to Miguel Vega-Castellote through the Training Programme for Academic Staff (FPU).

Conflicts of Interest: The authors declare that they have no known competing financial interests or personal relationships that could have appeared to influence the work reported in this paper.

Nomenclature

CCD	charged-coupled device
CE	conditional entropy
CN	combined nomenclature
DDSIMCA	data-driven soft independent modeling of class analogies
DT	decision tree
EU	European Union
FT	Fourier-transform
GH	global Mahalanobis distance
GP	Gaussian process

HSI	hyperspectral imaging
InGaAs	indium gallium arsenide
iPLS	interval partial least squares discriminant analysis
JCR	journal citation reports
KC-FFQ	Knowledge Centre for Food Fraud and Quality
KNN	k-nearest neighbor
LDA	linear discriminant analysis
LR	logistic regression
MLP	multilayer perceptron
MPLS	modified partial least squares
MSC	multiplicative scatter correction
MSI	multispectral imaging
NB	naive bayes
NIR	near-infrared
NMR	nuclear magnetic resonance
OCPLS	one-class partial least squares
OJEU	official journal of the European Union
PCA	principal components analysis
PCR	polymerase chain reaction
PDO	protected designation of origin
PGI	protected geographical indication
PLS	partial least squares
PLS-DA	partial least squares–discriminant analysis
QDA	quadratic discriminant analysis
RF	random forest
SG	Savitzky–Golay
Si	Silicon
SIMCA	soft independent modeling of class analogies
SNV	standard normal variate
SVM	support vector machine
VIP	variable importance in projection
VIS	visible

References

1. Official Journal of the European Union (OJEU). Regulation (EU) 2017/625 of the European Parliament and of the Council of 15 March 2017 on official controls and other official activities performed to ensure the application of food and feed law, rules on animal health and welfare, plant health and plant protection products, amending Regulations (EC) No 999/2001, (EC) No 396/2005, (EC) No 1069/2009, (EC) No 1107/2009, (EU) No 1151/2012, (EU) No 652/2014, (EU) 2016/429 and (EU) 2016/2031 of the European Parliament and of the Council, Council Regulations (EC) No 1/2005 and (EC) No 1099/2009 and Council Directives 98/58/EC, 1999/74/EC, 2007/43/EC, 2008/119/EC and 2008/120/EC, and repealing Regulations (EC) No 854/2004 and (EC) No 882/2004 of the European Parliament and of the Council, Council Directives 89/608/EEC, 89/662/EEC, 90/425/EEC, 91/496/EEC, 96/23/EC, 96/93/EC and 97/78/EC and Council Decision 92/438/EEC (Official Controls Regulation) Text with EEA relevance. OJ L 95/1, 7.4.2017. 2017. pp. 1–142. Available online: https://eur-lex.europa.eu/legal-content/EN/TXT/PDF/?uri=CELEX:32017R0625 (accessed on 1 May 2024).
2. Winkler, B.; Maquet, A.; Reeves-Way, E.; Siegener, E.; Cassidy, T.; Valinhas de Oliveira, T.; Verluyten, J.; Jelic, M.; Muznik, A. *Fighting Fraudulent and Deceptive Practices in the Agri-Food Chain*; Technical Report Implementation of Article 9(2) of Regulation (EU) 2017/625; Publications Office of the European Union: Luxembourg, 2023. [CrossRef]
3. Ulberth, F. Tools to combat food fraud—A gap analysis. *Food Chem.* **2020**, *330*, 127044. [CrossRef] [PubMed]
4. Knowledge Centre for Food Fraud and Quality of the European (KC-FFQ). Monthly Food Fraud Summary Reports. 2023. Available online: https://knowledge4policy.ec.europa.eu/food-fraud-quality/monthly-food-fraud-summary-reports_en (accessed on 28 June 2023).
5. Official Journal of the European Union (OJEU). Commission Implementing Regulation (EU) 2017/1925 of 12 October 2017 amending Annex I to Council Regulation (EEC) No 2658/87 on the tariff and statistical nomenclature and on the Common Customs Tariff. OJ L 282/1, 31.10.2017. 2017. pp. 1–958. Available online: https://eur-lex.europa.eu/legal-content/EN/TXT/PDF/?uri=CELEX:32017R1925 (accessed on 1 May 2024).
6. Buthelezi, N.M.D.; Tesfay, S.Z.; Ncama, K.; Magwaza, L.S. Destructive and non-destructive techniques used for quality evaluation of nuts: A review. *Sci. Hortic.* **2019**, *247*, 138–146. [CrossRef]
7. Eurostats. Crop Production in EU Standard Humidity. 2023. Available online: https://ec.europa.eu/eurostat/databrowser/view/APRO_CPSH1__custom_8035705/default/table?lang=en (accessed on 27 February 2024).

8. Luparelli, A.; Losito, I.; De Angelis, E.; Pilolli, R.; Lambertini, F.; Monaci, L. Tree nuts and peanuts as a source of beneficial compounds and a threat for allergic consumers: Overview on methods for their detection in complex food products. *Foods* **2022**, *11*, 728. [CrossRef] [PubMed]
9. Campmajó, G.; Saurina, J.; Núñez, O. Liquid chromatography coupled to high-resolution mass spectrometry for nut classification and marker identification. *Food Control* **2023**, *152*, 109834. [CrossRef]
10. Ellis, D.I.; Brewster, V.L.; Dunn, W.B.; Allwood, J.W.; Golovanov, A.P.; Goodacre, R. Fingerprinting food: Current technologies for the detection of food adulteration and contamination. *Chem. Soc. Rev.* **2012**, *41*, 5706–5727. [CrossRef] [PubMed]
11. Visciano, P.; Schirone, M. Food frauds: Global incidents and misleading situations. *Trends Food Sci. Technol.* **2021**, *114*, 424–442. [CrossRef]
12. Aslam, R.; Sharma, S.R.; Kaur, J.; Panayampadan, A.S.; Dar, O.I. A systematic account of food adulteration and recent trends in the non-destructive analysis of food fraud detection. *J. Food Meas. Charact.* **2023**, *17*, 3094–3114. [CrossRef]
13. Barreira, J.C.M.; Alves, R.C.; Casal, S.; Ferreira, I.C.F.R.; Oliveira, M.B.P.P.; Pereira, J.A. Vitamin E profile as a reliable authenticity discrimination factor between chestnut (*Castanea sativa* Mill.) cultivars. *J. Agric. Food Chem.* **2009**, *57*, 5524–5528. [CrossRef]
14. Esteki, M.; Farajmand, B.; Amanifar, S.; Barkhordari, R.; Ahadiyan, Z.; Dashtaki, E.; Mohammadlou, M.; Heyden, Y.V. Classification and authentication of Iranian walnuts according to their geographical origin based on gas chromatographic fatty acid fingerprint analysis using pattern recognition methods. *Chemom. Intell. Lab. Syst.* **2017**, *171*, 251–258. [CrossRef]
15. Rabadan, A.; Pardo, J.E.; Gómez, R.; Alvarruiz, A.; Álvarez-Ortí, M. Usefulness of physical parameters for pistachio cultivar differentiation. *Sci. Hortic.* **2017**, *222*, 7–11. [CrossRef]
16. Brüning, P.; Haase, I.; Matissek, R.; Fischer, M. Marzipan: Polymerase chain reaction-driven methods for authenticity control. *J. Agric. Food Chem.* **2011**, *59*, 11910–11917. [CrossRef] [PubMed]
17. Ballin, N.Z.; Mikkelsen, K. Polymerase chain reaction and chemometrics detected several Pinus species including *Pinus armandii* involved in pine nut syndrome. *Food Control* **2016**, *64*, 234–239. [CrossRef]
18. Caligiani, A.; Coisson, J.D.; Travaglia, F.; Acquotti, D.; Palla, G.; Palla, L.; Arlorio, M. Application of 1H NMR for the characterisation and authentication of "Tonda Gentile Trilobata" hazelnuts from Piedmont (Italy). *Food Chem.* **2014**, *148*, 77–85. [CrossRef] [PubMed]
19. Eksi-Kocak, H.; Mentes-Yilmaz, O.; Boyaci, I.H. Detection of green pea adulteration in pistachio nut granules by using Raman hyperspectral imaging. *Eur. Food Res. Technol.* **2016**, *242*, 271–277. [CrossRef]
20. Firmani, P.; Bucci, R.; Marini, F.; Biancolillo, A. Authentication of "Avola almonds" by near infrared (NIR) spectroscopy and chemometrics. *J. Food Compos. Anal.* **2019**, *82*, 103235. [CrossRef]
21. Taylan, O.; Cebi, N.; Yilmaz, M.T.; Sagdic, O.; Ozdemir, D.; Bulubaid, M. Rapid detection of green-pea adulteration in pistachio nuts using Raman spectroscopy and chemometrics. *J. Food Sci. Agric.* **2021**, *101*, 1699–1708. [CrossRef] [PubMed]
22. Schmitt, C.; Bastek, T.; Stelzer, A.; Schneider, T.; Fischer, M. Detection of peanut adulteration in food samples by nuclear magnetic resonance spectroscopy. *J. Agric. Food Chem.* **2020**, *68*, 14364–14373. [CrossRef] [PubMed]
23. Von Wuthenau, K.; Segelke, T.; Müller, M.-S.; Behlok, H.; Fischer, M. Food authentication of almonds (*Prunus dulcis* Mill.). Origin analysis with inductively coupled plasma mass spectrometry (ICP-MS) and chemometrics. *Food Control* **2022**, *134*, 108696. [CrossRef]
24. Von Wuthenau, K.; Müller, M.-S.; Cvancar, L.; Oest, M.; Fischer, M. Food authentication of almonds (*Prunus dulcis* mill.). Fast origin analysis with laser ablation inductively coupled plasma mass spectrometry and chemometrics. *J. Agric. Food Chem.* **2022**, *70*, 5237–5244. [CrossRef]
25. Teixeira, M.; Sousa, C. A review on the application of vibrational spectroscopy to the chemistry of nuts. *Food Chem.* **2019**, *277*, 713–724. [CrossRef]
26. Panda, B.K.; Mishra, G.; Ramirez, W.A.; Jung, H.; Singh, C.B.; Lee, S.-H.; Le, I. Rancidity and moisture estimation in shelled almond kernels using NIR hyperspectral imaging and chemometric analysis. *J. Food Eng.* **2022**, *318*, 110889. [CrossRef]
27. Pannico, A.; Schouten, R.E.; Basile, B.; Romano, R.; Woltering, E.J.; Cirillo, C. Non-destructive detection of flawed hazelnut kernels and lipid oxidation assessment using NIR spectroscopy. *J. Food Eng.* **2015**, *160*, 42–48. [CrossRef]
28. Torres, I.; Sánchez, M.T.; Vega-Castellote, M.; Pérez-Marín, D. Fraud detection in batches of sweet almonds by portable near-infrared spectral devices. *Foods* **2021**, *10*, 1221. [CrossRef]
29. Mohammadi-Moghaddam, T.; Razavi, S.M.A.; Taghizadeh, M. Applications of hyperspectral imaging in grains and nuts quality and safety assessment: A review. *Food Meas.* **2013**, *7*, 129–140. [CrossRef]
30. Gowen, A.A.; O'Donnell, C.P.; Cullen, P.J.; Downey, G.; Frias, J.M. Hyperspectral imaging an emerging process analytical tool for food quality and safety control. *Trends Food Sci. Technol.* **2007**, *18*, 590–598. [CrossRef]
31. Lu, Y.; Saeys, W.; Kim, M.; Peng, Y.; Lu, R. Hyperspectral imaging technology for quality and safety evaluation of horticultural products: A review and celebration of the past 20-year progress. *Postharvest Biol. Technol.* **2020**, *170*, 111318. [CrossRef]
32. Pérez-Marín, D.; Garrido-Varo, A. NIR Sensors for the In-Situ Assessment of Iberian Ham. *Compr. Foodomics* **2021**, *3*, 340–345. [CrossRef]
33. Momtaz, M.M.; Bubli, S.Y.; Khan, M.S. Mechanisms and health aspects of food adulteration: A comprehensive review. *Foods* **2023**, *12*, 199. [CrossRef]
34. Dimitrakopoulou, M.-E.; Vantarakis, A. Does Traceability Lead to Food Authentication? A Systematic Review from A European Perspective. *Food Rev. Int.* **2023**, *39*, 537–559. [CrossRef]

35. Faqeerzada, M.A.; Lohumi, S.; Joshi, R.; Kim, M.; Baek, I.; Cho, B.-K. Non-targeted detection of adulterants in almond powder using spectroscopic techniques combined with chemometrics. *Foods* **2020**, *9*, 876. [CrossRef]
36. Menevseoglu, A.; Entrenas, J.A.; Gunes, N.; Dogan, M.A.; Pérez-Marín, D. Machine learning-assisted near-infrared spectroscopy for rapid discrimination of apricot kernels in ground almond. *Food Control* **2024**, *159*, 110272. [CrossRef]
37. Biancolillo, A.; De Luca, S.; Bassi, S.; Roudier, L.; Bucci, R.; Magrì, A.D.; Marini, F. Authentication of an Italian PDO hazelnut ("Nocciola Romana") by NIR spectroscopy. *Environ. Sci. Pollut. Res.* **2018**, *25*, 28780–28786. [CrossRef]
38. Sammarco, G.; Dall'Asta, C.; Suman, M. Near infrared spectroscopy and multivariate statistical analysis as rapid tools for the geographical origin assessment of Italian hazelnuts. *Vib. Spectrosc.* **2023**, *126*, 103531. [CrossRef]
39. Amendola, L.; Firmani, P.; Bucci, R.; Marini, F.; Biancolillo, A. Authentication of Sorrento walnuts by NIR spectroscopy coupled with different chemometric classification strategies. *Appl. Sci.* **2020**, *10*, 4003. [CrossRef]
40. Nardecchia, A.; Presutto, R.; Bucci, R.; Marini, F.; Biancolillo, A. Authentication of the geographical origin of "Vallerano" chestnut by near infrared spectroscopy coupled with chemometrics. *Food Anal. Methods* **2020**, *13*, 1782–1790. [CrossRef]
41. Cortés, V.; Barat, J.; Talens, P.; Blasco, J.; Lerma-García, M.J. A comparison between NIR and ATR-FTIR spectroscopy for varietal differentiation of Spanish intact almonds. *Food Control* **2018**, *94*, 241–248. [CrossRef]
42. Arndt, M.; Rurik, M.; Drees, A.; Ahlers, C.; Feldmann, S.; Kohlbacher, O.; Fischer, M. Food authentication: Determination of the geographical origin of almonds (*Prunus dulcis* MILL.) via near-infrared spectroscopy. *Microchem. J.* **2021**, *160*, 105702. [CrossRef]
43. Corona, P.; Frangipane, M.T.; Moscetti, R.; Lo Feudo, G.; Castellotti, T.; Massantini, R. Chestnut cultivar identification through the data fusion of sensory quality and FT-NIR spectral data. *Foods* **2021**, *10*, 2575. [CrossRef] [PubMed]
44. Moscetti, R.; Berhe, D.H.; Agrimi, M.; Haff, R.P.; Liang, P.; Ferri, S.; Monarca, D.; Massantini, R. Pine nut species recognition using NIR spectroscopy and image analysis. *J. Food Eng.* **2021**, *292*, 110357. [CrossRef]
45. Genis, H.E.; Durna, S.; Boyaci, I.H. Determination of green pea and spinach adulteration in pistachio nuts using NIR spectroscopy. *LWT—Food Sci. Technol.* **2021**, *136*, 110008. [CrossRef]
46. Aykas, D.P.; Menevseoglu, A. A rapid method to detect green pea and peanut adulteration in pistachio by using portable FT-MIR and FT-NIR spectroscopy combined with chemometrics. *Food Control* **2021**, *121*, 107670. [CrossRef]
47. Vega-Castellote, M.; Pérez-Marín, D.; Torres, I.; Moreno-Rojas, J.M.; Sánchez, M.T. Exploring the potential of NIRS technology for the in situ prediction of amygdalin content and classification by bitterness of in-shell and shelled intact almonds. *J. Food Eng.* **2021**, *294*, 110406. [CrossRef]
48. Torres-Rodríguez, I.; Sánchez, M.T.; Entrenas, J.A.; Vega-Castellote, M.; Garrido-Varo, A.; Pérez-Marín, D. Hyperspectral Imaging for the detection of bitter almonds in sweet almond batches. *Appl. Sci.* **2022**, *12*, 4842. [CrossRef]
49. Miaw, C.S.W.; Martins, M.L.C.; Sena, M.M.; De Souza, S.V.C. Screening method for the detection of other allergenic nuts in cashew nuts using chemometrics and portable near-infrared spectrophotometer. *Food Anal. Methods* **2022**, *15*, 1074–1084. [CrossRef]
50. Rovira, G.; Miaw, C.S.W.; Martins, M.L.C.; Sena, M.M.; De Souza, S.V.C.; Ruisánchez, I.; Callao, M.P. In-depth chemometric strategy to detect up to four adulterants in cashew nuts by IR spectroscopic techniques. *Microchem. J.* **2022**, *181*, 107816. [CrossRef]
51. Feng, Z.; Li, W.; Cui, D. Detection of endogenous foreign bodies in Chinese hickory nuts by hyperspectral spectral imaging at the pixel level. *Int. J. Agric. Biol. Eng.* **2022**, *15*, 204–210. [CrossRef]
52. Arndt, M.; Rurik, M.; Drees, A.; Bigdowski, K.; Kohlbacher, O.; Fischer, M. Comparison of different sample preparation techniques for NIR screening and their influence on the geographical origin determination of almonds (*Prunus dulcis* MILL.). *Food Control* **2020**, *115*, 107302. [CrossRef]
53. Vega-Castellote, M.; Sánchez, M.T.; Torres, I.; Pérez-Marín, D. An innovative non-targeted control system based on NIR spectral information for detecting non-compliant batches of sweet almonds. *Spectrochim. Acta Part A Mol. Biomol. Spectrosc.* **2021**, *250*, 119407. [CrossRef]
54. Lösel, H.; Shakiba, N.; Wenck, S.; Tan, P.L.; Arndt, M.; Seifert, S.; Hackl, T.; Fischer, M. Impact of Freeze-Drying on the determination of the geographical origin of almonds (*Prunus dulcis* Mill.) by near-infrared (NIR) spectroscopy. *Food Anal. Methods* **2022**, *15*, 2847–2857. [CrossRef]
55. Netto, J.M.; Honorato, F.A.; Celso, P.G.; Pimentel, M.F. Authenticity of almond flour using handheld near infrared instruments and one class classifiers. *J. Food Compos. Anal.* **2023**, *115*, 104981. [CrossRef]
56. Rovira, G.; Miaw, C.S.W.; Martins, M.L.C.; Sena, M.M.; De Souza, S.V.C.; Callao, M.P.; Ruisánchez, I. One-class model with two decision thresholds for the rapid detection of cashew nuts adulteration by other nuts. *Talanta* **2023**, *253*, 123916. [CrossRef] [PubMed]
57. Ayvaz, H.; Temizkan, R.; Genis, H.E.; Mortas, M.; Genis, D.O.; Dogan, M.A.; Nazlim, B.A. Rapid discrimination of Turkish commercial hazelnut (*Corylus avallana* L.) varieties using Near-Infrared Spectroscopy and chemometrics. *Vib. Spectrosc.* **2022**, *119*, 103353. [CrossRef]
58. Shakiba, N.; Gerdes, A.; Holz, N.; Wenck, S.; Bachmann, R.; Schneider, T.; Seifert, S.; Fischer, M.; Hackl, T. Determination of the geographical origin of hazelnuts (*Corylus avellana* L.) by near-infrared spectroscopy and a low-level fusion with nuclear magnetic resonance (NMR). *Microchem. J.* **2022**, *174*, 107066. [CrossRef]
59. Huang, B.; Liu, J.; Jiao, J.; Lu, J.; Lv, D.; Mao, J.; Zhao, Y.; Zhang, Y. Applications of machine learning in pine nuts classification. *Sci. Rep.* **2022**, *12*, 8799. [CrossRef] [PubMed]
60. Gu, X.; Li, Z.; Li, L.; Ma, N.; Tu, K.; Song, L.; Pan, L. Multisource fingerprinting for the region identification of walnuts in Xinjiang combined with chemometrics. *J. Food Process Eng.* **2018**, *41*, e12687. [CrossRef]

61. Arndt, M.; Drees, A.; Ahlers, C.; Fischer, M. Determination of the geographical origin of walnuts (*Juglans regia* L.) using near-infrared spectroscopy and chemometrics. *Foods* **2020**, *9*, 1860. [CrossRef] [PubMed]
62. Faqeerzada, M.A.; Perez, M.; Lohumi, S.; Lee, H.; Kim, G.; Wakholi, C.; Joshi, R.; Cho, B.-K. Online application of a hyperspectral imaging system for the sorting of adulterated almonds. *Appl. Sci.* **2020**, *10*, 6569. [CrossRef]
63. Porep, J.U.; Kammerer, D.R.; Reinhold, C. On-line application of near infrared (NIR) spectroscopy in food production. *Trends Food Sci. Technol.* **2015**, *46*, 211–230. [CrossRef]
64. Liu, Y.; Pu, H.; Sun, D.-W. Hyperspectral imaging technique for evaluating food quality and safety during various processes: A review of recent applications. *Trends Food Sci. Technol.* **2017**, *69*, 25–35. [CrossRef]
65. Müller-Maatsch, J.; van Ruth, S.M. Handheld devices for food authentication and their applications: A review. *Foods* **2021**, *10*, 2901. [CrossRef]
66. Cortés, V.; Blasco, J.; Aleixos, N.; Cubero, S.; Talens, P. Monitoring strategies for quality control of agricultural products using visible and near-infrared spectroscopy: A review. *Trends Food Sci. Technol.* **2019**, *85*, 138–148. [CrossRef]
67. Pérez-Marín, D.C.; Garrido-Varo, A. Near-Infrared Spectroscopy and Chemometrics in Food and Agriculture. In *Encyclopedia of Analytical Chemistry*; Meyers, R.A., Ed.; John Wiley & Sons Ltd.: Hoboken, NJ, USA, 2023. [CrossRef]
68. Shenk, J.S.; Workman, J.; Westerhaus, M.O. Application of NIR spectroscopy to agricultural products. In *Handbook of Near Infrared Analysis*; Burns, D.A., Ciurczac, E.W., Eds.; Marcel Dekker: New York-Basel, NY, USA, 2008; pp. 347–386.
69. Zhu, C.; Fu, X.; Zhang, J.; Qin, K.; Wu, C. Review of portable near infrared spectrometers: Current status and new techniques. *J. Near Infrared Spectrosc.* **2022**, *30*, 51–66. [CrossRef]
70. Qin, J.; Chao, K.; Kim, M.S.; Lu, R.; Burks, T.F. Hyperspectral and multispectral imaging for evaluating food safety and quality. *J. Food Eng.* **2013**, *118*, 157–171. [CrossRef]
71. Rinnan, Å.; van de Berg, F.; Engelsen, S.B. Review of the most common pre-processing techniques for near-infrared spectra. *Trends Anal. Chem.* **2009**, *28*, 1201–1222. [CrossRef]
72. Boysworth, M.K.; Booksh, K.S. Aspects of multivariate calibration applied to near-infrared spectroscopy. In *Handbook of Near-Infrared Analysis*; Burns, D.A., Ciurczak, E., Eds.; Marcel Dekker: Basel, Switzerland, 2008; pp. 347–386.
73. Riedl, J.; Esslinger, S.; Fauhl-Hassek, C. Review of validation and reporting of non-targeted fingerprinting approaches for food authentication. *Anal. Chim. Acta* **2015**, *885*, 17–32. [CrossRef]
74. Xu, L.; Yan, S.-M.; Cai, C.-B.; Yu, X.-P. One-class partial least squares (OCPLS) classifier. *Chemom. Intell. Lab. Syst.* **2013**, *126*, 1–5. [CrossRef]
75. McGrath, T.F.; Haughey, S.A.; Patterson, J.; Fauhl-Hassek, C.; Donarski, J.; Alewijn, M.; van Ruth, S.; Elliot, C.T. What are the scientific challenges in moving from targeted to non-targeted methods for food fraud testing and how can they be addressed? Spectroscopy case study. *Trends Food Sci. Technol.* **2018**, *76*, 38–55. [CrossRef]
76. Sádecká, J.; Jakubíková, M.; Májek, P.; Kleinová, A. Classification of plum spirit drinks by synchronous fluorescence spectroscopy. *Food Chem.* **2016**, *196*, 783–790. [CrossRef] [PubMed]
77. Cortes, C.; Vapnik, V. Support-vector networks. *Mach. Learn.* **1995**, *20*, 273–297. [CrossRef]
78. Oliveri, P. Class-modelling in food analytical chemistry: Development, sampling, optimisation and validation issues—A tutorial. *Anal. Chim. Acta* **2017**, *982*, 9–19. [CrossRef]
79. De Maesschalck, R.; Candolfi, A.; Massart, D.L.; Heuerding, S. Decision criteria for soft independent modelling of class analogy applied to near infrared data. *Chemom. Intell. Lab. Syst.* **1999**, *47*, 65–77. [CrossRef]
80. Kemsley, E.K. Discriminant analysis of high-dimensional data: A comparison of principal components analysis and partial least squares data reduction methods. *Chemom. Intell. Lab. Syst.* **1996**, *33*, 47–61. [CrossRef]

beverages MDPI

Article

Production and Optimisation of Fermented Pumpkin-Based Mature Coconut Water Kefir Beverage Using Response Surface Methodology

Wee Yin Koh [1,*], Xiao Xian Lim [2], Ban Hock Khor [1], Babak Rasti [3], Thuan Chew Tan [2], Rovina Kobun [1] and Utra Uthumporn [2]

1 Faculty of Food Science and Nutrition, Universiti Malaysia Sabah, Jalan UMS, Kota Kinabalu 88400, Malaysia; khorbanhock@ums.edu.my (B.H.K.); rovinaruby@ums.edu.my (R.K.)
2 Food Technology Division, School of Industrial Technology, Universiti Sains Malaysia, Minden 11800, Malaysia; ayumi_xiaoxian@yahoo.com (X.X.L.); thuanchew@usm.my (T.C.T.); sapina@usm.my (U.U.)
3 Australasian Nanoscience and Nanotechnology Initiative, 8054 Monash University LPO, Clayton 3168, Australia; rasti_babak@yahoo.com
* Correspondence: weeyin@ums.edu.my; Tel.: +60-88-320000

Abstract: Fermentation of pumpkin puree and mature coconut water using water kefir grains is a potential method for producing a novel functional non-dairy-based probiotic drink. In the present study, response surface methodology based on Box–Behnken design (RSM-BBD) was used to optimise fermentation temperature and substrates' concentrations. The optimised fermentation temperature, pumpkin puree, and brown sugar concentrations of pumpkin-based mature coconut water kefir beverage (PWKC) were 27 °C, 20%, and 10% w/v, respectively. The optimised PWKC ($PWKC_{opt}$) obtained an overall acceptability (OA) score of 4.03, with a desirable *Lactobacillus* count (6.41 Log CFU/mL), 0.68% v/v lactic acid content, 31% of water kefir grains' biomass growth rate, and fermentation time (to reach pH 4.5) of 4.5 h. The optimized beverage, $PWKC_{opt}$, contained 3.26% proteins, 2.75% dietary fibre, 2186.33 mg/L of potassium, 180.67 mg/L phosphorus, and 137.33 mg/L calcium and had a total phenolic content of 89.93 mg GAE/100 mL, flavonoid content of 49.94 mg QE/100 mL, and carotenoid content of 33.24 mg/100 mL, with antioxidant activity (FRAP: 169.17 mM Fe(II)/100 mL, IC_{50} value of DPPH free radicals scavenging activity: 27.17 mg/mL). Water kefir microorganisms in $PWKC_{opt}$ remained stable for at least 56 days at 4 °C. Therefore, $PWKC_{opt}$ might potentially serve as a value-added product, offering a basis for sustainable development within both the coconut and pumpkin industries.

Keywords: acetic acid bacteria; Box–Behnken design; fermented pumpkin; lactic acid bacteria; mature coconut water; yeast

1. Introduction

The growing demand of consumers for foods that offer more than just basic sustenance has increased interest in the development of fermented functional foods containing probiotics. Probiotics have been defined as 'live microorganisms, that are believed to confer health benefits to the host when administrated in adequate amounts' by the FAO/WHO [1]. Traditionally, dairy products are the predominant vehicle for the delivery of probiotic bacteria. However, the high prevalence of lactose intolerance and the ongoing trend of veganism have led to an increase in the demand for non-dairy alternatives in the probiotic market. According to Euromonitor International [2], approximately 20% of consumers in the world's major markets intend to increase plant origin consumption while decreasing animal origin consumption.

Water kefir, a traditional fermented beverage crafted from sugary water using water kefir grains, offers a viable non-dairy alternative to conventional probiotic dairy products.

Water kefir grains are irregularly shaped and brittle, composed of a grain-like exopolysaccharide matrix made up of a symbiotic culture of lactic acid bacteria (LAB), acetic acid bacteria (AAB), and yeasts [3]. The historical origin of water kefir grains is not precisely known, but some of the earliest records indicate that 'Tibi grains' were plucked from the leaves of a Mexican cactus (*Optunia*) [4]. Water kefir brewing is a relatively mild fermentation process (24–72 h, 21–30 °C [3,5]) and hence the bioactive compounds such as vitamins, minerals, and organic acids present in the fermented water kefir brew can be preserved well. Water kefir grains can be retrieved from the fermented medium and used indefinitely by re-inoculating the grains in a new sugar–water medium [6], and therefore, water kefir production is relatively inexpensive. Low-energy water kefir-based fermentations can improve sensory, functional, and nutritional properties, increase shelf life, and generate value-added goods [3].

The consumption of fermented water kefir beverages has also been related to health benefits, such as gastrointestinal beneficial effects and anti-hypertensive, anti-bacterial, anti-microbial, anti-cancer, anti-oxidative, anti-allergenic, anti-asthmatic, anti-tumour, anti-inflammatory, anti-obesity, anti-hyperglycaemia, anti-carcinogenic, cholesterol-lowering, and immune modulation effects, and helps in minimising or preventing oedema, ulcers, and pathogen growth [3,7–10]. Traditionally, a table sugar or brown sugar solution is used as the substrate for water kefir fermentation [3]. However, a sugar solution has a relatively simple nutritional composition and reduces the growth rate and activity of water kefir over time. In order to ensure the health benefits of kefir, studies on fermentation using alternative matrices are necessary. Locally available, nutritious but highly perishable pumpkin and mature coconut water, which are often discarded as waste, have the potential to be used as fermentation substrates in the production of water kefir.

Pumpkin (*Cucurbita* L.), a squash fruit vegetable belonging to the *Cucurbitaceae* family [11], is a significant domestic crop grown worldwide [12], with global production of 22.81 million tons in 2022 [13]. The total production of pumpkin in Malaysia was 20,369 tons in 2022, making it one of the most important agricultural commodities in the country [14]. In the present study, pumpkin is selected as the fermentation substrate over different sources due to its richness in indigestible carbohydrates (such as rhamnose, glucose, arabinose, and galactose) with potential prebiotic properties [15]. Previous studies have demonstrated that the prebiotic potential of pumpkin could enhance the viability of probiotics under adverse food processing and simulated gastrointestinal conditions [16,17]. Pumpkin is also a good source of vitamins, minerals, and carotenoids, which include α- and ß-carotene, and lutein, and is low in fat, which is beneficial for the developed beverage as a functional food [15,18]. However, raw pumpkin is not ready-to-eat, and the processing of pumpkin from cutting to cooking is laborious due to its large volume and mass. Furthermore, freshly cut pumpkin is susceptible to enzymatic browning and microbial spoilage [19]. Fermentation with *Lactobacillus* has been found to reduce anti-nutrients (phytates, tannins, phenolic compounds, and oxalates) in pumpkin [20], produce antimicrobial ingredients to inhibit the growth of pathogenic microorganisms, and generate high antioxidant activity, which can inhibit enzymatic browning in pumpkin [21]. Pumpkin can be transformed into a ready-to-drink, value-added beverage product with prolonged shelf life, enhanced nutritional values, and improved digestibility through the fermentation process [20,21]. Previously, pumpkin has been used as a fermentation substrate in the production of fermented juices using *Lactobacillus casei* [18] and *Lactobacillus rhamnosus* [15]. However, to the best of our knowledge, using pumpkin as a substrate for water kefir fermentation appears to have received limited attention in scientific research.

Coconut (*Cocos nucifera* L.) is a multipurpose perennial crop that is widely grown in tropical regions around the world. Multifaceted uses of coconut have resulted in high global production of 62.41 million tons in 2022 [13]. Among the major coconut-producing countries, Malaysia ranked 11th, with a total production of 604,428 tons in 2022. Coconut is also the fourth essential industrialised crop in Malaysia, after rice, rubber, and oil palm [22]. Coconut water is suitable to be consumed directly as well as to be transformed into different beverages owing to its high quantities of sugar, amino acids, vitamins and minerals and low fat content [23]. Scientific evidence is growing to support the use of coconut water for health and therapeutic purposes. Coconut water's diverse applications can be attributed to its unique chemical composition, including carbohydrates, vitamins, minerals, amino acids, and phytochemicals [24]. Depending on the age of the coconut upon harvest, coconut water can be classified as young or mature. As a coconut grows, in particular, in the ninth month, the overall sugar level peaks at 2.9% [25], and therefore, young coconut water is generally sweeter than mature coconut water. While the maturity level does not notably affect the nutritional properties, the rise in fat content and decline in water content as the coconut matures can lead to a degradation in flavour [23]. As a result, mature coconut water is disposed of as waste most of the time during the manufacturing of coconut milk, chutney, chips, or cream, resulting in substantial environmental pollution and a significant loss of mature coconut water [26]. For instance, a total of 200,000 tons/year of mature coconut water is discharged down the drain in Thailand [27]. There have been attempts to make various value-added products such as therapeutic drinks [28], refreshment drinks [25], and vinegar [23] using mature coconut water. The flavour of mature coconut water can be overshadowed by the acid–sweetish flavour imparted by other additives [25], and hence, mature coconut water is drinkable after the fermentation process with improved organoleptic properties [28]. A functional beverage produced by fermenting mature coconut water with potential probiotic *Lactobacillus plantarum* DW12 yielded a probiotic count of 8.4 Log CFU/mL [27], surpassing the minimum therapeutic dose of probiotic cells, which is 6 Log CFU/mL. This suggests that mature coconut water is an appropriate medium for delivering probiotic strains [29]. Therefore, employing the water kefir fermentation process to create fermented mature coconut water beverages could mitigate off-flavours present in mature coconut water and potentially improve customer acceptance.

The usage of water with low buffering capacity and calcium concentrations is linked to reduced water kefir grain growth [30]. The calcium content of coconut water constantly rises throughout the maturation process [24]. The protein content of coconut water also increases as it matures, which could be linked to increased stored protein synthesis during the change of liquid watery endosperm into solid white coconut meat [24]. According to Guzel-Seydim et al. [31], the addition of whey protein isolates or modified whey protein to kefir grains could maintain the kefir grain microflora and enhance biomass production. As a result, it could be hypothesised that mature coconut water with high protein and calcium ion contents is more suited for water kefir fermentation. Young coconut water has been extensively studied, but the exploration of mature coconut water as a water kefir fermentation substrate has been relatively limited.

This study used a combination of mature coconut water and pumpkin puree as substrates in the development of a fermented pumpkin-based mature coconut water kefir beverage with water kefir grains as a starter. While a wide range of substrates have been utilised in the preparation of water kefir beverages, the practice of combining multiple substrates is not so common. Mature coconut water is naturally rich in sugars, vitamins, and minerals [23] while pumpkin puree is a good source of indigestible carbohydrates [15]. Utilisation of a combination of mature coconut water and pumpkin puree as substrates could offer a diverse range of nutrients for the kefir microbiota and consequently support the growth of kefir grains and enhance the fermentation process and fermented final product.

Traditional water kefir is manufactured by people in Southeast Asian countries on a household scale, using solid-state fermentation under non- or barely aseptic conditions, which poses a risk of pathogenic or possibly unwanted microbial contamination, requires a long fermentation period, has low acid production and inconsistent flavour, and hence is not applicable for industrial production [32]. Water kefir can be prepared via a consistent method to ensure quality and consistency in the final product. Therefore, in the present study, response surface methodology based on the Box–Behnken design (RSM-BBD) was used to optimise the fermentation temperature and the concentrations of substrates (pumpkin puree and brown sugar) to obtain a final fermented product (pumpkin-based mature coconut water kefir beverage (PWKC)) with minimum fermentation time (to reach pH 4.5), while achieving maximum overall acceptability (OA) score, water kefir grain biomass growth rate, lactic acid content, and *Lactobacillus* count. The chemical composition, physicochemical properties (total soluble solids, viscosity, pH value, and colour), antioxidative contents and activities, sugar (glucose and fructose) and ethanol contents, and viabilities of presumptive *Lactobacillus*, LAB, AAB, and yeast, sensorial properties, and shelf-life stability of optimised PWKC ($PWKC_{opt}$) were also studied.

2. Materials and Methods

2.1. Materials

Ripe pumpkins (*Cucurbita moschata* Duschene) and mature coconuts (12 months old, Malayan Coconut Tall) were bought from Comfort Organic Farm (Balik Pulau, Penang, Malaysia) and Anba Coconuts (Abu Siti Lane, Penang, Malaysia), respectively. The water kefir grains were purchased from My Kefir World, Kuala Lumpur, Malaysia.

2.2. Chemicals and Reagents

Brilliant green lactose bile broth, chloramphenicol, cycloheximide, de Man Rogosa Sharpe (MRS) agar, lactose broth, Folin–Ciocalteau reagent, modified deoxycholate-mannitol-sorbitol (mDMS) agar, peptone solution, plate count agar (PCA), Ringer's solution, sodium acetate buffer, and yeast extract glucose chloramphenicol (YGC) agar were purchased from Merck (Darmstadt, Germany). Dichloran Rose Bengal Chloramphenicol agar (DRBC) was obtained from Biokar Diagnostics (Beauvais, France). *Lactobacillus-selective* base agar was bought from Neogen Corporation (Lansing, MI, USA). Aluminium trichloride, ferric chloride hexahydrate, and 2,4,6-tri(2-pyridyl)-s-triazine (TPTZ) solution were purchased from Sigma-Aldrich (St. Louis, MO, USA). Ethanol and methanol were supplied by HmbG Chemicals (Hamburg, Germany). Sodium carbonate and hydrochloric acid were bought from Systerm ChemAR (Shah Alam, Malaysia) and R&M Chemicals (London, UK), respectively. All chemicals and reagents were of analytic grade and had not undergone any further purification.

2.3. Preparation of Fermentation Substrates

2.3.1. Preparation of Pumpkin Puree

The pumpkins were trimmed, washed, cut into slices, steam-pasteurised to aid in softening the fruits at 121 °C for 20 min in an autoclave (HV-25, Hirayama Manufacturing Co., Saitama, Japan), and subsequently blended into a puree. Once the puree had cooled to room temperature, it was packaged in sterile polyethylene zip-seal bags and stored at −20 °C until use. Before use as the beverage substrate, the pumpkin puree was thawed at 4 °C overnight.

2.3.2. Preparation of Mature Coconut Water

The mature coconuts were washed with distilled water containing 1% household bleach (Clorox, The Clorox Company, Oakland, CA, USA) and rinsed thoroughly with distilled water. After air-drying for 1 h, the coconut water was extracted by drilling the fruit mesocarp with a blade. The extracted mature coconut water was filtered using a muslin cloth and flash pasteurised (15 min at 121 °C) using a lab-scale autoclave (HV-25, Hirayama Manufacturing Co., Saitama, Japan), poured into jars, and cooled to 25 °C.

2.4. Activation of Water Kefir Grains

The mixture of mature coconut water (64% v/v), pumpkin puree (22% w/v), and brown sugar (9% w/v, Malayan Sugar Manufacturing (MSM) Prai Berhad, Seberang Perai, Malaysia) was pasteurised at 121 °C for 20 min using a lab-scale autoclave and left to cool to 25 °C. Subsequently, water kefir grains at 5% (w/v) were added, and the mixture was left to ferment at 32 °C for 12 h [33]. After fermentation, the water kefir grains were separated from the brew by filtering the mixture through a plastic fine mesh sieve (3 mm pore size).

2.5. Preparation of PWKC

Activated water kefir grains were used directly as inoculum at 5% *w*/*v* to ferment the PWKC beverage samples as shown in Table 1. The fermentation process was considered complete when the pH value of the sample reached 4.5. All the preparation procedures were conducted in a laminar airflow cabinet (Class II Type A2 Biological Safety Cabinet, Esco Labculture, Horsham, PA, USA) to maintain aseptic conditions. The fermented beverage samples were stored in a refrigerator (4 °C) until used.

Table 1. Box–Behnken design (BBD) arrangement of the independent variables (X_1, X_2, and X_3) and their observed responses including the fermentation time to reach pH 4.5 ($T_{pH4.5}$).

Run	X_1 (% *w*/*v*)	X_2 (% *w*/*v*)	X_3 (°C)	$T_{pH4.5}$ (h)	OA (Score)	Biomass Growth Rate (% G)	Lactic Acid (% *v*/*v*)	*Lactobacillus* Count (Log CFU/mL)
1	20.00 (−1)	0.00 (−1)	27.00 (0)	6.36	2.09	45.00	0.80	7.54
2	30.00 (+1)	0.00 (−1)	27.00 (0)	6.15	2.02	40.00	0.62	6.01
3	20.00 (−1)	10.00 (+1)	27.00 (0)	4.56	3.98	30.00	0.64	6.23
4	30.00 (+1)	10.00 (+1)	27.00 (0)	3.64	5.89	22.00	0.65	4.32
5	20.00 (−1)	5.00 (0)	22.00 (−1)	6.00	2.64	42.00	0.73	6.82
6	30.00 (+1)	5.00 (0)	22.00 (−1)	5.45	2.83	38.00	0.64	5.95
7	20.00 (−1)	5.00 (0)	32.00 (+1)	3.00	4.98	15.00	0.54	3.82
8	30.00 (+1)	5.00 (0)	32.00 (+1)	2.25	6.50	10.00	0.32	2.08
9	25.00 (0)	0.00 (−1)	22.00 (−1)	8.08	0.74	55.00	0.82	8.04
10	25.00 (0)	10.00 (+1)	22.00 (−1)	4.76	2.07	31.00	0.65	6.03
11	25.00 (0)	0.00 (−1)	32.00 (+1)	3.64	2.25	20.00	0.40	3.08
12	25.00 (0)	10.00 (+1)	32.00 (+1)	3.00	6.26	15.00	0.45	3.82
13	25.00 (0)	5.00 (0)	27.00 (0)	4.07	4.46	25.00	0.40	4.01
14	25.00 (0)	5.00 (0)	27.00 (0)	4.00	4.46	25.00	0.45	3.26
15	25.00 (0)	5.00 (0)	27.00 (0)	4.00	4.22	24.00	0.45	3.88
16	25.00 (0)	5.00 (0)	27.00 (0)	4.08	4.37	24.00	0.45	3.88
17	25.00 (0)	5.00 (0)	27.00 (0)	4.00	4.23	26.00	0.43	3.12

X_1 = pumpkin puree content (20–30% *w*/*v*); X_2 = brown sugar content (0–10% *w*/*v*); X_3 = fermentation temperature (22–32 °C).

2.6. Experimental Design

The three independent variables studied were concentrations of pumpkin puree (X_1; 20–30% *w*/*v*) and brown sugar (X_2; 0–10% *w*/*v*), and fermentation temperature (X_3; 22–32 °C). The independent variables were optimised using RSM-BBD with Design-Expert 13.0 software (DX13, Stat Ease Inc., Minneapolis, MN, USA), for PWKC production. The three coded levels of variables were −1, 0, and +1 (low, medium, and high) (Table 1). Seventeen experiments were executed following the design matrix, and their response values were obtained. The activated water kefir grains were used directly as inoculum at (5% *w*/*v*) to ferment the PWKC samples at different fermentation temperatures and brown sugar and pumpkin puree concentrations, as shown in Table 1. Five dependent variables, namely (i) fermentation time to reach pH 4.5 ($T_{pH4.5}$, h), (ii) overall acceptability score (OA), (iii) water kefir grains' biomass growth rate (% G), (iv) lactic acid content (%), and (v) *Lactobacillus* count (Log CFU/mL), were assessed.

Each variable's significance level (*p*-value) was established based on Student's *t*-test with $p < 0.05$ indicating a statistically significant difference. Analysis of variance (ANOVA) was employed to determine the significance of the generated models. DX13 software was used to determine the response (Y) of the second-order polynomial equation, the coefficient of determination (R^2), the predicted R-squared and adjusted R-squared, the coefficient of variance (CV), and the probability F-value. The functional (mathematical) relationship between the three independent variables and the dependent variable was presented by a second-degree polynomial equation (Equation (1)):

$$Y = \beta_0 + \beta_1 X_1 + \beta_2 X_2 + \beta_3 X_3 + \beta_{11} X_1^2 + \beta_{22} X_2^2 + \beta_{33} X_3^2 + \beta_{12} X_1 X_2 + \beta_{13} X_1 X_3 + \beta_{23} X_2 X_3 \quad (1)$$

where Y and X_i are the predicted response and independent variables (3 variables), respectively. β_0, β_i, β_{ii}, and β_{ij} are the constant, linear, quadratic, and interaction regression coefficients of the model computed by DX13.

The response surface and contour plots were used to study the effect of two independent variables on dependent variables. An experiment was then conducted using a combination of the optimum predicted values to validate an optimal model.

2.7. Microbiological Safety Analyses

Microbiological safety analyses were conducted following the methods described by Gómez-Aldapa et al. [34] and Cassani et al. [35] prior to hedonic assessment. The sample (50 mL) was diluted with quarter-strength ($\frac{1}{4}$) Ringer's solution (450 mL) and homogenised in a stomacher (1 min; Lab Blender, Seward 400, London, UK). The homogenised samples were diluted (until 10^6) using $\frac{1}{4}$ Ringer's solution and inoculated onto the corresponding agar or broth for enumeration of total aerobic mesophilic bacteria (PCA Agar, 35 °C for 48 h), filamentous fungi (DRBC agar, 30 °C for 120 h), coliform bacteria, faecal coliforms, and *Escherichia coli* presumptive identification and confirmation (lactose broth and brilliant green lactose bile broth at 37 °C for 48 h and 44.5 °C for 48 h, respectively). The total mesophilic aerobic bacteria and filamentous fungi counts were expressed as CFU/mL whereas coliform count was expressed as the most probable number (MPN)/100 mL.

2.8. Measurement of Responses

2.8.1. Fermentation Time

The fermentation time ($T_{pH4.5}$, h) was determined based on the amount of time for PWKC beverage samples to reach a pH value of 4.5. This pH value was selected as the fermentation endpoint of plant-based water kefir and probiotic yoghurt-like products [36,37]. Throughout the fermentation process, the pH value of PWKC samples was determined using a pH meter (Mettler-Toledo, SevenEasy, Griefensee, Switzerland) at 30 min intervals.

2.8.2. Overall Acceptability

Hedonic assessment of PWKC beverage samples was carried out in sensory booths, and sensory panellists were recruited according to ISO 11136 [38]. Fifty semi-trained panellists (25 males and 25 females, Food Technology undergraduates at School of Industrial Technology, Universiti Sains Malaysia, Penang, Malaysia) were recruited to evaluate the OA of the beverage samples using a 7-point hedonic scale, where 1 = dislike extremely, 4 = neither like nor dislike, and 7 = like extremely. All the panellists were trained on the products' sensory evaluation, and consent was obtained from the panellists. Purified water was provided to cleanse the palate between samples.

2.8.3. Water Kefir Grain Biomass Growth Rate

The biomass growth rate was determined using Equation (2) to evaluate the effects of culture conditions [39]:

$$G(\%) = \frac{X_{n+1} - X_n}{X_n} \times 100 \quad (2)$$

where G is the growth rate, X_n is the biomass weight after n hours (g), and X_{n+1} is the biomass weight after (n + 1) hours (g).

2.8.4. Lactic Acid Content

The lactic acid content of PWKC beverage samples was evaluated using a high-performance liquid chromatography (HPLC) equipped with a refractive index (RI) detector system (Waters 2414, Waters Corporation, Milford, MA, USA) [40]. The lactic acid content was then calculated based on the standard curve constructed.

2.8.5. Viability of Presumptive *Lactobacillus*

Presumptive *Lactobacillus* count of PWKC beverage samples was determined according to the method proposed by Laureys and De Vuyst [41], using the spread plate method on *Lactobacillus*-selective base agar supplemented with cycloheximide (0.1 g/L). The colonies were counted after being incubated at 30 °C for 6 days, and results were expressed as Log CFU per mL of beverage samples.

2.9. Characterisation of Optimised Fermented Pumpkin-Based Mature Coconut Water Kefir (PWKC$_{opt}$)

PWKC$_{opt}$ were placed in glass bottles, closed tightly with lids, and stored at 4 °C for further characterisation and shelf-life study in terms of physicochemical (total soluble solids (TSS), pH, viscosity, and colour) and sensory properties (appearance (colour), odour, taste (sourness), texture (consistency), and overall acceptability), and microbiological changes (*Lactobacillus*, lactic acid bacteria, acetic acid bacteria, and yeast counts), measured at days 0, 28, and 56 of storage. The antioxidative properties, proximate and elemental compositions of the PWKC$_{opt}$ were measured at the end of fermentation. Chromatographic analysis of sugars, ethanol, and organic acid contents were conducted during the fermentation process at 0, 2.25, and 4.50 h of fermentation.

2.9.1. Chemical Composition

The proximate composition of PWKC$_{opt}$, i.e., moisture content (AOAC 925.19), ash content (AOAC 923.03), total protein content (AOAC, 955.04), crude lipid (AOAC 960.39), total dietary fibre (AOAC 991.43), and mineral composition (AOAC 985.35), was determined according to AOAC methods [42].

2.9.2. Physicochemical Properties

The physicochemical properties (total soluble solids, viscosity, pH value, and colour) of PWKC$_{opt}$ were determined following the methods described by Hlangwani et al. [43].

Total Soluble Solids

Total soluble solids content of the PWKC$_{opt}$ was measured using a refractometer (PAL-BX/RI, Atago, Tokyo, Japan).

Viscosity

Viscosity of the PWKC$_{opt}$ was measured using a viscometer (DVII + Pro, Brookfield, Middleborough, MA, USA) with LV-4 spindle rotating at 12 rpm. The reading displayed in centipoise (cP) was recorded when the value shown remained constant, with no fluctuation.

pH Value

pH value of the PWKC$_{opt}$ was measured using a pH meter (Orion 4 star A211, Thermo Fisher Scientific, Waltham, MA, USA). The reading was taken once a constant value was observed.

Colour

Colour (CIE L*; lightness, CIE a*; redness, and CIE b*; yellowness) of the PWKC$_{opt}$ was analysed using a colourimeter (Model 45/0-L, HAL, New York, NY, USA).

2.9.3. Antioxidative Contents and Activities

The carotenoid content, total phenolic content, total flavonoid content, ferric reducing antioxidant power, and free radical scavenging activity of PWKC$_{opt}$ were examined using the methods proposed by Barkallah et al. [44] and Wang et al. [45]. PWKC$_{opt}$ was centrifuged (15,000× *g*, 10 min, 4 °C), filtered through a polytetrafluoroethylene (PTFE) membrane (pore size = 0.22 μm, Merck, Frankfurt, Germany), and stored in a screw-capped, amber-coloured glass bottle at −20 °C until analysis.

Total Carotenoid Content

Sample (1 mL) was centrifuged (3000× g, 10 min), sonicated (65 °C, 30 min) in ethanol (1 mL) and centrifuged (10,000× g, 5 min) again before its absorbance was taken (UV mini-1240, Shimadzu, Kyoto, Japan). The total carotenoid content of the sample was calculated according to Equation (3) [46] and expressed in mg/100 mL:

$$\text{Totalcarotenoid}(\text{mg/L}) = \frac{(1000 \times A_{470} - 2.86[15.65 \times A_{666} - 7.340 \times A_{653})] - 85.9[27.05 \times A_{653} - 11.21 \times A_{666}])}{245} \quad (3)$$

where A_{470}, A_{653}, and A_{666} are the absorbances taken at 470 nm, 653 nm, and 666 nm, respectively.

Total Phenolic Content (TPC)

The TPC of $PWKC_{opt}$ was measured using the Folin–Ciocalteau method. The sample in methanol (2.5 mL) was homogenised with Folin–Ciocalteau reagent (0.2 N, 2.5 mL) for 5 min followed by the addition of sodium carbonate (2.5 mL, 7.5% *w*/*v*) and incubation at 37 °C for 30 min. The absorbance of the mixture was then spectrophotometrically measured at 765 nm. TPC of the sample was estimated from the gallic acid calibration curve (0–100 mg/100 mL) and expressed as mg gallic acid equivalent (GAE) per 100 mL sample (mg GAE/100 mL).

Total Flavonoid Content (TFC)

Aluminium trichloride colourimetric method was used to determine the TFC of $PWKC_{opt}$. Sample (5 mL) was homogenised in aluminium trichloride methanol solution (5 mL, 2% *w*/*v* aluminium trichloride in methanol) and incubated at 37 °C for 30 min. The absorbance of the mixture was read at 415 nm, and the TFC of the sample was estimated based on the quercetin calibration curve (0–60 mg/100 mL) and quantified as mg quercetin equivalent (QE) per 100 mL sample.

Ferric Reducing Antioxidant Power (FRAP)

Sample (1 mL) was combined with FRAP working solution (20 mL, prepared by mixing sodium acetate buffer (300 mM, pH 3.6), TPTZ solution (10 mM) in 40 mM hydrochloric acid, and ferric chloride hexahydrate (20 mM,) at proportions of 10:1:1 ratio (*v*/*v*)) and incubated at 37 °C for 30 min in the dark. The absorbance of the mixture was read at 593 nm and reported as the millimole equivalent of ferrous sulphate per 100 mL sample (mM eq Fe (II)/100 mL) according to the calibration curve of ferrous sulphate in deionised water (0–30 mM/mL).

DPPH Radical Scavenging Ability (RSA)

A mixture of the sample (200 μL) and DPPH solution prepared in methanol (1 mL, 0.2 mM) was incubated at 37 °C for 30 min in the dark and the absorbance of the mixture was taken at 517 nm. The RSA of the sample was calculated according to Equation (4) and expressed as half the maximal inhibitory concentration value (IC_{50}, mg/mL), according to the regression equation (y = 12.246 ln(x) − 9.1955) obtained from the standard curve.

$$\text{Inhibition}(\%) = \left[\frac{A_B - A_S}{A_B}\right] \times 100 \quad (4)$$

where A_B and A_S are the absorbances of the blank (DPPH solution) and sample with DPPH solution, respectively.

2.9.4. Chromatographic Analysis

The sugar (glucose and fructose) content of $PWKC_{opt}$ was quantified using an HPLC equipped with an RI detector (Waters 2414, Waters Corporation, Milford, MA, USA) as previously described by Filip et al. [47], whereas organic acid (lactic, acetic, malic, and tartaric) contents were quantified using an HPLC with UV detector (Model 2487, Waters Corporation, Milford, MA, USA) [40]. Ethanol content (% v/v) was determined using a gas chromatograph (Shidmazu, GC-2010, Milan, Italy) fitted with a flame ionisation detector (Shidmazu, Milan, Italy) and equipped with a ZB-WAX film capillary column (length of 30 m × diameter of 0.25 mm, film thickness of 0.25 µm) (Phenomenex, Torrance, CA, USA) [48]. Identification and quantification of the ethanol, sugars, and acids were performed on samples collected at 0, 2.25, and 4.50 h of fermentation. The sample collected at 0 h was used as a control.

2.9.5. Viability of Presumptive *Lactobacillus*, LAB, AAB, and Yeast

To determine the viability of presumptive *Lactobacillus*, LAB, AAB, and yeast, a total of 10 mL of $PWKC_{opt}$ was homogenised with 90 mL of 1 g/L aqueous peptone solution, and necessary serial dilutions were performed. The diluted sample (100 µL) was spread on the *Lactobacillus*-selective base agar, MRS agar supplemented with cycloheximide (0.1 g/L,), mDMS agar supplemented with cycloheximide (0.1 g/L), and YGC agar supplemented with chloramphenicol (0.1 g/L,) using sterile beads followed by incubation at 30 °C for 48 to 72 h. The viability of presumptive *Lactobacillus*, LAB, AAB, and yeasts was expressed as Log CFU per mL [41,49].

2.9.6. Hedonic Test

Hedonic sensory testing was conducted in sensory booths according to the ISO 11136 standard [38]. A total of 50 untrained panellists (25 males and 25 females, Food Technology undergraduates at School of Industrial Technology, Universiti Sains Malaysia, Penang, Malaysia) were recruited to evaluate the colour, odour, taste (sourness), texture (consistency), and overall acceptability of $PWKC_{opt}$ using a 7-point hedonic scale. Purified water was served to the panellists, who were instructed to rinse their mouths during the sensory test.

2.10. *Identification of Isolate*

Genomic deoxyribonucleic acid (DNA) from the dominant cultivable species isolates in the optimised beverage was isolated with the NucleoSpin Tissue kit (Machery-Nagel, Duren, Germany) based on the instructions supplied by the manufacturer. The concentration and purity of the extracted DNA were determined using a UV spectrophotometer (UV mini-1240, Shimadzu, Kyoto, Japan). The pure genomic DNA was identified by 16S rRNA gene sequencing using universal primers (designed by the Centre for Chemical Biology, Universiti Sains Malaysia, Penang, Malaysia). The sequence similarity and closest phylogenetic relatives were then analysed and matched using the basic local alignment search tool (BLAST) at the National Center for Biotechnology Information (NCBI).

2.11. *Statistical Analysis*

The analysis of the experimental design and data were performed using Design-Expert 13.0 (DX13, Stat Ease Inc., Minneapolis, MN, USA). The fitness of the polynomial equations was evaluated using the analysis of variance (ANOVA), and the statistical significance of the regression coefficients was tested by Fisher's test (F-test). All analyses were performed in triplicate, and data from characterisation of the $PWKC_{opt}$ were statistically analysed using one-way ANOVA with Tukey's honestly significant difference (HSD) test for post hoc comparisons using the IBM SPSS Statistics software version 27 (IBM, Chicago, IL, USA). The results from hedonic test were analysed using the Kruskal–Wallis test. Differences with p-values less than 0.05 were regarded as statistically significant.

3. Results and Discussion

3.1. RSM Model Fitting

A total of seventeen experimental runs were made to model the empirical relationship between the key factor parameters using the Box–Behnken design (BBD) (Table 1). ANOVA was conducted to confirm the validity of the regression model and determine the effects of the key factor parameters, i.e., pumpkin puree concentration (X_1; % *w*/*v*), brown sugar concentration (X_2; % *w*/*v*), and fermentation temperature (X_3; °C), on the output response function.

The model for all the output response functions was significant (p-value < 0.05, Table 2). The coefficient of determination (R^2) for each model achieved a desirability of 0.99, indicating the model developed for PWKC fitted the experimental data. All models possessed low correlation variation, non-significant lack of fit (P_{lof} > 0.05), and an adequate precision level of ≥4, which indicated that the developed model could predict the response function and the effect of independent variables on dependent variables. The quadratic model was attempted, and the non-significant terms (p > 0.05) were excluded from the fitted second-degree polynomial equations. The regression equations presented the extent of the response as a function of the three independent variables and are interpreted in the following sub-sections.

Table 2. Estimated coefficients of the fitted second-order polynomial models for dependent variables.

Term	Regression Coefficients Estimated				
	$T_{pH4.5}$ (h)	OA (Score)	Biomass Growth Rate (% G)	Lactic Acid (% *v*/*v*)	*Lactobacillus* Count (Log CFU/mL)
			Intercept		
β_0	4.03	4.35	24.80	0.44	3.63
			Linear		
β_1	−0.30 *	0.44 *	−2.75 **	−0.06 **	−0.76 **
β_2	−1.03 *	1.39 *	−7.75 *	−0.03 ***	−0.54 **
β_3	−1.55 *	1.46 *	−13.25 *	−0.14 *	−1.76 *
			Interaction		
$\beta_1 \beta_2$	−0.18 **	0.49 *	−0.75 n.s.	0.05 ***	−0.10 n.s.
$\beta_1 \beta_3$	−0.05 n.s.	0.33 **	−0.25 n.s.	−0.03 n.s.	−0.22 n.s.
$\beta_2 \beta_3$	0.67 *	0.67 *	4.75 *	0.06 **	0.69 ***
			Quadratic		
β_{11}	0.23 **	0.28 ***	2.72 **	0.11 **	0.91 **
β_{22}	0.92 *	−1.13 *	6.72 *	0.13 *	1.48 **
β_{33}	−0.08 ***	−0.39 **	−1.27 n.s.	0.01 n.s.	0.13 n.s.
R^2	0.9993	0.9979	0.9961	0.9837	0.9796
CV%	1.34	3.05	3.91	5.23	8.37
Adq. Pre.	125.87	66.27	53.49	22.13	20.31
p_{lof}-value	0.1188 n.s.	0.5293 n.s.	0.1683 n.s.	0.1758 n.s.	0.4990 n.s.
p_m-value	<0.0001	<0.0001	<0.0001	<0.0001	<0.0001

Subscripts: 1 = pumpkin puree content (20–30%, *w*/*v*); 2 = brown sugar content (0–10% *w*/*v*); 3 = fermentation temperature (22–32 °C). CV%: coefficient variation (%); Adq. Pre.: adequate precision; p_{lof}-value: probability of F value for the lack of fit; p_m-value: probability of F value for the model. $T_{pH4.5}$: fermentation time to reach pH 4.5; OA: overall acceptability; % G: water kefir grain biomass growth rate. n.s. non-significant at p > 0.05. * p < 0.001, ** p < 0.01, *** p < 0.05.

3.2. Effect of Independent Variables on Dependent Variables of RSM Model

The model of second-order regression from the experimental data of the five output response functions: Y_1 = fermentation time to reach pH 4.5 ($T_{pH4.5}$), Y_2 = overall acceptability (OA) score, Y_3 = water kefir grain biomass growth rate, Y_4 = lactic acid content, and Y_5 = *Lactobacillus* count (Log CFU/mL) are displayed in Equations (5)–(9), respectively.

$$Y_1 = 4.03 - 0.30X_1 - 1.03X_2 - 1.55X_3 + 0.23X_1^2 + 0.92X_2^2 - 0.08X_3^2 - 0.18X_1X_2 + 0.67X_2X_3 \quad (5)$$

$$Y_2 = 4.35 + 0.44X_1 + 1.39X_2 + 1.46X_3 + 0.28X_1^2 - 1.13X_2^2 - 0.39X_3^2 + 0.49X_1X_2 + 0.33X_1X_3 + 0.67X_2X_3 \quad (6)$$

$$Y_3 = 24.80 - 2.75X_1 - 7.75X_2 - 13.25X_3 + 2.72X_1^2 + 6.72X_2^2 + 4.75X_2X_3 \quad (7)$$

$$Y_4 = 0.44 - 0.06X_1 - 0.03X_2 - 0.14X_3 + 0.11X_1^2 + 0.13X_2^2 + 0.05X_1X_2 + 0.06X_2X_3 \quad (8)$$

$$Y_5 = 3.63 - 0.76X_1 - 0.54X_2 - 1.76X_3 + 0.91X_1^2 + 1.48X_2^2 - 0.44X_1X_2 + 0.69X_2X_3 \quad (9)$$

where X_1 and X_2 are the concentrations of pumpkin puree and brown sugar (% w/v), respectively, and X_3 is the fermentation temperature (°C).

3.2.1. Fermentation Time ($T_{pH4.5}$)

Fermentation time is crucial in development of probiotic products, as it can directly influence both the safety and quality of the final product. An inadequate fermentation time could lead to the proliferation of pathogenic microorganisms, but a prolonged fermentation time could lead to over-acidification that could negatively affect the sensory properties of the product [50]. In the development of probiotic foods, pH 4.5 is frequently chosen to be the fermentation endpoint, as a pH value < 4.6 could inhibit the growth of putrefying and pathogenic bacteria, consequently increasing microbial stability during storage and the shelf life of the product [51]. As shown in Table 1, $T_{pH4.5}$ varied from 2.25 to 8.08 h. The shortest and longest fermentation times were obtained at 32 °C, with a medium containing 30% w/v pumpkin puree and 5% w/v brown sugar (run 8), and 22 °C, with a medium containing 25% w/v pumpkin puree and 0% w/v brown sugar (run 9), respectively. The $T_{pH4.5}$ obtained was similar to the result obtained in the studies conducted by Acevedo-Martínez et al. [52] and Fawzi et al. [53], which reported a fermentation time of 2.25–5.5 h and 8 h to reach a pH value of 4.5 in the fermentation of skim milk supplemented with different grape pomace extracts, using *Streptococcus thermophilus* and *Lactobacillus acidophilus* [52] and broken rice milk, using *Lactobacillus plantarum* ATCC 14917, *Lactobacillus casei* DSM 20011, and *Lactobacillus acidophilus* ATCC20552 [53], respectively.

All the independent variables showed a significant negative linear effect ($p < 0.001$) on $T_{pH4.5}$ (Table 2). The increment of the fermentation temperature and concentrations of pumpkin puree and brown sugar within the studied range resulted in decrements of $T_{pH4.5}$ (Figure 1a–c), with the fermentation temperature in its linear term showing the dominant effect (Figure 1b,c). Microflora in water kefir were principally mesophilic. Mesophilic temperature could lead to higher metabolic activity and facilitate a faster fermentation rate and a higher organic acid production, leading to a shorter fermentation time [54,55].

As shown in Table 2, brown sugar concentration demonstrated a positive interactional effect with fermentation temperature ($p < 0.001$) and a negative interactional effect with pumpkin puree concentration ($p < 0.01$) on $T_{pH4.5}$. As illustrated in Figure 1a,c, decrement of $T_{pH4.5}$ was more significant when fermentation temperature and pumpkin puree and brown sugar concentrations at the upper-end of the studied range were used. The indigestible carbohydrates, sugar in pumpkin puree [16,17], and sucrose [3] in brown sugar were proven to be excellent nutrient/carbon sources and growth media for LAB and yeast strains in water kefir grains. Bengoa et al. [54] and Laureys et al. [55] also reported that nutrient-rich fermentation media accompanied by a mesophilic temperature (25–37 °C) could facilitate faster water kefir microorganisms' fermentation rate.

The quadratic effect of brown sugar concentration positively impacted $T_{pH4.5}$ significantly ($p < 0.001$), where a gradual decrease of $T_{pH4.5}$ was observed when brown sugar concentration increased from 0–8% w/v, as shown in Figure 1a,c. When a higher sugar concentration was used, more carbon was available for fermenting microorganisms' growth, thus facilitating a faster metabolic rate and producing a higher amount of organic acid, and hence, reaching a pH value of 4.5 in a shorter time [55]. However, when more than 8% *w*/*v* brown sugar was used, $T_{pH4.5}$ was observed to increase slightly. A high substrate concentration could exert an osmotic pressure, which prevents the fermenting microorganisms from metabolising, thus prolonging the fermentation time [56]. Pumpkin puree concentration had a positive quadratic effect on $T_{pH4.5}$ ($p < 0.01$). When pumpkin puree concentration increased from 20–26% *w*/*v*, the $T_{pH4.5}$ decreased slightly. However, $T_{pH4.5}$ became almost constant when pumpkin puree concentration further increased, indicating that 26% *w*/*v* of pumpkin puree contributed to a minimum $T_{pH4.5}$. Pumpkin puree is rich in indigestible polysaccharides such as rhamnose, glucose, arabinose, and galactose and sugar, which could facilitate and support the growth of microorganisms [15]. The negative linear effect ($p < 0.001$) of fermentation temperature on $T_{pH4.5}$ was more significant than that of quadratic effect ($p < 0.05$) (Table 2). A sharp decrease of $T_{pH4.5}$ was observed when fermentation temperature was raised from 22 to 32 °C (Figure 1b,c). Water kefir microorganisms grew faster at the mesophilic temperature (upper end of the studied range), produced more acids, and, hence, reached the fermentation endpoint in a shorter time [54,55].

(a) (b) (c)

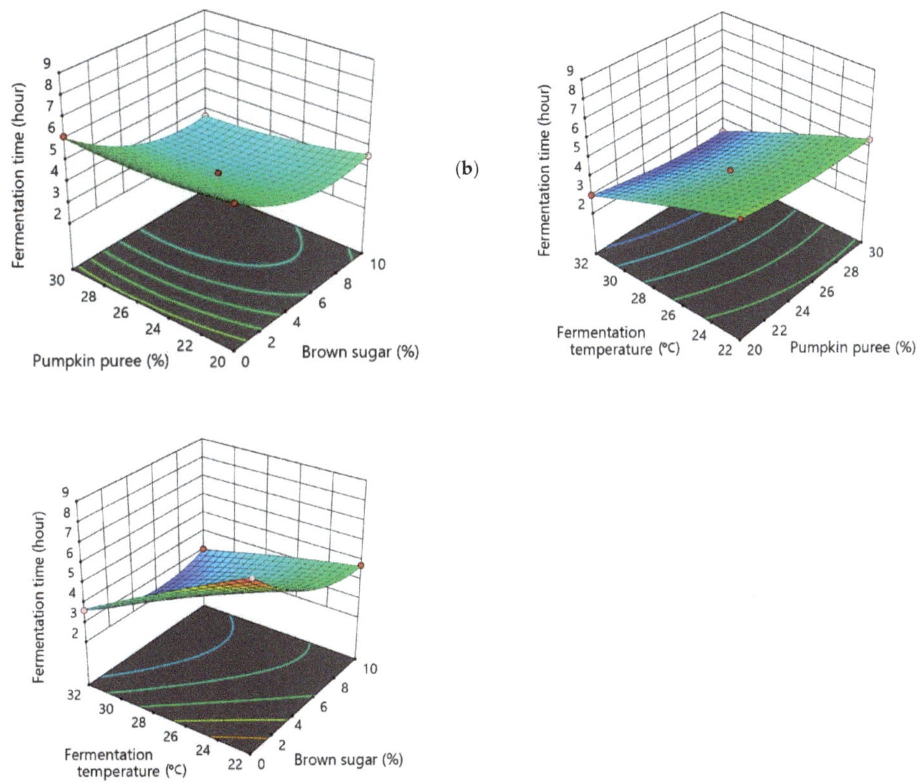

Figure 1. Response surface plots illustrating the interaction effect of (**a**) pumpkin puree and brown sugar content, (**b**) fermentation temperature and pumpkin puree content, and (**c**) fermentation temperature and brown sugar content on the fermentation time required to prepare pumpkin-based mature coconut water kefir beverage (PWKC).

3.2.2. Overall Acceptability (OA)

The adequate level of sensory acceptance of fermented food products has been considered an important factor due to its influence on the overall quality of the product [41,57].

The OA score of PWKC samples falls in the range of 0.74 to 6.50 on a 7-point hedonic Likert scale (Table 1). The PWKC sample containing 25% v/v pumpkin puree and 0% v/v brown sugar fermented at 22 °C (run 9) had the lowest OA score, whereas the PWKC sample containing 30% v/v pumpkin puree and 5% v/v brown sugar fermented at 32 °C scored the highest (run 8). The score obtained was comparable to kefir-like beverages derived from fruits or vegetables with an OA score of 1.5 to 5.0 on a 7-point hedonic Likert scale [56].

Fermentation temperature and pumpkin puree and brown sugar concentrations demonstrated a positive linear effect ($p < 0.001$) on the OA score of PWKC samples, indicating that the OA score can be enhanced by increasing the value of the independent variable as standalone events. The sensory attributes of fermented beverage products could be attributed to fermentation metabolites produced during the fermentation process, such as volatile compounds and organic acids. The production of metabolites during fermentation is depending on the fermentation condition, types and concentration of ingredients used in the production of fermented food products [58]. As illustrated in Figure 2a–c, a marked increase in OA score was observed as fermentation temperature and substrates' concentrations were elevated. When the fermentation temperature and substrates' concentration increased, the metabolism and growth rate of microorganisms during fermentation fasten, available substrates can be utilised to a greater extent, thus resulting in higher production of organoleptic compounds, which contributes to the improvement in the sensory acceptance of fermented beverage sample [59].

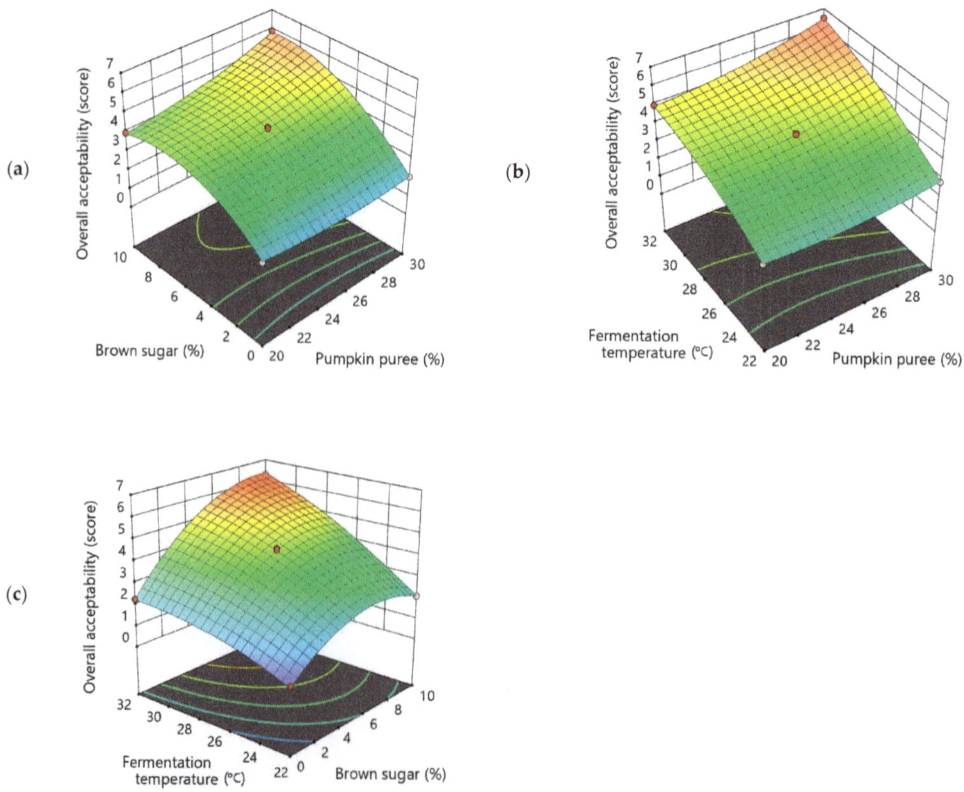

Figure 2. Response surface plots illustrating the interaction effect of (**a**) pumpkin puree and brown sugar content, (**b**) fermentation temperature and pumpkin puree content, and (**c**) fermentation temperature and brown sugar content on the overall acceptability (OA) score of pumpkin-based mature coconut water kefir beverage (PWKC).

The interactional effect between all the independent variables were significantly positive (Table 2, Figure 2a–c), revealing that by increasing fermentation temperature along with brown sugar or pumpkin puree concentration to the upper limit of the studied range, OA scores increased. Pumpkin puree with no additives has a bitter aftertaste [60], whereas mature coconut has a neutral taste, and hence both are unacceptable in terms of sensory properties [23]. Results demonstrated that the water kefir fermentation had improved the sensory qualities of the mature coconut water and pumpkin puree. Under mesophilic conditions with adequate nutrients, water kefir microorganisms, LAB, and yeast could produce several desirable organoleptically-active by-products such as lactic acid, acetic acid, ethanol, carbon dioxide, and an array of aromatic molecules [10]. Previous studies also proved that fermentation could eliminate unpleasant notes and resemble the organoleptic properties of pumpkin puree and mature coconut water [20,21,61].

Brown sugar concentration ($p < 0.001$) and fermentation temperature ($p < 0.01$) demonstrated significant negative quadratic effect, whereas pumpkin puree concentration exerted a moderately significant positive quadratic effect ($p < 0.05$) on the OA score (Table 2). OA score increased with the concentration of brown sugar (0–8% w/v) and became almost constant when more than 8% w/v of brown sugar was used (Figure 2a,c). The addition of sugar in the fermentation medium was effective in stimulating the growth of fermenting microorganisms. LAB and yeast ferment sugars to produce carbon dioxide, ethanol, and some aromatic compounds which give better sensorial properties to the fermented beverages [10]. Excessive sweetness and acidity can lead to a cloying or overly rich taste and a sharp tart flavour, respectively, and adversely affect the sensory experience. As observed in Figure 1b,c, OA scores increased when fermentation temperature increased from 22 to 32 °C. Water kefir microorganisms, which prefer to grow at mesophilic temperatures (25–37 °C), could produce more desirable metabolites and improve the sensory acceptance of the fermented product [54,55]. A gradual increase in OA score was noticed when 20–30% w/v of pumpkin puree was used (Figure 2a,b). Szydłowska et al. [15] reported that readily available nutrients in pumpkin puree favoured the growth and metabolic activities of LAB and yeast and increased the production of flavanols and esters that contribute to better organoleptic properties of the fermented beverages.

3.2.3. Water Kefir Grains Biomass Growth Rate

Culture medium for producing kefir grain biomass must be designed for high biomass output and reduction of production costs to develop economically viable cultures for industrial applications [62]. Water kefir grains' biomass growth rate is useful in studying the factors affecting the grains' growth. The minimum water kefir grain biomass growth rate obtained was 10%, which was found in the medium containing 30% w/v pumpkin puree and 5% w/v brown sugar fermented at 32 °C (run 8). Meanwhile, fermentation at 22 °C in a medium with 25% w/v pumpkin puree and 0% w/v brown sugar (run 9) resulted in the maximum water kefir grain biomass growth rate of 55% (Table 1). This outcome is consistent with observations of Pop et al. [63] that reported the biomass growth rate of kefir grains in skim milk medium ranged from 22.97 to 74.55%.

The water kefir grain biomass growth rate reduced when fermentation temperature and pumpkin puree and brown sugar concentrations rose as standalone events, with temperature in its linear term as the dominant effect (Figure 3a–c). This indicated that fermentation temperature at the lower end limit of the studied range favoured the kefir grains' biomass growth rate. This outcome is consistent with Pop et al. [63], who showed that the optimum temperature for kefir grains to grow was 25 to 28 °C, with the greatest growth rate recorded at 37.93 g/L for 24 h.

The factors influencing kefir grain biomass increment have received little attention. Guzel-Syedim et al. [64] reported that it is critical to utilise proper concentrations of different nutritional carbohydrate substrates to increase the growth rate of kefir grains. The quadratic term of substrates' concentrations demonstrated a significant positive effect on water kefir grains' biomass growth rate (Figure 3a–c). The water kefir grains' biomass growth rate decreased gradually when 22–26% w/v of pumpkin puree and 0–6% w/v brown sugar concentration were used. Subsequent increase of substrates' concentrations increased the water kefir grains' biomass growth rate. This demonstrated that increasing the amount of pumpkin puree and brown sugar to the top of the studied range could boost the kefir biomass.

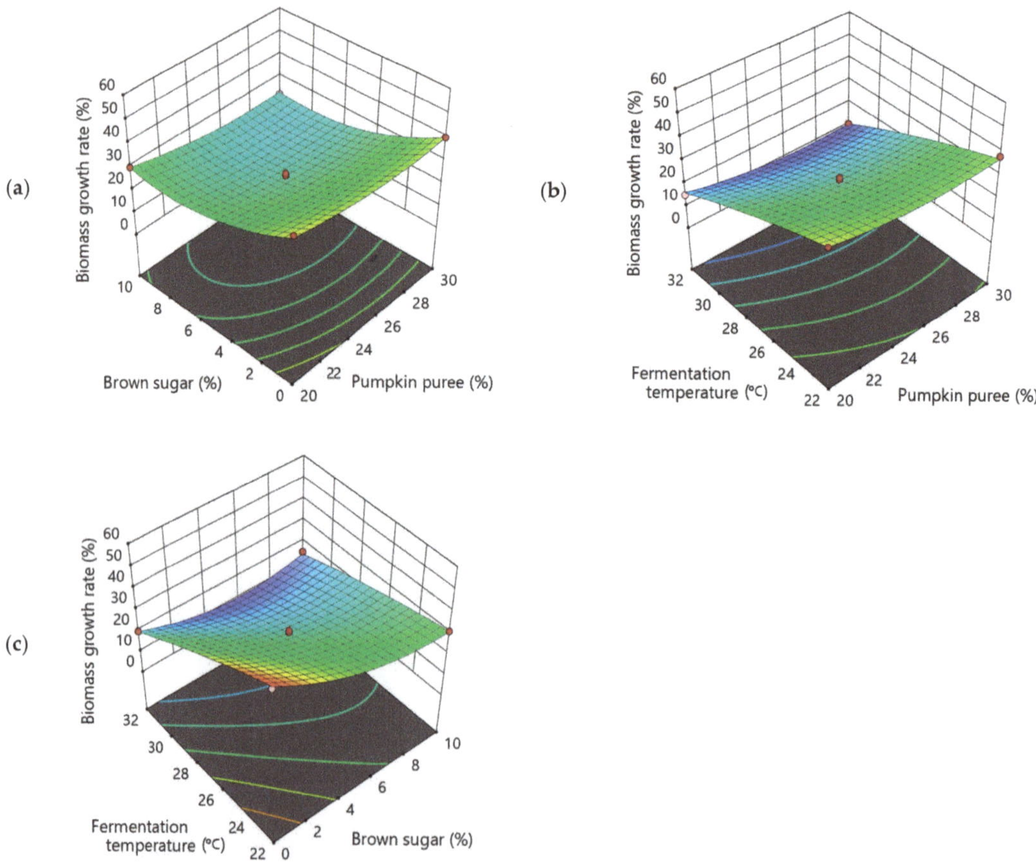

Figure 3. Response surface plots illustrating the interaction effect of (**a**) pumpkin puree and brown sugar content, (**b**) fermentation temperature and pumpkin puree content, and (**c**) fermentation temperature and brown sugar content on the water kefir grains' biomass growth rate.

Figure 3c demonstrates a positive interactional effect between temperature and brown sugar. The rate of biomass growth of the water kefir grains was accelerated by simultaneous increases in fermentation temperature and brown sugar content. The increase in fermentable sugar (glucose, sucrose, and fructose) and temperature enhanced the rate of metabolism for the microbial flora and led to a rise in the kefir grains' biomass growth.

3.2.4. Lactic Acid Content

The distinctive aroma and flavour of water kefir are mainly associated with the organic acid produced by the water kefir microorganisms during fermentation [65]. Lactic acid fermentation occurs during the production of water kefir by *Lactobacillus*, which breaks down sugars to produce energy in the form of adenosine triphosphate [66]. Lactic acid generated during the fermentation process gives the water kefir a tangy or sour flavour [10].

The production of lactic acid could also lead to the development of an acidic environment, which could inhibit the growth of spoilage microorganisms and consequently extend the shelf life of the fermented product [67]. As shown in Table 1, the minimum and maximum lactic acid contents in PWKC are 0.32% v/v (run 8: 30% w/v pumpkin puree, 5% w/v brown sugar, 32 °C) and 0.82% v/v (run 9: 25% w/v, 0% w/v, 22 °C), respectively (Table 1). Previously, Laureys and De Vuyst [68] also reported a lactic acid content of 0.5% v/v in honey beverages fermented by water kefir grains.

Fermentation temperature and pumpkin puree and brown sugar concentrations demonstrated negative linear effect ($p < 0.05$) on the lactic acid content of PWKC samples. Fermentation temperature in its linear term showed the dominant effect. Lactic acid content in PWKC samples decreased when the fermentation temperature increased from 22 to 32 °C (Figure 4b,c). The *Lactobacillus* sp. in PWKC might require specific nutrients and temperature to grow optimally. This behaviour was also reported by Alves et al. [69], who fermented insulin-supplemented coconut extract with water kefir grains. While certain microbial species never linked to water kefir grains cultivated in sugary water were discovered, the usual strains were not present [69]. This result could be supported by Darvishzadeh et al. [7] who studied Russian olive juice and found that the optimal fermentation temperature of the *Lactobacillus* sp. (*Lactobacillus paracasei*, *Lactobacillus parabuchneri*, *Lactobacillus kefiri*, and *Lactobacillus casei*) was 31.2 °C. Although the optimal growth temperature of water kefir grains ranges from 21 to 30 °C [3,5], some *Lactobacilli* are mesophilic (30–40 °C), with an upper limit of 40 °C [70], thereby restricting the *Lactobacilli* growth and consequently lowering the content of lactic acid in PWKC.

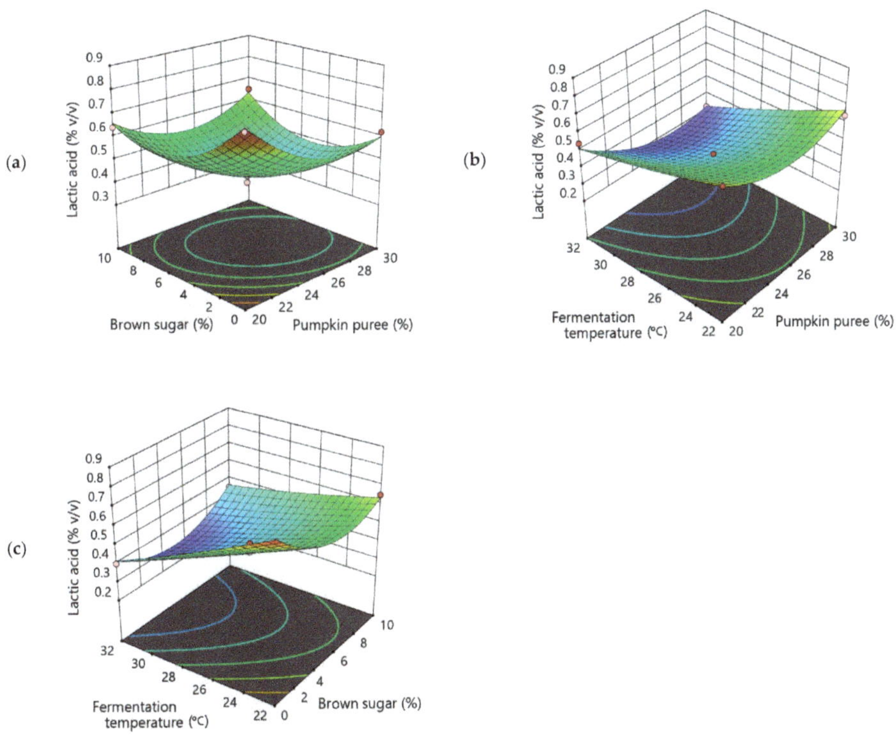

Figure 4. Response surface plots illustrating the interaction effect of (**a**) pumpkin puree and brown sugar content, (**b**) fermentation temperature and pumpkin puree content, and (**c**) fermentation temperature and brown sugar content on the lactic acid content of pumpkin-based mature coconut water kefir beverage (PWKC).

A positive interactional effect between the substrates' concentrations is shown in Figure 4a. When both the pumpkin puree and brown sugar concentrations increased, the lactic acid content of PWKC increased. At the upper limit of the pumpkin puree and brown sugar concentrations, more carbohydrates were available for *Lactobacillus* fermentation and lactic acid conversion. Pumpkin puree (rich in sucrose, glucose, and fructose [16,17]) and brown sugar (consisting mainly of sucrose [3]) are potential carbon sources for lactic acid fermentation. Dimitrovski et al. [18] also reported on the utilisation of glucose and fructose in pumpkin juice and the production of lactic acid by *Lactobacillus casei* 431. Bueno et al. [65], who fermented red pitaya and apple pulp using water kefir grains, also reported on the reduction of sucrose over the fermentation time, along with the production of lactic acid. As observed in Table 2 and Equation (5), the quadratic term of the substrates positively affected the lactic acid content in PWKC. When the concentrations of pumpkin puree and brown sugar increased from 20–26% w/v and 0–6% w/v, respectively, lactic acid content showed a slight decrease (Figure 4a–c). Further increase in the concentrations of both substrates increased the lactic acid content. The pumpkin puree and brown sugar concentrations below 6 and 26% w/v were inadequate to initiate the production of lactic acid.

As observed in Figure 1c, the lactic acid content of PWKC rose when higher concentration of brown sugar and fermentation temperature were used. At the top end of the studied range, the positive interactional effect between brown sugar concentration and fermentation temperature was more pronounced. Fermentation temperature is a crucial factor in determining the growth and fermentative activities of water kefir microorganisms [3,5]. Most of the water kefir *Lactobacilli* in PWKC grew better at a fermentation temperature at the upper limit of the studied range, and thus, when more fermentable nutrients (pumpkin puree and brown sugar) were available, *Lactobacilli* fermented at a faster rate and produced a greater amount of lactic acid. Laureys et al. [71] also found that the metabolism of certain microorganisms in the water kefir grains could be affected by the fermentation temperature and consequently reduced the lactic acid content.

3.2.5. *Lactobacillus* Count

Lactobacillus is the dominant bacterial genera in water kefir beverages [3,5]. The consumption of *Lactobacillus* has been associated with health benefits such as protective effects against pathogenic bacteria, immune response-modulating effects, allergy and cancer prevention effects, oxidative stress reduction effects, and anti-diabetic effects [72]. The count of *Lactobacilli* contained in the final fermented product at the time of consumption is crucial for determining its potential health benefits. Higher counts generally indicate a potentially more potent probiotic food product, as more bacteria are available to exert therapeutic effects on the host. The lowest and highest *Lactobacillus* counts of PWKC obtained were 2.08 Log CFU/mL (run 8: 30% w/v pumpkin puree, 5% w/v brown sugar, fermentation temperature of 32 °C) and 8.04 Log CFU/mL (run 9: 25% w/v, 0% w/v, 22 °C), respectively (Table 1). Our results are comparable to the study conducted by Ozcelik et al. [73], who obtained *Lactobacillus* spp. count varying between 5.72 and 7.94 Log CFU/mL in water kefir grain-fermented fruit juices.

Table 2 and Equation (6) show the negative linear effect ($p < 0.05$) of fermentation temperature and pumpkin puree and brown sugar concentrations on the *Lactobacillus* count. As can be seen in Figure 5b,c, fermentation temperature in its linear term dominantly affected *Lactobacillus* count. *Lactobacillus* count decreased when the fermentation temperature increased (Figure 5a,c), indicating that some of the *Lactobacillus* strains in PWKC preferred the fermentation temperature at the lower end limit of the studied range. This finding can be supported by a study conducted by Dadkhah et al. [74], who used kefir grains to ferment soy milk at 22 °C, and a high *Lactobacillus* count of 8.14 Log CFU/mL was obtained.

Brown sugar concentration in its quadratic term demonstrated a significant positive effect on *Lactobacillus* count (Figure 5a,c). *Lactobacillus* count decreased slightly when the brown sugar concentration increased from 0–6% w/v, but rose with further increase of brown sugar. Water kefir microbes prefer the sucrose in brown sugar compared to other substrates [3]. When sucrose is adequate, water kefir grains could grow at a faster rate, as more sucrose can be converted into water kefir grain dextran exopolysaccharide by glucansucrases of *Lentilactobacillus hilgardii*, thus resulting in a higher *Lactobacillus* count [6]. Figure 1a,b shows the positive quadratic effects of pumpkin puree concentration on the *Lactobacillus* count. When the pumpkin puree concentration increased from 20 to 26% w/v, the *Lactobacillus* count decreased. However, when the pumpkin puree concentration was further increased (26–30% w/v), the *Lactobacillus* count increased. The substitution of sucrose with glucose and fructose has been reported to reduce the water kefir grain growth rate as sucrose is necessary for dextran exopolysaccharide production [10]. Brown sugar concentration and fermentation temperature demonstrated a positive interactional effect ($p < 0.05$) on the viability of *Lactobacillus* (Table 2, Equation (6)). *Lactobacillus* count increased when the concentration of brown sugar and fermentation temperature rose simultaneously (Figure 5c). This indicated that fermentation temperature and brown sugar content at the upper limit of the measured parameter favoured *Lactobacillus* growth in PWKC. When more sugar is available, *Lactobacillus* cell division will be higher, thus resulting in a higher viable count.

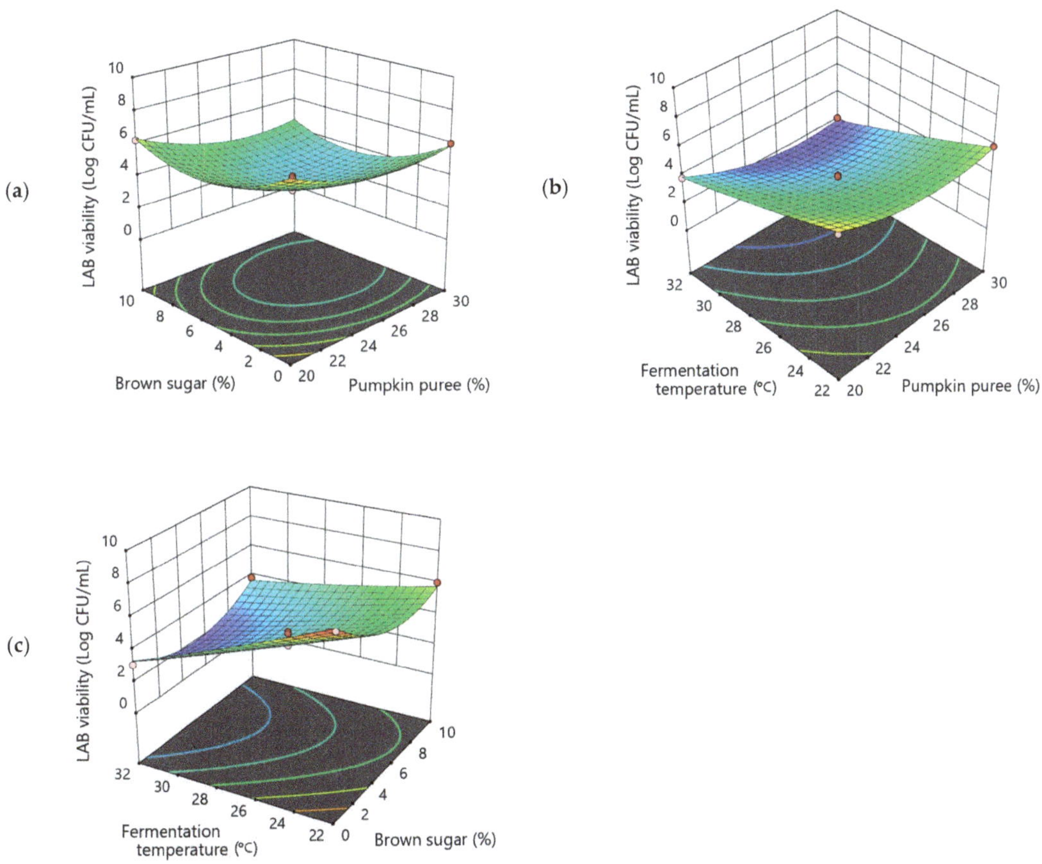

Figure 5. Response surface plots illustrating the interaction effect of (**a**) pumpkin puree and brown sugar content, (**b**) fermentation temperature and pumpkin puree content, and (**c**) fermentation temperature and brown sugar content on the *Lactobacillus* count of pumpkin-based mature coconut water kefir beverage (PWKC).

3.3. Optimisation

The optimal conditions for the $PWKC_{opt}$ obtained in the current study are X_1 = 20.00% w/v, X_2 = 9.99% w/v and X_3 = 27.19 °C. These optimal values were then rounded to 20% w/v pumpkin puree, 10% w/v brown sugar, and a 27 °C fermentation temperature within a range of 1% and 1 °C. Under these optimum conditions, all five responses of $PWKC_{opt}$ were validated by triplicate experimental runs (Table 3).

Table 3. Model predicted, experimental values and the intervals obtained from two-sided confirmation of point predicted by RSM at confidence level of 95% confidence level.

Response Variables	Goal	Predicted Mean [a]	95% PI Low [c]	Observed Value [b]	95% PI High [c]
$T_{pH4.5}$ (h)	Minimize	4.59	4.44	4.54 ± 0.06	4.74
OA (score)	Maximize	4.01	3.73	4.03 ± 0.12	4.30
Biomass growth rate (% G)	Maximize	29.68	26.92	31.00 ± 1.12	32.44
Lactic acid (% v/v)	Maximize	0.66	0.59	0.68 ± 0.03	0.73
Lactobacillus count (Log CFU/mL)	Maximize	6.31	5.32	6.41 ± 0.40	7.30

[a] Desirability for this result was 0.99. [b] Values are the means of three determinations ± standard error. [c] Prediction interval (PI). $T_{pH4.5}$: fermentation time to reach pH 4.5; OA: overall acceptability.

Overall desirability value is a measure that combines multiple response variables into a single metric and is always employed to assess the suitability of the prediction of the optimal value of input parameters corresponding to the output responses [75]. $PWKC_{opt}$ obtained in the current study possessed an overall desirability value of 0.99, which indicated that $PWKC_{opt}$ attained an ideal response value. For instance, the PWKC obtained under the optimal condition required a fermentation time of 4.54 h to reach pH 4.5, which corresponds to previous studies [52,53]. The optimal condition has resulted in a water kefir grain biomass growth rate of 31%, *Lactobacillus* count of 6.41 Log CFU/mL, and lactic acid content of 0.68% v/v, which are consistent with the result obtained in previous studies [63,68,73]. The optimal condition has optimised the water kefir grains' biomass growth, allowed the kefir microbiota to efficiently utilise the nutrients (such as carbohydrates, sugars, vitamins, and minerals) provided in mature coconut water and pumpkin puree, consequently resulting in a high *Lactobacillus* count and producing lactic acid in a concentration that would not adversely affect the sensory characteristics of the final product, PWKC, resulting in an OA score of 4.03 (on a 7-point hedonic Likert scale), which is acceptable and is comparable to other fruit- or vegetable-based kefir-like beverages [76].

3.4. Microbiological Safety Analyses

Microbiological safety analyses were performed on all 17 experimental runs of PWKC before the sensory evaluation test to ensure the absence of food-borne pathogens in the production of PWKC. There were no coliform bacteria or *E. coli* detected in all 17 PWKC samples throughout the entire storage period. These results confirmed that adequate hygienic measures were adopted during the preparation and processing of PWKC. The findings also revealed that PWKC presented a good microbial shelf stability for up to 56 days (4 °C).

Aerobic mesophilic bacteria (2.05×10^{10}–2.46×10^{10} CFU/mL) and filamentous fungi (1.00×10^2–1.95×10^2 CFU/mL) were detected in all 17 PWKC samples during storage, which agreed with previous literature for kefir [77]. While the mixture of mature coconut water, pumpkin puree, and brown sugar was heat-treated (121 °C for 20 min) to reduce microbial load before fermentation, the aerobic mesophilic bacteria and filamentous fungi counts observed in PWKC were attributed to the water kefir grains. LAB in water kefir grains constitute a significant number among the aerobic mesophilic genera. Water kefir grains also contain filamentous fungi living in symbiosis [77].

3.5. *Characterisation of the Optimised Pumpkin-Based Mature Coconut Water Kefir Brew (PWKC_{opt})*

3.5.1. Chemical Composition

The proximate composition of PWKC_{opt} is depicted in Table 4. Results revealed that the optimised beverage was 89.06% moisture, 2.81% ash, 3.26% protein, 2.75% dietary fibre, 0.14% fat, and 1.98% carbohydrates. According to the literature, pumpkin fruit consists of 91.6% moisture, 1% protein, 0.1% fat, 0.8% ash, 0.5% dietary fibre, and 6.5% carbohydrates [78], whereas mature coconut water is 95.97% moisture, 0.51% protein, 0.06% fat, 0.47% ash, 0.08% dietary fibre, and 2.91% carbohydrates [79]. When comparing the values, PWKC_{opt} was lower in moisture and carbohydrate content but higher in protein, fat, ash, and dietary fibre contents than pumpkin fruit and mature coconut water. During fermentation, kefir microorganisms metabolise nutrients for growth and produce organic compounds and at the same time reduce the moisture and carbohydrate contents. The enzymatic lipolytic and proteolytic activities of kefir microorganisms promote the release of fatty acids and amino acids and subsequently increase the fat and protein contents. Kefir microorganisms could promote an increase in mineral content under certain circumstances, hence resulting in higher ash content [69,80]. The findings indicated that water kefir fermentation could enhance the nutritive values of the beverage. According to Annex of Regulation (EC) No 1924/2006 [81], a beverage product can be claimed as 'fat-free' when the beverage product contains no more than 0.15 g of fat per 100 mL. Therefore, PWKC_{opt} formulated in this study can be characterised as a fat-free beverage product.

Table 4. Major constituents of the optimised fermented pumpkin-based mature coconut water kefir beverage (PWKC_{opt}).

Parameters	Values
Proximate composition (% wet bases)	
Moisture	89.06 ± 0.14
Ash	2.81 ± 0.05
Protein	3.26 ± 0.13
Dietary fibre	2.75 ± 0.07
Fat	0.14 ± 0.03
Carbohydrate	1.98 ± 0.20
Minerals (mg/L)	
Potassium (K)	2186.33 ± 1.53
Phosphorus (P)	180.67 ± 0.58
Magnesium (Mg)	207.07 ± 2.65
Calcium (Ca)	137.33 ± 2.08
Iron (Fe)	1.37 ± 0.03
Zinc (Zn)	0.23 ± 0.03
Copper (Cu)	0.56 ± 0.04
Selenium (Se)	0.47 ± 0.03
Manganese (Mn)	0.77 ± 0.03

Results are presented as mean ± standard deviation (n = 3).

Nine minerals were detected in PWKC_{opt}, with K being the element identified at the highest concentration (2186.33 mg/L), followed by Mg, P, Ca, Fe, Mn, Cu, Se, and Zn (Table 4). The minerals detected in PWKC_{opt} matched the mineral profile of pumpkin fruit reported by USDA [78]. Mineral elements such as K, Na, Ca, Zn, Fe, and Mg were also detected in mature coconut water [24]. K is the predominant mineral in pumpkin fruit (3400 mg/L) [78] and mature coconut water (1960–2400 mg/L) [24]. Results suggested that the mineral profile of PWKC_{opt} was highly influenced by the raw materials (pumpkin puree and mature coconut water) used. The differences between the mineral contents of PWKC_{opt} and raw materials were attributed to the proportions used in the formulations.

3.5.2. Physicochemical Properties

The physicochemical properties (TSS, pH value, viscosity, and colour) of $PWKC_{opt}$ were evaluated (Table 5). The TSS and pH values of $PWKC_{opt}$ were 6.87 °Brix and 4.53, respectively. The TSS of $PWKC_{opt}$ was lower compared to the TSS of pumpkin puree (11.7 °Brix) [82] and mature coconut water (4.0 °Brix) [83] available in the literature. During fermentation, LAB utilised the sugars in pumpkin puree and mature coconut water to produce organic acids, thus resulting in a lower TSS [69]. According to the literature, pumpkin puree and mature coconut water have pH values of 6.30 [82] and 5.50 [83], respectively. The lower pH value of $PWKC_{opt}$ was due to the production of organic acids during the LAB fermentation [10]. Viscosity is a crucial quality characteristic of a fermented product, as it would affect sensory qualities such as flavour release and mouthfeel. $PWKC_{opt}$ had a viscosity of 54.57 cP, which is considered a mildly thick (nectar-like) consistency by the National Dysphagia Diet (NDD). The viscosity of the $PWKC_{opt}$ was comparable with the viscosities of *Lactobacillus plantarum*-fermented fruit juice (31.34–63.86 cP) prepared by Aly et al. [84]. $PWKC_{opt}$ had L*, a*, and b* values of 37.21, 2.24, and 23.83, respectively, indicating its 'yellowy-orange' colour. The colour of the $PWKC_{opt}$ could corresponded to the high total carotenoids content (33.24 mg/100 mL) detected in the $PWKC_{opt}$.

Table 5. The physicochemical properties of the optimised fermented pumpkin-based mature coconut water kefir beverage ($PWKC_{opt}$) during 56 days storage period (4 °C).

Physicochemical Properties		Storage Time (Day)		
		0	28	56
TSS (°Brix)		6.87 ± 0.02 [a]	6.87 ± 0.02 [a]	6.84 ± 0.04 [a]
pH		4.53 ± 0.02 [a]	4.55 ± 0.05 [a]	4.43 ± 0.12 [a]
Viscosity (cP)		54.57 ± 0.02 [a]	54.57 ± 0.02 [a]	54.53 ± 0.05 [a]
Colour	L*	37.21 ± 0.01 [a]	37.23 ± 0.12 [a]	37.10 ± 0.90 [a]
	a*	2.24 ± 0.03 [a]	2.24 ± 0.01 [a]	2.15 ± 0.13 [a]
	b*	23.83 ± 0.02 [a]	23.83 ± 0.03 [a]	23.81 ± 0.11 [a]

Results are displayed as mean ± standard deviation (n = 3). Values in a row with different superscript letters are significantly different ($p < 0.05$).

3.5.3. Antioxidative Contents and Activities

$PWKC_{opt}$ was found to have 33.24 mg/100 mL carotenoids, 89.93 mg GAE/100 mL phenols, 49.94 mg QE/100 mL flavonoids, a FRAP antioxidant power of 169.17 mM Fe (II)/100 mL, and half the maximal inhibitory concentration of DPPH free radicals of 27.17 mg/mL (Table 6). Pumpkin fruit is known to contain bioactive phenolic compounds with antioxidant capacities, particularly polyphenols, flavonoids, and carotenoids [85]. For instance, the most common phenolic acids in pumpkin are chlorogenic acid, quercetin, caffeic acid, gallic acid, p-Coumaric acid, and ferulic acid. Flavonoids such as catechin, quercetin, and narginin can also be found in the pumpkin fruit [86]. Antioxidative properties of mature coconut water have also been reported [20,87]. During fermentation, LAB release polyphenol oxidase, which could depolymerise the conjugated, complex phenolic compounds into smaller molecules, thereby increasing the antioxidant content [88,89]. Studies by Ozcelik et al. [73] and Tu et al. [88] revealed that lactic acid fermentation could help in increasing the antioxidant activities of Cornelian cherry, hawthorn, red plum, rose-ship, and pomegranate juices [73] and soy whey [88], respectively. Fiorda et al. [9] also found that some of the LAB strains (*Lactobacillus satsumensis* and *Leuconostoc mesenteroides*) present in kefir microbiota can scavenge reactive oxygen species.

Table 6. Antioxidant properties of the optimised fermented pumpkin-based mature coconut water kefir beverage ($PWKC_{opt}$).

Antioxidative Contents and Activities	Values
Total carotenoid (mg/100 mL)	33.24 ± 0.29
Total phenolic (mg GAE/100 mL)	89.93 ± 0.35
Total flavonoid (mg QE/100 mL)	49.94 ± 0.34
FRAP (mM Fe (II)/100 mL)	169.17 ± 0.06
DPPH IC_{50} (mg/mL)	27.17 ± 0.07

Results are presented as mean ± standard deviation (n = 3). GAE: gallic acid equivalents; QE: quercetin equivalents; Fe (II): iron (II) sulphate equivalents; DPPH IC_{50}: half maximal inhibitory concentration of 2,2-diphenylpicryl hydrazine (DPPH) free radicals.

3.5.4. Sugar Content

Sucrose is the main carbon and energy source for the microorganism's growth and fermentation, whereas glucose and fructose are major carbohydrates that are mainly metabolised into organic acids and ethanol. The sucrose in $PWKC_{opt}$ is mainly from the brown sugar added initially to produce $PWKC_{opt}$. At the beginning of the water kefir fermentation, yeast hydrolysed sucrose into glucose and fructose and produced ethanol at the same time. The produced glucose and fructose are the carbon source for LAB, whereas ethanol is utilised by AAB [88]. The composition of glucose and fructose sugars in the $PWKC_{opt}$ was monitored during the fermentation process (Table 7). Prior to the fermentation process, the initial values of glucose and fructose detected in the non-fermented $PWKC_{opt}$ were 38.75 and 27.17 g/L, respectively. Both glucose and fructose contents decreased significantly during fermentation. This finding indicated that water kefir microorganisms were able to metabolise both glucose and fructose presented in the pumpkin puree and mature coconut water. The glucose consumption (22.03 g/L, 56.85% of consumption) was higher than that of fructose (12.30 g/L, 45.27% of consumption). Glucose is always the preferred substrate for water kefir microorganisms over fructose owing to its faster degradation rate [90].

Table 7. The composition of glucose and fructose sugars in the optimised fermented pumpkin-based mature coconut water kefir beverage ($PWKC_{opt}$) during the fermentation process.

Fermentation Time (h)	Sugar (g/L)	
	Glucose	Fructose
0	38.75 ± 0.09 ^{c}	27.17 ± 0.08 ^{c}
2.25	23.35 ± 0.05 ^{b}	19.10 ± 0.05 ^{b}
4.50	16.72 ± 0.07 ^{a}	14.87 ± 0.07 ^{a}

Results are displayed as mean ± standard deviation (n = 3). Values in a column with different superscript letters are significantly different ($p < 0.05$).

3.5.5. Organic Acids Content

Sucrose, glucose, and fructose are the major sugars present in pumpkin and mature coconut water. Probiotic strains (*Lactobacillus casei* 431, *Saccharomyces cerevisiae* D254) have been reported to metabolise these sugars through the glycolytic pathway and tricarboxylic acid cycle to produce organic acids in their respective fermented products [18,26] The four main organic acids identified in $PWKC_{opt}$ were lactic acid, acetic acid, malic acid, and tartaric acid (Table 8).

Lactic acid is one of the organic acids formed during lactic acid fermentation, which can help in inhibiting the growth of pathogenic microorganisms and preventing food spoilage [88]. Among the organic acids detected, lactic acid was the most concentrated in $PWKC_{opt}$. The lactic acid content in the $PWKC_{opt}$ increased prominently to 5.80 g/L after fermented for 2.25 h and further to 6.79 g/L when fermentation time reached 4.50 h. The high lactic acid content detected in $PWKC_{opt}$ corresponded to the high viability of LAB (8.57 Log CFU/mL) in $PWKC_{opt}$. LAB metabolised glucose and fructose in pumpkin puree and mature coconut water into lactic acid. This finding is supported by the results obtained in the previous section (sugar content) of this work, where the increment in lactic acid content was accompanied by the decrement in glucose and fructose contents during fermentation. Lactic acid was also found to be the main fermentation end-product in other studies [65,76].

Table 8. The content of organic acids in the optimised fermented pumpkin-based mature coconut water kefir beverage ($PWKC_{opt}$) during the fermentation process.

Fermentation Time (h)	Organic Acid (g/L)			
	Lactic Acid	Acetic Acid	Malic Acid	Tartaric Acid
0	0.00 ± 0.00	0.00 ± 0.00	0.20 ± 0.01^{b}	0.02 ± 0.02^{a}
2.25	5.80 ± 0.01^{a}	0.08 ± 0.02^{a}	0.17 ± 0.01^{a}	0.12 ± 0.03^{b}
4.50	6.79 ± 0.03^{b}	0.13 ± 0.01^{b}	0.16 ± 0.02^{a}	0.14 ± 0.02^{b}

Results are displayed as mean ± standard deviation (n = 3). Values in a column with different superscript letters are significantly different ($p < 0.05$).

Acetic acid plays an important role in producing the fruity flavour and aroma of water kefir beverages [88]. The acetic acid content in $PWKC_{opt}$ increased after fermentation, and higher ($p < 0.05$) acetic acid content was obtained from the 4.5 h-fermentation than that of 2.20 h. Increments of acetic acid could be attributed to the ethanol oxidation by acetic acid bacteria and the anaerobic degradation of lactic acid by heterofermentative LAB [5,73,88].

Malic and tartaric acids were quantified at 0.20 and 0.02 g/L, respectively, at the beginning of the fermentation. The malic and tartaric acids are the organic acids in pumpkin and coconut water [24,91,92]. Malic acid can also be produced during the fermentation of water kefir by yeast metabolism [76]. A fermentation time of 2.25 h was found to reduce the malic acid content but increase the tartaric acid content in $PWKC_{opt}$. LAB could decarboxylate malic acid into lactic acid and CO_2 and metabolise glucose to tartaric acid. Decrements in malic acid and increments in tartaric acid were also observed in the fermentation of coconut water and blueberry juice, using *Lactobacillus casei* L4 [40] and *Lactobacillus plantarum* [93], respectively.

3.5.6. Ethanol Content (EC)

Water kefir is an ethanol and carbon dioxide-containing beverage that has a 'floral' and 'fruity' taste, which are primarily attributed to the yeast sugar fermentation and alcohol dehydrogenase activity of *Lactobacillus* sp. [5]. The ethanol content (EC) in the $PWKC_{opt}$ rose remarkably ($p < 0.05$) during fermentation (Table 6). The increment of EC was accompanied by decrements in glucose and fructose contents (Table 9). Yeast cells in the beverage metabolised the available glucose and fructose into ethanol.

Table 9. The ethanol content in the optimised fermented pumpkin-based mature coconut water kefir beverage ($PWKC_{opt}$) during the fermentation process.

Fermentation Time (h)	Ethanol (g/L)
0	0.05 ± 0.00^{a}
2.25	0.51 ± 0.03^{b}
4.50	0.66 ± 0.02^{c}

Results are displayed as mean ± standard deviation (n = 3). Values in a column with different superscript letters are significantly different ($p < 0.05$).

The EC of $PWKC_{opt}$ was 0.66%, and our finding was in accordance with the EC of fruit/vegetable-based water kefir-like fermented beverages ranging between 0.1 and 5.0%, varying depending on the substrate used for the fermentation [94]. In the Association of Southeast Asian Nations (ASEAN) countries, the allowable maximum alcohol level in the final product for alcohol-containing halal food is only 1%. EC of the $PWKC_{opt}$ was less than 1%, and, hence, it can be considered a non-alcoholic beverage [95].

3.5.7. Viability of Presumptive Lactic Acid Bacteria (LAB), *Lactobacillus*, Acetic Acid Bacteria (AAB) and Yeast

Water kefir is known to contain a consortium of microorganisms which have potential health benefits [3]. As shown in Table 10, the viable counts of $PWKC_{opt}$ for LAB, AAB, and yeast are 8.57, 2.08, and 6.29 Log CFU/mL, respectively. Therefore, the $PWKC_{opt}$ complied with the established minimum viable counts (10^7 and 10^4 CFU/mL of viable microorganisms and yeast, respectively) of the Codex definition for kefir and can be characterised as a kefir beverage [96].

Table 10. The microbiological properties of optimised fermented pumpkin-based mature coconut water kefir beverage ($PWKC_{opt}$) during 56 days storage period (4 °C).

Microorganisms' Viability (Log CFU/mL)	Storage Time (Day)		
	0	28	56
Lactic acid bacteria	8.57 ± 0.06^a	8.54 ± 0.36^a	8.98 ± 0.56^a
Lactobacillus	6.41 ± 0.40^a	6.65 ± 0.58^a	6.75 ± 0.59^a
Acetic acid bacteria	2.08 ± 0.08^a	2.08 ± 0.06^a	2.03 ± 0.32^a
Yeast	6.29 ± 0.08^a	6.36 ± 0.23^a	6.52 ± 0.42^a

Results are displayed as mean ± standard deviation (n = 3). Values in a row with different superscript letters are significantly different ($p < 0.05$).

As shown in Table 10, the viable count of the $PWKC_{opt}$ for *Lactobacillus* sp. is 6.41 Log CFU/mL, demonstrating that *Lactobacillus* sp. is the dominant LAB species in the $PWKC_{opt}$. Our finding agreed with the literature, in which LAB in water kefir is dominantly *Lactobacillus* [5]. *Lactobacillus* sp. has the potential to modulate host immunity and inhibit the growth of foodborne pathogenic bacteria [10]. Ozcelik et al. [73] have reported similar results, where *Lactobacillus* sp. is also the predominant LAB species in fermented fruit juices after the kefir fermentation process. The other LAB in water kefir includes *Lactococcus* and *Leuconostoc* [5].

3.5.8. Sensory Evaluation

The sensorial quality of a probiotic product is of crucial importance because it has a collective role in influencing the consumers' preference for the product as well as their purchasing behaviour. The metabolism of probiotic bacteria has been reported to negatively affect the sensory characteristics of the product. Despite the potential health benefits, consumers are unwilling to compromise on the taste of probiotic foods [97]. Therefore, sensory evaluation was performed on the optimised beverage in this study. The hedonic mean scores for appearance (colour), odour, taste (sourness), texture (consistency), and overall acceptability of the optimised beverage are shown in Figure 6. In general, the overall acceptability of sensory panellists on $PWKC_{opt}$ was acceptable (4.03 out of a 7-point hedonic Likert scale). The highest score was found for the appearance (colour) attribute (4.27), indicating that the colour of $PWKC_{opt}$ was the main quality attribute, which influenced the overall acceptance. The sensory panellists gave the lowest score for odour (3.93) among other sensory attributes, indicating that the panellists neither liked nor disliked the sour odour of $PWKC_{opt}$. Overall, despite the sour odour, $PWKC_{opt}$ was well accepted by the panellists. It is common for a fermented beverage to have a sour odour due to the formation of fermentation by-products.

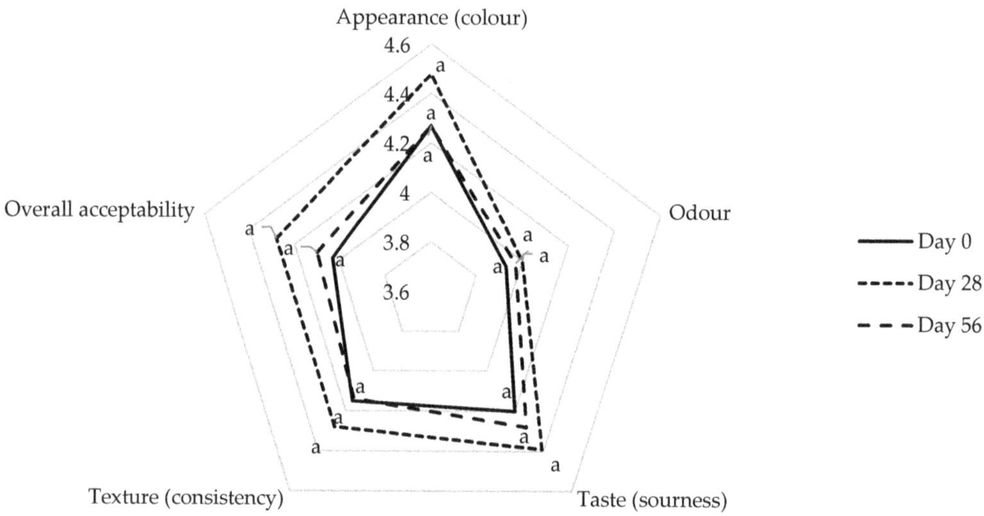

Figure 6. The sensory properties of optimised fermented pumpkin-based mature coconut water kefir beverage during 56 days storage period (4 °C). Results are displayed as mean ± standard deviation (n = 50). Similar superscript letters represent no significant difference ($p > 0.05$) between the attributes of different storage day.

3.5.9. Shelf-Life Study

The shelf-life of $PWKC_{opt}$ in terms of physicochemical (TSS, pH, viscosity, and colour), microbiological (LAB, *Lactobacillus*, AAB, and yeast), and sensorial properties, was conducted on days 0, 28, and 56 during refrigerated storage at 4 °C. In general, the changes of $PWKC_{opt}$ in its physicochemical (Table 5), microbiological (Table 10), and sensorial (Figure 2) properties during refrigerated storage were not statistically significant ($p > 0.05$).

Both the TSS and pH values of $PWKC_{opt}$ were observed to remain constant (TSS: day 0 = 6.87 ± 0.02 °Brix, day 28 = 6.87 ± 0.02 °Brix, day 56 = 6.84 ± 0.04 °Brix, $p > 0.05$; pH value: day 0 = 4.53 ± 0.02, day 28 = 4.55 ± 0.05, day 56 = 4.43 ± 0.012, $p > 0.05$) throughout the storage period (4 °C, 56 days). Results demonstrated that the lactic culture (*Lactobacillus* sp. K8) in $PWKC_{opt}$ could have become metabolically inactive during the refrigerated storage. A low temperature of 4 °C can reduce the cellular fluidity, subsequently, decrease the metabolic rate, and induce the dormant state in most of the probiotic cells [98]. Acevedo-Martínez et al. [52] also reported on the metabolic inactivity of the probiotic strain (*Lactobacillus casei*) in fermented mango juice stored at 4 °C. The TSS and sugar contents of the fermented mango juice were maintained over the 4 weeks of refrigerated storage [52]. The viscosity of $PWKC_{opt}$ ranged from 54.57 ± 0.01 (day 0) to 54.57 ± 0.02 (day 28) and 54.53 ± 0.05 cP (day 56) during the storage. The high stability in the viscosity of the beverage could be attributed to the water-soluble pumpkin pectinic polysaccharides consisting of glucose, rhamnose, arabinose, galactose, and galacturonic acid [15,99]. The polymeric properties of the polysaccharides created a repulsive force counterbalancing attractive van der Waals forces acting on a particle approaching another particle, stabilizing the colloid of liquid, and hence maintaining the viscosity of the beverage [100]. No difference was found in the colour (L*: day 0 = 37.21 ± 0.01, day 28 = 37.23 ± 0.12, day 56 = 37.10 ± 0.90, $p > 0.05$; a*: day 0 = 2.24 ± 0.03, day 28 = 2.24 ± 0.01, day 56 = 2.15 ± 0.13, $p > 0.05$; b*: day 0 = 23.83 ± 0.02, day 28 = 23.83 ± 0.03, day 56 = 23.81 ± 0.11, $p > 0.05$) of $PWKC_{opt}$ during the refrigerated storage time. The result indicated that the natural pigments and polyphenols in $PWKC_{opt}$ were well preserved. This might be attributed to the *Lactobacillus*'s ability to entrap residual oxygen in the beverage, which in turn prevents the occurrence of the auto-oxidation process that is responsible for the colour degradation [89].

According to Codex Alimentarius, for a product to be considered kefir, it should contain at least 10^7 CFU/mL of viable microorganisms and 10^4 CFU/mL of viable yeast [96]. The viability of LAB (8.57 ± 0.06–8.98 ± 0.56 Log CFU/mL), *Lactobacillus* (6.41 ± 0.40–6.75 ± 0.59 Log CFU/mL), AAB (2.08 ± 0.08–2.03 ± 0.32 Log CFU/mL), and yeast (6.29 ± 0.08–6.52 ± 0.42 Log CFU/mL) remained stable over 56 days of storage. The preservation of high count of *Lactobacillus*, LAB, AAB, and yeast in $PWKC_{opt}$ during storage was thought to be attributed to the availability of nutrients in pumpkin puree. The nutrients available in pumpkin purees, such as dietary fibre, vitamins, minerals, and antioxidants sustained the growth and viability of the water kefir microorganisms in $PWKC_{opt}$ [15]. Similar to the shelf-life study by Genevois et al. [101], it was found that the viability of *Lactobacillus casei* supplemented with powdered pumpkin peel and pulp was constant for 21 days of storage at 8 and 22 °C with counts of more than 10^6 CFU/g. At the end of the storage, $PWKC_{opt}$ contained 8.98 ± 0.56 Log CFU/mL LAB and 6.52 ± 0.42 Log CFU/mL yeast, which complied with the minimum level of total viable counts of microorganisms and yeast in kefir [96]. In addition, according to FAO/WHO [1], a minimum viable probiotics dose of 10^7 CFU/mL in probiotic food at the time of consumption could exert health benefits to the host. Therefore, $PWKC_{opt}$ could confer probiotic effects to the host with a shelf-life of at least 56 days.

After 56 days of storage at 4 °C, panellists did not detect any significant or undesirable changes ($p > 0.05$) in the sensorial properties of $PWKC_{opt}$ during storage (Figure 2). The sensory deterioration of probiotic-fermented products is always correlated with the proteolytic activity of bacteria, high acid production, and degradation of fat and protein. Therefore, the maintenance of sensory profiles indicated that $PWKC_{opt}$ possessed a desirable physicochemical characterisation and considerable proteolysis pattern, which is an acceptable and stable product [102].

3.6. Identification of Selected Isolate

The 16S rRNA gene sequences of dominant cultivable species isolates in $PWKC_{opt}$ was perceived to be a close relative to *Lactobacillus mali*, *Lactobacillus satsumensis*, and *Lactobacillus hordei* (96% similarity) under the accession number of NR_112691.1, LC311746.1, and NR_044394.1, respectively (Table 11). This result proposed that dominant cultivable species isolates belong to the genus *Lactobacillus*, supported by similar phenotypic characteristics to *Lactobacillus* species.

Table 11. Closely-related relatives of dominant cultivable species isolates in the optimised fermented pumpkin-based mature coconut water kefir ($PWKC_{opt}$) were determined based on almost complete 16S rRNA gene sequence.

Closely-Related Relatives (Type Strains)	Max Score	Query Cover	E Value	Similarity
Lactobacillus satsumensis	2215	94%	0	96%
Lactobacillus mali	2215	94%	0	96%
Lactobacillus hordei	2207	92%	0	96%
Lactobacillus oeni	2189	95%	0	95%
Lactobacillus uvarum	2141	94%	0	95%
Lactobacillus aquaticus	2141	94%	0	95%
Lactobacillus sucicola	2130	95%	0	94%
Lactobacillus capillatus	2126	95%	0	94%
Lactobacillus mobilis	2098	93%	0	94%
Lactobacillus vini	2098	93%	0	94%

4. Conclusions

Developing a kefir beverage using non-dairy substrates such as mature coconut water and pumpkin puree is a novel approach for probiotic beverage production. This study was conducted to determine the optimal fermentation conditions (fermentation temperature, concentrations of pumpkin puree and brown sugar) to minimize fermentation time (to reach pH 4.5) while maximising the overall acceptability (OA) score, water kefir grains' biomass growth rate, lactic acid content, and *Lactobacillus* count of the fermented pumpkin-based mature coconut water kefir. The optimised fermentation condition included a fermentation temperature of 27 °C, 20% w/v of pumpkin puree, and 10% w/v of brown sugar, and optimised kefir drink ($PWKC_{opt}$) obtained had an overall acceptability (OA) score of 4.03, 6.41 Log CFU/mL of *Lactobacillus* count, 0.68% v/v lactic acid content, 31% of water kefir grains' biomass growth rate, and fermentation time (to reach pH 4.5) of 4.5 h. The fermentation process had a positive effect on the nutritional value, physicochemical, antioxidative, and microbiological properties of the beverage, including increases in protein, dietary fibre, minerals, antioxidant activities, organic acid content, and viable count of probiotic microorganisms. $PWKC_{opt}$ is stable for at least 56 days when stored under refrigerated conditions (4 °C). Therefore, $PWKC_{opt}$ has the potential to function as a value-added probiotic beverage product.

However, there are some limitations in this work that should be acknowledged. First, the viabilities of *Lactobacillus*, LAB, acetic acid bacteria, and yeast reported in this work reflected only their estimated counts. Future research should identify and confirm the counts through characterisation tests such as microscopic examination, catalase test, and Gram stain test.

Author Contributions: Project administration, resources, funding acquisition, supervision, investigation, methodology, data curation, and visualisation, W.Y.K.; writing—original draft, and formal analysis, W.Y.K. and X.X.L.; writing—review and editing, W.Y.K., X.X.L., B.H.K., B.R., T.C.T., R.K. and U.U. All authors have read and agreed to the published version of the manuscript.

Funding: This work was supported by the SPLB Scheme (SLB2228) provided by Universiti Malaysia Sabah (UMS), Malaysia.

Data Availability Statement: Data are contained within the article.

Conflicts of Interest: The authors declare no conflicts of interest.

References

1. FAO/WHO. *FAO/WHO Joint Working Group Report on Drafting Guidelines for the Evaluation of Probiotics in Food (30 April 2002 and 1 May 2002)*; Scientific Research Publishing: London, ON, Canada, 2002.
2. Tom, R. The Rising Tide of Non-Animal Proteins. Euromonitor International. 2022. Available online: https://www.euromonitor.com/article/the-rising-tide-of-non-animal-proteins (accessed on 28 March 2024).
3. Lynch, K.M.; Wilkinson, S.; Daenen, L.; Arendt, E.K. An update on water kefir. *Int. J. Food Microbiol.* **2021**, *345*, 109128. [CrossRef] [PubMed]
4. Lutz, L. Recherches Biologiques Sur la Constitution du Tibi. *Bull. Soc. Mycol. Fr.* **1899**, *15*, 68–72.
5. Egea, M.B.; Santos, D.C.d.; Oliveira Filho, J.G.d.; Ores, J.d.C.; Takeuchi, K.P.; Lemes, A.C. A review of non-dairy kefir products: Their characteristics and potential human health benefits. *Crit. Rev. Food Sci. Nutr.* **2022**, *62*, 1536–1552. [CrossRef] [PubMed]
6. Pendón, M.D.; Bengoa, A.A.; Iraporda, C.; Medrano, M.; Garrote, G.L.; Abraham, A.G. Water kefir: Factors affecting grain growth and health-promoting properties of the fermented beverage. *J. Appl. Microbiol.* **2021**, *133*, 162–180. [CrossRef] [PubMed]
7. Darvishzadeh, P.; Orsat, V.; Martinez, J.L. Process optimization for development of a novel water kefir drink with high antioxidant activity and potential probiotic properties from Russian olive fruit (*Elaeagnus angustifolia*). *Food Bioprocess Technol.* **2021**, *14*, 248–260. [CrossRef]
8. Rodrigues, K.L.; Caputo, L.R.G.; Carvalho, J.C.T.; Evangelista, J.; Schneedorf, J.M. Antimicrobial and healing activity of kefir and kefiran extract. *Int. J. Antimicrob.* **2005**, *25*, 404–408. [CrossRef] [PubMed]
9. Fiorda, F.A.; de Melo Pereira, G.V.; Thomaz-Soccol, V.; Medeiros, A.P.; Rakshit, S.K.; Soccol, C.R. Development of kefir-based probiotic beverages with DNA protection and antioxidant activities using soybean hydrolyzed extract, colostrum and honey. *LWT* **2016**, *68*, 690–697. [CrossRef]
10. Cai, Y.; Sounderrajan, A.; Serventi, L. Water kefir: A review of its microbiological profile, antioxidant potential and sensory quality. *Acta Sci. Nutr. Health* **2020**, *4*, 10–17. [CrossRef]
11. Ninčević Grassino, A.; Rimac Brnčić, S.; Badanjak Sabolović, M.; Šic Žlabur, J.; Marović, R.; Brnčić, M. Carotenoid content and profiles of pumpkin products and by-products. *Molecules* **2023**, *28*, 858. [CrossRef]
12. Aggarwal, S.; Mohite, A.; Sharma, N. The maturity and ripeness phenomenon with regard to the physiology of fruits and vegetables: A review. *Bull. Transilv. Univ. Brasov. Ser. II For. Wood Ind. Agric. Food Eng.* **2018**, *11*, 77–88.

13. The Food and Agriculture Organization Corporate Statistical Database (FAOSTAT). 2024. Available online: https://www.fao.org/faostat/en/#home (accessed on 28 March 2024).
14. Kamarubahrin, A.F.; Haris, A.B.; Daud, S.N.B.M.; Kefeli, Z.B.; Ahmad, N.B.; Muhamed, N.A.B.; Shukor, S.B.A. The potential of pumpkin (*Cucurbita moschata* Duschene) as commercial crop in Malaysia. *Pertanika J. Sch. Res. Rev.* **2018**, *4*, 1–10.
15. Szydłowska, A.; Kołożyn-Krajewska, D. Development of potentially probiotic and synbiotic pumpkin frozen desserts. *CyTA-J. Food* **2019**, *17*, 251–259. [CrossRef]
16. Lokuge, G.M.; Vidanarachchi, J.; Thavarajah, P.; Siva, N.; Thavarajah, D.; Liyanage, R.; Balasooriya, N.; Alwis, J. Prebiotic carbohydrate profile and in vivo prebiotic effect of pumpkin (*Cucurbita maxima*) grown in Sri Lanka. *J. Natl. Sci. Found. Sri Lanka* **2018**, *46*, 477–485. [CrossRef]
17. Nguyen, T.H.; Kim, Y.; Kim, J.-S.; Jeong, Y.; Park, H.M.; Kim, J.W.; Kim, J.-E.; Kim, H.; Paek, N.-S.; Kang, C.-H. Evaluating the cryoprotective encapsulation of the lactic acid bacteria in simulated gastrointestinal conditions. *Biotechnol. Bioproc. Eng.* **2020**, *25*, 287–292. [CrossRef]
18. Dimitrovski, D.; Dimitrovska-Vetadjoka, M.; Hristov, H.; Doneva-Shapceska, D. Developing probiotic pumpkin juice by fermentation with commercial probiotic strain *Lactobacillus casei* 431. *J. Food Process. Preserv.* **2021**, *45*, e15245. [CrossRef]
19. Kiharason, J.W.; Isutsa, D.K. Shelf-life of pumpkin fruit slices, flour and blended products. *Int. J. Food Sci. Biotechnol.* **2019**, *4*, 14.
20. Kumari, N.; Sindhu, S.C.; Rani, V.; Kumari, A.; Kumari, V. Fermentation Improves Mineral Bioavailability of Pumpkin Seed Flour. *Ann. Biol.* **2021**, *37*, 252–255.
21. Roger, T.; Ezo'o Ezo'o, J.; Essia Ngang, J.J. mprovement of sanitary quality of the digue by use of lactic starters. *Am. J. Microbiol. Res.* **2022**, *10*, 6–16. [CrossRef]
22. Ahmad, R.K.; Sulaiman, S.A.; Yusup, S.; Dol, S.S.; Inayat, M.; Umar, H.A. Exploring the potential of coconut shell biomass for charcoal production. *Ain Shams Eng. J.* **2022**, *13*, 101499. [CrossRef]
23. Xu, S.; Ma, Z.; Chen, Y.; Li, J.; Jiang, H.; Qu, T.; Zhang, W.; Li, C.; Liu, S. Characterization of the flavor and nutritional value of coconut water vinegar based on metabolomics. *Food Chem.* **2022**, *369*, 130872. [CrossRef]
24. Kumar, M.; Saini, S.S.; Agrawal, P.K.; Roy, P.; Sircar, D. Nutritional and metabolomics characterization of the coconut water at different nut developmental stages. *J. Food Compos. Anal.* **2021**, *96*, 103738. [CrossRef]
25. Chauhan, O.; Archana, B.; Singh, A.; Raju, P.; Bawa, A. A refreshing beverage from mature coconut water blended with lemon juice. *J. Food Sci. Technol.* **2014**, *51*, 3355–3361. [CrossRef] [PubMed]
26. Zhang, G.; Li, X.; Chen, W.; Chen, P.; Jin, X.; Chen, W.; Chen, H. Organic acid content, antioxidant capacity, and fermentation kinetics of matured coconut (*Cocos nucifera*) water fermented by *Saccharomyces cerevisiae* D254. *Int. J. Food Eng.* **2018**, *14*, 20170331. [CrossRef]
27. Kantachote, D.; Ratanaburee, A.; Hayisama-ae, W.; Sukhoom, A.; Nunkaew, T. The use of potential probiotic *Lactobacillus plantarum* DW12 for producing a novel functional beverage from mature coconut water. *J. Funct. Foods* **2017**, *32*, 401–408. [CrossRef]
28. Preetha, P.P.; Devi, V.G.; Rajamohan, T. Mature coconut water exhibits antidiabetic and antithrombotic potential via L-arginine-nitric oxide pathway in alloxan induced diabetic rats. *J. Basic Clin. Physiol. Pharmacol.* **2015**, *26*, 575–583. [CrossRef] [PubMed]
29. Skryplonek, K.; Dmytrów, I.; Mituniewicz-Małek, A. Probiotic fermented beverages based on acid whey. *J. Dairy Sci.* **2019**, *102*, 7773–7780. [CrossRef]
30. Laureys, D.; Aerts, M.; Vandamme, P.; De Vuyst, L. The buffer capacity and calcium concentration of water influence the microbial species diversity, grain growth, and metabolite production during water kefir fermentation. *Front. Microbiol.* **2019**, *10*, 2876. [CrossRef] [PubMed]
31. Guzel-Seydim, Z.B.; Gökırmaklı, Ç.; Greene, A.K. A comparison of milk kefir and water kefir: Physical, chemical, microbiological and functional properties. *Trends Food Sci. Technol.* **2021**, *113*, 42–53. [CrossRef]
32. Kazou, M.; Grafakou, A.; Tsakalidou, E.; Georgalaki, M. Zooming into the microbiota of home-made and industrial kefir produced in Greece using classical microbiological and amplicon-based metagenomics analyses. *Front. Microbiol.* **2021**, *12*, 64. [CrossRef]
33. Prasad, B.N.; Sivamani, S. Fate of acidity and alcohol during fermentation of Salalah and Sri Lanka coconut water. *Eur. J. Eng. Technol.* **2021**, *9*, 1–9.
34. Gómez-Aldapa, C.A.; Rangel-Vargas, E.; Bautista-De León, H.; Castro-Rosas, J. Presence of non-O157 Shiga toxin-producing *Escherichia coli*, enterotoxigenic *E. coli*, enteropathogenic *E. coli* and *Salmonella* in fresh beetroot (*Beta vulgaris* L.) juice from public markets in Mexico. *J. Sci. Food Agric.* **2014**, *94*, 2705–2711. [CrossRef] [PubMed]
35. Cassani, L.; Tomadoni, B.; Moreira, M.d.R.; Ponce, A.; Agüero, M.V. Optimization of inulin: Oligofructose proportion and non-thermal processing to enhance microbiological and sensory properties of fiber-enriched strawberry juice. *LWT* **2017**, *80*, 446–455. [CrossRef]
36. Hecer, C.; Ulusoy, B.; Kaynarca, D. Effect of different fermentation conditions on composition of kefir microbiota. *Int. Food Res. J.* **2019**, *26*, 401–409.
37. dos Santos Filho, A.L.; Freitas, H.V.; Rodrigues, S.; Abreu, V.K.G.; de Oliveira Lemos, T.; Gomes, W.F.; Narain, N.; Pereira, A.L.F. Production and stability of probiotic cocoa juice with sucralose as sugar substitute during refrigerated storage. *LWT* **2019**, *99*, 371–378. [CrossRef]
38. *ISO 11136:2014*; Sensory Analysis-Methodology-General Guidance for Conducting Hedonic Tests with Consumers in a Controlled Area. International Organization for Standardization (ISO): Geneva, Switzerland, 2014.

39. Pop, C.R.; Salanţă, L.; Rotar, A.M.; Semeniuc, C.A.; Socaciu, C.; Sindic, M. Influence of extraction conditions on characteristics of microbial polysaccharide kefiran isolated from kefir grains biomass. *J. Food Nutr. Res.* **2016**, *55*, 121–130.
40. Giri, S.S.; Sukumaran, V.; Sen, S.S.; Park, S.C. Use of a potential probiotic, *Lactobacillus casei* L4, in the preparation of fermented coconut water beverage. *Front. Microbiol.* **2018**, *9*, 1976. [CrossRef] [PubMed]
41. Laureys, D.; Van Jean, A.; Dumont, J.; De Vuyst, L. Investigation of the instability and low water kefir grain growth during an industrial water kefir fermentation process. *Appl. Microbiol. Biotechnol.* **2017**, *101*, 2811–2819. [CrossRef]
42. AOAC. *Official Methods of Analysis*, 18th ed.; The Association of Official Analytical Chemists: Gaithersburg, MD, USA, 2005.
43. Hlangwani, E.; Doorsamy, W.; Adebiyi, J.A.; Fajimi, L.I.; Adebo, O.A. A modeling method for the development of a bioprocess to optimally produce *umqombothi* (a South African traditional beer). *Sci. Rep.* **2021**, *11*, 20626. [CrossRef]
44. Barkallah, M.; Dammak, M.; Louati, I.; Hentati, F.; Hadrich, B.; Mechichi, T.; Ayadi, M.A.; Fendri, I.; Attia, H.; Abdelkafi, S. Effect of *Spirulina platensis* fortification on physicochemical, textural, antioxidant and sensory properties of yogurt during fermentation and storage. *LWT* **2017**, *84*, 323–330. [CrossRef]
45. Wang, L.; Zhang, H.; Lei, H. Phenolics profile, antioxidant activity and flavor volatiles of pear juice: Influence of lactic acid fermentation using three *Lactobacillus* strains in monoculture and binary mixture. *Foods* **2021**, *11*, 11. [CrossRef]
46. Hentati, F.; Barkallah, M.; Ben Atitallah, A.; Dammak, M.; Louati, I.; Pierre, G.; Fendri, I.; Attia, H.; Michaud, P.; Abdelkafi, S. Quality characteristics and functional and antioxidant capacities of algae-fortified fish burgers prepared from common barbel (*Barbus barbus*). *Biomed Res. Int.* **2019**, *2019*, 2907542. [CrossRef]
47. Filip, M.; Vlassa, M.; Coman, V.; Halmagyi, A. Simultaneous determination of glucose, fructose, sucrose and sorbitol in the leaf and fruit peel of different apple cultivars by the HPLC–RI optimized method. *Food Chem.* **2016**, *199*, 653–659. [CrossRef] [PubMed]
48. Bonello, F.; Cravero, M.C.; Asproudi, A.; Lottero, M.R.; Piras, F.; Damasco, G.; Demelas, L.; Petrozziello, M.; Lovicu, G. Exploring the aromatic complexity of Sardinian red wines obtained from minor and rare varieties. *Eur. Food Res. Technol.* **2021**, *247*, 133–156. [CrossRef]
49. Gamba, R.R.; Caro, C.A.; Martínez, O.L.; Moretti, A.F.; Giannuzzi, L.; De Antoni, G.L.; Peláez, A.L. Antifungal effect of kefir fermented milk and shelf life improvement of corn arepas. *Int. J. Food Microbiol.* **2016**, *235*, 85–92. [CrossRef] [PubMed]
50. Grujović, M.Ž.; Mladenović, K.G.; Semedo-Lemsaddek, T.; Laranjo, M.; Stefanović, O.D.; Kocić-Tanackov, S.D. Advantages and disadvantages of non-starter lactic acid bacteria from traditional fermented foods: Potential use as starters or probiotics. *Compr. Rev. Food Sci. Food Saf.* **2022**, *21*, 1537–1567. [CrossRef]
51. M'hir, S.; Mejri, A.; Atrous, H.; Ayed, L. Optimization of parameters using response surface methodology to develop a novel kefir-like functional beverage from cheese whey enriched with myrtle juice. *J. Chem.* **2021**, *2021*, 2984470. [CrossRef]
52. Acevedo-Martínez, E.; Gutiérrez-Cortés, C.; García-Mahecha, M.; Díaz-Moreno, C. Evaluation of viability of probiotic bacteria in mango (*Mangifera indica* L. Cv."Tommy Atkins") beverage. *Dyna* **2018**, *85*, 84–92. [CrossRef]
53. Fawzi, N.Y.; Abdelghani, D.Y.; Abdel-azim, M.A.; Shokier, C.G.; Youssef, M.W.; El-Rab, M.K.G.; Gad, A.I.; Abou-Taleb, K.A. The ability of probiotic lactic acid bacteria to ferment Egyptian broken rice milk and produce rice-based yoghurt. *Ann. Agric. Sci.* **2022**, *67*, 107–118. [CrossRef]
54. Bengoa, A.A.; Llamas, M.G.; Iraporda, C.; Dueñas, M.T.; Abraham, A.G.; Garrote, G.L. Impact of growth temperature on exopolysaccharide production and probiotic properties of *Lactobacillus paracasei* strains isolated from kefir grains. *Food Microbiol.* **2018**, *69*, 212–218. [CrossRef]
55. Laureys, D.; Aerts, M.; Vandamme, P.; De Vuyst, L. Oxygen and diverse nutrients influence the water kefir fermentation process. *Food Microbiol.* **2018**, *73*, 351–361. [CrossRef]
56. M'hir, S.; Rtibi, K.; Mejri, A.; Ziadi, M.; Aloui, H.; Hamdi, M.; Ayed, L. Development of a novel whey date beverage fermented with kefir grains using response surface methodology. *J. Chem.* **2019**, *2019*, 1218058. [CrossRef]
57. Viana, R.O.; Magalhães-Guedes, K.T.; Braga, R.A.; Dias, D.R.; Schwan, R.F. Fermentation process for production of apple-based kefir vinegar: Microbiological, chemical and sensory analysis. *Braz. J. Microbiol.* **2017**, *48*, 592–601. [CrossRef] [PubMed]
58. Mota-Gutierrez, J.; Botta, C.; Ferrocino, I.; Giordano, M.; Bertolino, M.; Dolci, P.; Cannoni, M.; Cocolin, L. Dynamics and biodiversity of bacterial and yeast communities during fermentation of cocoa beans. *Appl. Environ. Microbiol.* **2018**, *84*, e01164-18. [CrossRef]
59. Guimarães, J.T.; Balthazar, C.F.; Scudino, H.; Pimentel, T.C.; Esmerino, E.A.; Ashokkumar, M.; Freitas, M.Q.; Cruz, A.G. High-intensity ultrasound: A novel technology for the development of probiotic and prebiotic dairy products. *Ultrason. Sonochem.* **2019**, *57*, 12–21. [CrossRef] [PubMed]
60. Rosdiana, R.K.; Fitrilia, T. Proximate composition and organoleptic properties of pumpkin (*Cucurbita moschata*)-based complementary foods with several types of nuts. *Indones. J. Appl. Res.* **2022**, *3*, 23–31.
61. Segura-Badilla, O.; Lazcano-Hernández, M.; Kammar-García, A.; Vera-López, O.; Aguilar-Alonso, P.; Ramírez-Calixto, J.; Navarro-Cruz, A.R. Use of coconut water (*Cocus nucifera* L.) for the development of a symbiotic functional drink. *Heliyon* **2020**, *6*, e03653. [CrossRef] [PubMed]
62. Teijeiro, M.; Pérez, P.F.; De Antoni, G.L.; Golowczyc, M.A. Suitability of kefir powder production using spray drying. *Food Res. Int.* **2018**, *112*, 169–174. [CrossRef]

63. Pop, C.; Apostu, S.; Salanţă, L.; Rotar, A.M.; Sindic, M.; Mabon, N.; Socaciu, C. Influence of different growth conditions on the kefir grains production used in the kefiran synthesis. *Bull. Univ. Agric. Sci. Vet. Med. Cluj-Napoca Food Sci. Technol.* **2014**, *71*, 147–153. [CrossRef] [PubMed]
64. Guzel-Seydim, Z.; Kok-Tas, T.; Ertekin-Filiz, B.; Seydim, A. Effect of different growth conditions on biomass increase in kefir grains. *J. Dairy Sci.* **2011**, *94*, 1239–1242. [CrossRef]
65. Bueno, R.S.; Ressutte, J.B.; Hata, N.N.; Henrique-Bana, F.C.; Guergoletto, K.B.; de Oliveira, A.G.; Spinosa, W.A. Quality and shelf life assessment of a new beverage produced from water kefir grains and red pitaya. *LWT* **2021**, *140*, 110770. [CrossRef]
66. Altuntas, S.; Hapoglu, H. *Kefir-Type Drinks from Whey Non-Alcoholic Beverages*; Elsevier: Amsterdam, The Netherlands, 2019; pp. 185–226.
67. Gut, A.M.; Vasiljevic, T.; Yeager, T.; Donkor, O.N. Kefir characteristics and antibacterial properties-potential applications in control of enteric bacterial infection. *Int. Dairy J.* **2021**, *118*, 105021. [CrossRef]
68. Laureys, D.; De Vuyst, L. Microbial species diversity, community dynamics, and metabolite kinetics of water kefir fermentation. *Appl. Environ. Microbiol.* **2014**, *80*, 2564–2572. [CrossRef]
69. Alves, V.; Scapini, T.; Camargo, A.F.; Bonatto, C.; Stefanski, F.S.; de Jesus, E.P.; Diniz, L.G.T.; Bertan, L.C.; Maldonado, R.R.; Treichel, H. Development of fermented beverage with water kefir in water-soluble coconut extract (*Cocos nucifera* L.) with inulin addition. *LWT* **2021**, *145*, 111364. [CrossRef]
70. Xu, Z.; Lu, Z.; Soteyome, T.; Ye, Y.; Huang, T.; Liu, J.; Harro, J.M.; Kjellerup, B.V.; Peters, B.M. Polymicrobial interaction between *Lactobacillus* and *Saccharomyces cerevisiae*: Coexistence-relevant mechanisms. *Crit. Rev. Microbiol.* **2021**, *47*, 386–396. [CrossRef] [PubMed]
71. Laureys, D.; Leroy, F.; Vandamme, P.; De Vuyst, L. Backslopping time, rinsing of the grains during backslopping, and incubation temperature influence the water kefir fermentation process. *Front. Microbiol.* **2022**, *1487*, 871550. [CrossRef] [PubMed]
72. Slattery, C.; Cotter, P.D.; W. O'Toole, P. Analysis of health benefits conferred by *Lactobacillus* species from kefir. *Nutrients* **2019**, *11*, 1252. [CrossRef]
73. Ozcelik, F.; Akan, E.; Kinik, O. Use of Cornelian cherry, hawthorn, red plum, roseship and pomegranate juices in the production of water kefir beverages. *Food Biosci.* **2021**, *42*, 101219. [CrossRef]
74. Dadkhah, S.; Pourahmad, R.; Assadi, M.M.; Moghimi, A. Kefir production from soymilk. *Ann. Biol. Res.* **2011**, *2*, 293–299.
75. Premjit, Y.; Mitra, J. Optimization of electrospray-assisted microencapsulation of probiotics (*Leuconostoc lactis*) in soy protein isolate-oil particles using Box-Behnken experimental design. *Food Bioprocess Technol.* **2021**, *14*, 1712–1729. [CrossRef]
76. Paredes, J.L.; Escudero-Gilete, M.L.; Vicario, I.M. A new functional kefir fermented beverage obtained from fruit and vegetable juice: Development and characterization. *LWT* **2022**, *154*, 112728. [CrossRef]
77. de Oliveira, A.F.; Silva, I.S.; Ferreira, A.M.; Velázquez-Moyado, J.A.; Bezerra, R.M.; Carvalho, J.C.T.; Ferreira, I.M.; Florentino, A.C. Mercury tolerance of *Penicillium* sp. isolated from kefir grains. *Ciênc. Nat.* **2018**, *40*, e73. [CrossRef]
78. US Department of Agriculture (USDA). Pumpkin, Raw. 2018. Available online: https://fdc.nal.usda.gov/fdc-app.html#/food-details/168448/nutrients (accessed on 28 March 2024).
79. Othaman, M.A.; Sharifudin, S.A.; Mansor, A.; Kahar, A.; Long, K. Coconut water vinegar: New alternative with improved processing technique. *J. Eng. Sci. Technol.* **2014**, *9*, 293–302.
80. Pinto, L.C.; de Oliveira, T.P.; de Souza, R.; Santos, N.B.F.; Santos, L.F.P.; de Assis Santos, A.; dos Santos, T.X.; Santos, C.T.; de Jesus Nunes, C.; Costa, I.B. Probiotic kefir-fermented beverage-based *Colocasia esculenta* L.: Development, characterization, and microbiological stability during chilled storage. *J. Food Process. Preserv.* **2021**, *45*, e15113. [CrossRef]
81. EC (European Commission). Commission Regulation (EC) No 1924/2006 of 20 December 2006 on Nutrition and Health Claims Made on Foods. Available online: http://data.europa.eu/eli/reg/2006/1924/oj (accessed on 28 March 2024).
82. García-Parra, J.; González-Cebrino, F.; Ramírez, R. Volatile compounds of a pumpkin (*Cucurbita moschata*) purée processed by high pressure thermal processing. *J. Sci. Food Agric.* **2020**, *100*, 4449–4456. [CrossRef] [PubMed]
83. Tan, T.-C.; Easa, A.M. The evolution of physicochemical and microbiological properties of green and mature coconut water (*Cocos nucifera*) under different storage conditions. *J. Food Meas. Charact.* **2021**, *15*, 3523–3530. [CrossRef]
84. Aly, E.; Darwish, A.A.; Tawfek, M. Quality characteristics of sweet whey-based fruits beverages fermented with *Lactobacillus plantarum*. *Egypt. J. Food Sci.* **2019**, *47*, 141–254. [CrossRef]
85. Kulczyński, B.; Gramza-Michałowska, A. The profile of carotenoids and other bioactive molecules in various pumpkin fruits (*Cucurbita maxima* Duchesne) cultivars. *Molecules* **2019**, *24*, 3212. [CrossRef]
86. Enneb, S.; Drine, S.; Bagues, M.; Triki, T.; Boussora, F.; Guasmi, F.; Nagaz, K.; Ferchichi, A. Phytochemical profiles and nutritional composition of squash (*Cucurbita moschata* D.) from Tunisia. *S. Afr. J. Bot.* **2020**, *130*, 165–171. [CrossRef]
87. Rodsamran, P.; Sothornvit, R. Bioactive coconut protein concentrate films incorporated with antioxidant extract of mature coconut water. *Food Hydrocoll.* **2018**, *79*, 243–252. [CrossRef]
88. Tu, C.; Azi, F.; Huang, J.; Xu, X.; Xing, G.; Dong, M. Quality and metagenomic evaluation of a novel functional beverage produced from soy whey using water kefir grains. *LWT* **2019**, *113*, 108258. [CrossRef]
89. abadi Sherahi, M.H.; Shahidi, F.; Yazdi, F.T.; Hashemi, S.M.B. Effect of *Lactobacillus plantarum* on olive and olive oil quality during fermentation process. *LWT* **2018**, *89*, 572–580. [CrossRef]

90. Laureys, D.; Leroy, F.; Hauffman, T.; Raes, M.; Aerts, M.; Vandamme, P.; De Vuyst, L. The type and concentration of inoculum and substrate as well as the presence of oxygen impact the water kefir fermentation process. *Front. Microbiol.* **2021**, *12*, 628599. [CrossRef] [PubMed]
91. Limareva, N.; Donchenko, L.; Malaknov, V.; Semenova, E. Functional beverages containing pectin from different raw material. *IOP Conf. Ser. Earth Environ. Sci.* **2019**, *337*, 012013. [CrossRef]
92. Zhou, C.-L.; Mi, L.; Hu, X.-Y.; Zhu, B.-H. Evaluation of three pumpkin species: Correlation with physicochemical, antioxidant properties and classification using SPME-GC–MS and E-nose methods. *J. Food Sci. Technol.* **2017**, *54*, 3118–3131. [CrossRef]
93. Li, S.; Tao, Y.; Li, D.; Wen, G.; Zhou, J.; Manickam, S.; Han, Y.; Chai, W.S. Fermentation of blueberry juices using autochthonous lactic acid bacteria isolated from fruit environment: Fermentation characteristics and evolution of phenolic profiles. *Chemosphere* **2021**, *276*, 130090. [CrossRef] [PubMed]
94. Fiorda, F.A.; de Melo Pereira, G.V.; Thomaz-Soccol, V.; Rakshit, S.K.; Pagnoncelli, M.G.B.; de Souza Vandenberghe, L.P.; Soccol, C.R. Microbiological, biochemical, and functional aspects of sugary kefir fermentation-A review. *Food Microbiol.* **2017**, *66*, 86–95. [CrossRef]
95. Jamaludin, M.A.; Amin, A.; Othman, R.; Fadzlillah, N.A.; Kartika, B. Technical review on vinegar fermentation process and physiochemical properties of vinegar product based on Shariah and scientific perspectives. In Proceedings of the 3rd International Halal Conference (INHAC 2016), Selangor, Malaysia, 21–22 November 2016; pp. 491–499.
96. Codex Alimentarius, FAO/WHO. *Standard for Fermented Milks*; CODEX STAN 243-2003; Codex Alimentarius Commission: Rome, Italy, 2011.
97. Gámbaro, A.; McSweeney, M.B. Sensory methods applied to the development of probiotic and prebiotic foods. In *Advances in Food and Nutrition Research*; Elsevier: Amsterdam, The Netherlands, 2020; pp. 295–337.
98. Puttarat, N.; Thangrongthong, S.; Kasemwong, K.; Kerdsup, P.; Taweechotipatr, M. Spray-drying microencapsulation using whey protein isolate and nano-crystalline starch for enhancing the survivability and stability of *Lactobacillus reuteri* TF-7. *Food Sci. Biotechnol.* **2021**, *30*, 245–256. [CrossRef] [PubMed]
99. de Souza, T.S.; Kawaguti, H.Y. Cellulases, hemicellulases, and pectinases: Applications in the food and beverage industry. *Food Bioprocess Technol.* **2021**, *14*, 1446–1477. [CrossRef]
100. Wang, S.; Lu, A.; Zhang, L.; Shen, M.; Xu, T.; Zhan, W.; Jin, H.; Zhang, Y.; Wang, W. Extraction and purification of pumpkin polysaccharides and their hypoglycemic effect. *Int. J. Biol. Macromol.* **2017**, *98*, 182–187. [CrossRef]
101. Genevois, C.; Pieniazek, F.; Messina, V.; Flores, S.; de Escalada Pla, M. Bioconversion of pumpkin by-products in novel supplements supporting *Lactobacillus casei*. *LWT* **2019**, *105*, 23–29. [CrossRef]
102. Alirezalu, K.; Inácio, R.S.; Hesari, J.; Remize, F.; Nemati, Z.; Saraiva, J.A.; Barba, F.J.; Sant'Ana, A.S.; Lorenzo, J.M. Nutritional, chemical, syneresis, sensory properties, and shelf life of Iranian traditional yoghurts during storage. *LWT* **2019**, *114*, 108417. [CrossRef]

separations MDPI

Article

Decoding the Volatile Profile of White Romanian Fetească Wines

Fulvia-Ancuța Manolache [1], Denisa-Eglantina Duță [1], Gabriela Daniela Criveanu-Stamatie [1], Teodora-Alexandra Iordache [1] and Maria-Cristina Todașcă [2,*]

[1] National Research & Development Institute of for Food Bioresources, IBA Bucharest, 6th Dinu Vintila Street, 021102 Bucharest, Romania; fulvia.manolache@bioresurse.ro (F.-A.M.); denisa.duta@bioresurse.ro (D.-E.D.); gabriela.stamatie@bioresurse.ro (G.D.C.-S.); teodora.manasia@bioresurse.ro (T.-A.I.)

[2] "C. Nenitescu" Department of Organic Chemistry, Faculty of Chemical Engineering and Biotechnology, National University of Science and Technology Politehnica, 1–7 Gh. Polizu Street, 011061 Bucharest, Romania

* Correspondence: cristina.todasca@upb.ro

Abstract: The wine's volatilome, most of the time, defines not only its aroma, but also, its major attributes. In the case of wines, the authentication process has become imperative, in light of increased production of alcoholic beverages; consequently, reliable analytical methods have served for it. Therefore, the goal of this research was to establish the global volatile profile of traditional Romanian white wines from Fetească varieties (Fetească albă, Fetească regală) in order to identify its unique characteristics by means of a headspace solid-phase microextraction coupled with gas chromatography analysis (HS-SPME/GC-MS) and *e*-Nose devoted techniques. Statistics was also employed aimed at differentiating the analyzed wine by varietal groups. Consequently, 23 volatile compounds were detected and quantified in 39 Fetească white wine samples originating from various production areas (Muntenia, Oltenia, Transylvania, Banat and Dobrogea), then further classified according to their odor thresholds in five aromatic classes (floral, fruity, sweet, lactic (cheesy) and other). In addition, statistics (Principal Component Analysis (PCA), Hieratical Clustering Analysis (HCA)) were used aiming to differentiate the analyzed varietal groups. The outcomes have pointed out the existence of distinct clusters connected with ethyl esters or alcohol composition and production year, depending on each examined variety.

Keywords: volatile organic compounds; *e*-Nose; fingerprint; PCA discrimination; HS-SPME/GC-MS

1. Introduction

Traditional wines are part of the cultural heritage of every EU country. In the case of grape varieties that are cultivated widely, there are specific expectations in terms of taste and color of the obtained wines (for instance, Cabernet Sauvignon wines, which meet specific taste and color requirements regardless of the geographical origin). Local grape varieties used for traditional winemaking, which are cultivated only in limited areas, are still in need of steadiness in the characteristics of the final product.

Currently, the global wine market is growing, and so is wine production, showing the need for continuous authentication of local, traditional, specific wines. Oenologists and researchers have endeavored to prove the genuineness of conventional wines or trademarks wines (e.g., PDO, CDO or PGI), mainly based on the distinctive fragrance/aroma [1,2].

Wine's aroma is one of the most predominant qualitative traits, with volatile compounds leading to the promotion of some molecules further binding to olfactory receptors, finally affecting the consumer's acceptance [3]. The aroma of wine can be influenced by several factors, including grape variety, climatic conditions (temperature and precipitation), harvesting time or biochemical processes that occur during fermentation/aging/maturation processes. According to the literature, over 800 compounds have been identified in wines, including alcohols, esters, phenols, lactones and aldehydes, with a large variety of these standing for the pursued objective [4].

The HS-SPME/GC-MS technique proved to be a reliable tool for detecting and identifying specific wine components, such as volatile aroma compounds (terpenes, alcohols, ethyl esters, etc.) [5,6]. This technique allows for the analysis of the volatile profile of wines, providing information on the composition and characteristics of the integrated aroma compounds [7]. It has been used to analyze the volatile compounds in different wine varieties and regions, highlighting the sensory attributes and volatiles that contribute to the diversity of wines [8]. Overall, the HS-SPME/GC-MS assay is a valuable analytical technique for the detection and identification of specific wine components, allowing for a better understanding of the aroma profile and sensory attributes of wines. HS/SPME, coupled with GC/MS and statistical analyses, such as analysis of variance, linear discriminant analysis (LDA), PCA, and factor analysis (FA), were employed by Karabagias et al. (2021) to analyze the data obtained from the volatile profile of eight Greek white wines (dry and demi-dry from different growing regions). The findings of the study underscored the effectiveness of volatile compound identification and characterization in clear facilitating the differentiation among the examined wine samples [9].

Lin Zhang et al. (2023) characterized 99 commercially available Chinese wine samples, comprising Cabernet Sauvignon (CS), Syrah (S), and Merlot (M) using HS-SPME/GC-MS. The primary focus was on delineating the volatile fingerprint. Examination of a total of 54 constituents, both qualitatively and quantitatively, revealed significant disparities among specific regions. Notably, ethyl acetate emerged as the most prevalent ethyl ester in CS samples, within the concentration range of 40.8 to 264.3 mg/L. Phenylethyl alcohol, with concentrations ranging from 21.7 to 76.0 mg/L, significantly contributed to defining the overall aroma profile of CS wines. Additionally, various statistical methodologies, including PCA and sensory description assays, complemented the study, highlighting the considerable diversity among the examined samples [10].

In a separate study, Rubén Del Barrio-Galán et al. (2021) tried to identify the volatiles and other specific compounds, such as total organic acids, glycerol, and low-molecular-weight phenolic compounds, present in 73 white and rosé wines originating from diverse Spanish regions with Protected Designations of Origin (PDO). Analysis was conducted using HPLC-DAD and HS-SPME/GC-FID techniques. Principal Component Analysis and least significant differences, employing the ANOVA tool, were indispensable tools in identifying the variables contributing most to the differentiation of wines, based on their PDOs. This analysis revealed distinct differences among the studied wines originating from various PDO areas, despite their geographical proximity [11].

Another import technique used for the volatilome characterization of wine is electronic nose technology. This technique has been used to assess volatile aromatic compounds in agriculture and food applications, including determining products' maturity stage, pest and disease attacks, or for classification purposes [12]. An electronic nose, also known as an *e*-Nose, plays a significant role in wine analysis, offering a modern and efficient alternative to traditional methods. In the wine industry, electronic noses have shown potential for the identification and classification of wine varieties or quality [13]. Electronic noses are designed to detect and analyze volatile organic compounds (VOCs) present in wines. These VOCs contribute to the aroma and flavor profile of the wine. They have been used to analyze and identify distinct varieties of wine, showing significant differences between samples and classifying them according to chemical analysis [14]. Additionally, electronic noses have been used to detect 2,4,6-trichloroanisole (TCA) in wine, which causes cork taint, and to classify cork samples with low TCA concentrations [15].

Electronic nose technology has the potential to be implemented in the wine industry for routine quality assurance purposes. Winemakers could use electronic noses to assess the aging potential and quality of their wines throughout the production process, enabling them to make informed decisions about blending, bottling, and storage conditions.

The objective of this study was to highlight the volatile profile of two traditional Romanian white wines using HS-SPME/GC-MS and electronic nose techniques. Fetească regală (FR) and Fetească albă (FA) wine samples were selected from different wine producing regions of Romania, covering all growing regions, specific for each variety of grapes. This sampling strategy aimed to provide a comprehensive representation of the various geographic regions where these grape varieties are cultivated. Based on the current knowledge, no research has been conducted on Romanian wines covering the selected winemaking areas. In this study, the electronic nose provided a digital fingerprint of the volatile compounds present in the analyzed wine samples.

The selection of HS-SPME GC-MS and *e*-Nose methods used in this paper for wine sample analysis is supported by a wealth of prior research, which has yielded favorable results [7,16,17].

2. Materials and Methods

2.1. Wine Samples

The research study has focused on two native grape varieties, specifically, Fetească albă (FA) and Fetească regală (FR). A total of 39 wine samples originating from various wine-producing regions across Romania were investigated. Namely, 16 samples of FA wine were analyzed, sourced from vineyards from Muntenia, Oltenia, Moldova and Transylvania, whereas the 23 samples of FR wine were cultivated in Muntenia, Oltenia, Moldova, Transylvania, Dobrogea and Banat. These wines were procured from local wineries, and from different production years (Table 1).

Table 1. The winemaking region of Fetească wines analyzed in this study.

Group	Growing Area	Samples Details	Sample Codification	Year of Production	pH *	Alc.% **
Fetească albă	Muntenia	Dealu Mare (PH)	FA_LA_2014	2014	3.29	12.5
		D.O.C—C.M.D Ștefănești	FA_MAR _2017	2017	3.31	12.5
		Urlați (PH)	FA_BAS_2017	2017	3.51	13.0
		Dealu Mare (PH)	FA_NMD_2019	2019	3.39	10.0
		Dealu Mare (PH)	FA_ART_2019	2019	3.54	13.5
		Dealu Mare (PH)	FA_SAH_DM_2020	2020	3.73	14.3
	Oltenia	D.O.C—C.M.D Banu	FA_BM_7ART_2017	2017	3.35	13
		Mărăcine, Craiova	FA_SEG_2017	2017	3.54	12
		D.O.C—C.M.D, Dolj Drăgășani, Vâlcea	FA_BAU_2020	2020	3.27	13
	Moldova	Cotnari (IS)	FA_COL_CT_2011	2011	3.54	13.5
		Dealurile Moldovei (IS)	FA_STR_IS_2019	2019	3.59	12.5
		D.O.C—C.M.D, Cotnari (IS)	FA_COT_2019	2019	3.22	12.0
		D.O.C—C.M.D, Bivolari (IS)	FA_HER_2019	2019	3.57	13.2
	Transylvania	D.O.C—C.M.D, Lechința (BN)	FA_LCH_2018	2018	3.37	12.7
		Valea Ascunsă, Teaca (BN)	FA_LCT_2018	2018	3.17	12.5
		D.O.C—C.M.D, Lechința (BN)	FA_LIL_2020	2020	3.29	12.5
Fetească regală	Transylvania	Dealurile Crișanei (AL)	FR_RAT_2014	2014	3.19	13.9
		D.O.C—C.M.D, Târnave (AL)	FR_JID_2018	2018	3.28	12.0
		D.O.C—C.M.D, Jelba (BN)	FR_LCH_2018	2018	3.29	12.4
		D.O.C—C.M.D, Valea Ascunsă, Teaca (BN)	FR_LCT_2018	2018	3.05	13.5
		D.O.C—C.M.D, Târnava (MS)	FR_TAR_2019	2019	3.36	13.0
		D.O.C—C.M.D, Lechința (BN)	FR_LIL_2019	2019	3.14	12.5

Table 1. *Cont.*

Group	Growing Area	Samples Details	Sample Codification	Year of Production	pH *	Alc.% **
	Muntenia	D.O.C—C.M.D, Lechința (BN)	FR_LCH_2019	2019	3.12	13
		D.O.C—C.M.D, Dealu Mare, Urlați (PH)	FR_URL_2015	2015	3.06	13.5
		D.O.C—C.M.D, Mizil (PH) D.O.C—Dealu Mare, Gura Vadului (PH)	FR_SRF_2017 FR_BUD_2018	2017 2018	3.23 3.05	12 13
		Gura Vadului (PH)	FR_TOH_2018	2018	3.44	11
		D.O.C—C.M.D, Ceptura (PH)	FR_CEP_2019	2019	3.37	13
		Gura Vadului (PH)	FR_DM_2020	2020	3.34	13
		Ștefănești (AG)	FR_MRC_2020	2020	3.22	12
		Dealu Mare, Urlați (PH)	FR_BLG_2020	2020	3.6	13
	Moldova	Bucium (IS)	FR_GRM_2015	2015	3.23	12
		Huși (VS)	FR_AVER_2020	2020	3.01	12.5
	Oltenia	Dealurile Olteniei	FR_STB_2018	2018	2.92	13.5
		Dealurile Segarcea, Dolj	FR_MRM_2018	2018	3.48	12.2
	Dobrogea	Valea lui Traian (CT)	FR_GIT_2019	2019	3.22	13.5
	Banat	Babadag (TL)	FR_HMG_2019	2019	3.43	13.5
		Miniș-Măderat (AR)	FR_MARC_2020	2020	3.47	13.5
		Recaș (TM)	FR_HND_2021	2021	3.30	11.5

* Measured with a pH-meter, ** Alcohol content indicated on the label.

The selected samples cover all winemaking areas where Fetească variety of white wines are produced (Figure 1).

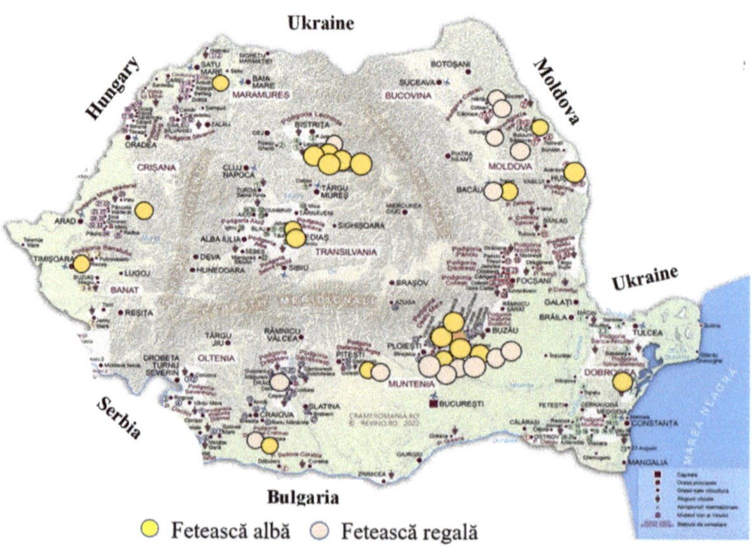

Figure 1. Samples collected in the study highlighting wine regions (picture modified from [18]). Reproduced with permission of Revino Romania SRL; available online: https://www.crameromania.ro/ (accessed on 26 January 2024).

2.2. Chemicals and Reagents

The reagents and materials which were used for the GC-MS analysis are: 2-octanol, the reference solution, from Sigma Aldrich (Saint Louis, MO, USA); Sodium chloride (>99% purity) from LachNer (Neratovice, Czech Republic); TG-Wax MS capillary column (30 m × 0.25 mm × 0.25 μm) from Thermo Fisher Scientific (Waltham, MA, USA); The grey SPME fiber 50/30 μm divinylbenzene/carboxen/polydimethylsiloxane (DVB/CAR/PDMS) was bought from Merck, Supelco (Darmstadt, Germany).

2.3. Methods

2.3.1. HS-SPME/GC-MS Analysis

HS-SPME/GC-MS has been extensively used to identify the volatile profile of wine samples.

a. Wine sample preparation. Volatile compounds were analyzed according to the method of Hanganu A. et.al (2012) with some modifications. Prior to the analysis, the wine samples were stored at a temperature of +4 °C. An aliquot of wine (10 mL) was added to a 20 mL glass vial containing NaCl (3 g); an internal standard solution of 2-octanol (1 mg/mL) was further added. The vial with the mixture was agitated on a shaker with controlled temperature, 40 °C at 200 rpm for 30 min. The SPME fiber, divinylbenzene/carboxen/polydimethylsiloxane (DVB/CAR/PDMS), was inserted into the headspace of the sample vial and kept under constant stirring of the sample, 200 rpm, 30 min, for continuous adsorbtion. Thus, the adsorption of volatile compounds on the fiber was achieved. Different SPME fibers, PDMS, PDMS–DVB, CAR–PDMS and DVB/CAR/PDMS were previously tested by Jordi Torrens et al. in a volatile compound analysis of white and red wines, and the DVB/CAR/PDMS fiber proved to be the best option for the analysis of volatile compounds in wines. This way, adapting the parameters and conditions to the specific equipment, the same fiber was used for the identification of the volatile profile of the analyzed white wines [19].

b. Chromatographic Analysis

Following the incubation and extraction operations, the volatile compounds adsorbed on the fiber were thermally desorbed in the GC-MS injector (FOCUS GC), and were further analyzed using the mass spectrometer POLARISQ, an ion-trap mass spectrometer. The gas chromatograph was equipped with a TG-Wax MS capillary column (30 m × 0.25 mm × 0.25 μm). All injections were performed in split mode following a 1:10 split ratio, using He as carrier gas, at a constant rate of 1 mL/min. The mass range, m/z 25–250, Full Scan Mode in EI (electron ionization) mode, served at compounds identification.

The initial temperature of the FOCUS GC oven was 40 °C for 5 min, then it ramped up to 240 °C, with 5 °C/min and held at 240 °C for 10 min. The desorption and separation of the compounds lasted 55 min. The transfer line was maintained at 240 °C

Identification of the aroma substances was carried out using the detected m/z (details about volatile organic compounds detected via GC-MS analysis, matching with NIST library are shown in the supplementary materials Figure S1), and corroborated the data found in the literature and NIST library [11].

2.3.2. Odor Threshold Value (OTV) and Aroma Series

The odor threshold value (OTV) represents the smallest content of a chemical in the vapory phase possible to be detected by smell. The OTVs measurements were conducted by dividing the determined compound concentration in the wine (using GC-MS assay) by the concentration corresponding to its odor threshold recovered from literature sources; see Equation (1).

$$\mathrm{OTV} = \mathrm{C}/\mathrm{T} \quad (1)$$

where, C is the concentration of volatile determined into the analyzed samples, using GC-MS analysis; and T is the odor threshold reported in the specialized literature sources.

The aroma series comprises a group of volatiles, similar in terms of sensorial descriptors, with their mathematical value being the average of OTVs of aroma constituents [20].

2.3.3. Electronic Nose (*e*-Nose) Technique Sampling

The global volatile composition of white wine samples was analyzed, using the α-Prometheus multi-sensor system for food quality control (*e*-Nose) (FOX 4000, Alpha M.O.S., Toulouse, France) equipped with 18 metal oxide sensors.

Sample preparation consisted of 0.05 g of each sample weighted into 10 mL vials. Vials were capped and incubated at 35 °C in the oven of the equipment for 300 s, under agitation at 250 rpm, to engender liquid vapors to generate the volatile headspace. A total of 1000 μL of the headspace was injected, with a speed of 1000 μL/s, into the measuring chamber of the electronic nose.

Three replications were performed for each wine type, and a statistical analysis was applied to the recorded signals. The data underwent analysis using the Alpha Soft version 2.1 software. Weekly sensor diagnostics checks were conducted utilizing the manufacturer-provided sensor diagnostics kit.

2.3.4. Statistical Analysis

Statistical analysis was carried out using the statistical package XLSTAT, an add-in for Microsoft Excel. PCA and HCA were performed on wine's compositional data, generated from the HS-SPME/GC-MS analysis.

3. Results and Discussion

3.1. HS-SPME/GC-MS

Following the analysis of the 39 white wine samples from FA and FR varieties, a total number of 23 volatile compounds have been identified and quantified: ethyl acetate; butanoic acid, ethyl ester; isovaleric acid, ethyl ester; isopentyl alcohol, acetate; hexanoic acid, ethyl ester; lactic acid, ethyl ester; octanoic acid, ethyl ester; decanoic acid, diethyl ester; butandioic acid, diethyl ester; acetic acid, penthyl ester; propyl alcohol; isobutyl alcohol; isopentyl alcohol; phenylethyl alcohol; 2, 3-butandiol; glycerin; hexanoic acid; octanoic acid; sorbic acid; decanoic acid; ionone; and furfural and acetic acid. Figure 2 presents the gas chromatogram (relative abundance function time) of the volatile compounds extracted from Fetească albă wine produced at Dealu Mare, and in Table 2, the names of volatile compound identified in this sample are presented.

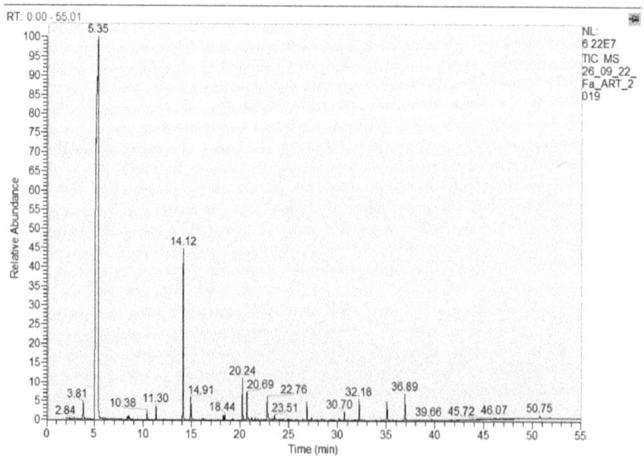

Figure 2. Representation of a GC-MS chromatogram from Fetească Albă sample, SAH_DM.

Table 2. Volatile compounds extracted from Fetească albă sample (SAH_DM_ 2020).

No.crt.	Compounds Name	RT	Molecular Formula	Molecular Weight, g/mol	Integrated Ions
1.	Ethyl acetate	4.02	$C_4H_8O_2$	88.11	43/61/45
2.	Butanoic acid ethyl ester	8.5	$C_6H_{12}O_2$	116.16	71/43/29
3.	Propyl alcohol	8.64	C_3H_8O	60.09	31/29/27
4.	Isobutyl alcohol	10.41	$C_4H_{10}O$	74.13	43/41/42
5.	Acetic acid, pethyl ester	11.39	$C_7H_{14}O_2$	130.19	43/70/42
6.	Isopentyl alcohol	14.15	$C_5H_{12}O$	88.15	29/42/57
7.	Hexanoic acid ethyl ester	14.95	$C_8H_{16}O_2$	144.21	88/29/43
8.	Lactic acid ethyl ester	18.3	$C_5H_{10}O_3$	118.13	45/29/27
9.	2-octanol	20.25	$C_8H_{18}O$	130.23	45/55/41
10.	Octanoic acid ethyl ester	20.7	$C_{10}H_{20}O_2$	172.27	88/101/57
11.	Acetic acid	21.52	$C_2H_4O_2$	60.05	43/45/60
12.	Ionone	23.25	$C_{13}H_{20}O$	192.30	121/93/136
13.	2–3 butandiol	23.52	$C_4H_{10}O_2$	90.12	45/43/57
14.	Decanoic acid, ethyl ester	25.8	$C_{12}H_{24}O_2$	200.32	88/101/29
15.	Butandioic acid diethyl ester	26.85	$C_8H_{14}O_4$	174.20	101/29/129
16.	Hexanoic acid	30.7	$C_6H_{12}O_2$	116.16	60/73/41
17.	Phenylethyl alcohol	32.19	$C_8H_{10}O$	122.16	91/92/65
18.	Octanoic acid	35.05	$C_8H_{16}O_2$	144.21	60/73/43

RT—retention time, min

The volatile compounds identified in all of the 39 analyzed samples were divided in several chemical classes, including esters, alcohols, aldehydes/ketones and organic/carboxylic acids. The number of identified volatile compounds in Fetească variety wines is comparable to those in the literature sources. As proof, in the study performed by Delia Elena Urcan et al. (2015) [21], the authors have identified in Fetească albă and regală varieties, by means of HPLC, 24 bound and free volatile compounds. On the same grape varieties as those we have employed, Colibaba Cintia et al. (2014) [22], using GC, have scored a number of 31 compounds (including terpenes). In addition, George Adrian Cojocaru and Arina Oana (2019) [23] have identified, working on Fetească regală produced by different pre-treatments of the must, 23 volatile compounds, including monoterpenes.

Alcohols, aromatic and aliphatic ones, are usually synthesized via the anabolic pathway of sugar metabolism. Some of these compounds contribute to defining/completing the fruity and flowery notes (e.g., 2-phenylethanol can express aromas such as honey or roses); meanwhile, others promote unpleasant odors (e.g., 2/3-methyl-1-butanol draws undesirable smells and aromas), depending on the concentration retrieved in the wine [24,25]. From the comparative analysis of FA and FR wine samples, the highest amount of alcohols were noted in the Muntenia region, namely isobutyl and isopentyl alcohols, and in the case of FR wines, in addition to Muntenia, in the Transylvania area. 2,3–butanediol and phenylethyl alcohol distinguished themselves among the six determined alcohols, by their contribution to the buttery or floral sent (rose like) [26], being spotted into the same two mentioned regions.

Esters play a fundamental role in defining wine destination by favoring fruity and floral flavors. They are formed, in a small quantity, during grape ripening, and the majority during the fermentation processes (alcoholic and malolactic) [24]. According to the odor descriptors previously reported (Table 3), ethyl butyrate may be responsible for the sour fruit, strawberry, or sweet fruit aroma. The flavor of green apple, fruit, strawberry, and anise was considered to originate from the contribution of ethyl hexanoate. Ethyl octanoate yielded the pineapple, pear and floral aromas, and ethyl lactate helped in pointing out the role of lactic and raspberry odors in defining the global aroma profile [27]. Additionally, ethyl isovalerate and amyl acetate have been observed in Transylvania, Muntenia, Moldova, and even Oltenia (only amyl acetate), in FA samples.

Ethyl isovalerate and amyl acetate were associated with banana or sweet fruit smells. Among the ethyl esters found in FR and FA, the most abundant were hexanoic, octanoic acids ethyl esters, closely followed to butandioic acid diethyl ester.

The presence of carboxylic acids can be detected, in a small part, in the must prior to the alcoholic fermentation; meanwhile, the rest or the highest quantity of these are formed during the metabolic processes conducted by microorganisms [28]. Referring to the volatile fatty acids retrieved for the wine samples, hexanoic acid, followed by octanoic acid, were widespread among the selected areas. The distribution of sorbic acid (pH corrector acid), among FR and FA samples was marked in Moldova and Muntenian regions; meanwhile, decanoic acid had a more visible contribution, particularly in FR wines in Muntenia, Dobrogea, and Moldova.

Acetic acid measured at a maximum concentration level of 243 mg/L (the normal measured level, in wines, is around 300 mg/L), independent of white wine category or regions, indicated a positive influence, especially in reacting with esters producing acetate esters in defining the final quality of wines [29].

Furfural was determined in the Transylvania, Oltenia and Moldova regions for both FA and FR; the same as ionone, which has a more pronounced influence.

In view of making data easier to visualize and compare, a heatmap representation was plotted. The correlations established between the variable compounds, using the heatmap (Figure 3), revealed that, although the identified chemical families are almost the same for both varieties of Fetească wine (FA, FR), there are no visible differences between the samples collected from different growing regions.

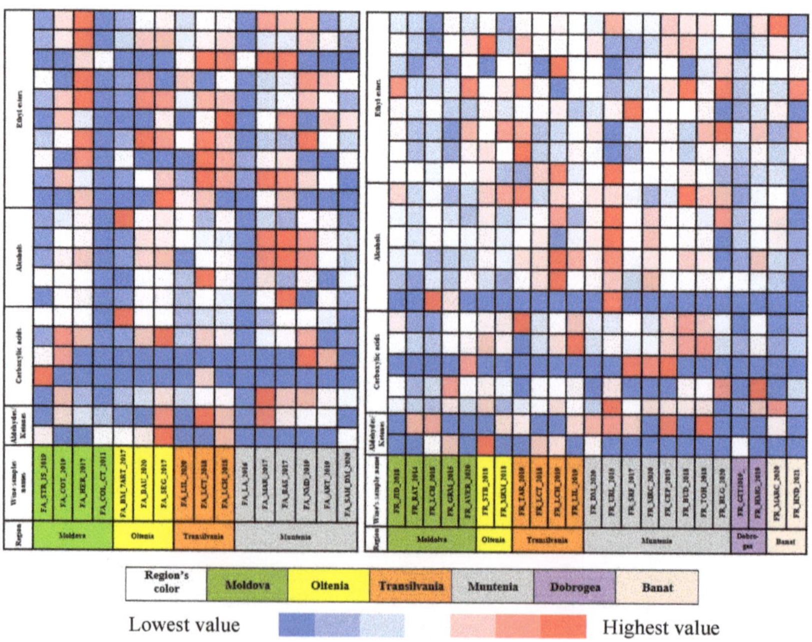

Figure 3. Heatmap representation of Fetească albă and regală wine samples.

3.2. Aroma Correlation between White Wine Samples

Each identified volatile compound imprints a different characteristic on the overall aroma profile of white wines. Thus, in order to assess the influence of the aroma compounds in the studied Fetească white wines, the OTVs were determined. However, a volatile can express multiple aromas, making it difficult to be single-classified; therefore, these aromatic compounds (similar in terms of sensorial descriptors) were further grouped in five aromatic series: floral, fruity, sweet, lactic (cheesy) and other [20].

The gathered data was assembled into a bipartite network consisting of two nodes: the five aromatic series and their corresponding volatile organic compounds, covering five classes, using yEd graph Editor (Figure 4). Further, for the construction of an aroma network dedicated to each category of the analyzed white wines, the OTVs were summed together based on shared aroma characteristics, employing Microsoft Excel, 2013 (Figure 5) [2].

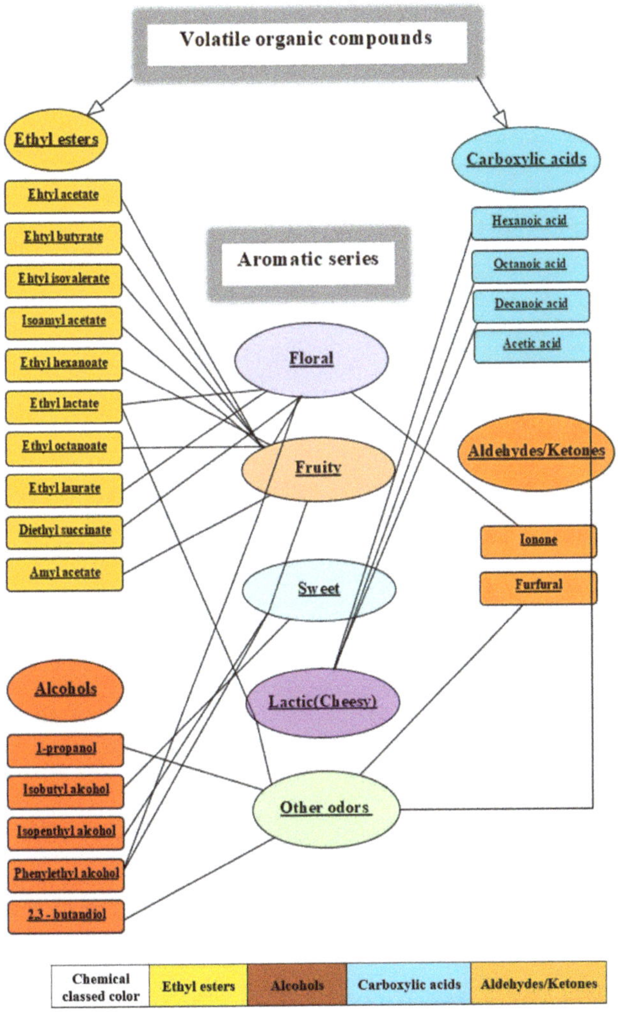

Figure 4. The correlation between aroma descriptors and corresponding volatile compounds identified in the white wine samples.

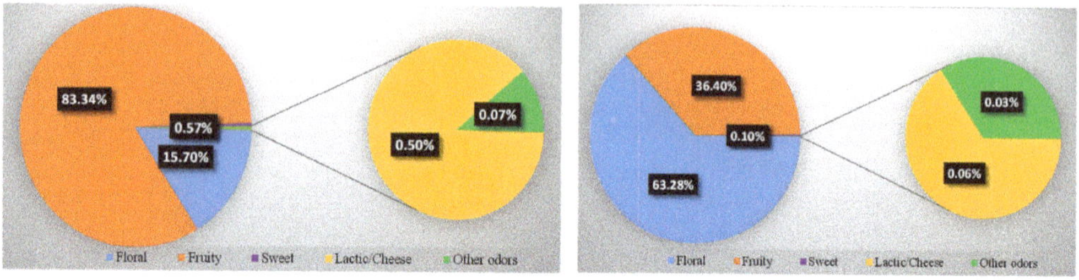

Figure 5. The influence of aroma series on the volatile profile of FA wines (**left**) and FR wines (**right**).

For FA wines, the fruity scent of volatile organic compounds gives the most of the wine's global aroma, shortly followed by floral, in a smaller percentage (16%). Ethyl esters and alcohols are those responsible for this fruity predominant note, their concentration being much higher than their odor thresholds reported by various scientific papers. These findings are similar to the results reported by Sílvia Petronilho et al. (2020). In accordance with the graphical representation of aroma series for FR wines, the floral scent of volatile organic compounds is responsible for the global profile of this wine category. Ethyl octanoate exhibited the biggest OVT, among the fifth compounds responsible for the fruity aroma [2].

Table 3. Detailed data regarding the sensorial attributes of the identified volatiles compounds.

No.	Wine's Volatile Compounds		Wine's Odor Descriptors	Odor Thresholds
	Compound's Chemical Classes	Volatile Compound's Name		
1.	Ethyl esters	Ethyl acetate	Aromatic, Brandy, Grape [30]	5 ppb to 5 ppm [31]
2.		Butanoic acid ethyl ester (ethyl butanoate/ethyl butyrate)	Sour fruit, banana and strawberry flavors, floral and fruity aromas [32]	0.1 to 18 ppb [31]
3.		Isovaleric acid, ethyl ester (ethyl isovalerate)	Apple, Fruit, Pineapple, Sour [3]	0.01 to 0.4 ppb [31]
4.		Isopentyl alcohol, acetate (1-butanol, 3-methyl-, acetate/isoamyl acetate)	Banana, Apple, Glue, Pear [30]	17 μg/L [3]
5.		Hexanoic acid ethyl ester (ethyl hexanoate)	Apple Peel, Brandy, Fruit Gum, Overripe Fruit, Pineapple [30]	0.3 to 5 ppb [31]
6.		Lactic acid, ethyl ester (ethyl lactate)	Cheese, Floral, Fruit, Pungent, Rubber [30]	0.15 ppm [33]
7.		Octanoic acid ethyl ester (ethyl octanoate)	Apricot, Brandy, Fat, Floral, Pineapple [30]	0.002 ppm [33]
8.		Decanoic acid diethyl ester (ethyl laurate)	Floral, Fruity [31]	50 ppm [31]
9.		Butandioic acid diethyl ester (diethyl succinate)	Cotton, Fabric, Floral, Fruit, Wine [30]	10 ppm [31]
10.		Acetic acid, penthyl ester (amyl acetate)	Apple, Banana, Glue, Pear [30]	0.0052 ppm [34]
11.	Alcohols	Propyl alcohol (1-propanol)	Alcohol, ripe fruit flavors [32]	5.7 to 40 ppm [31]
12.		Isobutyl alcohol (2-methyl-1-propanol)	Sweet, Musty [31]	360 ppb to 3.3 ppm [31]
13.		Isopentyl alcohol (isoamyl alcohol/3-methyl-1-butanol)	Mellow, Astringent, whisky-characteristic [32]	250 ppb to 4.1 ppm [31]
14.		Phenylethyl alcohol (2-phenylethanol)	Fruit, Honey, Lilac, Rose, Wine [30]	0.015 ppb to 3.5 ppm [31]
15.		2–3 butandiol	Butter, creamy [27,35]	120 ppm

Table 3. *Cont.*

No.	Wine's Volatile Compounds		Wine's Odor Descriptors	Odor Thresholds
	Compound's Chemical Classes	Volatile Compound's Name		
16.	Carboxylic acids	Hexanoic acid (Caproic acid)	Cheese, Oil, Pungent, Sour [30]	93 ppb to 10 ppm [31]
17.		Octanoic acid (Caprylic acid)	Cheese, Fat, Grass, Oil [30]	10 ppm [31]
18.		Decanoic acid (Capric acid)	Dust, Fat, Grass [3]	1.6 ppm
19.		Acetic acid	Sour and pungent vinegar odor [32]	15 ppm [31]
20.	Aldehydes/ Ketones	Ionone	Violet, Wood, Floral [30,33]	0.9 ppb
21.		Furfural	Almond, Baked Potatoes, Bread, Burnt, Spice [3]	65 ppm

3.3. Electronic Nose (e-Nose) Results

The electronic nose analysis presents pattern recognition for the detection of volatile compounds present in wine samples.

In the FOX equipment, there are three types of sensors currently used: P and T Sensors implemented in the Chambers A and B and LY Sensors used in the Chamber C. If the sensitive layer is placed on a tube of aluminum, the sensor's name possesses the prefix T (example: T30/1, T70/2, etc.) and if the sensitive layer is placed on a plan substrate, the sensor's name possesses the prefix P (example: P30/1, P10/1, P10/2). The LY sensors are metal oxide sensors based on chromium titanium oxide (Cr_2-xTi_xO_{3+y}) and on Tsungsten oxide (WO_3). Wine samples were analyzed using all of the 18 sensors [36]. Each sensor is sensitive and reacts to certain compounds in wine.

Figure 6 illustrates the response of 18 sensors from a sample of Fetească albă: LY2/LG: LY2/GLY2/AA; LY2/Gh; LY2/gCTI: LY2/gCT: T30/1: P10/1: P10/2: P40/1: T70/2; PA2: P30/1: P40/2: P30/2; T40/2: 140/1: TA2 [37].

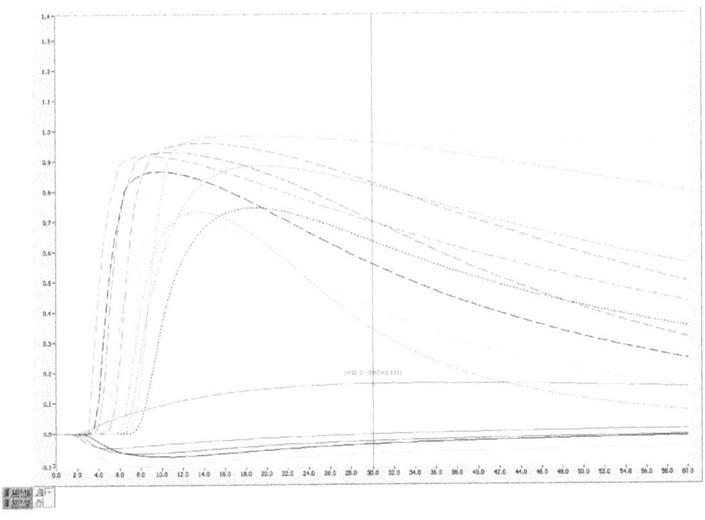

Figure 6. Raw sensor signals of Fetească albă wine samples.

Following the acquisition of the sensor responses for each wine sample, a comprehensive data library was created for all analyzed samples. This data library comprises a table detailing the values obtained from each sensor, aggregating all results of sensors.

Subsequently, statistical analysis, specifically Principal Component Analysis (PCA), was conducted utilizing the dataset from the data library.

To simplify the high dimensionality of the data set, and to understand the variability among the wine samples, all responses of sensors obtained for each analysis of wine using the *e*-Nose were submitted to a PCA to generate a visual representation of the discrimination between the wine samples (Figure 7).

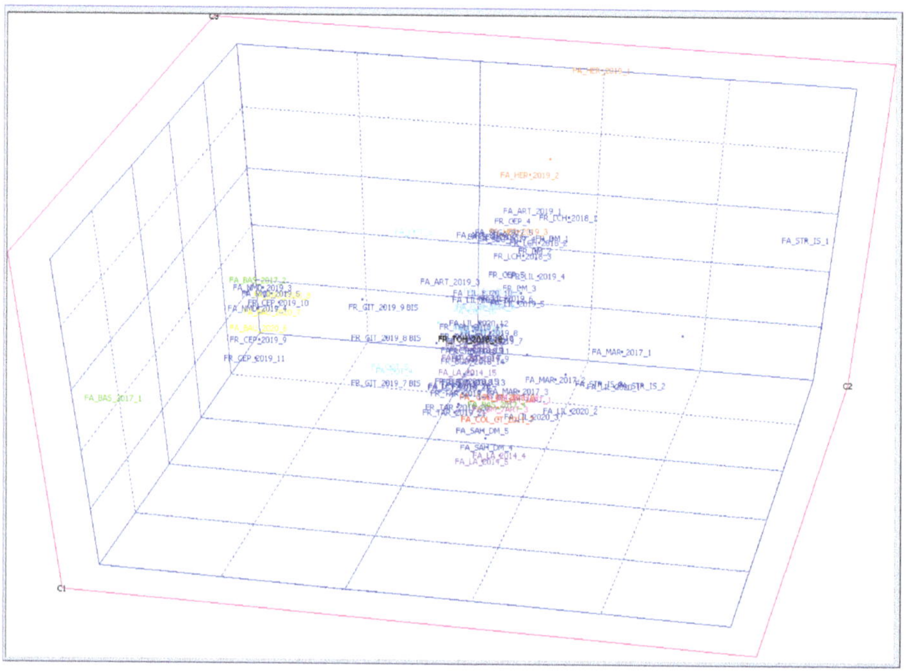

Figure 7. PCA analysis of result obtained with *e*-Nose of analyzed samples.

In the PCA visualization of the electronic nose data, the observed sample lacks major fluctuations in analyzed volatile compounds.

We can conclude that the Fetească albă and Fetească regală samples analyzed present the same characteristics of volatile compounds, even though they were obtained in different winemaking regions and different years. These results could help the producers of these traditional Romanian wines in the process of obtaining them since, currently, the traditional wines are produced in accordance with oenologist experience and taste, because they have no reliable database and lists with the expected characteristics for each variety. Consequently, for one grape variety, several distinct wines are produced from one region to another and, sometimes, from one producer to another in the same region. Therefore, especially for traditional wines, a general list with the expected varietal characteristic in relation to the obtained wine is mandatory. This study aims to establish the main varietal characteristics and, in this way, to help the producers to identify and to highlight the main characteristics of Fetească white wines. Also, the results obtained from the sensors of the *e*-Nose can help establish the authenticity of these wines.

3.4. Statistical Analysis of Wine Samples Based on HS SPME/GC-MS Data

A characterization of unicity in the case of each studied Fetească variety was aimed by means of PCA. The representation of the two main principal components F1/F2 is shown in Figure 8. The analyzed wine samples tend to cluster in accordance with the wine variety

(Fetească albă or Fetească regală), but the system variability is really high, considering the broad area from which the analyzed wine sample were collected and the different production year. The PCA result shows that we have a small explanation of the variation from the first three principal components; therefore, the variation from one sample to another it is significant. This result supports the claim for a clear list of characteristics specific for each Fetească variety. Despite this, several conclusions could be extracted; for example: Fetească albă variety tends to contain more esters in comparison to the Fetească regală variety. The Fetească regală variety contains more glycerin in comparison to Fetească albă variety.

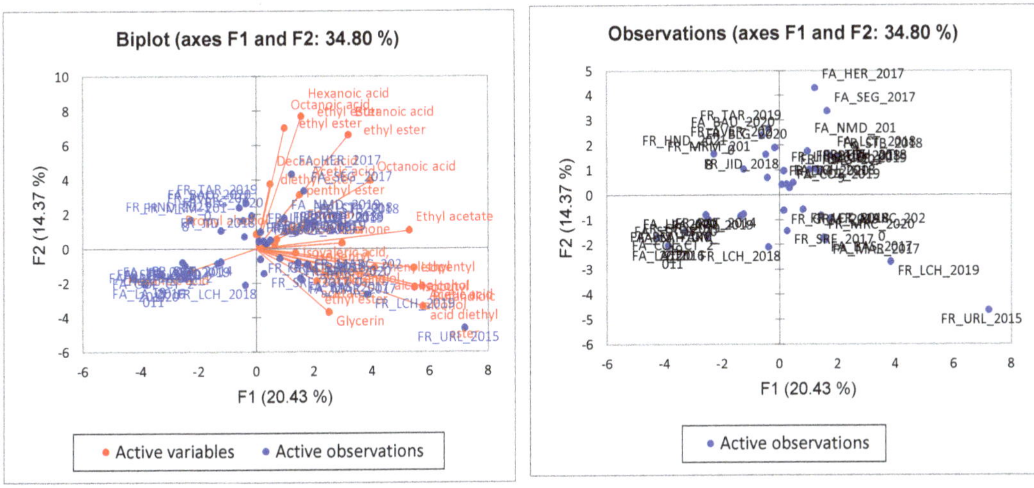

Figure 8. Biplot representation for F1/F2 resulted after PCA analysis of FR and FA samples.

In Figure 9, the PCA analysis was applied to samples from Fetească regală (FR) variety. In this case, the first four principal components explain 70% of the system variation, which has an important value when samples from all producing regions are included in the study. Of course, if we consider only the samples for each region of the country, the results in terms of variation explanation are improved, but the idea was to try to find the general characteristics of the wine variety across all production areas. In the case of individual variety characterization, the system variability explanation was improved. Therefore, it can be noticed in Figure 9 that Fetească regală samples tend to differentiate based on the production year, in accordance with the ester compositions (2019 and 2020) and in accordance with alcohol concentrations (2018).

In Figure 10, the results for the PCA analysis applied to the Fetească albă (FA) wines is illustrated. In this case, the system variance is explained better than in the case of Fetească regală wines, with this proving a more comparable processing approach of the producers in the case of this Fetească variety. In the case of Fetească albă wines, the clustering tendency is generated by the carboxylic acid composition (2019 and 2020), and the esters and alcohols composition for the other cultivars (2018 and 2017). Those results are in accordance with the observations made by the *e*-Nose.

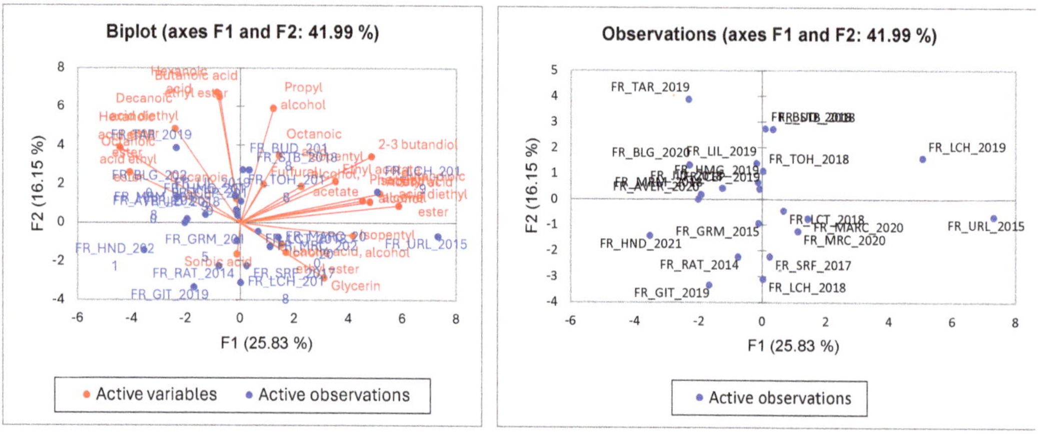

Figure 9. Biplot representation for F1/F2 resulted after PCA analysis of FR wine samples.

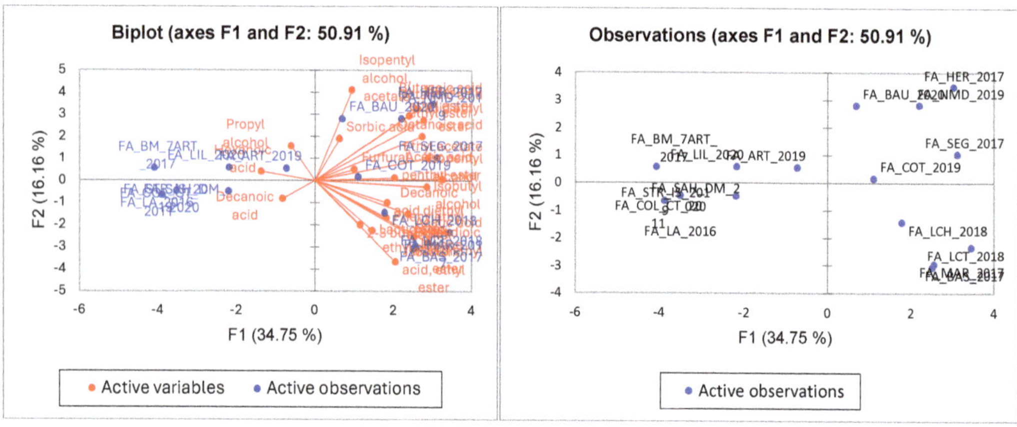

Figure 10. Biplot representation for F1/F2 resulted after PCA analysis of FA wine samples.

The Hierarchical Clustering Analysis of the FA wine samples, presented in Figure 11, shows four separate clusters corresponding to different samples processing, and not in accordance with the different geographical regions. Therefore, the influence of the oenological techniques tends to differentiate the samples more than the wine growing area.

The same results are shown in Figure 12 by the FR samples, which clustered in five distinct clusters, which supports the results obtained by PCA.

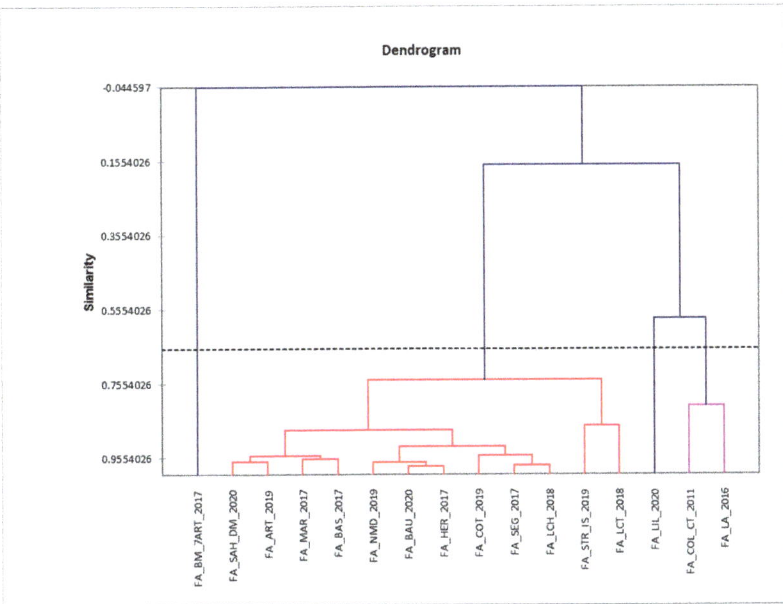

Figure 11. Dendrogram of the FA wine samples analyzed.

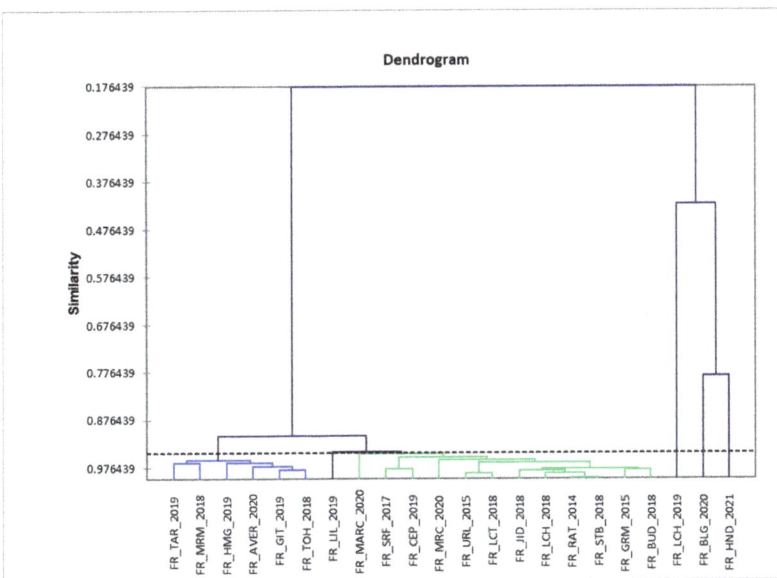

Figure 12. Dendrogram of the FR wine samples analyzed.

4. Conclusions

A total of 39 samples of white wine from Romanian Fetească varieties were analyzed, and 23 volatile compounds were quantified using the HS-SPME/GC-MS and *e*-Nose techniques in order to reveal the authenticity attributes. In this regard, the volatile profile defined by the flavor compounds was depicted. The 23 chemical compounds detected were classified into four classes: esters, alcohols, aldehydes/ketones and organic/carboxylic acids. According to the odor threshold detailed in the literature, the compounds were further divided in five aroma series, in this way being easily to visualize the imprint of each category on Fetească albă and Fetească regală wine samples. The data collected via *e*-Nose software has scored the authenticity of these wines bearing almost the same characteristics of volatile compounds, despite the different winemaking regions and production years. Principal Component Analysis finally confirmed *e*-Nose outputs that, despite the great variability of samples (e.g., different growing regions, different production years), the samples tend to cluster according to wine variety and the processing techniques.

The results obtained in the present study are a useful tool when a specific handling techniques will be applied in the white Fetească grape variety processing to obtain wines with similar individualities, which could be easily identify by oenologist when the wine quality is established. In the meanwhile, those results are important for when the variety authentication of Fetească white wines is targeted.

Supplementary Materials: The following supporting information can be downloaded at: https://www.mdpi.com/article/10.3390/separations11050141/s1, Figure S1: Spectrum of volatile organic compounds detected via GC-MS analysis, matching with NIST library.

Author Contributions: Conceptualization M.-C.T. and F.-A.M.; T.-A.I. and F.-A.M. writing—original draft preparation; formal analysis, F.-A.M., D.-E.D. and G.D.C.-S.; writing—review and editing, M.-C.T., F.-A.M. and T.-A.I.; project administration, F.-A.M. All authors have read and agreed to the published version of the manuscript.

Funding: This research was funded by the Ministry of Research, Innovation and Digitalization, contract 22N/2019, grant number PN 19 02 04 02, "Experimental research regarding the analysis of some traditional Romanian wines by spectral methods and chromatography's in order to establish the unique characteristics for uniqueness" and contract 17 PFE (2021–2024).

Data Availability Statement: The data presented in this study are available on request from the corresponding author.

Conflicts of Interest: The authors declare no conflicts of interest.

References

1. Khakimov, B.; Bakhytkyzy, I.; Fauhl-Hassek, C.; Engelsen, S.B. Non-Volatile Molecular Composition and Discrimination of Single Grape White of Chardonnay, Riesling, Sauvignon Blanc and Silvaner Using Untargeted GC–MS Analysis. *Food Chem.* **2022**, *369*, 130878. [CrossRef] [PubMed]
2. Petronilho, S.; Lopez, R.; Ferreira, V.; Coimbra, M.A. Supplementary Material Revealing the Usefulness of Aroma Networks to Explain Wine Aroma Properties: The Case Study of Portuguese Wines by Chemical Families, Odor Threshold, Content, and Odor Active Value (OAV). *Molecules* **2020**, *25*, 272. [CrossRef] [PubMed]
3. Han, S.; Yang, J.; Choi, K.; Kim, J.; Adhikari, K.; Lee, J. Chemical Analysis of Commercial White Wines and Its Relationship with Consumer Acceptability. *Foods* **2022**, *11*, 603. [CrossRef] [PubMed]
4. Luzzini, G.; Slaghenaufi, D.; Ugliano, M. Volatile Compounds in Monovarietal Wines of Two Amarone Della Valpolicella Terroirs: Chemical and Sensory Impact of Grape Variety and Origin, Yeast Strain and Spontaneous Fermentation. *Foods* **2021**, *10*, 2474. [CrossRef] [PubMed]
5. Jiang, K.; Xu, K.; Wang, J.; Meng, F.; Wang, B. Based on HS-SPME-GC-MS Combined with GC-O-MS to Analyze the Changes of Aroma Compounds in the Aging Process of Citri Reticulatae Pericarpium. *Food Biosci.* **2023**, *54*, 102798. [CrossRef]
6. Gu, W.; Wei, Y.; Fu, X.; Gu, R.; Chen, J.; Jian, J.; Huang, L.; Yuan, C.; Guan, W.; Hao, X. HS-SPME/GC×GC-TOFMS-Based Flavoromics and Antimicrobial Properties of the Aroma Components of Zanthoxylum Motuoense. *Foods* **2023**, *12*, 2225. [CrossRef]
7. Rossi, L.; Foschi, M.; Biancolillo, A.; Maggi, M.A.; D'Archivio, A.A. Optimization of HS-SPME-GC/MS Analysis of Wine Volatiles Supported by Chemometrics for the Aroma Profiling of Trebbiano d'Abruzzo and Pecorino White Wines Produced in Abruzzo (Italy). *Molecules* **2023**, *28*, 1534. [CrossRef] [PubMed]
8. Piergiovanni, M.; Carlin, S.; Lotti, C.; Vrhovsek, U.; Mattivi, F. Development of a Fully Automated Method HS-SPME-GC-MS/MS for the Determination of Odor-Active Carbonyls in Wines: A "Green" Approach to Improve Robustness and Productivity in the Oenological Analytical Chemistry. *J. Agric. Food Chem.* **2024**, *72*, 1995–2007. [CrossRef]

9. Karabagias, I.K.; Karabagias, V.K.; Badeka, A.V. Volatilome of White Wines as an Indicator of Authenticity and Adulteration Control Using Statistical Analysis. *Aust. J. Grape Wine Res.* **2021**, *27*, 269–279. [CrossRef]
10. Zhang, L.; Liu, Q.; Li, Y.; Liu, S.; Tu, Q.; Yuan, C. Characterization of Wine Volatile Compounds from Different Regions and Varieties by HS-SPME/GC-MS Coupled with Chemometrics. *Curr. Res. Food Sci.* **2023**, *6*, 100418. [CrossRef]
11. Del Barrio-Galán, R.; Valle-Herrero, H.d.; Bueno-Herrera, M.; López-de-la-Cuesta, P.; Pérez-Magariño, S. Volatile and Non-Volatile Characterization of White and Rosé Wines from Different Spanish Protected Designations of Origin. *Beverages* **2021**, *7*, 49. [CrossRef]
12. Fuentes, S.; Viejo, C.G. Novel Use of E-Noses for Digital Agriculture, Food, and Beverage Applications. In *Nanotechnology-Based E-Noses: Fundamentals and Emerging Applications*; Gupta, R.K., Nguyen, T.A., Bilal, M., Ahmadi, M., Eds.; Woodhead Publishing: Shaston, UK, 2023; pp. 415–432.
13. Celdrán, A.C.; Oates, M.J.; Molina Cabrera, C.; Pangua, C.; Tardaguila, J.; Ruiz-Canales, A. Low-Cost Electronic Nose for Wine Variety Identification through Machine Learning Algorithms. *Agronomy* **2022**, *12*, 2627. [CrossRef]
14. Erwanto, D.; Wahyudi, D.; Fatkhur Rizal, R. Sistem Electronic Nose Untuk Deteksi Aroma Pada Fasilitas Kamar Mandi Berbasis IoT. *J. Zetroem* **2023**, *5*, 43–50. [CrossRef]
15. Meléndez, F.; Arroyo, P.; Gómez-Suárez, J.; Palomeque-Mangut, S.; Suárez, J.I.; Lozano, J. Portable Electronic Nose Based on Digital and Analog Chemical Sensors for 2,4,6-Trichloroanisole Discrimination. *Sensors* **2022**, *22*, 3453. [CrossRef]
16. Pati, S.; Tufariello, M.; Crupi, P.; Coletta, A.; Grieco, F.; Losito, I. Quantification of Volatile Compounds in Wines by HS-SPME-GC/MS: Critical Issues and Use of Multivariate Statistics in Method Optimization. *Processes* **2021**, *9*, 662. [CrossRef]
17. Jin, X.; Wu, S.; Yu, W.J.; Xu, X.; Huang, M.; Tang, Y.; Yang, Z. Wine Authentication Using Integration Assay of MIR, NIR, E-Tongue, HS-SPME-GC-MS, and Multivariate Analyses: A Case Study for a Typical Cabernet Sauvignon Wine. *J. AOAC Int.* **2019**, *102*, 1174–1180. [CrossRef]
18. Harta A3 Podgorii Spate Romana 2023. Available online: https://www.crameromania.ro/upload/HartaA3podgoriispate%20romana%202023.png (accessed on 26 January 2024).
19. Torrens, J.; Riu-Aumatell, M.; López-Tamames, E.; Buxaderas, S. Volatile Compounds of Red and White Wines by Headspace-Solid-Phase Microextraction Using Different Fibers. *J. Chromatogr. Sci.* **2004**, *42*, 310–316. [CrossRef]
20. Dumitriu, G.D.; Sánchez-Suárez, F.; Peinado, R.A.; Cotea, V.V.; de Lerma, N.L.; Gabur, I.; Simioniuc, V. Metabolomics of Red Wines Aged Traditionally, with Chips or Staves. *Foods* **2024**, *13*, 196. [CrossRef]
21. Urcan, D.E.; Lung, M.L.; Giacosa, S.; Torchio, F.; Ferrandino, A.; Vincenzi, S.; Rió Segade, S.; Pop, N.; Rolle, L. Phenolic Substances, Flavor Compounds, and Textural Properties of Three Native Romanian Wine Grape Varieties. *Int. J. Food Prop.* **2016**, *19*, 76–98. [CrossRef]
22. Colibaba, C.; Cotea, V.; Liliana, R.; Marius, N.; Vararu, F.; Luchian, C. Studies of the Influence of Some Conditioning Treatments on Some Volatile Compounds in Fetească Albă Wines. *Sci. Pap. Ser. Hortic.* **2014**, *55*, 181–186.
23. Cojocaru, G.A.; Antoce, A.O. Influence of Glutathione and Ascorbic Acid Treatments during Vinification of Feteasca Regala Variety and Their Antioxidant Effect on Volatile Profile. *Biosensors* **2019**, *9*, 140. [CrossRef] [PubMed]
24. Liu, S.; Lou, Y.; Li, Y.; Zhao, Y.; Laaksonen, O.; Li, P.; Zhang, J.; Battino, M.; Yang, B.; Gu, Q. Aroma Characteristics of Volatile Compounds Brought by Variations in Microbes in Winemaking. *Food Chem.* **2023**, *420*, 136075. [CrossRef] [PubMed]
25. Petretto, G.; Urgeghe, P.P.; Cabizza, R.; Del Caro, A. Evaluation of Volatile and Chemical Profile of Sherry-like White Wine Vernaccia Di Oristano from Sardinia by Comprehensive Targeted and Untargeted Approach. *Eur. Food Res. Technol.* **2023**, *249*, 1887–1897. [CrossRef]
26. Cordente, A.G.; Nandorfy, D.E.; Solomon, M.; Schulkin, A.; Kolouchova, R.; Francis, I.L.; Schmidt, S.A. Aromatic Higher Alcohols in Wine: Implication on Aroma and Palate Attributes during Chardonnay Aging. *Molecules* **2021**, *26*, 4979. [CrossRef] [PubMed]
27. Tao, Y.-S.; Li, H. Active Volatiles of Cabernet Sauvignon Wine from Changli County. *Health* **2009**, *1*, 176–182. [CrossRef]
28. Avram, V.; Floare, C.G.; Hosu, A.; Cimpoiu, C.; Măruţoiu, C.; Moldovan, Z. Characterization of Romanian Wines by Gas Chromatography–Mass Spectrometry. *Anal. Lett.* **2015**, *48*, 1099–1116. [CrossRef]
29. The Role of Acetic Acid in Wine. Available online: https://www.calwineries.com/learn/wine-chemistry/wine-acids/acetic-acid (accessed on 1 February 2024).
30. Flavor Ingredient Library. Available online: https://www.femaflavor.org/flavor-library (accessed on 15 January 2024).
31. Chemical Book. Available online: https://www.chemicalbook.com/ (accessed on 15 January 2024).
32. Cao, W.; Shu, N.; Wen, J.; Yang, Y.; Jin, Y.; Lu, W. Characterization of the Key Aroma Volatile Compounds in Nine Different Grape Varieties Wine by Headspace Gas Chromatography–Ion Mobility Spectrometry (HS-GC-IMS), Odor Activity Values (OAV) and Sensory Analysis. *Foods* **2022**, *11*, 2767. [CrossRef] [PubMed]
33. Jiang, B.; Zhang, Z. Volatile Compounds of Young Wines from Cabernet Sauvignon, Cabernet Gernischet and Chardonnay Varieties Grown in the Loess Plateau Region of China. *Molecules* **2010**, *15*, 9184–9196. [CrossRef] [PubMed]
34. Amyl Acetate. Available online: https://pubchem.ncbi.nlm.nih.gov/#query=amyl%20acetate (accessed on 15 January 2024).
35. Hao, N.; Liu, J.; Wang, F.; Cao, J.; Chen, Q.; He, L.; Guan, X.; Liu, S.; Shi, K. Malolactic Fermentation Performance of Indigenous Oenococcus Oeni Strains from Shaanxi Wine Region (China) and Their Mutants on Pinot Noir and Chardonnay Wines. *Lwt* **2023**, *185*, 115170. [CrossRef]

36. Plawiak, P.; Maziarz, W. Comparison of Artificial Intelligence Methods on the Example of Tea Classification Based on Signals from E-Nose Sensors. *Adv. Signal Process.* **2013**, *1*, 19–32. [CrossRef]
37. Pławiak, P.; Rzecki, K. Approximation of Phenol Concentration Using Computational Intelligence Methods Based on Signals from the Metal-Oxide Sensor Array. *IEEE Sens. J.* **2015**, *15*, 1770–1783. [CrossRef]

foods

MDPI

Article

Characterization of Key Odorants in Lushan Yunwu Tea in Response to Intercropping with Flowering Cherry

Yinxiang Gao [1,†], Zhiyong Lei [1,†], Jigang Huang [2], Yongming Sun [3], Shuang Liu [2], Liping Yao [1], Jiaxin Liu [2], Wenxin Liu [1], Yanan Liu [1,*] and Yan Chen [2,*]

1 Institute of Jiangxi Oil-Tea Camellia, Jiujiang University, Jiujiang 332005, China; 4120001@jju.edu.cn (Y.G.)
2 Jiujiang Agricultural Technology Extension Center, Jiujiang 332000, China
3 Jiangxi Institute of Red Soil and Germplasm Resources, Nanchang 330046, China
* Correspondence: liuyanan81211@163.com (Y.L.); jjzz8028@163.com (Y.C.)
† These authors contributed equally to this work.

Abstract: Lushan Yunwu tea (LSYWT) is a famous green tea in China. However, the effects of intercropping tea with flowering cherry on the overall aroma of tea have not been well understood. In this study, headspace solid-phase microextraction (HS-SPME) coupled with gas chromatography–mass spectrometry (GC–MS) was used for analysis. A total of 54 volatile compounds from eight chemical classes were identified in tea samples from both the intercropping and pure-tea-plantation groups. Principal component analysis (PCA), orthogonal partial least-squares discriminant analysis (OPLS-DA), and odor activity value (OAV) methods combined with sensory evaluation identified cis-jasmone, nonanal, and linalool as the key aroma compounds in the intercropping group. Benzaldehyde, α-farnesene, and methyl benzene were identified as the main volatile compounds in the flowering cherry using headspace solid-phase microextraction/gas chromatography–mass spectrometry (HS-SPME/GC–MS). These findings will enrich the research on tea aroma chemistry and offer new insights into the product development and quality improvement of LSYWT.

Keywords: Lushan Yunwu tea; flowering cherry; volatile compounds; key aroma compounds; HS-SPME/GC–MS

1. Introduction

The tea plant (*Camellia sinensis*) is a shrub or tree distributed in tropical and subtropical regions that belongs to the genus *Camellia* in the Camellia family. The forests of Southwest China are the center of origin of tea plants [1]. Tea leaves are a popular source of tea. The growth and quality of tea leaves are affected by soil nutrition, environmental factors, cultivation methods, and other factors [2,3]. Traditional tea plantations have gradually encountered problems such as soil acidification, increased pest and disease pressure, and decreased yields [4–6]. The intercropping model of growing tea trees and other species can maximize land resource benefits, enhance the resistance of tea trees to adversity, and improve the biodiversity of tea gardens, thereby enhancing their comprehensive benefits for ecology, landscapes, and economies [7–10].

Intercropping models have been widely used in tea cultivation to improve the quality of tea. Recent studies found that many intercropping plants, including herbaceous plants, shrubs, and woody plants, can promote the growth of tea plants and improve their quality [1,11,12]. Chestnut–tea intercropping can reduce the environment and soil temperatures and increase the air and soil moisture in tea plantations [1]. Intercropping tea with soybean, corn, and rubber can also retain soil moisture and reduce surface erosion [13,14]. Soybean–tea intercropping can affect the content of available nitrogen in the soil and significantly increase the content of ammonium nitrogen [15]. Zhang et al. [16] found that volatile compounds released by interplanting *Cassia tora* and *Leonurus artemisia* in tea plantations can repel green tea leafhoppers. Chinese chestnut–tea intercropping was found to decrease the content of tea polyphenols and the bitter taste of tea and increase the content of amino acids and the freshness of green tea [1]. Soybean–tea intercropping was found to increase the content of non-ester-type catechins, total catechins, and theanine; decrease the content of ester-type catechins; and improve tea quality [17]. Recent studies showed that pea–tea intercropping can improve tea quality by regulating amino acid metabolism and flavonoid biosynthesis [12].

Studies showed that interplanting tea with other plants can promote the biosynthesis of aroma compounds, thereby improving the aroma of tea. Intercropping chestnut trees with tea increases the aroma of green tea by increasing phenylalanine-related metabolites and jasmine lactones [1]. Yang et al. [18] found that the intercropping of *Prunus salicina lindl.* in tea gardens improved the chestnut and floral aroma of green tea. The planting patterns of tea trees intercropped with other plants can be traced back to the Eastern Jin Dynasty in ancient China [11]. During the Song Dynasty, "Tea Garden" recorded a tea plant company with plums belonging to the Rosaceae family. In a study on companion plants in ancient, forest, and terraced tea gardens, Rosaceae was found to be one of the dominant families [11]. Flowering cherry (*Cerasus* sp.) belongs to the genus *Prunus* of the Rosaceae family. It is a famous ornamental woody plant that generally flowers in April and is cultivated worldwide [19].

Lushan Yunwu tea (LSYWT) is one of the top ten famous green teas in China, originating from Lushan Mountain, Jiangxi Province, China [20]. LSYWT has a history of more than 1800 years. As a new cultivation model, intercropping tea with flowering cherry has received little attention regarding its influence on tea aroma. In this study, 54 volatile compounds were identified using HS-SPME/GC–MS from the tea samples of the intercropping group (EG) and the pure tea tree group (CG). Through principal component analysis (PCA), orthogonal partial least-squares discriminant analysis (OPLS-DA), and odor activity value (OAV) methods, combined with sensory evaluation, we found that the EG green tea exhibited a floral aroma, while the CG did not possess a floral aroma. We also identified volatile compounds in flowering cherry using HS-SPME/GC–MS analysis and investigated the potential factors contributing to the production of key fragrance compounds, such as EG. The results of this study provided information on the aroma-active compounds in Lushan Yunwu green tea intercropped with flowering cherry. The findings of our study provide meaningful guidance for aroma quality control in Lushan Yunwu green tea.

2. Materials and Methods

2.1. Plant Materials and Growth Conditions

Fresh tea leaves from the cultivar "Population" with one leaf and one bud were collected from the Hailu Tea Expo Garden (116°3′38″ E, 29°34′114″ N) in Jiujiang, Jiangxi on 2 April 2023. The line spacing of the tea trees was 1.5 m, crowns were 70–90 cm, and plant height was 1–1.2 m. The spacing between the flowering cherry trees was 4 m, row spacing was 6 m, the crown breadth was 4–5 m, and the tree height was 4.5–5.5 m. Two sample plots were set up, one of which involved interplanting flowering cherry trees with tea trees. Tea trees within a 3 m radius of a well-growing cherry blossom tree were selected, and one bud and one leaf were randomly collected from each tea tree. The second sample comprised pure tea trees. Tea trees with good growth were selected, and one bud and leaf were randomly collected from each tea tree. LSYWT was produced using traditional processing methods, including fixing, rolling, and roasting, by an inheritor of the intangible cultural heritage of a tea master. The tea samples were stored at −40 °C until further analysis was performed.

2.2. Sensory Assessment

The tea sensory evaluation adopted a quantitative descriptive analysis (QDA) method [21]. The team members systematically evaluated the typical aroma types and descriptive vocabulary of green tea before performing a quantitative descriptive analysis. The evaluation panel consisted of four males and five females, aged 20–44 years. During the preparatory phase, the aroma attributes of the samples were discussed until all members agreed on their properties. Flowery, chestnut, tender, fresh, clear, and grass were selected as indicators for the quantitative description. The reference system was established and training was conducted according to the method of Wang et al. [22] until the members were familiar with the odor characteristics and intensity of these five aroma attributes [23].

Quantitative descriptive analysis was used to evaluate the intensity of various aroma characteristics with an intensity scale ranging from 1 to 9, where 1 represents the weakest aroma and 9 represents the strongest aroma. Tea samples (3 g) were added to 150 mL of boiling water and the tea infusion was obtained after 4 min. Subsequently, the tea soup was poured into a matching bowl and the cylindrical cup and matched bowl were randomly numbered for blind review. The evaluation panel repeated the process for each sample thrice, and the results were averaged. Sensory evaluation data collection and use were in accordance with a Jiujiang University Human Ethics application, ID Number: JJUM20240090.

2.3. Chemicals

β-Myrcene (≥90.0%) and cis-3-Hexenyl hexanoate (98%) were purchased from Macklin Biochemical Co. Ltd. (Shanghai, China). Benzylalcohol (≥99.5%), Linalool (98%), Nonanal (96%), α-Terpineol (>95.0%), Geraniol (≥99.0%), cis-Jasmone (98%), α-Farnesene (98%), and methyl palmitate (99%) were obtained from Aladdin Biochemical Co. Ltd. (Shanghai, China). Linalool oxide (pyranoid) and trans-Nerolidol were purchased from Yuanye Bio-Technology Co. Ltd. (Shanghai, China). *n*-Alkane mixtures of C7–C40 were supplied by Sigma-Aldrich Trading Co. Ltd. (Shanghai, China) and used to determine retention indices (RI). Ethyl decanoate was used as the internal standard and purchased from Aladdin Biochemical Co. Ltd. (Shanghai, China). Ultra-pure water was obtained using a Milli-Q purification system (Millipore, Bedford, MA, USA). Distilled water was purchased from the Hangzhou Wahaha Group Co. Ltd. (Hangzhou, China).

2.4. Extraction of Volatile Compounds from Tea Using HS-SPME

Volatile compounds were extracted using HS-SPME with a divinylbenzene/carboxen/polydimethylsiloxane coating fiber (50/30 μm; 1 cm) purchased from Supelco (Bellefonte, PA, USA) [24]. All tea samples were ground, passed through 40 mesh, and sealed for future use. Briefly, ground tea powder (1.5 g) was accurately weighed into a 20 mL headspace flask. Then, NaCl (0.5 g) was added, followed by 20 μL of ethyl decanoate (20 μg/mL) as an internal standard, and 5.25 mL of distilled water. Subsequently, the autosampler was stabilized to ensure injection stability. The flask was incubated at 70 °C for 30 min with shaking (5 s on and 2 s off). The coating fiber was then inserted into the headspace of the flask for extraction at 70 °C for 30 min. Finally, the fiber was inserted into the injector port of the gas chromatography–mass spectrometer (GC–MS) for desorption at 250 °C for 5 min, after which GC–MS separation and identification were performed [24].

2.5. Gas Chromatography–Mass Spectrometry Analysis

GC–MS was performed using a Thermo TSQ 8000 spectrometer equipped with a TriPlus RSH three-in-one automatic sampler TPR (Thermo Fisher Scientific, Waltham, MA, USA). A TG-5MS gas chromatography column with 30 m × 0.25 mm × 0.25 μm film thickness (Thermo Fisher, USA) was used for separation and high-purity helium (≥99.999%) was used as carrier gas. The constant flow rate was 1.0 mL/min, and the injector temperature was 280 °C. The mass spectrometry conditions were as follows: electron bombardment ion source (EI) temperature of 300 °C, mass interface temperature of 280 °C, and ionization voltage of 70 eV. The mass-scanning range was 33 to 550 m/z. The initial oven temperature was maintained at 40 °C for 2 min and then increased at 5 °C/min to 85 °C, where it was held for 2 min. Subsequently, the temperature was increased at a rate of 2 °C/min to 110 °C and held for 2 min, then increased at a rate of 4 °C/min to 220 °C, where it was held for 2 min, and finally increased at 5 °C/min to 250 °C and held for 10 min.

2.6. Identification and Quantitation of Volatiles

Peak identification was performed by searching the mass spectra in the National Institutes of Standards and Technology Mass Spectrometry (NIST MS) library and comparing their retention indices (RIs) with published data using the same capillary column [25]. An *n*-alkane standard (10 μg/mL) was injected under the same GC conditions as the external standard. The volatile compounds were preliminarily identified using the similarities between the RI results, mass spectra, and published data. The relative concentrations of the released volatiles were calculated using the following formula:

$$Wi = \frac{A_i \times W_s}{A_s}$$

Wi is the content of the target volatile concentration to be measured (μg/L), A_i is the peak area of the target volatile concentration to be measured, W_s is the concentration of the internal standard (μg/L), and A_s is the peak area of the internal standard in the sample.

2.7. OAV Calculation

OAVs were used to assess the contributions of volatile compounds to the aroma of tea samples [26]. OAVs were calculated using the following equation:

$$OAV_i = \frac{C_i}{OT_i}$$

C_i is the absolute content of volatile compounds and OT_i is the odor threshold of volatile compounds in water.

2.8. Statistical Analysis

PCA and OPLS-DA were performed using SIMCA14.1 (Umetrics, Umea, Sweden). Significant differences between samples were assessed using one-way analysis of variance (ANOVA) with SPSS (version 20.0; SPSS Inc., Chicago, IL, USA). TBtools (version 2.056, accessed on 20 January 2024, https://github.com/CJ-Chen/TBtools) was used for heat mapping and hierarchical analysis. Data significance analysis was performed using the Student's *t*-test and significant differences ($p < 0.05$) are denoted by *.

3. Results and Discussion

3.1. Aroma Characteristics of Teas in the Intercropping Group (EG) and Pure Tea Tree Group (CG)

Aroma is an important indicator for evaluating tea quality [27]. In this study, we found differences between the aroma profiles of EG and CG green tea using sensory descriptive analysis. As shown in the radar plot in Figure 1, the "flowery" aroma of EG green tea was significantly enhanced ($p < 0.01$), whereas the "fresh and clear" aroma was significantly weakened ($p < 0.01$); the "grass", "chestnut", and "tender" aromas were not significantly different ($p > 0.05$).

3.2. Identification and Quantification of Volatile Compounds in EG and CG Green Tea

To further explore why the floral aromas of the EG tea samples were more prominent than those of the CG samples, HS-SPME and GC–MS were used to analyze the volatile compounds in the EG and CG green teas grown under the two different cultivation modes, and 54 volatile compounds were identified (Table 1). Among them, 46 and 29 volatiles were identified in EG and CG green tea, respectively. Based on their chemical structures, these volatile compounds were divided into eight different chemical classes: 16 alcohols, 7 esters, 6 ketones, 2 aldehydes, 11 olefins, 2 aromatics, 1 heterocycle, and 9 hydrocarbons. To further analyze the volatile components of EG and CG green tea, we analyzed the percentage contents of 54 volatile components in EG and CG green tea (Figure 2). VIP values of the 54 volatile components are shown in Figure S1. The primary volatile components in EG and CG green tea were alcohols and esters, which accounted for 32.41%

and 21.90% of the total volatile content in EG green tea, and 30.70% and 16.97% in CG green tea, respectively. Additionally, the hydrocarbon content in EG green tea (13.42%) was lower than that in CG green tea (17.77%). The ketone, aldehyde, and aromatic contents in EG green tea (16.14%, 6.15%, 5.08%) were higher than those in CG green tea (7.69%, 0.00%, and 0.00%, respectively). The main volatile compounds in EG green tea were cis-3-hexenylhexanoate (60.12 µg/L), cis-jasmone (46.04 µg/L), methyl salicylate (39.71 µg/L), linalool oxide (pyranoid) (36.42 µg/L), linalool (34.58 µg/L), and nonanal (30.45 µg/L). The main volatile compounds in CG green tea were linalool (76.47 µg/L), cis-3-hexenylhexanoate (67.93 µg/L), trans-nerolidol (48.91 µg/L), pentadecane (41.71 µg/L), hexadecane (34.84 µg/L), and tetradecane (32.81 µg/L).

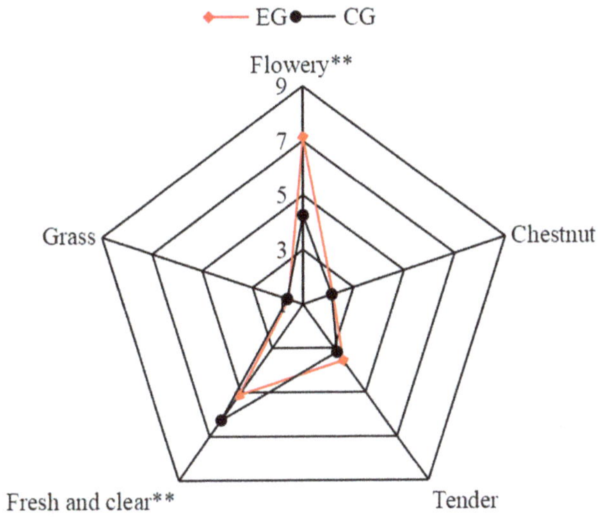

Figure 1. Radar plot for comparative descriptive aroma profiles of the EG and CG green teas. (EG—teas from the intercropping group; CG—teas from the pure tea tree group. ** indicates $p < 0.01$).

Table 1. Volatile compounds identified in EG and CG green tea (µg/L).

	RT	Name	CAS	RI [a]	RI [b]	ID	EG	CG
1	7.57	Styrene	100-42-5	897	893	MS, RI	2.52 ± 0.6	2.87 ± 0.38
2	8.8	Ethylene glycol monoisobutyl ether	4439-24-1	940	947	MS, RI	25.22 ± 5.49	-
3	9.03	3-Ethyl-3-methylheptane	17302-01-1	959	953	MS, RI	3.09 ± 0.59	-
4	10.34	β-Myrcene	123-35-3	992	991	MS, RI, STD	-	0.85 ± 0.16
5	11.67	D-sylvestrene	1461-27-4	1030	1027	MS, RI	-	4.85 ± 0.37
6	12.14	Benzyl alcohol	100-51-6	1040	1036	MS, RI, STD	0.77 ± 0.89	-
7	12.49	1-Ethyl-1H-pyrrole-2-carboxaldehyde	2167-14-8	1048	1046	MS, RI	2.27 ± 0.74	-
8	12.85	Decane, 3-methyl-	13151-34-3	1061	1071	MS, RI	2.98 ± 0.69	-
9	13.3	cis-Linalool oxide (furanoid)	5989-33-3	1073	1074	MS, RI	12.99 ± 3.14	13.9 ± 1.89
10	13.95	(E)-Linalol oxide (furanoid)	34995-77-2	1090	1886	MS, RI	2.72 ± 5.44	15.91 ± 0.96
11	14.16	Undecane	1120-21-4	1095	1100	MS, RI	2.83 ± 0.31	-
12	14.38	Linalool	78-70-6	1101	1098	MS, RI, STD	34.58 ± 3.73	76.47 ± 6.14
13	14.58	Nonanal	124-19-6	1105	1102	MS, RI, STD	30.45 ± 5.61	-

Table 1. *Cont.*

	RT	Name	CAS	RI [a]	RI [b]	ID	EG	CG
14	14.61	2-Nonen-1-ol	22104-79-6	1105	1105	MS, RI	-	28.57 ± 6.49
15	15.36	Phenylethyl Alcohol	60-12-8	1122	1116	MS, RI	0.88 ± 0.31	-
16	17.69	(Z)-Linalool oxide (pyranoid)	14009-71-3	1170	1174	MS, RI	5.04 ± 1.64	-
17	17.85	Linalool oxide (pyranoid)	14049-11-7	1175	1178	MS, RI, STD	36.42 ± 5.53	31.75 ± 3.51
18	18.43	cis-3-Hexenyl butyrate	16491-36-4	1187	1187	MS, RI	4.94 ± 1.15	-
19	18.72	α-Terpineol	98-55-5	1193	1190	MS, RI, STD	3.29 ± 0.6	14.12 ± 1.32
20	19.05	Methyl salicylate	119-36-8	1200	1192	MS, RI	39.71 ± 7.82	29.55 ± 6.78
21	22.03	Geraniol	106-24-1	1255	1255	MS, RI, STD	14.64 ± 5.51	-
22	22.24	2,6-Octadien-1-ol, 3,7-dimethyl-	624-15-7	1259	1260	MS, RI	5.68 ± 0.8	0.66 ± 0.32
23	24.45	Indole	120-72-9	1300	1295	MS, RI	25.52 ± 2.54	-
24	26.67	γ-Ionone	79-76-5	1338	1340	MS, RI	0.09 ± 0.12	-
25	27.36	α-Cubebene	17699-14-8	1350	1352	MS, RI	6.77 ± 0.89	-
26	28.87	α-Copaene	3856-25-5	1376	1376	MS, RI	1.89 ± 0.25	2.17 ± 0.33
27	29.17	cis-3-Hexenyl hexanoate	31501-11-8	1382	1380	MS, RI, STD	60.12 ± 10.38	67.93 ± 6.69
28	30.21	Tetradecane	629-59-4	1400	1400	MS, RI	-	32.81 ± 6.97
29	30.24	cis-Jasmone	488-10-8	1400	1394	MS, RI, STD	46.04 ± 6.37	-
30	32.18	γ-Elemene	29873-99-2	1434	1434	MS, RI	3.5 ± 0.64	-
31	32.93	Cyclopentane, nonyl-	2882-98-6	1447	1451	MS, RI	2.89 ± 1.53	-
32	33.32	trans-Geranylacetone	3796-70-1	1454	1453	MS, RI	-	1.63 ± 0.32
33	33.54	cis-β-Farnesene	28973-97-9	1458	1457	MS, RI	3.13 ± 0.38	7.29 ± 0.66
34	33.77	2,6,10-Trimethyltridecane	3891-99-4	1462	1461	MS, RI	10.61 ± 1.26	-
35	34.87	Valencene	4630-07-3	1481	1476	MS, RI	3.92 ± 3.74	-
36	35.24	Trans-β-Ionone	79-77-6	1487	1486	MS, RI	4.18 ± 0.48	8.41 ± 0.66
37	35.97	Pentadecane	629-62-9	1500	1500	MS, RI	26.28 ± 19.46	41.71 ± 4.58
38	36.52	α-Farnesene	502-61-4	1510	1508	MS, RI, STD	-	5.65 ± 2.59
39	37.67	Dihydroactinidiolide	17092-92-1	1530	1532	MS, RI	-	3.02 ± 0.7
40	38.47	α-Calacorene	21391-99-1	1544	1542	MS, RI	1.42 ± 0.34	0.43 ± 0.17
41	39.63	trans-Nerolidol	40716-66-3	1565	1564	MS, RI, STD	22.81 ± 4.33	48.91 ± 24.79
42	39.97	Pentadecane, 3-methyl-	2882-96-4	1570	1570	MS, RI	3.59 ± 1.27	-
43	41.48	Diethyl Phthalate	84-66-2	1598	1603	MS, RI	0.89 ± 0.97	2.02 ± 1.35
44	41.63	Hexadecane	544-76-3	1600	1600	MS, RI	19.33 ± 3.3	34.84 ± 4.15
45	43.75	2,6-Bis(1,1-dimethylethyl)-4-(1-oxopropyl)phenol	14035-34-8	1639	1640	MS, RI	2.41 ± 1.45	-
46	43.99	τ-Muurolol	19912-62-0	1643	1642	MS, RI	5.63 ± 1.42	3.39 ± 0.42
47	44.24	Torreyol	19435-97-3	1648	1645	MS, RI	0.16 ± 0.1	-
48	44.39	Methyl jasmonate	1211-29-6	1650	1652	MS, RI	9.58 ± 1.34	-
49	44.68	α-Cadinol	481-34-5	1656	1653	MS, RI	1.49 ± 0.87	-
50	45.79	Cadalene	483-78-3	1676	1674	MS, RI	1.48 ± 0.23	-
51	47.11	Heptadecane	629-78-7	1700	1700	MS, RI	2.63 ± 0.65	4.96 ± 0.49
52	54.33	Neophytadiene	504-96-1	1838	1837	MS, RI	-	2.76 ± 0.32
53	54.67	Caffeine	50-8-2	1847	1840	MS, RI	33.13 ± 9.42	2.76 ± 0.34
54	58.75	Methyl palmitate	112-39-0	1928	1926	MS, RI, STD	1.21 ± 0.28	0.68 ± 0.11

MS—identification based on the NIST 2017 mass spectral database; [a] RI—retention index of each volatile on the TG-5MS column (30 m × 0.25 mm × 0.25 μm); [b] RI—retention index based on the NIST database. STD—means the compound was identified by standards. All data are expressed as mean ± S.D. (n = 5).

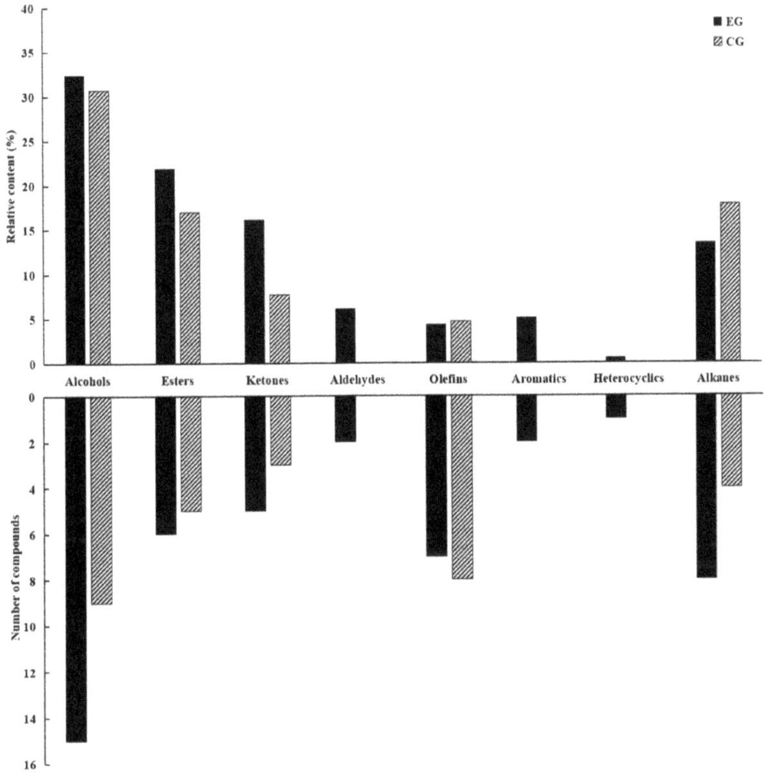

Figure 2. Volatile profiles of EG and CG green tea determined by SPME-GC–MS. The upper part of the figure shows the relative content of each chemical category, and the lower part shows the number of different compounds in each category.

The aroma is an important factor in evaluating the quality of tea, and the formation of the overall aroma of tea is crucial to the type and content level of volatile compounds. In this study, there were 46 types of volatile compounds in EG green tea and 29 types in CG green tea. The difference in volatile components between the EG and CG green teas may be one of the reasons why EG green tea presents a floral scent, whereas CG green tea does not. Among these volatile compounds, cis-jasmone, methyl salicylate, and nonanal had the highest content levels in EG green tea. As reported in the existing literature, these compounds are components of floral and fruit odors. The differences in the proportions and content levels of volatile compounds may be one of the reasons why the experimental group's tea was floral in odor, whereas the control group was not. Xu et al. [28] found significant differences in the varieties and proportions of volatile compounds (alcohols, aldehydes, esters, terpenes, alkenes, and ketones) in different cultivation environments, which is consistent with the results of this experiment.

3.3. Key Volatile Component Analysis in EG

Figure 3 shows a heat map of the 54 volatile compounds, and PCA was performed based on the quantitative results of the aroma components. Principal components PC1 and PC2 individually accounted for 76.3% and 10.9% of the total variance (Figure 4). PCA showed a clear separation between the different groups, and the results were consistent with those of heat map cluster analysis. To further reveal the influences of cherry blossoms on key aroma compounds in tea, OPLS-DA was performed on EG and CG green tea in this study (Figure 5). The fitting index of the independent variable was $R^2X = 0.862$, that of the dependent variable was $R^2Y = 0.99$, and that of the prediction was $Q^2 = 0.969$, indicating that the model had a good cumulative explanatory degree, predictive ability, and stability. Nine key volatile compounds were screened using the OPLS-DA model (VIP > 1, $p < 0.05$): cis-jasmone, linalool, indole, geraniol, ethylene glycol monoisobutyl ether, tetradecane, nonanal, 2-nonen-1-ol, and hexadecane.

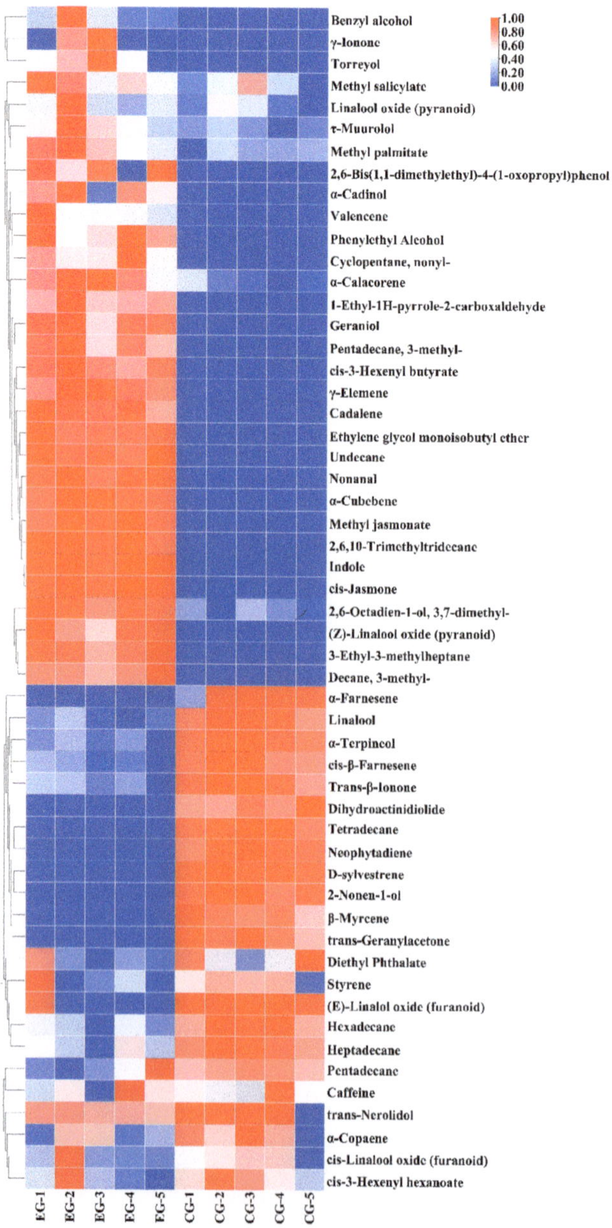

Figure 3. Heat map analysis of the concentrations of the main volatile compounds in tea samples.

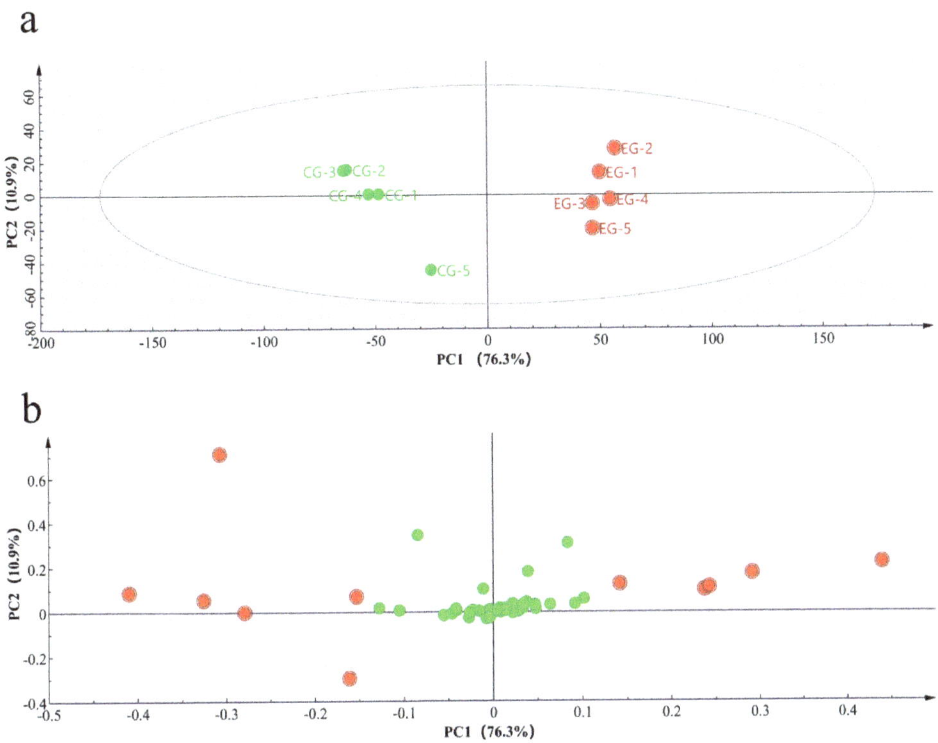

Figure 4. Multivariate analysis of volatile compound contents between the EG and CG green tea samples. (**a**) PCA score plot and (**b**) loading plots of volatile compounds (red and green dots represent different compounds; red dot represent the most differential compounds, VIP > 1).

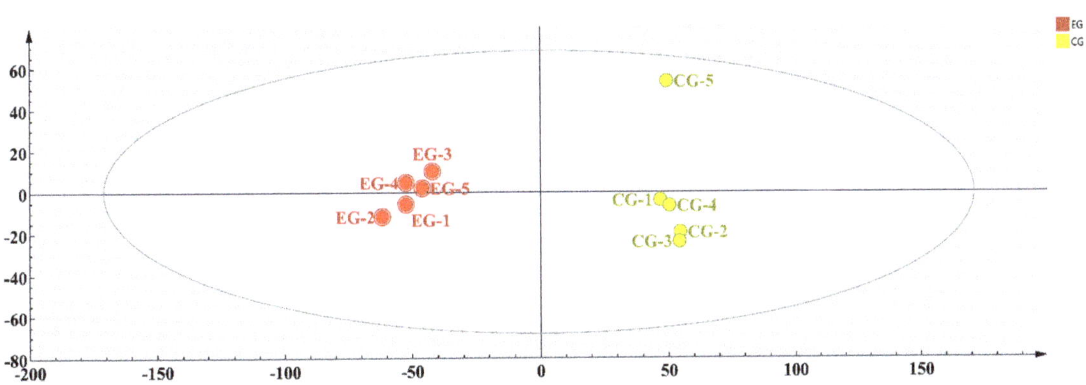

Figure 5. The OPLS-DA model in the EG and CG green tea samples ($R^2X = 0.862$, $R^2Y = 0.99$, and $Q^2 = 0.969$).

Cis-jasmone is one of the characteristic aromas of green tea [29] and one of the main components of the plant's direct and indirect stress-resistance mechanisms [30]. Indole in tea is produced from tryptophan, which presents a floral aroma at low concentrations and enhances the overall aroma of green tea within a certain concentration range, thus contributing to the chestnut aroma of green tea [31,32]. Geraniol and linalool are released from geranyl pyrophosphate precursors by geraniol synthase and linalool synthase, respectively, and linalool is further oxidized to form dehydrolinalool and various other linalool oxides [33]. Linalool has floral and citrus fruit aromas, whereas geraniol has a rose aroma [33]. Liu et al. [31] showed that geraniol is an important source of floral aroma. Han et al. [34] found that geraniol, linalool, cis-jasmone, and indol are the key aroma compounds in green tea, which is in agreement with the findings of this study. Nonanal is the key aroma compound in green tea and it is positively correlated with the floral aroma intensity of tea [35,36].

3.4. Key Aroma Compounds Identified by OAV in the EG Group

The OAV for each compound was calculated in order to assess the contribution of the volatiles to the overall aroma. Based on these nine key volatile compounds, three key aroma compounds, cis-jasmone, nonanal, and linalool, were identified (OVA > 1) (Table 2). Among these compounds, two showed OAVs > 10, indicating their substantial contributions to the overall aroma profile of EG green tea. Cis-jasmone had the highest OAV (177.07), followed by nonanal (OAV = 30.45). Cis-jasmone and nonanal were aroma compounds unique to EG. Cis-jasmone and nonanal may have contributed to the floral appearance of the EG.

Table 2. Aroma compounds with OAVs identified in EG and CG green tea samples.

Odorants	Odor Description	VIP	OT (μg/L)	OAV
cis-Jasmone *	Flowery	3.19	0.26	177.07
Linalool *	Floral, lavender	2.88	6	2.17
Tetradecane *	Alkane	2.29	10,000	0.0032
Nonanal *	Sweet Orange aroma	2.12	1	30.45
2-Nonen-1-ol *	-	1.98	-	-
Indole *	Floral, animal-like	1.76	40	0.63
Ethylene glycol monoisobutyl ether *	-	1.72	-	-
Hexadecane *	Alkane	1.08	13,000,000	0
Geraniol *	Rose	1.05	40	0.37

OT—odor threshold in water based on the literature [37,38]; OAV—odor activity value; VIP—variable importance in projection. Data significance analysis was performed using Student's *t*-test, and a significant difference ($p < 0.05$) is denoted by *.

Nine key aromatic compounds were identified using the OPLS-DA model (VIP > 1, $p < 0.05$), and the contents of these nine compounds were significantly different between the EG and CG teas. Recent studies showed that high concentrations of compounds do not necessarily produce a strong aroma. Tea typically contains hundreds of volatile compounds; however, not all volatile components play a significant role in its aroma quality [26,30,39,40]. Wang et al. [26] showed that only a few compounds determined the overall aroma quality of tea, and when some key compounds were missing, the overall aroma quality of tea would be significantly affected. The aroma of tea depends not only on the type and content of volatile compounds but also on their OAVs [41]. We used OAVs to evaluate the contribution of volatile compounds to the EG green tea and selected three key aromatic compounds (VIP > 1, $p < 0.05$, OAV > 1): cis-jasmone, linalool, and nonanal. Linalool has a citrus-like and flowery aroma [27,41]. Su et al. [42] showed that linalool is the key volatile compound of green tea and that it plays an important role in the aroma of green tea. Nonanal is widely found in green tea and has a sweet, fruity flavor [37]. Notably, cis-jasmone and nonanal were the key volatile compounds responsible for the unique floral flavor of EG green tea.

Cis-jasmone and nonanal were the main components of EG green tea. The aroma quality of tea is determined not only by its content of compounds but also by the activity value of its aroma. The unique floral-aroma compound cis-jasmone in EG green tea had not only a high content but also a high OAV of 177.07. This indicates that cis-jasmone is a characteristic aroma compound of EG green tea and contributes significantly to its overall aroma. cis-jasmone is a compound with a jasmine scent that is commonly found in green tea and produces a pleasant floral jasmine scent at lower concentrations. Zhu et al. [43] found that cis-jasmone is a characteristic aroma compound of Xihu Longjing green tea, which is consistent with the findings of this study. Katsuno et al. [32] demonstrated that cis-jasmone contributes to the formation of floral fragrances. Cis-jasmone, linalool, and nonanal, the three key aromatic compounds in EG, promote the formation of floral fragrances [44].

Cis-jasmone and nonanal are the unique aroma-characteristic compounds of EG green tea. To explain this phenomenon, we studied the growth environment of the EG. Tea plants are exposed to various natural stresses, such as light, temperature, and pests. When facing stress, tea plants release volatile organic compounds to reduce the adverse effects of the stress [45]. The EG tea trees were intercropped with flowering cherry trees. When cherry trees bloom, flowers compete for scarce resources by continuously releasing volatile compounds. During resource competition, tea trees secrete large amounts of volatile compounds to defend against stress.

3.5. Analysis of Volatile Compounds in Flowering Cherry Trees

The results showed that EG green tea produced floral volatile compounds, such as cis-jasmone and nonanal, and that the content of other key floral volatile compounds changed. To explain these phenomena, HS-SPME/GC–MS was used to analyze the volatile compounds present in flowering cherry flowers. The detailed results are presented in Table 3. The results showed that benzaldehyde, α-farnesene, methyl benzoate, 3, 5-dimethoxytoluene had high relative contents, and that benzaldehyde accounted for 86.12% of all aroma components.

Table 3. The volatile compounds present in flowering cherry flowers.

Name	CAS	Relative Content (%)
Benzaldehyde	100-52-7	86.12
α-Farnesene	502-61-4	4.49
Methyl benzoate	93-58-3	3.15
Orcinol dimethyl ether	4179-19-5	2.01
Anisic aldehyde	123-11-5	0.81
Phenylethyl Alcohol	60-12-8	0.75
3,5-Dimethoxybenzaldehyde	7311-34-4	0.53
Benzene, 1-methoxy-2-(methoxymethyl)-	21998-86-7	0.1
Benzoic acid, ethyl ester	93-89-0	0.07
Methyl salicylate	119-36-8	0.07

Allelopathy was first proposed by Molisch in 1937. During growth and development, plants change their surrounding microecological environment by releasing specific metabolites that affect the growth and development of other plants [46]. Allelopathy includes promotion and inhibition [47]. Allelopathic compounds can inhibit seed germination and slow or destroy normal plant growth. Therefore, the use of allelopathy to remove weeds is an effective method [48]. Studies showed that many allelopathic compounds can stimulate plant growth. The leaf extracts of *Pinus sylvestris* and *Broussonetia papyrifera* have been shown to promote the growth of *Amygdalus pedunculata* seedlings [49]. Research has shown that leaf extracts from *Hippophae rhamnoides*, *Hedysarum mongolicum*, and *Sabina vulgaris* significantly promote the seed germination and seedling growth of *Amygdalus pedunculata* Pall [50]. Plant allelochemicals are mainly secondary metabolites distributed in the roots, stems, leaves, flowers, fruits, and seeds of plants. Plant secondary metabolites include phenols, alkaloids, organic acids, terpenoids, and glycosides.

We analyzed the volatile compounds present in the flowers of the flowering cherry. Benzaldehyde had the highest concentration, followed by α-farnesene. Benzaldehyde is the simplest aromatic aldehyde found in nature, and it is present not only in >50% of plant families but also in insects and non-insect arthropods [51]. Benzaldehyde acts as a pollinator attractant, fragrance volatiler, and an antifungal compound in plants [52]. Benzaldehyde was found to be one of the main volatiles in sweet cherry flowers, which is consistent with the results of Zhang et al. [53]. Recent studies found that in the interplanting system between oil tea and peanuts, oil tea releases benzaldehyde as an allelopathic compound that inhibits the germination of peanut seeds [54]. α-farnesene has a citrusy odor and is involved in signal transduction between tea plants; it promotes metabolite synthesis in adjacent undamaged tea leaves [55]. Studies showed that α-farnesene acts as a signal to activate the expression of β-1, 3-glucanase (CsBGL) gene in adjacent undamaged tea leaves [56]. Methyl benzoate has a fruity odor and is a common floral volatile component found in more than 80 plants [57,58]. We speculated that benzaldehyde, α-farnesene, and methyl benzoate in flowering cherry may have allelopathic interactions with tea trees as major allelopathic compounds, which affect the metabolism of tea trees and prompt EG green tea trees to produce key volatile compounds with unique flower flavors (Figure 6). The results showed that a system of tea tree intercropping with flowering cherry has the potential to improve the tea quality.

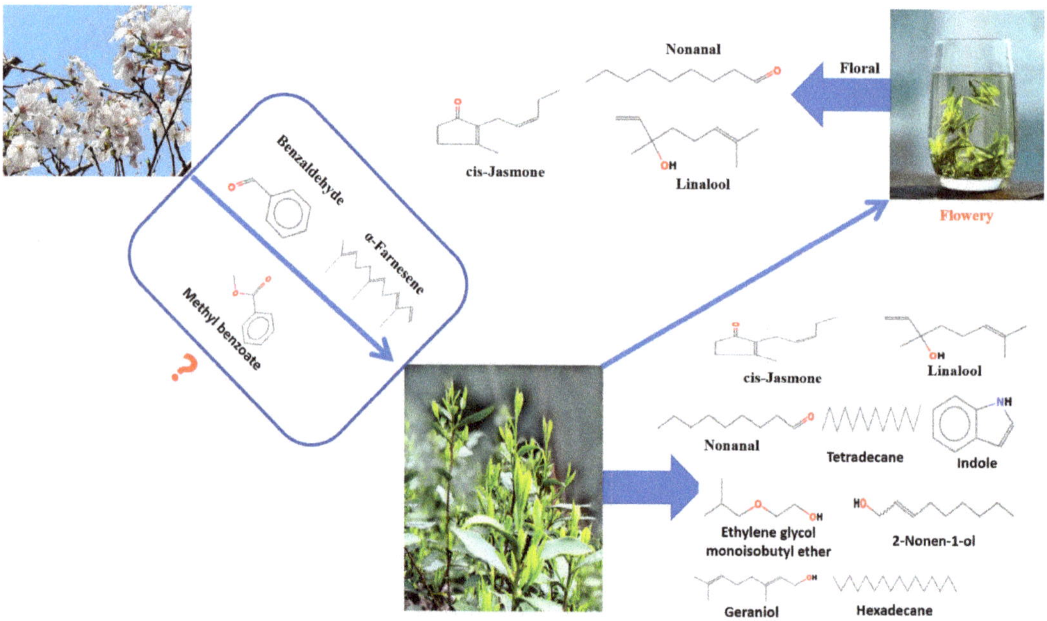

Figure 6. Model of Lushan Yunwu tea intercropping with flowering cherry leading to floral tea. We speculated that benzaldehyde, α-farnesene, and methyl benzoate in flowering cherry may have allelopathic interactions with tea trees as major allelopathic compounds, which affect the metabolism of tea trees and prompt leaves to produce key volatile compounds with unique flower flavors.

There were significant differences between the volatile compounds of tea under the two cultivation models, indicating that the intercropping model had a significant influence on the secondary metabolism of tea. These results enhance our understanding of the improvement in tea quality created by the model with tea intercropping with the flowering cherry, and they provide a reference for the future popularization and application of tea intercropping in the flowering cherry model.

4. Conclusions

In this study, HS-SPME/GC–MS was used to identify volatile compounds of EG and CG green tea, and 54 volatile compounds were identified. The 54 volatile compounds were identified using PCA, OPLS-DA, and OAV. Among these compounds, cis-jasmone, linalool, and nonanal were found to be the key aromatic compounds responsible for EG green tea's floral scent production. It was speculated that benzaldehyde, α-farnesene, and methyl benzene in flowering cherry trees acted as major allelopathic compounds, affecting the metabolism of tea trees through allelopathy. Cis-jasmone, linalool, and nonanal promoted the formation of a floral aroma and improved the overall aroma quality of tea. Further exploration is needed to determine whether other flowering plants affect the aroma quality of tea.

Supplementary Materials: The following supporting information can be downloaded through the link https://www.mdpi.com/article/10.3390/foods13081252/s1: Figure S1, VIP values of the 54 volatile compounds of EG and CG green tea samples.

Author Contributions: Conceptualization, Y.G. and Z.L.; investigation, Z.L.; methodology, Z.L., J.H., S.L. and J.L.; writing—original draft preparation, Y.G., Y.L. and Z.L.; writing—review and editing, Y.L., Y.G., Y.S. and Y.C.; data curation, W.L. and L.Y. All authors have read and agreed to the published version of the manuscript.

Funding: This research was funded by the Key Projects of Science and Technology Cooperation in Jiangxi Province (20212BDH80011), the Modern Agricultural Industrial Technology System of Jiangxi Province (JXARS-02), the National Modern Industrial Technology System (CARS-19), and the Natural Science Basic Research Program of Jiujiang (S2022KXJJ001).

Institutional Review Board Statement: Sensory evaluation data collection and use were in accordance with Jiujiang University Human Ethics application ID Number: JJUM20240090. (Approval Date: 7 April 2024).

Informed Consent Statement: All sensory evaluators agreed to participate and signed the in-formed consent form.

Data Availability Statement: The original contributions presented in the study are included in the article/supplementary material, further inquiries can be directed to the corresponding authors.

Acknowledgments: Thanks to Li Jie of Tea College of Jiujiang Vocational University for her guidance and help in tea sensory evaluation.

Conflicts of Interest: The authors declare no conflicts of interest.

References

1. Wu, T.; Zou, R.; Pu, D.; Lan, Z.Q.; Zhao, B.Y. Non-targeted and targeted metabolomics profiling of tea plants (*Camellia sinensis*) in response to its intercropping with Chinese chestnut. *BMC Plant Biol.* **2021**, *21*, 55. [CrossRef] [PubMed]
2. Huang, Z.; Cui, C.H.; Cao, Y.J.; Dai, J.H.; Cheng, X.Y.; Hua, S.W.; Wang, W.T.; Duan, Y.; Petropoulos, E.; Wang, H.; et al. Tea plant-legume intercropping simultaneously improves soil fertility and tea quality by changing *Bacillus* species composition. *Hortic. Res.* **2022**, *9*, uhac046. [CrossRef] [PubMed]
3. Wang, T.; Duan, Y.; Liu, G.D.; Shang, X.W.; Liu, L.F.; Zhang, K.X.; Li, J.Q.; Zou, Z.W.; Zhu, X.J.; Fang, W.P. Tea plantation intercropping green manure enhances soil functional microbial abundance and multifunctionality resistance to drying-rewetting cycles. *Sci. Total Environ.* **2022**, *810*, 151282. [CrossRef] [PubMed]
4. Sun, Y.M.; Li, X.F.; Yu, S.Q.; Zhang, K.; Zhang, J.; Yu, P.L.; Wang, Y.G.; Ye, C. Comparison on effects of different weed control measures in tea garden. *J. South. Agric.* **2017**, *48*, 1832–1837.
5. Ye, J.H.; Wang, Y.H.; Wang, Y.C.; Hong, L.; Jia, X.L.; Kang, J.Q.; Lin, S.X.; Wu, Z.Y.; Wang, H.B. Improvement of soil acidification in tea plantations by long-term use of organic fertilizers and its effect on tea yield and quality. *Front. Plant Sci.* **2022**, *13*, 1055900. [CrossRef] [PubMed]
6. Pokharel, S.S.; Yu, H.; Fang, W.P.; Parajulee, M.N.; Chen, F.J. Intercropping cover crops for a vital ecosystem service: A review of the biocontrol of insect pests in tea agroecosystems. *Plants* **2023**, *12*, 2361. [CrossRef] [PubMed]
7. Yang, H.B.; Sheng, Z.L.; Xie, K.; Wu, X.H.; Deng, M. Seasonal regulation of different cultivation mode on ecological environment and tea quality in hilly tea plantation. *Southwest China J. Agric. Sci.* **2015**, *28*, 1559–1563.
8. Liu, Y.; Sun, L.L.; Liao, H. Effects of nutrient management on soil fertility and tea quality in Anxi tea plantation. *Acta Pedol. Sin.* **2020**, *57*, 917–927.
9. Cao, Q.M.; Wang, Y.P.; Qin, J.W.; Yi, W.P.; Zhang, J.H.; Xiao, T.B. Effects of compound planting on soil properties and tea quality in tea garden of Hainan island. *Southwest China J. Agric. Sci.* **2020**, *33*, 1504–1509.
10. Shi, F.; Huang, H.J.; Chen, Y.T.; Chen, L.L. Effects of intercropping functional plants on the ecosystem functions and services in tea garden. *J. Tea Sci.* **2022**, *42*, 151–168.
11. Wu, H.; Long, X.F.; Geng, Y.F. Companion plants of tea: From ancient to terrace to forest. *Plants* **2023**, *12*, 3061. [CrossRef] [PubMed]
12. Ma, Q.P.; Song, L.C.; Niu, Z.H.; Qiu, Z.Y.; Sun, H.W.; Ren, Z.H.; Wu, H.H.; Wang, Y.; Mei, H.L.; Li, X.H.; et al. Pea-tea intercropping improves tea quality through regulating amino acid metabolism and flavonoid biosynthesis. *Foods* **2022**, *11*, 3746. [CrossRef]

13. Kateb, H.E.; Zhang, H.F.; Zhang, P.C.; Mosandl, R. Soil erosion and surface runoff on different vegetation covers and slope gradients: A field experiment in Southern Shanxi Province, China. *Catena* **2013**, *105*, 1–10. [CrossRef]
14. Liu, W.J.; Luo, Q.P.; Lu, H.J.; Wu, J.E.; Duan, W.P. The effect of litter layer on controlling surface runoff and erosion in rubber plantations on tropical mountain slopes, SW China. *Catena* **2017**, *149*, 167–175. [CrossRef]
15. Duan, Y.; Shen, J.Z.; Zhang, X.L.; Wen, B.; Ma, Y.C.; Wang, Y.H.; Fang, W.P.; Zhu, X.J. Effects of soybean-tea intercropping on soil-available nutrients and tea quality. *Acta Physiol. Plant.* **2019**, *41*, 140. [CrossRef]
16. Zhang, Z.Q.; Zhou, C.; Xu, Y.Y.; Huang, X.Q.; Zhang, L.X.; Mu, W. Effects of intercropping tea with aromatic plants on population dynamics of arthropods in Chinese tea plantations. *J. Pest Sci.* **2017**, *90*, 227–237. [CrossRef]
17. Sun, L.; Dong, X.; Wang, Y.; Maker, G.; Agarwal, M.; Ding, Z.T. Tea-Soybean intercropping improves tea quality and nutrition uptake by inducing changes of rhizosphere bacterial communities. *Microorganisms* **2022**, *10*, 2149. [CrossRef] [PubMed]
18. Yang, L.R.; Jiang, B.; Jiao, W.W.; Zhong, X.X.; Liu, K.Y.; Liu, L.; Li, L.X.; Peng, K.L.; Luo, X.P. Influence of tea-*Prunus salicina lindl.* intercropping on growth of tea trees and the quality of autumn green tea. *J. South. Agric.* **2023**, *10*, 1–15.
19. Nie, C.; Zhang, Y.; Zhang, X.; Xia, W.; Sun, H.; Zhang, S.; Li, N.; Ding, Z.; Lv, Y.; Wang, N. Genome assembly, resequencing and genome-wide association analyses provide novel insights into the origin, evolution and flower colour variations of flowering cherry. *Plant J.* **2023**, *114*, 519–533. [CrossRef] [PubMed]
20. Yan, X.; Xie, Y.; Chen, J.; Yuan, T.; Leng, T.; Chen, Y.; Xie, J.; Yu, Q. NIR spectrometric approach for geographical origin identification and taste related compounds content prediction of Lushan Yunwu tea. *Foods* **2022**, *11*, 2976. [CrossRef] [PubMed]
21. Wu, H.T.; Chen, Y.Y.; Feng, W.Z.; Shen, S.S.; Wei, Y.M.; Jia, H.Y.; Wang, Y.J.; Deng, W.W.; Ning, J.M. Effects of three different withering treatments on the aroma of white tea. *Foods* **2022**, *11*, 2502. [CrossRef] [PubMed]
22. Wang, B.; Shu, N.; Lu, A.X.; Liao, X.L.; Yan, J.N.; Xie, G.H.; Tong, H.R. Aroma formation of green tea effected by different pan-fire temperature. *Food Ferment. Ind.* **2020**, *46*, 197–203.
23. Wei, Y.M.; Yin, X.C.; Wu, H.T.; Zhao, M.J.; Huang, J.L.; Zhang, J.X.; Li, T.H.; Ning, J.M. Improving the flavor of summer green tea (*Camellia sinensis* L.) using the yellowing process. *Food Chem.* **2022**, *388*, 132982. [CrossRef] [PubMed]
24. Su, D.; He, J.J.; Zhou, Y.Z.; Li, Y.L.; Zhou, H.J. Aroma effects of key volatile compounds in Keemun black tea at different grades: HS-SPME-GC-MS, sensory evaluation, and chemometrics. *Food Chem.* **2022**, *373*, 131587. [CrossRef] [PubMed]
25. Sun, L.T.; Dong, X.; Ren, Y.L.; Agarwal, M.; Ren, A.; Ding, Z.T. Profiling real-time aroma from green tea infusion during brewing. *Foods* **2022**, *11*, 684. [CrossRef] [PubMed]
26. Wang, M.Q.; Ma, W.J.; Shi, J.; Zhu, Y.; Lin, Z.; Lv, H.P. Characterization of the key aroma compounds in Longjing tea using stir bar sorptive extraction (SBSE) combined with gas chromatography-mass spectrometry (GC-MS), gas chromatography-olfactometry (GC-O), odor activity value (OAV), and aroma recombination. *Food Res. Int.* **2020**, *130*, 108908. [CrossRef] [PubMed]
27. Zhai, X.T.; Zhang, L.; Granvogl, M.; Ho, C.T.; Wan, X.C. Flavor of tea (*Camellia sinensis*): A review on odorants and analytical techniques. *Compr. Rev. Food Sci. Food Saf.* **2022**, *21*, 3867–3909. [CrossRef] [PubMed]
28. Xu, K.; Tian, C.Y.; Zhou, C.Z.; Zhu, C.; Weng, J.J.; Sun, Y.; Lin, Y.L.; Lai, Z.X.; Guo, Y.Q. Non-targeted metabolomics analysis revealed the characteristic non-volatile and volatile metabolites in the *Rougui* Wuyi rock tea (*Camellia sinensis*) from different culturing regions. *Foods* **2022**, *11*, 1694. [CrossRef] [PubMed]
29. Wang, J.; Li, X.H.; Wu, Y.; Qu, F.F.; Liu, L.; Wang, B.Y.; Wang, P.Q.; Zhang, X.F. HS-SPME/GC-MS reveals the season effects on volatile compounds of green tea in high-latitude region. *Foods* **2022**, *11*, 3016. [CrossRef] [PubMed]
30. Zhu, J.C.; Niu, Y.W.; Xiao, Z.B. Characterization of the key aroma compounds in Laoshan green teas by application of odour activity value (OAV), gas chromatography-mass spectrometry-olfactometry (GC-MS-O) and comprehensive two-dimensional gas chromatography mass spectrometry (GC × GC-qMS). *Food Chem.* **2021**, *339*, 128136. [CrossRef] [PubMed]
31. Liu, N.F.; Shen, S.S.; Huang, L.F.; Deng, G.J.; Wei, Y.M.; Ning, J.M.; Wang, Y.J. Revelation of volatile contributions in green teas with different aroma types by GC-MS and GC-IMS. *Food Res. Int.* **2023**, *169*, 112845. [CrossRef] [PubMed]
32. Katsuno, T.; Kasuga, H.; Kusano, Y.; Yaguchi, Y.; Tomomura, M.; Cui, J.L.; Yang, Z.Y.; Baldermann, S.; Nakamura, Y.; Ohnishi, T.; et al. Characterisation of odorant compounds and their biochemical formation in green tea with a low temperature storage process. *Food Chem.* **2014**, *148*, 388–395. [CrossRef] [PubMed]
33. Ho, C.T.; Zheng, X.; Li, S.M. Tea aroma formation. *Food Sci. Hum. Well.* **2015**, *4*, 9–27. [CrossRef]
34. Han, Z.X.; Rana, M.M.; Liu, G.F.; Gao, M.J.; Li, D.X.; Wu, F.G.; Li, X.B.; Wan, X.C.; Wei, S. Green tea flavour determinants and their changes over manufacturing processes. *Food Chem.* **2016**, *212*, 739–748. [CrossRef] [PubMed]
35. Liu, P.P.; Zheng, P.C.; Gong, Z.M.; Feng, L.; Gao, S.W.; Wang, X.P.; Teng, J.; Zheng, L.; Liu, Z.H. Comparing characteristic aroma components of bead-shaped green teas from different regions using headspace solid-phase microextraction and gas chromatography–mass spectrometry/olfactometry combined with chemometrics. *Eur. Food Res. Technol.* **2020**, *246*, 1703–1714. [CrossRef]
36. Yin, X.; Huang, J.A.; Huang, J.; Wu, W.L.; Tong, T.; Liu, S.J.; Zhou, L.Y.; Liu, Z.H.; Zhang, S.G. Identification of volatile and odor-active compounds in Hunan black tea by SPME/GC-MS and multivariate analysis. *LWT-Food Sci. Technol.* **2022**, *164*, 113656. [CrossRef]
37. Wang, H.J.; Ouyang, W.; Yu, Y.Y.; Wang, J.J.; Yuan, H.B.; Hua, J.J.; Jiang, Y.W. Analysis of non-volatile and volatile metabolites reveals the influence of second-drying heat transfer methods on green tea quality. *Food Chem. X* **2022**, *14*, 100354. [CrossRef]
38. Guo, X.Y.; Ho, C.T.; Schwab, W.; Wan, X.C. Effect of the roasting degree on flavor quality of large-leaf yellow tea. *Food Chem.* **2021**, *347*, 129016. [CrossRef] [PubMed]
39. Hu, W.W.; Wang, G.G.; Lin, S.X.; Liu, Z.J.; Wang, P.; Li, J.Y.; Zhang, Q.; He, H.B. Digital evaluation of aroma intensity and odor characteristics of tea with different types-based on OAV-splitting method. *Foods* **2022**, *11*, 2204. [CrossRef] [PubMed]

40. Ma, L.J.; Gao, M.M.; Zhang, L.Q.; Qiao, Y.; Li, J.X.; Du, L.P.; Zhang, H.L.; Wang, H. Characterization of the key aroma-active compounds in high-grade Dianhong tea using GC-MS and GC-O combined with sensory-directed flavor analysis. *Food Chem.* **2022**, *378*, 132058. [CrossRef]
41. Zeng, L.; Watanabe, N.; Yang, Z. Understanding the biosyntheses and stress response mechanisms of aroma compounds in tea (*Camellia sinensis*) to safely and effectively improve tea aroma. *Crit. Rev. Food Sci. Nutr.* **2019**, *59*, 2321–2334. [CrossRef] [PubMed]
42. Su, W.Y.; Gao, S.Y.; Zhan, S.J.; Wu, Q.; Chen, G.M.; Han, J.Z.; Lv, X.C.; Rao, P.F.; Ni, L. Evaluation of volatile profile and in vitro antioxidant activity of fermented green tea infusion with *Pleurotus sajor-caju* (Oyster Mushroom). *Front. Nutr.* **2022**, *9*, 865991. [CrossRef] [PubMed]
43. Zhu, Y.; Lv, H.P.; Dai, W.D.; Guo, L.; Tan, J.F.; Zhang, Y.; Yu, F.L.; Shao, C.Y.; Peng, Q.H.; Lin, Z. Separation of aroma components in Xihu Longjing tea using simultaneous distillation extraction with comprehensive two-dimensional gas chromatography-time-of-flight mass spectrometry. *Sep. Purif. Technol.* **2016**, *164*, 146–154. [CrossRef]
44. Wang, Q.S.; Qin, D.D.; Jiang, X.H.; Fang, K.X.; Li, B.; Wang, Q.; Pan, C.D.; Ni, E.D.; Li, H.J.; Chen, D.; et al. Characterization of the aroma profiles of Guangdong black teas using non-targeted metabolomics. *Foods* **2023**, *12*, 1560. [CrossRef] [PubMed]
45. Zeng, L.; Jin, S.; Xu, Y.Q.; Granato, D.; Fu, Y.Q.; Sun, W.J.; Yin, J.F.; Xu, Y.Q. Exogenous stimulation-induced biosynthesis of volatile compounds: Aroma formation of oolong tea at postharvest stage. *Crit. Rev. Food Sci. Nutr.* **2022**, *28*, 1–11. [CrossRef] [PubMed]
46. Schandry, N.; Becker, C. Allelopathic plants: Models for studying plant-interkingdom interactions. *Trends Plant Sci.* **2020**, *25*, 176–185. [CrossRef] [PubMed]
47. Han, M.; Yang, H.; Huang, H.; Du, J.; Zhang, S.; Fu, Y. Allelopathy and allelobiosis: Efficient and economical alternatives in agroecosystems. *Plant Biol.* **2024**, *26*, 11–27. [CrossRef] [PubMed]
48. Khamare, Y.; Chen, J.; Marble, S.C. Allelopathy and its application as a weed management tool: A review. *Front. Plant Sci.* **2022**, *13*, 1034649. [CrossRef] [PubMed]
49. Wang, X.Q.; Zhang, R.Q.; Wang, J.X.; Di, L.; Wang, H.B.; Sikdar, A. The effects of leaf extracts of four tree species on amygdalus pedunculata seedlings growth. *Front. Plant Sci.* **2021**, *11*, 587579. [CrossRef] [PubMed]
50. Wang, X.Q.; Wang, J.X.; Zhang, R.Q.; Huang, Y.; Feng, S.L.; Ma, X.; Zhang, Y.Y.; Sikdar, A.; Roy, R. Allelopathic effects of aqueous leaf extracts from four shrub species on Seed germination and initial growth of *Amygdalus pedunculata* Pall. *Forests* **2018**, *9*, 711. [CrossRef]
51. Schiestl, F.P. The evolution of floral scent and insect chemical communication. *Ecol. Lett.* **2010**, *13*, 643–656. [CrossRef] [PubMed]
52. Huang, X.Q.; Li, R.Q.G.; Fu, J.X.; Dudareva, N. A peroxisomal heterodimeric enzyme is involved in benzaldehyde synthesis in plants. *Nat. Commun.* **2022**, *13*, 1352. [CrossRef] [PubMed]
53. Zhang, H.M.; Yan, H.G.; Li, Q.; Lin, H.; Wen, X.P. Identification of VOCs in essential oils extracted using ultrasound- and microwave-assisted methods from sweet cherry flower. *Sci. Rep.* **2021**, *11*, 1167. [CrossRef] [PubMed]
54. Wen, H.; Dan, P.P.; Liu, T.; Li, Z.Q.; Chen, X.Y.; Cao, Y.N.; Li, Y.; Yan, W.D. Allelopathic mechanisms in *Camellia oleifera-Arachis hypogaea* L. intercropping. *J. Agric. Food Chem.* **2023**, *71*, 19434–19444. [CrossRef] [PubMed]
55. Zeng, L.; Liao, Y.; Li, J.; Zhou, Y.; Tang, J.; Dong, F.; Yang, Z.Y. α-Farnesene and ocimene induce metabolite changes by volatile signaling in neighboring tea (*Camellia sinensis*) plants. *Plant Sci.* **2017**, *264*, 29–36. [CrossRef] [PubMed]
56. Wang, X.W.; Zeng, L.T.; Liao, Y.Y.; Li, J.L.; Tang, J.C.; Yang, Z.Y. Formation of α-farnesene in tea (*Camellia sinensis*) leaves induced by herbivore-derived wounding and its effect on neighboring tea plants. *Int. J. Mol. Sci.* **2019**, *20*, 4151. [CrossRef] [PubMed]
57. Dudareva, N.; Murfitt, L.M.; Mann, C.J.; Gorenstein, N.; Kolosova, N.; Kish, C.M.; Bonham, C.; Wood, K. Developmental regulation of methyl benzoate biosynthesis and emission in snapdragon flowers. *Plant Cell.* **2000**, *12*, 949–961. [CrossRef] [PubMed]
58. Yue, Y.C.; Wang, L.; Li, M.Y.; Liu, F.; Yin, J.L.; Huang, L.J.; Zhou, B.; Li, X.Y.; Yu, Y.Y.; Chen, F.; et al. A BAHD acyltransferase contributes to the biosynthesis of both ethyl benzoate and methyl benzoate in the flowers of *Lilium oriental* hybrid 'Siberia'. *Front. Plant Sci.* **2023**, *14*, 1275960. [CrossRef] [PubMed]

foods

MDPI

Article

Development, Validation, and Application of High-Performance Liquid Chromatography with Diode-Array Detection Method for Simultaneous Determination of Ginkgolic Acids and Ginkgols in *Ginkgo biloba*

Isaac Duah Boateng [1,2,†], Fengnan Li [1,†] and Xiao-Ming Yang [1,*]

1 School of Food & Biological Engineering, Jiangsu University, Zhenjiang 212013, China; boatengisaacduah@gmail.com (I.D.B.); lifengnan0118@163.com (F.L.)
2 Certified Group, 199 W Rhapsody Dr, San Antonio, TX 78216, USA
* Correspondence: zjyxmbs2003@ujs.edu.cn
† These authors contributed equally to this work.

Abstract: *Ginkgo biloba* leaves (GBLs), which comprise many phytoconstituents, also contain a toxic substance named ginkgolic acid (GA). Our previous research showed that heating could decarboxylate and degrade GA into ginkgols with high levels of bioactivity. Several methods are available to measure GA in GBLs, but no analytical method has been developed to measure ginkgols and GA simultaneously. Hence, for the first time, an HPLC-DAD method was established to simultaneously determine GA and ginkgols using acetonitrile (0.01% trifluoroacetic acid, v/v) as mobile phase A and water (0.01% trifluoroacetic acid, v/v) as mobile phase B. The gradient elution conditions were: 0–30 min, 75–90% phase A; 30–35 min, 90–90% phase A; 35–36 min, 90–75% phase A; 36–46 min, 75–75% phase A. The detection wavelength of GA and ginkgol were 210 and 270 nm, respectively. The flow rate and injection volume were 1.0 mL/min and 50 μL, respectively. The linearity was excellent ($R^2 > 0.999$), and the RSD of the precision, stability, and repeatability of the total ginkgols was 0.20%, 2.21%, and 2.45%, respectively, in six parallel determinations. The recoveries for the low, medium, and high groups were 96.58%, 97.67%, and 101.52%, respectively. The limit of detection of ginkgol C13:0, C15:1, and C17:1 was 0.61 ppm, 0.50 ppm, and 0.06 ppm, respectively. The limit of quantification of ginkgol C13:0, C15:1, and C17:1 was 2.01 ppm, 1.65 ppm, and 0.20 ppm, respectively. Finally, this method accurately measured the GA and ginkgol content in ginkgo leaves and ginkgo tea products (ginkgo black tea, ginkgo dark tea, ginkgo white tea, and ginkgo green tea), whereas principal component analysis (PCA) was performed to help visualize the association between GA and ginkgols and five different processing methods for GBLs. Thus, this research provides an efficient and accurate quantitative method for the subsequent detection of GA and ginkgols in ginkgo tea.

Keywords: *Ginkgo biloba*; *Ginkgo biloba* leaves; method validation; ginkgolic acids; ginkgols; ginkgo tea

1. Introduction

Ginkgo biloba (GB) is widely recognized as a prominent medicinal plant due to its bioactive constituents, and the extracts derived from this botanical genus are frequently employed in the therapeutic management of many medical conditions, such as arteriosclerosis and rheumatism. The GB (Figure 1A) has been used for food and traditional herbal treatment in Japan, Korea, and China for centuries [1–4]. Numerous bioactive compounds, including flavone glycosides, ginkgolides, and bilobalides, have been extracted from *Ginkgo biloba* leaves (GBLs) (Figure 1B) and have demonstrated various physiological functions. These effects include mitigating anoxia and ischemia in the eye, brain, and heart, alleviating arteriosclerosis and rheumatism, suppressing platelet-activating factors, and regulating neurotransmitters [5–7]. Moreover, the global application of the extract of *Ginkgo biloba* (EGb) in the pharmaceutical, cosmetic, and functional food industries has been substantial, with sales >USD10 billion since 2017 [8].

Figure 1. *Ginkgo biloba* tree (**A**), leaves (**B**), and structures of ginkgolic acids (**C**) and ginkgols (**D**).

Nonetheless, due to gingkolic acids (Figure 1C) being cytotoxic, genotoxic, mutagenic, neurotoxic, and allergenic, EGb utilization has been limited, as the United States, Chinese, and European Pharmacopoeias, in 2020, authorized that the concentration of gingkolic acids (GAs) in the GBE should be <5 μg/g [4,7,9]. According to our previous study [2], upon heating, GA undergoes decarboxylation, resulting in the formation of ginkgols (3-alkylphenols), which exhibit several pharmacological properties, including antibacterial, antioxidant, and anti-apoptotic actions [10]. These effects have been seen in vitro on tumor cells [3]. Moreover, in our previous study [2], it was observed that ginkgols (Figure 1D) had exceptional thermal stability and demonstrated anticancer effects when ginkgol C17:1, C15:1, and C13:0 were isolated. Since heating can decarboxylate and degrade GA into ginkgol, measuring ginkgol at the same time as GA would be more conducive to understanding the degradation of GA.

Methods for detecting ginkgolic acids, including the UV, HPLC, and LC-MS methods, have been researched [11]. The detection of GA in EGb and GBLs in various pharmacopeias is mainly based on the HPLC method. In Jianbiao et al.'s [12] research, the authors found that the GA content detected via the European Pharmacopoeia version 7.0 method [13] was much higher than the method shown in the Chinese Pharmacopoeia 2010 version [14]. Therefore, the 2015 version of the Chinese Pharmacopoeia changed the mobile phase to 0.1% trifluoroacetic acid (TFA)–acetonitrile solution and 0.1% TFA-aqueous solution [15]. The primary differences between the Chinese and the European Pharmacopoeia version 7.0 detection method concern the chromatographic column, detection wavelength, and trifluoroacetic acid. Compared with the 2015 version of the Chinese Pharmacopoeia, the 2020 version [16] mainly strictly controls the GA content in EGb from <10 μg/g to <5 μg/g. However, the detection method has not been further updated.

Since ginkgols naturally exist in GBLs and may also be generated through the degradation of GA during processing [2], measuring ginkgols at the same time as GA can effectively monitor the degradation of GA. Although Weihong et al. [17] and Liu [18] have developed a detection method for ginkgolic acids, the mobile phase is consistent with the 2010 version of the Chinese Pharmacopoeia method [14]. With the update to the Pharmacopoeia method, a new method for the simultaneous detection of ginkgolic acids and ginkgols should be developed. Nonetheless, there is no literature on a developed method that can simultaneously determine ginkgolic acids and ginkgols. Furthermore, there is no literature regarding the validation of this method and the application of the developed method on gingko leaves and gingko tea products.

Therefore, this article establishes an accurate, reproducible, and precise HPLC method for simultaneously measuring ginkgols and ginkgolic acids and conducted methodological verification to detect ginkgols. Additionally, this method was employed to quantify the GA and ginkgols in various processed GBL teas and freeze-dried GBLs. Principal component analysis (PCA) was performed to visualize the association between GA and ginkgols and five different processing methods for GBLs. We believe that this method can produce commercial quality-control applications for GBL products.

2. Materials and Methods

2.1. Materials and Chemicals

Fresh GBLs were picked in Pizhou City, Jiangsu Province, on 29 May 2021. After selection, the GBLs were frozen at −20 °C for 24 h, freeze-dried at 50 °C (FreeZone Triad Cascade, Labconco Corporation, Kansas City, MO, USA) for 72 h, and ground into powder. The powdered GBLs were passed through a 40-mesh sieve and stored at −20 °C.

The standard mixture of GA homologs (their % of concentration were C13:0 (20.7%), C15:0 (3.3%), C15:1 (51.6%), and C17:1 (21.1%), and C17:2 (3.3%)) with a purity ≥98% was prepared in our lab based on our previous research [2]. The ginkgolic acids C13:0, C15:1, and C17:1 standard (purity ≥ 98%) were purchased from Shanghai Yuanye Company (Shanghai, China). The standards of ginkgols C13:0, C15:1, and C17:1 were synthesized by our laboratory [2], with purity >98%. Methanol and acetonitrile (HPLC grade) were purchased from Tedia Company (Shanghai, China), while the trifluoroacetic acid (HPLC grade) was bought from Shanghai McLean Reagent Company (Shanghai, China). The glacial acetic acid, petroleum ether, ethyl acetate, and other reagents and chemicals (analytical grade) were bought from Sinopharm Chemical Reagent Co., Ltd. (Shanghai, China).

2.2. Experimental Design

2.2.1. Comparison of GA Detection Methods between European Pharmacopoeia 7.0 and Chinese Pharmacopoeia 2015 Edition

European Pharmacopoeia 7.0 edition [13] and Chinese Pharmacopoeia 2015 edition [15] were applied to detect GA. The chromatographic column used for testing the Chinese Pharmacopoeia method was an Agilent Eclipse XDB-C18 column (250 × 4.6 mm, 5 μm), and the chromatographic conditions were as follows: mobile phase (A) 0.1% trifluoroacetic acid (TFA) acetonitrile solution (B) 0.1% TFA aqueous solution; gradient elution: mobile phase A: 0–30 min, 75–90%, 30–50 min: 90–90%, 50–51 min, 90–75%, 51–55 min, 75–75%. The detection wavelength, chromatographic packing, injection volume, flow rate, and column temperature were 310 nm, C18, 50 μL, 1.0 mL/min, and 35 °C, respectively.

Regarding the European Pharmacopoeia 7.0 edition [13], the column for the European Pharmacopoeia method was an Ultimate XB-C8 column ((250 × 4.6 mm, 5 μm) from Yuexu Technology Company (Shanghai, China) and the chromatographic conditions were as follows: mobile phase (A) 0.01% trifluoroacetic acid acetonitrile solution (B) 0.01% trifluoroacetic acid aqueous solution, gradient elution: mobile phase A: 0–30 min, 75–90%, 30–35 min: 90–90%, 35–36 min, 90–75%, 36–46 min, 75–75%. The detection wavelength, chromatographic packing, injection volume, flow rate, and column temperature were 210 nm, C8, 50 μL, 1.0 mL/min, and 35 °C, respectively.

2.2.2. GA and Ginkgol Extraction from GBLs

Li et al.'s [7] protocol was applied for extraction. Freeze-dried GBL powder (0.5 g) was placed in a 50 mL centrifuge tube, 20 mL of methanol was added, and the total weight was recorded. The mixture was subjected to ultrasonication (40 °C, 100 W) for 45 min. The solution was cooled to room temperature, and methanol was added to recompensate the lost weight. The solution was centrifuged at 4669× *g* for 5 min at 4 °C. After diluting the supernatant 5×, the solution passed through a 0.45 μm filter. The solution was injected into the HPLC system (Agilent 1260, Agilent Corporation, Santa Clara, CA, USA) per the chromatographic conditions in Section 2.2.1. The peak area was recorded, and the contents of ginkgolic acid C13:0, C15:0, C15:1, C17:1, and C17:2 were determined. The total ginkgolic acid (TGA) was calculated as the sum of C13:0, C15:0, C15:1, C17:1, and C17:2 [7].

2.3. UV Absorption Spectrum for Ginkgolic Acids and Ginkgols

The GA and ginkgol standards were dissolved in acetonitrile solution to a concentration of 0.01 mg/ml and scanned within the 200–400 nm range to obtain the UV absorption, using a UV-visible spectrophotometer (UV-2450, Shimadzu Co., Ltd., Kyoto, Japan), to acquire a spectrum for GA and ginkgols [2].

2.4. Method Validation for GA and Ginkgols

The method of GA determination using European Pharmacopoeia version 7.0 has been validated [13]. Hence, in this study, we developed a reproducible method for detecting ginkgols in GBLs. Sample preparation was performed per the method stated in Section 2.2.2.

2.4.1. Chromatographic Conditions and System Suitability Test

The chromatographic column was the Ultimate XB-C8 column (250 × 4.6 mm, 5 μm, Yuexu Technology Company, Shanghai, China). The HPLC conditions are shown in Section 2.2.1, and the ginkgols' detection wavelength was set to 270 nm.

2.4.2. Preparation of Test Samples and Reference Solution

Detailed preparation of test samples is shown in Section 2.2.2. Regarding reference solution preparation, 5.66 mg of ginkgol C13:0, 5.30 mg of ginkgol C15:1, and 13.88 mg of ginkgol C17:1 were diluted with 100 mL methanol (100%) to obtain 0.0566 mg/mL, 0.0530 mg/mL, and 0.1388 mg/mL of ginkgol C13: 0, C15:1, and ginkgol 17:1 stock solution, respectively. Afterwards, each stock solution was diluted with methanol into gradient concentrations of ginkgol C13:0 (0.10608 mg/mL), C15:1 (0.1435 mg/mL), and C17:1 (0.1864 mg/mL) standard solutions.

2.4.3. Linear Range, Limit of Detection (LOD), and Limit of Quantification (LOQ)

The prepared gradient concentration standard solution was injected per the chromatographic conditions in Section 2.2.1. The peak area was documented, and a standard curve was drawn with the peak area as the ordinate (Y) and the concentration as the abscissa (X), with a linear regression performed to estimate linearity. The LOD was estimated as 3× the signal-to-noise (S/N) ratio [19,20]. The LOQ was determined using the protocol of Ahuja and Dong [21].

2.4.4. Precision, Stability, and Repeatability

For the precision test, following the procedure mentioned above (Section 2.4.2, reference solution), ginkgol stock solution (C13:0, C15:1, C17:1, mixed in a ratio of 3:4:3 $v/v/v$) was injected into the HPLC system per the chromatographic conditions in Section 2.4.1. The injection was performed 6 consecutive times, and the peak area was documented and substituted into the standard curve to estimate the ginkgol content and relative standard deviation (RSD) value.

For the stability test, the same test solution was injected once every 4 h based on the chromatographic conditions above. The injection was performed 6 consecutive times, and the peak area was noted and substituted into the standard curve to determine the ginkgol content and RSD value [20,22].

Finally, for the repeatability test, 6 parts of the test solution (3 different concentrations, each repeated 3 times; the test solution was repeated 6 times) were prepared, the samples were injected according to the chromatographic conditions, and the peak area was documented and substituted into the standard curve to compute the ginkgol content and RSD value [19].

2.4.5. Recovery Rate

The recovery rateprocedure required 3 × 3 parts of the test solution (3 different concentrations, each repeated 3 times). The ginkgol standard solution (C13:0, C15:1, C17:1, in a ratio of 1:1:3 $v/v/v$) was added to the GBLs in proportions of 100%, 120%, and

150%, respectively. After diluting the solution 5 times, 9 portions of the test solution were prepared. The sample was injected according to the chromatographic conditions above (Section 2.2.1). The peak area was documented and substituted into the standard curve to determine the ginkgol content and RSD value [23].

2.5. Usage of This Proposed Method on Processed GBL Products

The proposed method was applied to ginkgo tea and freeze-dried ginkgo leaves to detect ginkgols and ginkgolic acids simultaneously. The GBLs (1 kg) were divided into 5 equal parts to make ginkgo tea [24].

I. To make ginkgo black tea (BT), the GBLs were kneaded in a kneading machine with a rotation speed of 55 rpm (voltage 380 V, motor speed 1400 rpm) for 10 min and fermented in an incubator at 40 °C and 80% humidity for 8 h.
II. To make ginkgo dark tea (DT), the GBLs were dried in an oven at 150 °C for 5 min, kneaded at 55 rpm for 10 min, and fermented in an incubator at 40 °C and 80% humidity for 72 h.
III. When making ginkgo white tea (WT), the GBLs were spread flat and withered at 20 °C for 72 h.
IV. To make ginkgo green tea (GT), the ginkgo leaves were dried at 150 °C for 5 min and kneaded at 55 rpm for 10 min. All the above-mentioned semi-finished ginkgo teas were dried in an oven at 70 °C until their moisture content was approximately 7%.
V. For the freeze-dried (FD) control group, GBLs were frozen at −20 °C for 24 h and freeze-dried at −50 °C for 72 h. They were then crushed, passed through a 40-mesh sieve, and stored at −20 °C.

In the extraction procedure for ginkgols and ginkgolic acids in ginkgo tea products, Li et al.'s [7] protocol was applied. The GBT powder (0.5 g) was placed in a 50 mL centrifuge tube. Methanol (20 mL) was added, and the total weight was recorded. This was subjected to ultrasonication (40 °C, 40 kHz, 100 W, pulse off-and-on time of 2 s:10 s) for 45 min, after which the solution was cooled and methanol was added to balance the lost weight. The solution was centrifuged at 4669× g for 5 min at 4 °C. After diluting the supernatant 5×, the solution passed through a 0.45 μm filter.

2.6. Data Analysis

All experiments were performed in triplicate. Results are expressed as mean ± standard deviation. One-way analysis of variance (ANOVA) was performed on the ginkgol and ginkgolic acid homolog results using SPSS 26.0 software (IBM, Chicago, USA) and Tukey's test. $p < 0.05$ showed that the difference was statistically significant. Origin 2019b software (OriginLab®, Northampton, MA, USA) was used for the principal component analysis of the ginkgo tea products and other graphical representations.

3. Results and Discussion

3.1. Comparison of GA Detection Methods between European Pharmacopoeia 7.0 and Chinese Pharmacopoeia 2015 Edition

The chromatograms of the GA homolog standards and samples measured via the GA detection methods are shown in Figure 2A–D. The chromatographic conditions of the two detection methods achieved good baseline separation for the five monomers of the homologous mixture of GA and each monomer of GA in GBLs (Figure 2A–D). Comparing the analysis time, the European Pharmacopoeia method completed peak elution within 25 min (Figure 2A,C), and the analysis time was shorter, while the Chinese Pharmacopoeia method required approximately 44 min to complete peak elution (Figure 2B,D). The peak width/peak height was evaluated [25]. These findings showed that the peak width/peak height of C13:0, C15:1, and C17:1 in the Chinese Pharmacopoeia method for the detection of GA were 0.2746/32.41, 0.2778/65.08, 0.5900/10.45 min/mAu, respectively. For the European Pharmacopoeia, the peak width/peak height of C13:0, 15:1, and C17:1 were 0.2514/209.3, 0.2554/454.9, and 0.2849/148.0 min/mAu, respectively.

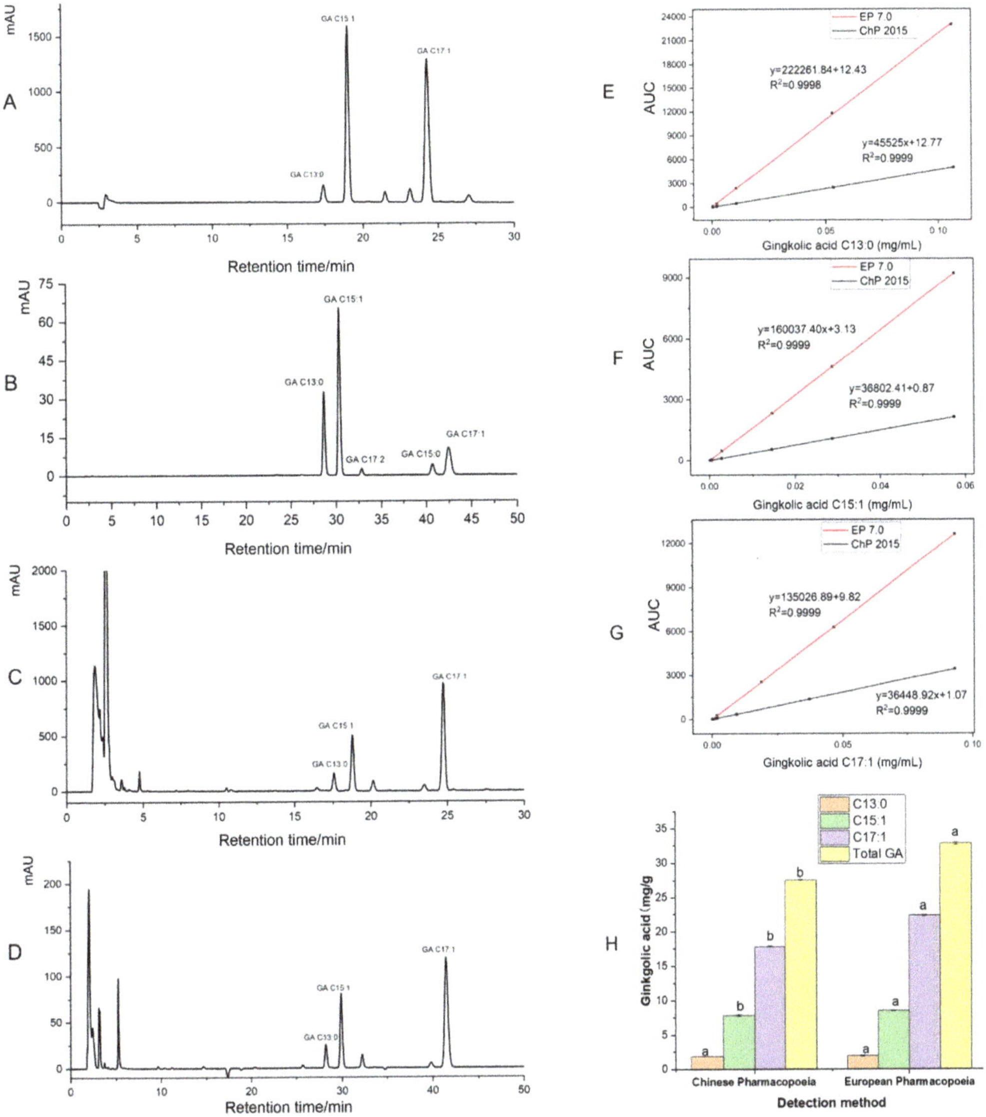

Figure 2. (**A–D**) Chromatogram of ginkgolic acid standard (**A**) and freeze-dried ginkgo leaf (**C**) using European Pharmacopoeia method, ginkgolic acid standard (**B**) and freeze-dried ginkgo leaf (**D**) using Chinese Pharmacopoeia method. (**E–G**) Chinese Pharmacopoeia and European Pharmacopoeia methods to detect ginkgolic acid C13:0 (**E**), C15:1 (**F**), and C17:1 (**G**) standard curves. (**H**) The content of ginkgolic acid in ginkgo leaves was detected via different detection methods. **Note:** different letters indicate significant differences in the same compound, $p < 0.05$, EP 7.0 using Tukey test. EP 7.0, European Pharmacopoeia 7.0 version method; ChP 2015, Chinese Pharmacopoeia 2015 version method.

Ginkgolic acid C13:0, C15:1, and C17:1 standard solutions (shown in Section 2.1) of different concentrations were injected, and the peak areas were measured to obtain the fitting of ginkgolic acid C13:0, C15:1, and C17:1 using the two detection methods. The standard curves are shown in Figure 2E–G. It can be seen that when the concentration of the standard GA was the same, the response value measured using the European Pharmacopoeia was much higher than the response value measured using the Chinese Pharmacopoeia. This is because 210 nm was the strong absorption of the phenyl ring of GA, and its intensity was much higher than 310 nm [23]. The absorption of the carboxyl group on the phenyl ring of GA was at 210 nm, and the detection sensitivity was higher. It is well known that the n---pi*band measured from COOH is of very low sensitivity. Therefore, the European Pharmacopoeia method is more suitable for detection at low concentrations than the Chinese Pharmacopoeia method. The European Pharmacopoeia method is more sensitive in determining the GA content in EGb and more suitable for detecting the GA limit standard in EGb. For non-polar compounds, such as ginkgo phenolic acids, both C18 and C8 columns are suitable for separation [26]; however, it is easier to use a C8 column than a C18 column [27]. It requires a small sample, there is no sample cleanup, the C8 RP-HPLC separation is fine, the chromatogram is clean, the sensitivity is sufficient, and it uses a readily available low-cost UV detector [26]. Therefore, the C8 column was decided upon for subsequent experiments.

The same GA in GBLs was measured using the Chinese Pharmacopoeia 2015 edition methods and the European Pharmacopoeia edition 7.0, as shown in Figure 2H. It was found that the C15:1 and C17:1 contents of GA and the total GA (TGA) measured using the European Pharmacopoeia method were significantly higher than those found using the Chinese Pharmacopoeia method. According to the test results, there were significant differences ($p < 0.05$) between the test results for the same sample (Figure 2H). The test results using the method specified in the European Pharmacopoeia were ~19.25% higher than those in the Chinese Pharmacopoeia. The main difference was due to the large difference in the results for ginkgolic acid C15:1 and C17:1 (Figure 2H), resulting in a high GA content measured in the European Pharmacopoeia. Comparing the two detection methods, the European Pharmacopoeia version 7.0 detection method had high sensitivity and a short peak time; the Chinese Pharmacopoeia method had slightly lower sensitivity and a longer peak time, especially since the retention time of C17:1 was close to 44 min (Figure 2B). Considering that the peak time was too long, this would have led to the peak broadening and affected the sensitivity [28]. Therefore, this study chose the method specified in the European Pharmacopoeia version 7.0 to develop a method to detect ginkgols and GA simultaneously.

The standard mixture of GA homologous series and the homologous mixture of ginkgol was mixed and injected into the HPLC according to the European Pharmacopoeia 7.0 method. The chromatogram is shown in Figure 3A. As shown in Figure 3A, the retention times of the four monomers of ginkgolic acid, C13:0, C15:1, C15:0, and C17:1, were 17.46, 18.65, 23.281 and 24.51 min, respectively. The retention times of the four monomers of ginkgols, C15:1, C17:2, C15:0, and C17:1, are 21.07, 21.16, 22.29, and 27.27 min, respectively. However, ginkgol C13:0 and ginkgolic acid C17:2 overlapped (Figure 3A) and were not separated.

Further comparisons were made between the chromatograms of ginkgol C13:0 injected alone (Figure 3B) and the GA mixture (Figure 3C). It was found that ginkgol C13:0 (Rt = 19.911 min, Figure 3B) and ginkgolic acid C17:2 (Rt = 19.979 min, Figure 3C) were very close; that is, the peak area at the retention time of 19.9 min may be the sum of the peak areas of ginkgol C13:0 and ginkgolic acid C17:2. The detection of other monomers of GA was not affected. The proportion of ginkgolic acid C17:2 in GA homologs was very small, and the content in GBLs was also very low (almost negligible). Neither the Chinese Pharmacopoeia [15] nor the European Pharmacopoeia [13] include ginkgolic acid C17:2 in the measurement of total GA. Ginkgol C13:0 is the main component of ginkgol homologs and cannot be ignored [3]. Therefore, we researched the absorption spectra of GA and ginkgols to select the wavelength for detecting ginkgol.

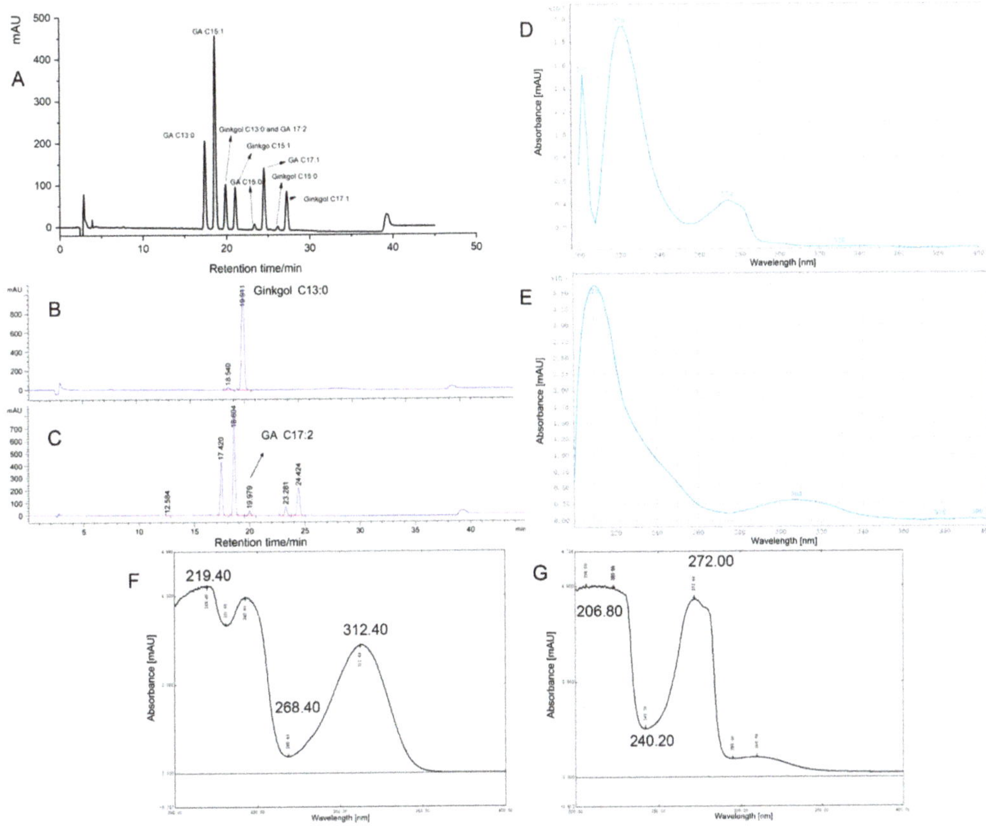

Figure 3. (**A**) Chromatogram of the European Pharmacopoeia method for standard ginkgolic acid mixed with ginkgol. (**B**,**C**) Chromatogram of ginkgol C13:0. (**B**) Ginkgolic acid mixture (**C**) detected at 210 nm using the European Pharmacopoeia method. (**D**,**E**) The UV spectra of ginkgol (**D**) and ginkgolic acid (**E**) in 0.01% trifluoroacetic acid acetonitrile solution–0.01% trifluoroacetic acid aqueous solution. (**F**,**G**) UV spectrum of ginkgolic acid (**F**) and ginkgol (**G**) in acetonitrile solution.

3.2. Selection of Wavelength for Detecting Ginkgol and GA

Figure 3D–G show the UV absorption spectra of GA and ginkgol in various solutions. As can be seen from Figure 3F, using acetonitrile solution, the two maximum absorption peaks of GA were located at 219 nm and 312 nm and the maximum UV absorption peaks of ginkgol were located at 206 nm and 272 nm (Figure 3G). Ginkgols had a maximum absorption of ~272 nm, while GA had almost no absorption at 268 nm. The UV scan of ginkgol and GA in the mobile phase of European Pharmacopoeia version 7.0 is shown in Figure 3D,E. Ginkgol had medium to strong absorption at 274 nm (Figure 3D), while GA had almost no absorption at 274 nm (Figure 3E). Therefore, based on detecting GA at 210 nm, ginkgol could be detected at 270 nm to avoid ginkgolic acid C17:2 influencing the detection of ginkgol C13:0.

To verify the feasibility of this idea, the GA homologous mixed standard was injected and monitored at wavelengths of 210 nm, 270 nm, and 310 nm (Figure S1). Figure S1A–C are the chromatograms of five homologs of ginkgolic acid at 310 nm, 270 nm, and 210 nm, respectively.

Figure S1A,C show five sharp, symmetrical absorption peaks for GA, with consistent retention times. Since the benzene ring of ginkgolic acid has strong absorption at 210 nm, its absorption intensity is much higher than that of the carboxyl group on the benzene ring at 310 nm [23]. Therefore, the response value in Figure S1C is ~10× that in Figure S1A (regarding peak height). At the same time, in the 270 nm chromatogram of the GA mixed standard (Figure S1B), except for a small impurity peak at 17.416 min (Rt of ginkgolic acid C13:0 = 17.42 min), no other peaks appeared. Ginkgol C13:0 had the earliest peak among the five ginkgols (Rt = 19.91 min). The retention times of the remaining four ginkgol monomers were 21.07, 21.16, 22.29, and 27.27 min, respectively. Therefore, the detection wavelength of ginkgol was changed and set at 270 nm, as there would be no interference from GA.

Ginkgol C13:0, C15:1, and C17:1 standards and ginkgolic acid C13:0, C15:1, C15:0, and C17:1 standards were mixed and injected, and their chromatograms at 270 nm and 210 nm were attained (Figure 4A,B). The chromatogram of ginkgol C13:0, C15:1, and C17:1 at 270 nm is shown in Figure 4A, and no interference from GA is seen. Figure 4B shows the chromatograms of ginkgol C13:0, C15:1, and C17:1 and ginkgolic acid C13:0, C15:1 and C17:1 at 210 nm. The baseline between the monomers of ginkgol and GA is complete.

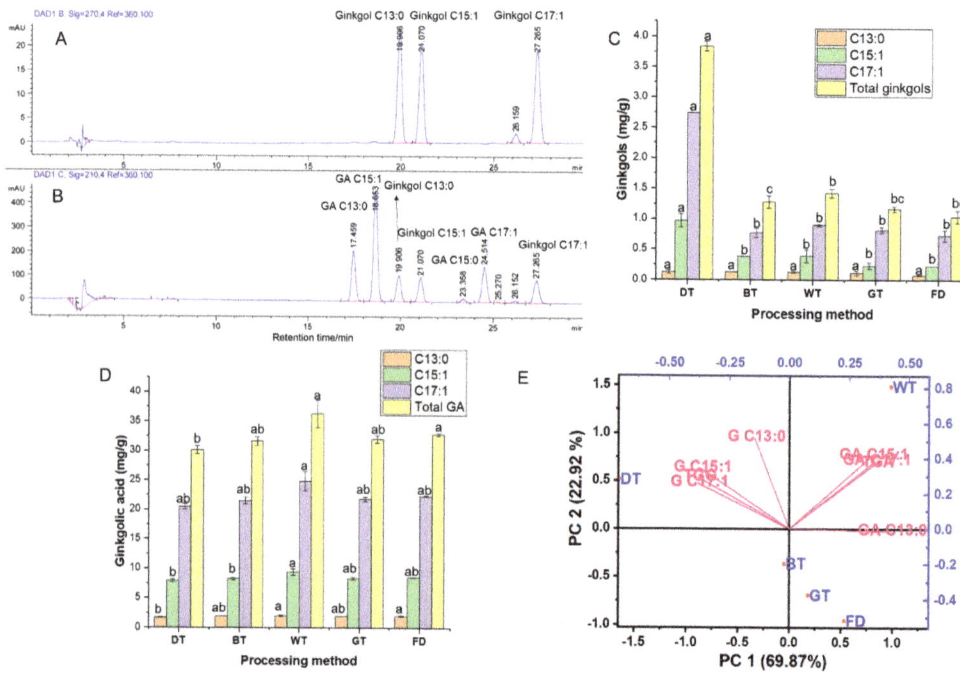

Figure 4. (**A**,**B**) Chromatogram of ginkgol (**A**) and the mixture of ginkgol and ginkgolic acid (**B**) detected using the European Pharmacopoeia method. (**C**) Ginkgol and ginkgolic acid (**D**) content in ginkgo leaf tea and freeze-dried ginkgo leaves. (**E**) Three-dimensional biplot from principal component analysis. **Note.** Different letters indicate significant differences in the same compound, $p < 0.05$ using the Tukey test. BT, black tea, WT, white tea, GT, green tea, DT, dark tea, and FD, freeze-dried GBLs.

3.3. Methodological Validation

3.3.1. Linear Range, Limit of Detection, and Limit of Quantification

To determine the method's accuracy, standard calibration curves were determined by injecting standard solutions at six concentration levels [23]. A standard curve was drawn, and the linear regression equation for ginkgol is shown in Table S1. The linearity of the method pertains to the extent of analyte concentration within which the method is applicable [29]. The standard curves of each ginkgol standard product showed an excellent linear relationship with an $R^2 < 0.999$ and a good linearity at a similar concentration range [30]. The limit of detection (LOD), which refers to the minimum concentration that can be reliably quantified using a certain analytical method [30], was determined (Table S1). The findings showed an LOD of 0.61, 0.5, and 0.06 ppm for C13:0, C15:1, and C17:1, respectively. The limit of quantification (LOQ) stands for the smallest amount or the lowest concentration of a substance that can be determined through a given analytical procedure with established accuracy, precision, and uncertainty [21]. The LOQ of ginkgol C13:0, C15:1, and C17:1 was 2.01 ppm, 1.65 ppm, and 0.20 ppm, respectively.

3.3.2. Precision and Repeatability Experiments

The method's precision was determined by injecting ginkgol standard stock solution (C13:0, C15:1, C17:1, 3:4:3 mixture) into the HPLC system using the above chromatographic conditions for six consecutive injections. As shown in Table 1, the RSD of the measured ginkgol C13:0, C15:1, and C17:1 standards was 0.27%, 0.33%, and 0.28%, respectively. The RSD of the total ginkgol content was 0.20%. The RSD values met the acceptable threshold, indicating that the approach demonstrated a rather high level of precision [23,31].

Table 1. Method validation for precision, stability, and repeatability of test results.

Standard Solution Precision Test	Standard Solution	Ginkgol C13:0 (mg/mL)	Ginkgol C15:1 (mg/mL)	Ginkgol C17:1 (mg/mL)	Total Ginkgol (mg/mL)
	1	0.0244	0.0295	0.0430	0.0969
	2	0.0245	0.0297	0.0431	0.0972
	3	0.0245	0.0297	0.0431	0.0973
	4	0.0246	0.0297	0.0432	0.0975
	5	0.0245	0.0295	0.0433	0.0974
	6	0.0246	0.0298	0.0430	0.0974
	Average value	0.0245	0.0297	0.0431	0.0973
	RSD	0.27%	0.33%	0.29%	0.20%
Sample solution stability test	Time (h)	Ginkgol C13:0 (mg/g)	Ginkgol C15:1 (mg/g)	Ginkgol C17:1 (mg/g)	Total ginkgol (mg/g)
	0	0.0259	0.3344	0.6511	1.0110
	4	0.0276	0.3351	0.6328	0.9981
	8	0.0278	0.3542	0.6570	1.0389
	12	0.0265	0.3540	0.6518	1.0324
	16	0.0285	0.3572	0.6390	1.0247
	20	0.0275	0.3591	0.5866	0.9731
	Average value	0.0273	0.3490	0.6364	1.0130
	RSD	3.06%	2.93%	3.73%	2.21%
Sample solution repeatability test	Sample	Ginkgol C13:0 (mg/g)	Ginkgol C15:1 (mg/g)	Ginkgol C17:1 (mg/g)	Total ginkgol (mg/g)
	1	0.0272	0.3572	0.6390	1.0247
	2	0.0264	0.3598	0.6459	1.0321
	3	0.0249	0.3518	0.6505	1.0272
	4	0.0244	0.3477	0.6252	0.9973
	5	0.0254	0.3367	0.6115	0.9735
	6	0.0251	0.3370	0.6138	0.9736
	Average value	0.0256	0.3484	0.6310	1.0047
	RSD	3.76%	2.58%	2.40%	2.45%

Repeatability experiments were conducted by taking six test samples and injecting them into the HPLC system per the above chromatographic conditions. As shown in Table 1, the RSDs of the measured ginkgol C13:0, C15:1, and C17:1 were 3.76%, 2.58%, and 2.40%, respectively. The total ginkols' RSD was 2.45%, and the method stability was high. The RSD values were within the acceptable criteria and indicated good method [23,31].

3.3.3. Stability Experiment

The same test solution was injected once every 4 h into the HPLC system, per the above chromatographic conditions, for six consecutive injections. As shown in Table 1, the RSD of the measured ginkgol C13:0, C15:1, and C17:1 was 3.06%, 2.93%, and 3.73%, respectively. The RSD of the total ginkgol content was 2.21%. The RSD values met the acceptable threshold, indicating that the approach demonstrated a rather good level of stability [23,31].

3.3.4. Recovery

An experiment was conducted to validate the accuracy and precision of the suggested approach for detecting ginkgols in *Ginkgo biloba* leaves [30]. Samples of the below 9 test solutions were injected into the HPLC, the data were recorded, and the results were calculated according to the standard curve. The recovery rates of ginkgol C13:0, C15:1, and C17:1 are shown in Table 2. The recoveries of the measured ginkgol C13:0 content (low, medium, and high sample groups) were 95.82%, 96.41%, and 95.60%, respectively, and the RSDs were 3.93%, 2.90%, and 3.72%, respectively. The recoveries of the measured ginkgol C15:1 content (low, medium, and high sample groups) were 95.82%, 96.41%, and 93.63%, respectively, and the RSDs were 3.24%, 2.86%, and 2.97%, respectively. The recoveries of the measured ginkgol C17:1 content (low, medium, and high loading groups) were 97.45%, 97.49%, and 103.94%, respectively, and the RSDs were 1.81%, 3.44%, and 1.94%, respectively. The acceptable recovery range for most residue analysis technique guidelines is typically between 70% and 120% [32]. Thus, our results (Table 2) show that our recovery range (~94 to 98%) is acceptable.

Table 2. The recovery rate experiment results for ginkgol C13:0, C15:1, C17:1 in *Ginkgo biloba* leaves.

C13:0	Sample	Sample Amount (g)	C13:0 (mg)	Added Amount (mg)	Theoretical Value (mg)	Actual Value (mg)	Total Recovery Rate (%)	Average Value (%)	RSD (%)
	1	0.5014	0.0147	0.0556	0.0703	0.0643	91.50		
	2	0.4985	0.0146	0.0556	0.0702	0.0669	95.29	95.82	3.93
	3	0.5015	0.0147	0.0556	0.0703	0.0708	100.67		
	4	0.5046	0.0148	0.0667	0.0815	0.0798	97.91		
	5	0.5010	0.0147	0.0667	0.0814	0.0804	98.83	96.41	2.90
	6	0.5011	0.0147	0.0667	0.0814	0.0753	92.48		
	7	0.5018	0.0147	0.0834	0.0981	0.0927	94.49		
	8	0.5020	0.0147	0.0834	0.0981	0.0985	100.40	95.60	3.72
	9	0.4999	0.0146	0.0834	0.0980	0.0901	91.91		
C15:1	**Sample**	**Sample amount (g)**	**C15:1 (mg)**	**Added amount (mg)**	**Theoretical value (mg)**	**Actual value (mg)**	**Total recovery (%)**	**Average value (%)**	**RSD (%)**
	1	0.5014	0.1756	0.0795	0.2551	0.2501	98.01		
	2	0.4985	0.1746	0.0795	0.2541	0.2414	95.01	94.52	3.24
	3	0.5015	0.1757	0.0795	0.2552	0.2311	90.56		
	4	0.5046	0.1768	0.0954	0.2722	0.2621	96.32		
	5	0.5010	0.1755	0.0954	0.2709	0.2777	102.50	98.52	2.86
	6	0.5011	0.1755	0.0954	0.2709	0.2621	96.75		
	7	0.5018	0.1758	0.1193	0.2950	0.2966	100.55		
	8	0.5020	0.1759	0.1193	0.2951	0.2768	93.80	96.61	2.97
	9	0.4999	0.1751	0.1193	0.2944	0.2811	95.50		

Table 2. *Cont.*

C17:1	Sample	Sample amount (g)	C17:1 (mg)	Added amount (m)	Theoretical value (mg)	Actual value (mg)	Total recovery (%)	Average value (%)	RSD (%)
	1	0.5014	0.3153	0.3470	0.6623	0.6371	96.21		
	2	0.4985	0.3134	0.3470	0.6604	0.6601	99.95	97.45	1.81
	3	0.5015	0.3153	0.3470	0.6623	0.6371	96.20		
	4	0.5046	0.3173	0.4164	0.7337	0.6831	93.11		
	5	0.5010	0.3150	0.4164	0.7314	0.7176	98.11	97.49	3.44
	6	0.5011	0.3151	0.4164	0.7315	0.7406	101.25		
	7	0.5018	0.3155	0.5205	0.8360	0.8683	103.86		
	8	0.5020	0.3156	0.5205	0.8361	0.8900	106.44	103.94	1.94
	9	0.4999	0.3143	0.5205	0.8348	0.8475	101.52		
Total ginkgol	**Sample**	**Sample amount (g)**	**Total ginkgol (mg)**	**Added amount (mg)**	**Theoretical sum (mg)**	**Actual sum (mg)**	**Total recovery rate (%)**	**Average value (%)**	**RSD (%)**
	1	0.5014	0.5056	0.4821	0.9877	0.9515	96.34	96.58	1.40
	2	0.4985	0.5027	0.4821	0.9848	0.9685	98.34		
	3	0.5015	0.5057	0.4821	0.9878	0.9390	95.06		
	4	0.5046	0.5088	0.5785	1.0873	1.0250	94.27	97.67	2.46
	5	0.5010	0.5052	0.5785	1.0837	1.0757	99.26		
	6	0.5011	0.5053	0.5785	1.0838	1.0780	99.46		
	7	0.5018	0.5060	0.7232	1.2291	1.2518	102.32	101.52	1.56
	8	0.5020	0.5062	0.7232	1.2293	1.2653	102.93		
	9	0.4999	0.5041	0.7232	1.2272	1.2187	99.31		

Finally, ginkgol C13:0, C15:1, and C17:1 were added to obtain the total ginkgol content. The calculations of the total ginkgo acid content of each sample are shown in Table 2. The recovery rate of ginkgol (low, medium, and high sample addition groups) was 96.58%, 97.67%, and 101.52%, respectively, and the RSDs were 1.40%, 2.46%, and 1.56%, respectively. The RSD values met the acceptable threshold, indicating that the approach demonstrated a rather high level of precision [31].

3.4. Application of the Developed Method to Investigate Ginkgols and GA in Processed GBLs

The developed method was employed to quantify the GA and ginkgols in various processed GBL teas and freeze-dried GBLs. The HPLC chromatogram is found in Figure 5, showing a baseline separation. The contents of GA homologs C13:0, C15:1, and C17:1 account for more than 90% of the total ginkgolic acid [24], and our results (Figure 4C,D) attest to that. It can be seen from Figure 4D that the total ginkgolic acid content of dark tea (30.07 mg/g) was significantly lower than those of the other three ginkgo teas and freeze-dried GBLs; the ginkgolic acid content of white tea was the highest among all tested samples, which may be due to withering [24]. The process leads to the accumulation of intracellular oxygen free radicals and the loss of selective permeability in the cell membrane [33]. Eventually, the cell contents burst, and ginkgolic acid flows out of the cells. After the tea is made, the ginkgolic acid content is relatively high. It should be noted that GA in all the GBTs was >5 ppm, thus requiring combined multiple physical processing technologies in GBT processing; hybrid processing is increasing [34]. In a study by Li et al. [7], the pulsed-light irradiation of GBLs improved the retention of the main bioactive compounds and the inactivation of polyphenol oxidase (PPO) and polyphenol peroxidase (POD) in GBLs compared to the control. Hence, combining new physical processing technology with fermentation technology used for GBTs may provide a way to reduce the GA content and retain bioactive ingredients, such as ginkgo flavonoids and lactones, while processing GBTs.

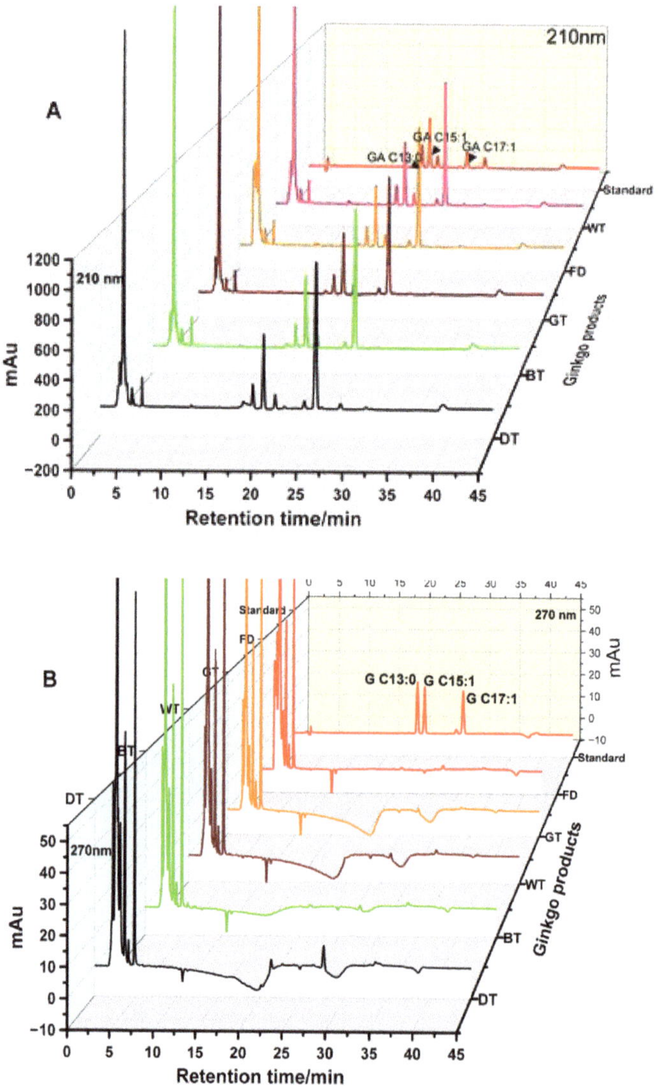

Figure 5. HPLC chromatogram of (**A**) ginkgolic acids and (**B**) ginkgols in various ginkgo products.

Three main types of ginkgols (ginkgol C13:0, C15:1, and C17:1) were measured in freeze-dried GBLs and ginkgo tea. Figure 4C shows the measurement results for three kinds of ginkgols and the total ginkgol content in ginkgo tea. Figure 4C shows that the total ginkgol content in dark tea was the highest and was significantly different from the content in freeze-dried GBLs and the other three ginkgo teas ($p < 0.05$). The total ginkgol content of dark tea was ~3.84 mg/g dt, which was ~2.76 mg higher than that of freeze-dried GBLs (Figure 4C), while the ginkgolic acid of dark tea was ~2.80 mg lower than that of freeze-dried GBLs (Figure 4D). The increased content of ginkgol is similar to the content degraded by ginkgolic acid. It is speculated that the combination of fixation [2] and fermentation [35] during dark tea production causes ginkgolic acid degradation, thus increasing ginkgol content. No similar phenomenon was found in other ginkgo teas, such as green and lightly fermented black tea. Compared with freeze-dried GBLs (Figure 4C,D), their ginkgolic acid content decreased and their ginkgol content increased slightly, but there was no significant difference ($p > 0.05$). The contents of ginkgolic acid and ginkgol in withered white tea were marginally higher than those in freeze-dried GBLs, possibly related to the outflow of cell contents caused by withering [24].

3.5. Principal Component Analysis (PCA)

PCA was performed (using the raw values of the ginkgolic acids and ginkgol homologs in Figure 4C,D) to visualize the association between GA and ginkgols and five different processing methods for GBLs (Figure 4E). PCA is an unsupervised statistical approach that uses the reduction of the dimensions of the principal components to analyze variations across sample groups [7]. Figure 4E shows the biplot of the first two principal components, PC1 and PC2, representing 92.79% of the total variability. A cumulative principal component greater than 85% is high enough to explain the total variance in the dataset [36,37]. Thus, our selection of PC1-2 is justifiable for accurately differentiating bioactive compounds among groups, and there is no need to perform orthogonal partial least squares discriminant analysis (OPLS-DA), used to mitigate environmental and systematic errors. PC1 (69.87%) describes the most sample variation. It also means that PC1 explained most related parameters, whereas PC2 and PC3 described lesser correlated parameters. In the biplot (Figure 4E), black tea (BT), white tea (GT), and freeze-dried GBLs appeared on PC1's positive side, while dark tea (DT) was on the negative side of PC1. This means that by using our developed method, we can classify ginkgo tea products based on the content of ginkgolic acid and ginkgols. In Figure 4B, FD shares the same characteristics as BT and GT, meaning that their processing methods have similar ginkgols and ginkgolic acid content. Furthermore, it was observed that gingkols were on the negative side of PC1, whereas ginkgolic acids were on the positive side of PC1. This proves that the increased content of ginkgol obtained by decarboxylating ginkgolic acids to ginkgols (3-alkylphenols) is due to processing treatments and aligns with the findings of Yang et al. [2].

The eigenvalues were calculated (Table S2) to understand which processing method and compound (various ginkgolic acids and ginkgols) contributed significantly to the PCA. The findings (Table S2) showed that PC1 was contributed considerably by GA C13:0 (0.4222) and ginkgol C17:1 (−0.3812), whereas PC2 was significantly contributed by ginkgol C13:0 (0.5181) and ginkgolic acid C15:1 (0.4168) based on their Eigenvector coefficients, and these align with the biplot (Figure 4E). Pearson's correlations were used to establish and characterize the relationships between the ginkgolic acid and ginkgols, and the results (Table S3) attested to the PCA findings (Figure 4E), thus indicating that the processing method affected ginkgolic acids and ginkgols.

4. Conclusions

This research used the methods of the Chinese Pharmacopoeia 2015 edition and the European Pharmacopoeia edition 7.0 to measure and compare the ginkgolic acid contents in GBLs. It was found that the accuracy of the European Pharmacopoeia method was ~19.25% higher than the Chinese Pharmacopoeia method and its sensitivity was higher than that of the Chinese Pharmacopeia. An HPLC-DAD method was established to simultaneously determine GA and ginkgols.. The detection wavelengths of GA and ginkgol were 210 and 270 nm, respectively. The flow rate and injection volume were 1.0 mL/min and 50 μL, respectively. The linearity had $R^2 > 0.999$, and the RSDs for the precision, stability, and repeatability of total ginkgols were 0.20%, 2.21%, and 2.45%, respectively, in six parallel determinations. The recoveries of low, medium, and high groups were 96.58%, 97.67%, and 101.52%, respectively, and were within the acceptable criteria. The LOD of ginkgol C13:0, C15:1, and C17:1 were 0.61 ppm, 0.50 ppm, and 0.06 ppm, respectively, whereas the LOQ of ginkgol C13:0, C15:1, and C17:1 were 2.01 ppm, 1.65 ppm, and 0.20 ppm, respectively. Finally, this method was able to accurately measure the GA and ginkgol content in five different ginkgo products that can be used for the simultaneous determination of the GA and ginkgol contents in various processed ginkgo leaves (freeze-dried ginkgo leaves, ginkgo black tea, ginkgo dark tea, ginkgo white tea, and ginkgo green tea). Future studies could evaluate the use of the European Pharmacopoeia column with the Chinese Pharmacopoeia eluent and gradient, and vice versa.

Supplementary Materials: The following supporting information can be downloaded at: https://www.mdpi.com/article/10.3390/foods13081250/s1, Figure S1: Chromatogram of ginkgolic acid mixed standard at 310nm (A), 270 nm (B) and 210 nm (C) by the European Pharmacopoeia methods; Table S1: Coefficients of the eigenvectors were taken from different principal components analysis for compounds; Table S2: Coefficients of the eigenvectors were taken from different principal components analysis for compounds; Table S3: Pearson correlation showing the relationship between ginkgolic acids and ginkgols.

Author Contributions: Conceptualization, X.-M.Y.; Methodology, F.L.; Software, I.D.B.; Validation, F.L.; Formal analysis, I.D.B. and F.L.; Investigation, F.L.; Resources, X.-M.Y.; Data curation, F.L.; Writing—original draft, I.D.B. and F.L.; Writing—review & editing, X.-M.Y.; Visualization, I.D.B.; Supervision, X.-M.Y.; Project administration, X.-M.Y.; Funding acquisition, X.-M.Y. All authors have read and agreed to the published version of the manuscript.

Funding: The Jiangsu Market Supervision and Administration Bureau (KJ21125110) and the National Natural Science Foundation of China (grant no. 81372404) supported this study. The APC was funded by Isaac Duah Boateng.

Institutional Review Board Statement: Not applicable.

Informed Consent Statement: Not applicable

Data Availability Statement: The original contributions presented in the study are included in the article/Supplementary Materials, further inquiries can be directed to the corresponding author.

Conflicts of Interest: Author Isaac Duah Boateng was employed by the company Certified Group. The role of the company was a scientist. He declares there is no conflicts of interest, as he analyzed, validated, and reviewed the work when he was in the University, instead of the company. The remaining authors declare that the research was conducted in the absence of any commercial or financial relationships that could be construed as a potential conflict of interest.

List of Abbreviations

GB	*Ginkgo biloba*
EGb	Extract of *Ginkgo biloba*
BT	Black tea
GT	Green tea
GA	Ginkgolic acid
TFA	Trifluoroacetic acid
LOQ	Limit of quantification
TGA	Total ginkgolic acid
HPLC	High-performance liquid chromatography
LC-MS	Liquid chromatography–mass spectrometry
GBLs	*Ginkgo biloba* leaves
DT	Dark tea
WT	White tea
FD	Freeze-dried
UV	Ultra-violet
LOD	Limit of detection
RSD	Relative standard deviation
PCA	Principal component analysis

References

1. Boateng, I.D.; Soetanto, D.A.; Li, F.; Yang, X.; Li, Y. Separation and Purification of Polyprenols from *Ginkgo biloba* L. Leaves by Bulk Ionic Liquid Membrane and Optimizing Parameters. *Ind. Crops Prod.* **2021**, *170*, 113828. [CrossRef]
2. Yang, X.M.; Wang, Y.F.; Li, Y.Y.; Ma, H. Le Thermal Stability of Ginkgolic Acids from *Ginkgo biloba* and the Effects of Ginkgol C17:1 on the Apoptosis and Migration of SMMC7721 Cells. *Fitoterapia* **2014**, *98*, 66–76. [CrossRef] [PubMed]

3. Li, F.; Boateng, I.D.; Yang, X.; Li, Y. Extraction, Purification, and Elucidation of Six Ginkgol Homologs from *Ginkgo biloba* Sarcotesta and Evaluation of Their Anticancer Activities. *Molecules* **2022**, *27*, 7777. [CrossRef] [PubMed]
4. Chen, H.Y.; Cheng, K.C.; Hsu, R.J.; Hsieh, C.W.; Wang, H.T.; Ting, Y. Enzymatic Degradation of Ginkgolic Acid by Laccase Immobilized on Novel Electrospun Nanofiber Mat. *J. Sci. Food Agric.* **2020**, *100*, 2705–2712. [CrossRef] [PubMed]
5. Ren, Q.; Chen, J.; Ding, Y.; Cheng, J.; Yang, S.; Ding, Z.; Dai, Q.; Ding, Z. In Vitro Antioxidant and Immunostimulating Activities of Polysaccharides from *Ginkgo biloba* Leaves. *Int. J. Biol. Macromol.* **2019**, *124*, 972–980. [CrossRef]
6. Liu, X.-G.; Lu, X.; Gao, W.; Li, P.; Yang, H. Structure, Synthesis, Biosynthesis, and Activity of the Characteristic Compounds from *Ginkgo biloba* L. *Nat. Prod. Rep.* **2022**, *3*, 474–511. [CrossRef]
7. Li, F.; Boateng, I.D.; Chen, S.; Yang, X.M.; Soetanto, D.A.; Liu, W. Pulsed Light Irradiation Improves Degradation of Ginkgolic Acids and Retainment of Ginkgo Flavonoids and Terpene Trilactones in *Ginkgo biloba* Leaves. *Ind. Crops Prod.* **2023**, *204*, 117297. [CrossRef]
8. Boateng, I.D. A Critical Review of Current Technologies Used to Reduce Ginkgotoxin, Ginkgotoxin-5′-Glucoside, Ginkgolic Acid, Allergic Glycoprotein, and Cyanide in *Ginkgo biloba* L. Seed. *Food Chem.* **2022**, *382*, 132408. [CrossRef]
9. Chinese Pharmacopoeia Committee. *Ginkgo Leaves Extract*; China Medical Science and Technology Press: Beijing, China, 2020; Volume 1.
10. Boateng, I.D. Ginkgols and Bilobols in *Ginkgo biloba* L. A Review of Their Extraction and Bioactivities. *Phyther. Res.* **2023**, *37*, 3211–3223. [CrossRef]
11. Liu, L.; Wang, Y.; Zhang, J.; Wang, S. Advances in the Chemical Constituents and Chemical Analysis of *Ginkgo biloba* Leaf, Extract, and Phytopharmaceuticals. *J. Pharm. Biomed. Anal.* **2021**, *193*, 113704. [CrossRef]
12. Jianbiao, Y.; Ling, F.; Wang, R. Comparative Study on the Limit Detection Methods of Ginkgolic Acid in Ginkgo Leaf Extracts from the Chinese Pharmacopoeia and the European Pharmacopoeia. *J. Pharm. Anal.* **2012**, *32*, 2055–2059.
13. European Pharmacopoeia Commission. *European Pharmacopoeia*, 7th ed.; EDQM: Strasbourg, France, 2012.
14. Chinese Pharmacopoeia Commission. *Pharmacopoeia of the People's Republic of China*; China Medical Science and Technology Press: Beijing, China, 2010.
15. Chinese Pharmacopoeia Commission. *Chinese Pharmacopoeia*; China Medical Science Press: Beijing, China, 2015.
16. Chinese Pharmacopoeia Commission. *Pharmacopoeia of the People's Republic of China*; China Medical Science and Technology Press: Beijing, China, 2020.
17. Weihong, T.; Zhaobang, S.; Wang, C. Isolation and Identification of Alkylphenol Compounds in Ginkgo Leaves. *Chem. Ind. For. Prod.* **2001**, *4*, 1–6.
18. Liu, P. *Isolation, Identification and Analysis Method Establishment of Ginkgol Monomers*; Jiangsu University: Zhenjiang, China, 2016.
19. Tsochatzis, E.D.; Alberto Lopes, J.; Gika, H.; Kastrup Dalsgaard, T.; Theodoridis, G. Development and Validation of an UHPLC-QTOF-MS Method for the Quantification of Cyclic Polyesters Oligomers in Pasta by Applying a Modified QuEChERS Clean-Up. *Food Chem.* **2021**, *347*, 129040. [CrossRef] [PubMed]
20. European Commission. *Analytical Quality Control and Method Validation Procedures for Pesticide Residues Analysis in Food and Feed*; European Commission: Brussels, Belgium, 2017.
21. Ahuja, S.; Dong, M.W. *Handbook of Pharmaceutical Analysis by HPLC*; Elsevier Inc.: Amsterdam, The Netherlands, 2005; Volume 6.
22. Barimah, A.O.; Chen, P.; Yin, L.; El-seedi, H.R.; Zou, X.; Guo, Z. SERS Nanosensor of 3-Aminobenzeneboronic Acid Labeled Ag for Detecting Total Arsenic in Black Tea Combined with Chemometric Algorithms. *J. Food Compos. Anal.* **2022**, *110*, 104588. [CrossRef]
23. Lee, H.; Lim, H.; Yang, J.; Hong, J. Rapid Determination of Ginkgolic Acids in *Ginkgo biloba* Leaf Using Online Column Switching High-Performance Liquid Chromatography-Diode Array Detection and Confirmation by Liquid Chromatography-Tandem Mass Spectrometry. *Bull. Korean Chem. Soc.* **2013**, *34*, 3629–3634. [CrossRef]
24. Li, F.; Boateng, I.D.; Yang, X.M.; Li, Y.; Liu, W. Effects of Processing Methods on Quality, Antioxidant Capacity, and Cytotoxicity of *Ginkgo biloba* Leaf Tea Product. *J. Sci. Food Agric.* **2023**, *103*, 4993–5003. [CrossRef]
25. Kadjo, A.F.; Dasgupta, P.K.; Su, J.; Liu, S.; Kraiczek, K.G. Width Based Quantitation of Chromatographic Peaks: Principles and Principal Characteristics. *Anal. Chem.* **2017**, *89*, 3884–3892. [CrossRef]
26. van Beek, T.A.; Montoro, P. Chemical Analysis and Quality Control of *Ginkgo biloba* Leaves, Extracts, and Phytopharmaceuticals. *J. Chromatogr. A* **2009**, *1216*, 2002–2032. [CrossRef]
27. VanBeek, T.A. *Ginkgo Biloba*, 1st ed.; CRC Press: Boca Raton, FL, USA, 2000.
28. Wilson, R.E.; Groskreutz, S.R.; Weber, S.G. Improving the Sensitivity, Resolution, and Peak Capacity of Gradient Elution in Capillary Liquid Chromatography with Large-Volume Injections by Using Temperature-Assisted On-Column Solute Focusing. *Anal. Chem.* **2016**, *88*, 5112–5121. [CrossRef]
29. Wageed, M.; Mahdy, H.M.; Kalaba, M.H.; Kelany, M.A.; Soliman, M. Development of LC-MS/MS Analytical Method for the Rapid Determination of Diquat in Water and Beverages. *Food Chem.* **2023**, *438*, 137869. [CrossRef]
30. Barimah, A.O.; Guo, Z.; Agyekum, A.A.; Guo, C.; Chen, P.; El-Seedi, H.R.; Zou, X.; Chen, Q. Sensitive Label-Free Cu_2O/Ag Fused Chemometrics SERS Sensor for Rapid Detection of Total Arsenic in Tea. *Food Control* **2021**, *130*, 108341. [CrossRef]
31. European Medicines Agency ICH Q2 (R1): Validation of Analytical Procedures: Text and Methodology. *Int. Conf. Harmon.* **2005**, *2*, 1–15.

32. European Commission. Guidance Document on Pesticide Analytical Methods for Risk Assessment and Post-Approval Control and Monitoring Purposes SANTE/2020/12830. Sante/2020/12830 2020, 1–50. Available online: https://food.ec.europa.eu/document/download/022e0538-7fcd-4fa3-8b99-f3ba2afc4c3d_en?filename=pesticides_ppp_app-proc_guide_res_mrl-guidelines-2020-12830.pdf (accessed on 6 March 2024).
33. Chen, X.M.; Ma, Z.; Kitts, D.D. Effects of Processing Method and Age of Leaves on Phytochemical Profiles and Bioactivity of Coffee Leaves. *Food Chem.* **2018**, *249*, 143–153. [CrossRef]
34. Krakowska-Sieprawska, A.; Kiełbasa, A.; Rafińska, K.; Ligor, M.; Buszewski, B. Modern Methods of Pre-Treatment of Plant Material for the Extraction of Bioactive Compounds. *Molecules* **2022**, *27*, 730. [CrossRef]
35. Wang, Y.; Tao, Y.; Zhang, X.; Shao, S.; Han, Y.; Chu, D.T.; Xie, G.; Ye, X. Metabolic Profile of Ginkgo Kernel Juice Fermented with Lactic Aicd Bacteria: A Potential Way to Degrade Ginkgolic Acids and Enrich Terpene Lactones and Phenolics. *Process Biochem.* **2019**, *76*, 25–33. [CrossRef]
36. Boasiako, T.A.; Yinka, A.A.; Yuqing, X.; Boateng, I.D.; Ma, Y. Tri-Cultured Lactic-Acetic Acid Co-Fermentation Improves Stored Jujube Puree Functionality, Physicochemical, Volatile Compounds, and Sensory Characteristics. *Food Biosci.* **2024**, *57*, 103534. [CrossRef]
37. Adainoo, B.; Thomas, A.L.; Krishnaswamy, K. Correlations between Color, Textural Properties and Ripening of the North American Pawpaw (*Asimina Triloba*) Fruit. *Sustain. Food Technol.* **2023**, *1*, 263–274. [CrossRef]

separations
MDPI
Article

Detection and Comparison of Volatile Organic Compounds in Four Varieties of Hawthorn Using HS-GC-IMS

Lijun Zhu [1,†], Feilin Ou [1,†], Yun Xiang [1], Bin Wang [1], Yingchao Mao [2], Lingfeng Zhu [2], Qun Zhang [2,*] and Chang Lei [1,3,*]

1 State Key Laboratory of Chinese Medicine Powder and Medicine Innovation in Hunan (Incubation), Science and Technology Innovation Center, Hunan University of Chinese Medicine, Changsha 410208, China; 202104030124@stu.hnucm.edu.cn (L.Z.); oufeilin@stu.hnucm.edu.cn (F.O.); 003872@hnucm.edu.cn (Y.X.); 004146@hnucm.edu.cn (B.W.)
2 Hunan Agricultural Product Processing Institute, Hunan Academy of Agricultural Sciences, Changsha 410125, China; mycjgs@hunaas.cn (Y.M.); zhulingfeng1988@163.com (L.Z.)
3 Yunnan Key Laboratory of Screening and Research on Anti-Pathogenic Plant Resources from Western Yunnan, Dali 671000, China
* Correspondence: zqun208@163.com (Q.Z.); leichang@hnucm.edu.cn (C.L.)
† These authors contributed equally to this work.

Abstract: Hawthorn is a type of natural food with significant medicinal and nutritional properties; it has been listed in the "Both Food and Drug" list by the Chinese Ministry of Health Item List since 1997. However, hawthorn varieties have complex origins, and there are significant differences in the content, type, and medicinal efficacy of the chemically active ingredients in different varieties of hawthorn. This leads to the phenomenon of mixed varieties and substandard products being passed off as high-quality. In this work, by using headspace gas chromatography–ion mobility spectrometry (HS-GC-IMS), we identified and analyzed volatile organic compounds (VOCs) in four varieties of hawthorn, establishing their characteristic fingerprints. As a result, a total of 153 peaks were detected, and 139 VOCs were also identified. As shown by the fingerprint profiles, the different hawthorn samples contained different VOCs. Meanwhile, by using principal component analysis (PCA), Euclidean distance, and partial least-squares discriminant analysis (PLS-DA), the relationship between the VOCs found in the different varieties of hawthorn was revealed. This study developed a simple, fast, accurate, and sensitive method for identifying, tracking, and evaluating hawthorn varieties.

Keywords: hawthorn; HS-GC-IMS; volatile organic compounds; PCA; Euclidean distance; PLS-DA

1. Introduction

Hawthorn is the dried ripe form of *Crataegus pinnatifida* Bge. var *major* N.E. Br., or *Crataegus pinnatifida* Bge. [1], and it is known as a "nutritious fruit" due to its richness in bioactive substances [2]. Hawthorn is not only a food product but also a medicinal plant that has many active ingredients and health benefits. In 1997, the Chinese Ministry of Health included it on its "Both Food and Drug" list, which encompasses drugs, dietary supplements, and foods. Hawthorn mainly contains flavonoids, organic acids, proanthocyanidins, triterpenes, pectin, vitamins, minerals, and other chemical components [3,4]. It has anti-atherosclerosis effects, can lower blood lipids and lower blood pressure, has antioxidant effects, improves liver damage, and has other pharmacological effects [5–7]. Due to its unique flavor and health benefits, hawthorn has become widely used in health foods, snacks, additives, and teas [8–12], such as hawthorn preserves, hawthorn pectin, hawthorn seed oil, hawthorn functional drinks, hawthorn plum, etc.

The variety of sources for hawthorns are varied and complex. In addition to Shanlihong (*Crataegus pinnatifida* Bge. Var. *major* N. E. Br.) and Shanzha (*Crataegus pinnatifida* Bge.), recorded in the 2020 edition of the "Chinese Pharmacopoeia", the most commonly used edible hawthorn varieties in the domestic market include "wild hawthorn" (*Crataegus cuneata* Sieb. et Zicc), *Malus doumeri* (Bois.) Chev. Shanlihong, and Shanzha, which are mostly grown in the north; therefore, they are commonly called "Northern Hawthorn", whereas wild hawthorn varieties mostly grow in the south; therefore, they are commonly known as "Nanshanzha". *Malus doumeri* (Bois.) Chev mostly grows in Guangdong, Guangxi, and other places, and it is commonly known as "Guangshanzha" [13]. These four varieties differ greatly in terms of source varieties, chemical composition, and pharmacological effects due to a series of factors, e.g., their origin, harvest period, variety, growth environment, soil, fertilizer, harvest, processing, and storage. Nanshanzha promotes qi, disperses blood stasis, astringes, and stops diarrhea; Guangshanzha regulates qi, strengthens the spleen, and digests stagnant food. The content and types of organic acid components in Northern Hawthorn, Nanshanzha, and Guangshanzha are very different. Using Nanshanzha and Guangshanzha instead of Northern Hawthorn will have a greater impact on research results and efficacy. The quality of the different varieties of Hawthorn is irregular, and so the market prices vary. This has led to common phenomena such as mixing production areas and passing off substandard products, resulting in greatly reduced efficacy and possibly even drug use and food safety accidents. However, in the 2020 edition, the Chinese Pharmacopoeia only uses acid–base titrations and indicators to evaluate the quality of hawthorn.

This method is not suitable for the qualitative and quantitative analysis of organic acid components, and it cannot reflect the overall quality of the medicinal materials and thus cannot be used for a reasonable evaluation; this makes it impossible to carry out a reasonable quality evaluation and control. Some researchers have also used empirical identification, microscopic identification, high-performance liquid chromatography (HPLC), molecular identification (such as DNA barcoding technology), and other techniques to identify hawthorn [14,15]. Although these methods provide more choices for identification, these approaches have their limitations. For example, an empirical identification is subjective, microscopic identification is not specific, HPLC identification is time-consuming and cumbersome, and molecular identification is relatively complex. Therefore, it is important to find a fast, easy, and green way to differentiate different varieties of hawthorn.

In addition to identifying food, volatile organic compounds (VOCs) also play an important role in determining a food's nutritional and sensory properties. At present, analyses of VOCs in food comprise two parts: sensory analysis and instrumental analysis. The sensory analysis of results is a kind of subjective sensory perception that analyzes the results. An instrumental analysis is a molecular analysis that is objective. However, the traditional odor and taste identification methods, which are highly subjective and empirical, can no longer match modern development. Moreover, most flavor substances and volatile odor substances are in the ppb range in food. Therefore, the detection of differentiation needs to be more sensitive and reliable. In the past, more and more researchers have used instrumental analysis techniques to detect VOCs in food, including gas chromatography–mass spectrometry (GC-MS), chromatography–olfactometry–mass spectrometry (GC-O-MS), headspace gas chromatography–ion mobility spectrometry (HS-GC-IMS), and electronic noses (E-noses) [16]. These methods can complement sensory evaluations by providing more objective information about the substance being tested at the molecular level. There are a number of gas chromatography and mass spectrometry methods available, but GC-MS has significant limitations in distinguishing isomeric molecules, especially cyclic isomers, due to its complex sample processing methods and high mass spectrometry resolution [17]. Aroma and flavor analysis can also be performed using GC-O-MS, but it shares the same disadvantages as GC-MS [18]. An E-nose is a new type of aroma detection technology that provides rapid detection, but it also suffers from low accuracy, sensor drift, and poor repeatability, as well as a high sensitivity to the surrounding environment [19].

A recent development in HS-GC-IMS technology incorporates the advantages of both GC and IMS, combining high separation capabilities with high resolution and high sensitivity. It is equipped with a static headspace sampling device to detect trace organic components emitted from liquid or solid samples. As an emerging technology, this method is simple, fast, non-destructive, there is no need for sample pre-processing, and it has good reproducibility. It is used in the identification of foods, medicinal varieties [20,21], and food aroma analysis [22]. The impact of different drying methods and temperatures on food [23,24], component analyses of food during different harvest periods and storage periods [25,26], and other aspects have played an important role in this method. There have been very few studies conducted on HS-GC-IMS for the identification of hawthorn, and this difference is often one of the key factors contributing to the differences in quality between traditional Chinese medicines. The aim of this study was to identify and analyze the VOCs of different varieties of hawthorn using HS-GC-IMS, principal component analysis (PCA), Euclidean distance, and partial least-squares discriminant analysis (PLS-DA) methods. The characteristic VOCs in the different varieties of hawthorn were displayed in a visual form, providing technical support for the rapid analysis of VOCs and the identification of different varieties of hawthorn. Simultaneously, this study enriched the study of flavor compounds of hawthorn.

2. Materials and Methods

2.1. Medicinal Materials

Four kinds of hawthorn powder were purchased from the National Institute for Food and Drug Control, Beijing, China: Shanlihong (the dried ripe fruit of *Crataegus pinnatifida* Bge. Var. *major*, No. 121138-201206, named SZ-01), Shanzha (the dried ripe fruit of *Crataegus pinnatifida* Bge., No. 121626-201803, named SZ-02), Guangshanzha (the dried ripe fruit of *Cantonese Crataegus*, No. 120943-201903, named GSZ), and Nanshanzha (the dried ripe fruit of *South Crataegus*, No. 121055-201704, named NSZ).

2.2. Sample Preparation

For each kind of hawthorn powder, 1 g was accurately weighed and placed into a 20 mL headspace bottle and incubated at 80 °C for 20 min, and then the samples were injected. Each sample was measured in three parallel groups.

2.3. Headspace Sampling Conditions

The temperature of incubation was 80 °C. The incubation time was 20 min. The injection volume was 500 μL. The splitless injection method was used. The incubation speed was 500 rotations per min, and the injection needle temperature was 85 °C.

2.4. Chromatographic Conditions

In this study, we used a FlavourSpec® gas-phase ion mobility spectrometer (G.A.S, Dortmund, Germany), a CTC-PAL 3 static headspace automatic sampling device (CTC Analytics AG, Switzerland), and a 20 mL headspace bottle (Shandong Haineng Scientific Instrument Co., Ltd., Shandong, China); an MXT-WAX capillary chromatography column (15 m × 0.53 mm × 1 μm, Restek Company of the United States, Bellefonte, PA, USA) was also used. The temperature was 60 °C. The carrier gas was high-purity N_2 (purity ≥99.999%). The initial flow rate was 2.00 mL/min, which was maintained for 2 min. Then, it was linearly increased to 10.00 mL/min and then 100.00 mL/min within 10 min, and this was maintained for 40 min. The running time for the chromatography was 60 min, and the injection port temperature was 80 °C.

2.5. IMS Conditions

A tritium source (3H) was used; the flow rate was 75 mL/min; the power of the electric field was 500 V/cm; the voltage of the drift gas was 99.999%; the length of the migration tube was 53 mm; the electric field strength was 500 V/cm; the temperature of the migration tube was 45 °C; and the drift gas was high-purity nitrogen (purity = 99.999%).

2.6. Data Analysis

In order to characterize the VOCs in the samples, the GAS software (G.A.S, Dortmund, Germany, version 2.0.0), the built-in NIST retention index database, and the IMS drift time database were used. To compare VOCs between the samples, VOCal data processing software plug-ins, such as Reporter (Version 11.x), Gallery Plot (Version 1.1.0.2), and Dynamic PCA (version 0.0.3), were used to generate three-dimensional (3D) spectrums, two-dimensional (2D) spectrums, difference spectrums, fingerprints, and PCA maps, respectively. The PLS-DA VIPs were calculated using the SIMCA software (version 14.0) (Umea, Sweden).

3. Results and Discussion

3.1. GC-IMS Analysis of VOCs in Different Samples

We compared the spectral differences between the samples and the characteristic VOCs of the Shanlihong, Shanzha, Guangshanzha, and Nanshanzha samples using the VOCal software's Reporter plug-in. The X axis represents the ion migration time, the Y axis represents the retention time of the gas chromatograph, and the Z axis represents the peak intensity used for quantification. A comparison of the VOC contents of the Shanlihong, Shanzha, Guangshanzha, and Nanshanzha varieties is shown in Figure 1. In Figure 2, a top view is used for comparison due to the inconvenience of observation. After normalizing the data, the red vertical line at 1.0 represents the reactive ion peak (RIP peak). The gas chromatography retention time is represented by the ordinate. The dots on either side of the RIP represent the VOCs. The color represents the concentration of the substance; white represents a lower concentration, and red represents a higher concentration. In general, the darker the color, the greater the concentration. The samples from the Shanlihong, Shanzha, Guangshanzha, and Nanshanzha varieties differed in terms of their VOCs, as shown in Figure 2.

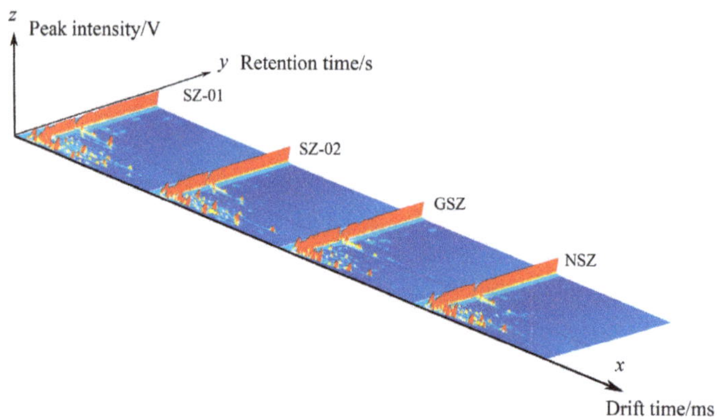

Figure 1. Three-dimensional spectra of hawthorn VOCs.

Using the spectrum of the Shanlihong (SZ-01) variety as a reference, we can visually compare the differences in the VOCs. From the spectra of the Shanzha, Guangshanzha, and Nanshanzha samples, we were able to determine their VOCs. The differences in organic matter are shown in Figure 3. In the figure, when the measured components of Shanlihong, Shanzha, Guangshanzha, and Nanshanzha were consistent, they are shown in white after deduction. Red and blue indicate that the concentrations of the measured components in the Shanzha, Guangshanzha, and Nanshanzha varieties were higher and lower than in the Shanlihong variety, respectively. With the difference comparison chart, it is easier to identify the differences between the four samples.

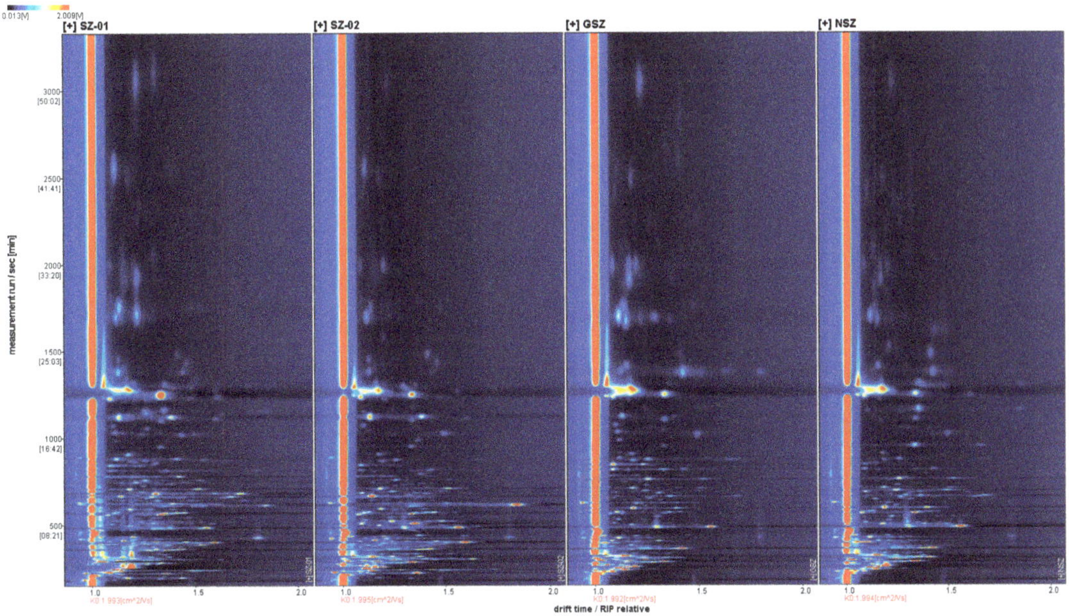

Figure 2. Two-dimensional spectra of hawthorn VOCs.

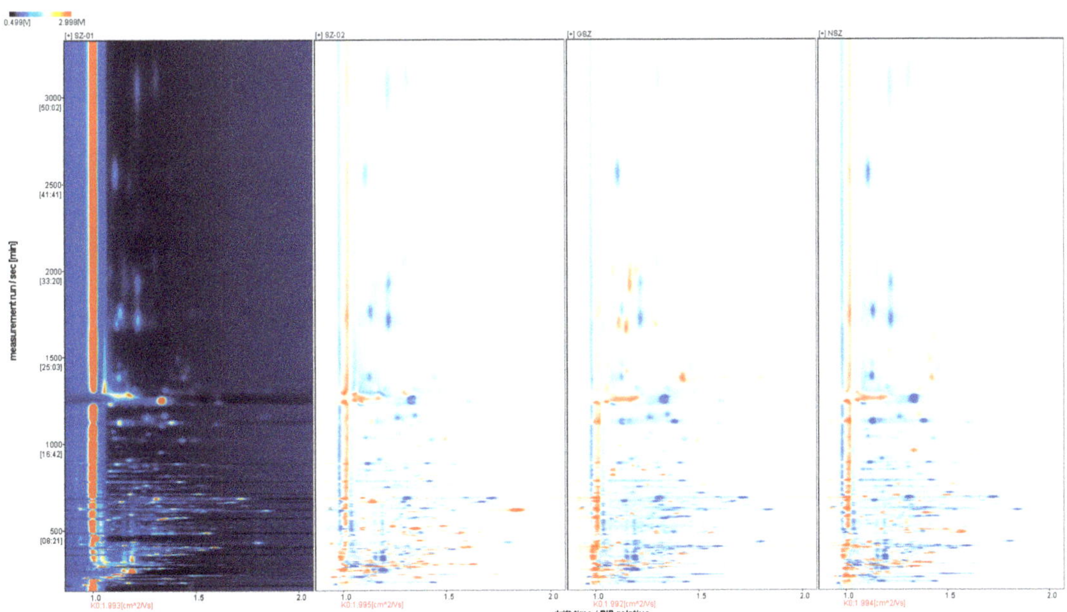

Figure 3. A comparison chart of the differences between varieties of hawthorn.

3.2. Comparison of the Fingerprints of VOCs

To compare the specific VOCs from the hawthorn samples, the Gallery Plot plug-in of VOCal was used to select the peaks for fingerprint comparison (Figure 4). Each row in the figure represents a sample; the rightmost side is the name of the sample, each column is a compound, and the bottom is the qualitative result of the samples. Some substances are followed by M and D, which indicate monomers and dimers of the same substance, respectively. The color indicates the concentration of the compounds, with darker colors indicating a higher concentration of the compounds, and black colors indicating a very low or undetectable concentration.

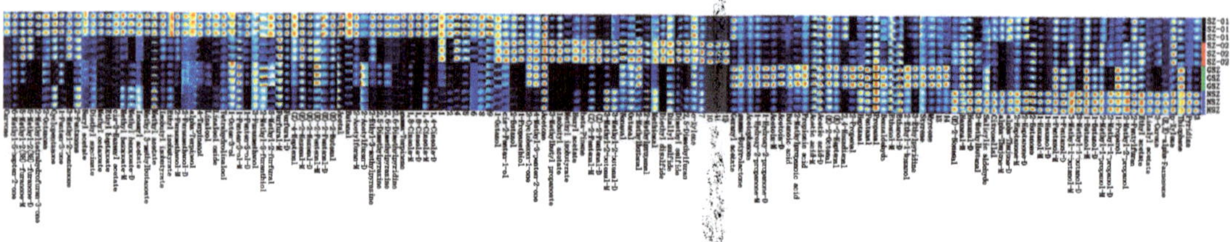

Figure 4. Fingerprints of VOCs of different varieties of hawthorn.

A comparison of the VOCs of the different hawthorn samples is shown in Figure 4. The contents of carvone, 6-methyl-5-hepten-2-one, 5-methyl-2(3H)-furanone, 2-methyltetrahydrofuran-3-one, 2-octanone, cyclopentanone, 1-penten-3-one, 3-methyl-2-pentanone, 2-pentanone, neryl acetate, diethyl succinate, methyl octanoate, ethyl enanthate, phytyl acetate, methyl caproate, isopentyl acetate, methyl 3-methylbutyrate, methyl acetate, isobutyl isobutyrate, ethyl isobutyrate, 2-furanmethanol, α-terpineol, 4-terpineol, linalool, linalool oxide, tetrahydrolinalool, 1-octen-3-ol, 1-Penten-3-ol, 3-furanmethanol, 2-methyl-3-furanthiol, 5-methyl-2-furfural, furfural, Z-4-heptenal, E-2-hexenal, E-2-pentenal, E-2-butenal, acetal, butyraldehyde, 2-acetylfuran, 2-ethyl-3-methylpyrazine, 2,5-dimethylpyrazine, 2,6-dimethylpyrazine, 2,6-dimethylpyridine, γ-terpinene, 1,8-cineole, 1,4-cineole, and 46 other substances were high in SZ-01.

The contents of 1-octanol, Z-2-penten-1-ol, 2-butanol, 1-propanethiol, 2-cyclohexen-1-one, 4-methyl-3-pentene-2-one, acetone, isoamyl propionate, isobutyl acetate, methyl isobutyrate, hexyl acetate, β-pinene, Z-2-pentenal, 3-methyl-2-butene aldehyde, nonanal, octanal, 2-methylbutyraldehyde, 2-methylpropionaldehyde, methylthiopropionaldehyde, diethyl disulfide, diallyl sulfide, dimethyl sulfide, 2,5-dimethylfuran, o-xylene, and 24 other substances were high in SZ-02.

The contents of geranyl acetate, γ-butyrolactone, acetophenone, 1-hydroxy-2-propanone, acetoin, butyric acid, 2-methyl propionic acid, propionic acid, acetic acid, E-2-octenal, E-2-heptenal, 2,4-hexadienal, 2-propenal, heptanal, valeraldehyde, propionaldehyde, benzaldehyde, ethanol, 2-ethyl-1-hexanol, 3-ethylpyridine, styrene, myrcene, and 22 other substances were high in GSZ.

The contents of E-2-nonenal, hexanal, 3-methyl butyraldehyde, salicylaldehyde, citronellal, α-thujone, 2-heptanone, 2-butanone, 1-hexanol, 1-pentanol, 3-methyl-1-butanol, 1-butanol alcohol, 2-methyl-1-propanol, 1-propanol, 2-methyl-2-propanol, 2-pentylfuran, ethyl acetate, linalyl acetate, 3-carene, (E, E), α-farnesene, p-xylene, thiophene, pyridine, and 23 other substances were high in NSZ.

3.3. *Identifying VOCs in Different Samples*

Our qualitative analysis was conducted using the NIST and IMS databases built into the VOCal software (Version 0.4.03, GAS Deutschland, Dortmund, Germany), as well as the Reporter, Gallery Plot, and Dynamic PCA plugins. Ultimately, we detected 153 peaks and identified 139 VOCs (Tables 1 and 2) from all the samples, including 31 aldehydes, 29 alcohols, 20 ketones, 20 esters, 16 terpenes, 5 acids, 4 furans, 3 pyridines, 3 pyrazines, 1 thiophene, and 7 other compounds.

Table 1. Results of the VOC analysis of different varieties of hawthorn.

No	Compounds	CAS	Molecular Formula	RI	Rt/s	Dt/ms	Comment
1	Carvone	C99490	$C_{10}H_{14}O$	1918.9	3096.478	1.31142	
2	Geranyl acetate	C105873	$C_{12}H_{20}O_2$	1911.1	3045.325	1.22205	
3	Neryl acetate	C141128	$C_{12}H_{20}O_2$	1866.5	2765.327	1.22205	
4	2-Furanmethanol	C98000	$C_5H_6O_2$	1829	2549.945	1.1134	Monomers
5	2-Furanmethanol	C98000	$C_5H_6O_2$	1829.5	2552.637	1.37275	Dimers
6	Acetophenone	C98862	C_8H_8O	1817.6	2488.023	1.17824	
7	alpha-Terpineol	C98555	$C_{10}H_{18}O$	1771.8	2253.795	1.2238	
8	Diethyl succinate	C123251	$C_8H_{14}O_4$	1727.1	2046.253	1.3028	
9	gamma-Butyrolactone	C96480	$C_4H_6O_2$	1713.6	1987.498	1.08941	
10	4-Terpinenol	C562743	$C_{10}H_{18}O$	1697.2	1917.95	1.2218	
11	5-Methyl-2-furfural	C620020	$C_6H_6O_2$	1656.9	1758.178	1.13623	
12	Propionic acid	C79094	$C_3H_6O_2$	1639.3	1692.39	1.11686	
13	Linalool	C78706	$C_{10}H_{18}O$	1646.4	1718.705	1.2218	
14	(E)-2-Nonenal	C18829566	$C_9H_{16}O$	1577.8	1481.866	1.41555	
15	2-Acetylfuran	C1192627	$C_6H_6O_2$	1546.6	1385.272	1.12833	Monomers
16	2-Acetylfuran	C1192627	$C_6H_6O_2$	1546.6	1385.29	1.44692	Dimers
17	1-Octanol	C111875	$C_8H_{18}O$	1562.1	1432.612	1.46191	
18	Acetic acid	C64197	$C_2H_4O_2$	1505.1	1266.629	1.05757	Monomers
19	Acetic acid	C64197	$C_2H_4O_2$	1505.6	1267.84	1.15537	Dimers
20	Furfural	C98011	$C_5H_4O_2$	1494.8	1238.763	1.08968	Monomers
21	Furfural	C98011	$C_5H_4O_2$	1494.8	1238.763	1.33199	Dimers
22	alpha-Thujone	C546805	$C_{10}H_{16}O$	1464.9	1161.223	1.35243	Monomers
23	alpha-Thujone	C546805	$C_{10}H_{16}O$	1464.4	1160.011	1.85457	Dimers
24	Linalool oxide	C60047178	$C_{10}H_{18}O_2$	1460.6	1150.319	1.26339	
25	5-Methyl-2(3H)-furanone	C591128	$C_5H_6O_2$	1450.2	1124.876	1.13055	Monomers
26	5-Methyl-2(3H)-furanone	C591128	$C_5H_6O_2$	1450.1	1124.55	1.38267	Dimers
27	(E)-2-Octenal	C2548870	$C_8H_{14}O$	1441.7	1104.36	1.33829	
28	2-Ethyl-3-methylpyrazine	C15707230	$C_7H_{10}N_2$	1445.7	1113.924	1.60293	
29	2-Cyclohexen-1-one	C930687	C_6H_8O	1428.1	1072.483	1.09996	
30	Tetrahydrolinalool	C78693	$C_{10}H_{22}O$	1422.1	1058.669	1.2709	
31	2,4-Hexadienal	C142836	C_6H_8O	1412.3	1036.355	1.10818	
32	Methyl octanoate	C111115	$C_9H_{18}O_2$	1405.1	1020.416	1.43527	
33	1-Hexanol	C111273	$C_6H_{14}O$	1373.7	953.473	1.33336	Monomers
34	6-Methyl-5-hepten-2-one	C110930	$C_8H_{14}O$	1352	909.719	1.17836	
35	2,5-Dimethylpyrazine	C123320	$C_6H_8N_2$	1340	886.579	1.12013	
36	Ethyl heptanoate	C106309	$C_9H_{18}O_2$	1340.4	887.259	1.4113	
37	cis-2-Penten-1-ol	C1576950	$C_5H_{10}O$	1344.3	894.746	0.94228	
38	(E)-2-Heptenal	C18829555	$C_7H_{12}O$	1335.8	878.412	1.25863	
39	1-Hydroxy-2-propanone	C116096	$C_3H_6O_2$	1319.7	848.466	1.07763	Monomers
40	1-Hydroxy-2-propanone	C116096	$C_3H_6O_2$	1319.3	847.786	1.23187	Dimers
41	cis-3-Hexenyl acetate	C3681718	$C_8H_{14}O_2$	1325.6	859.355	1.31057	
42	Acetoin	C513860	$C_4H_8O_2$	1304.2	820.562	1.08236	Monomers
43	2-Octanone	C111137	$C_8H_{16}O$	1302.3	817.159	1.36093	
44	2-Methyltetrahydrofuran-3-one	C3188009	$C_5H_8O_2$	1283.6	785.853	1.42389	
45	2,6-Dimethylpyrazine	C108509	$C_6H_8N_2$	1337.9	882.495	1.5372	
46	1-Pentanol	C71410	$C_5H_{12}O$	1270.7	765.435	1.2602	Monomers
47	1-Pentanol	C71410	$C_5H_{12}O$	1270.7	765.435	1.51989	Dimers
48	gamma-Terpinene	C99854	$C_{10}H_{16}$	1262.3	752.504	1.21456	
49	2,6-Dimethylpyridine	C108485	C_7H_9N	1261.8	751.823	1.44749	
50	2-Pentylfuran	C3777693	$C_9H_{14}O$	1246.9	729.364	1.25233	
51	(E)-2-Hexenal	C6728263	$C_6H_{10}O$	1235.8	713.03	1.18308	Monomers
52	(E)-2-Hexenal	C6728263	$C_6H_{10}O$	1235.8	713.03	1.51202	Dimers
53	3-Methyl-1-butanol	C123513	$C_5H_{12}O$	1225.9	698.738	1.2366	Monomers
54	1,8-Cineole	C470826	$C_{10}H_{18}O$	1217.1	686.342	1.30893	Monomers
55	1,8-Cineole	C470826	$C_{10}H_{18}O$	1217.6	687.087	1.71988	Dimers
56	Diethyl disulfide	C110816	$C_4H_{10}S_2$	1204.2	668.466	1.14134	
57	Methyl hexanoate	C106707	$C_7H_{14}O_2$	1203.6	667.721	1.2929	Monomers

Table 1. *Cont.*

No	Compounds	CAS	Molecular Formula	RI	Rt/s	Dt/ms	Comment
58	2-Heptanone	C110430	$C_7H_{14}O$	1198.1	660.273	1.26375	Monomers
59	Methyl hexanoate	C106707	$C_7H_{14}O_2$	1203.1	666.976	1.6747	Dimers
60	1,4-Cineole	C470677	$C_{10}H_{18}O$	1189	645.376	1.32059	Monomers
61	1,4-Cineole	C470677	$C_{10}H_{18}O$	1189	645.376	1.73154	Dimers
62	1-Penten-3-ol	C616251	$C_5H_{10}O$	1181.1	628.245	0.93878	Monomers
63	1-Penten-3-ol	C616251	$C_5H_{10}O$	1181.1	628.245	1.37013	Dimers
64	3-Methylbutyl propanoate	C105680	$C_8H_{16}O_2$	1175.2	615.582	1.84229	
65	1-Butanol	C71363	$C_4H_{10}O$	1165.1	594.727	1.18798	Monomers
66	1-Butanol	C71363	$C_4H_{10}O$	1165.5	595.472	1.39054	Dimers
67	Diallyl sulfide	C592881	$C_6H_{10}S$	1157.3	579.085	1.13114	
68	(E)-2-Pentenal	C1576870	C_5H_8O	1154.4	573.307	1.10787	Monomers
69	p-Xylene	C106423	C_8H_{10}	1149.6	563.911	1.07888	
70	Cyclopentanone	C120923	C_5H_8O	1151.1	566.878	1.33287	
71	(E)-2-Pentenal	C1576870	C_5H_8O	1154.4	573.262	1.35804	Dimers
72	4-Methyl-3-penten-2-one	C141797	$C_6H_{10}O$	1133	532.755	1.11615	
73	beta-Pinene	C127913	$C_{10}H_{16}$	1128.1	523.853	1.239	
74	Isoamyl acetate	C123922	$C_7H_{14}O2$	1124.2	516.93	1.29836	
75	(Z)-2-Pentenal	C1576869	C_5H_8O	1120.6	510.501	1.09683	Monomers
76	(Z)-2-Pentenal	C1576869	C_5H_8O	1121.1	511.49	1.34943	Dimers
77	2-Methyl-1-propanol	C78831	$C_4H_{10}O$	1114.8	500.61	1.18379	Monomers
78	Hexanal	C66251	$C_6H_{12}O$	1105.2	484.291	1.29698	Monomers
79	Hexanal	C66251	$C_6H_{12}O$	1106.1	485.774	1.56477	Dimers
80	(E)-2-Butenal	C123739	C_4H_6O	1071.3	437.31	1.19897	
81	1-Propanol	C71238	C_3H_8O	1061.4	424.947	1.11753	
82	2-Butanol	C78922	$C_4H_{10}O$	1054.5	416.54	1.15342	
83	1-Penten-3-one	C1629589	C_5H_8O	1049.1	410.111	1.30802	
84	Isobutyl isobutyrate	C97858	$C_8H_{16}O_2$	1065	429.398	1.80496	
85	3-Methyl-2-pentanone	C565617	$C_6H_{12}O$	1046.5	407.144	1.45296	
86	Isobutyl acetate	C110190	$C_6H_{12}O_2$	1044	404.177	1.59514	
87	Methyl 3-methylbutanoate	C556241	$C_6H_{12}O_2$	1032.4	390.824	1.52612	
88	Pentanal	C110623	$C_5H_{10}O$	1010.2	366.592	1.42812	
89	2-Pentanone	C107879	$C_5H_{10}O$	1005.9	362.141	1.36462	
90	Acetal	C105577	$C_6H_{14}O_2$	914.3	294.39	1.02919	
91	2,5-Dimethylfuran	C625865	C_6H_8O	954.4	321.337	1.37323	
92	Ethyl acetate	C141786	$C_4H_8O_2$	903.7	287.695	1.33952	
93	Methyl acetate	C79209	$C_3H_6O_2$	862.7	263.024	1.19002	
94	Acetone	C67641	C_3H_6O	841.7	251.249	1.11087	
95	Propanal	C123386	C_3H_6O	824	241.717	1.14459	
96	Butanal	C123728	C_4H_8O	890.1	279.284	1.26624	
97	1-Propanethiol	C107039	C_3H_8S	794.4	226.578	1.35711	
98	Dimethyl sulfide	C75183	C_2H_6S	800	229.381	0.96577	
99	2-Propenal	C107028	C_3H_4O	872.3	268.631	1.05811	
100	2-Methylbutanal	C96173	$C_5H_{10}O$	912.6	293.302	1.16364	
101	2-Butanone	C78933	C_4H_8O	922.3	299.57	1.24652	
102	Ethanol	C64175	C_2H_6O	951.8	319.548	1.05719	Monomers
103	Ethanol	C64175	C_2H_6O	952.3	319.865	1.12819	Dimers
104	Methyl isobutyrate	C547637	$C_5H_{10}O_2$	925.6	301.79	1.44837	
105	Thiophene	C110021	C_4H_4S	1035.4	394.247	1.04484	
106	2-Methyl-2-propanol	C75650	$C_4H_{10}O$	939.4	311.032	1.34116	
107	Heptanal	C111717	$C_7H_{14}O$	1200.7	663.817	1.35886	
108	2-Methyl-1-propanol	C78831	$C_4H_{10}O$	1115.5	501.77	1.37666	Dimers
109	3-Carene	C13466789	$C_{10}H_{16}$	1139.7	545.163	1.22154	
110	2-Methylpropanal	C78842	C_4H_8O	833.6	246.832	1.28257	
111	3-Methylbutanal	C590863	$C_5H_{10}O$	934.7	307.854	1.40845	
112	Butanoic acid	C107926	$C_4H_8O_2$	1694.3	1906.285	1.16893	
113	2-Methylpropanoic acid	C79312	$C_4H_8O_2$	1631.8	1665.286	1.15156	
114	Nonanal	C124196	$C_9H_{18}O$	1406.2	1022.81	1.49511	

Table 1. *Cont.*

No	Compounds	CAS	Molecular Formula	RI	Rt/s	Dt/ms	Comment
115	3-Ethylpyridine	C536787	C_7H_9N	1390	987.657	1.11846	
116	Acetoin	C513860	$C_4H_8O_2$	1304.7	821.482	1.32632	Dimers
117	1-Hexanol	C111273	$C_6H_{14}O$	1374.4	954.885	1.65336	Dimers
118	2-Methyl-3-furanthiol	C28588741	C_5H_6OS	1323.2	854.833	1.14631	
119	Octanal	C124130	$C_8H_{16}O$	1303.5	819.364	1.428	
120	3-Methyl-1-butanol	C123513	$C_5H_{12}O$	1226.1	699.033	1.48637	Dimers
121	3-Methyl-2-butenal	C107868	C_5H_8O	1219.5	689.643	1.09452	Monomers
122	3-Methyl-2-butenal	C107868	C_5H_8O	1219.5	689.643	1.35947	Dimers
123	2-Heptanone	C110430	$C_7H_{14}O$	1198.2	660.366	1.62302	Dimers
124	Pyridine	C110861	C_5H_5N	1172.8	610.581	1.24515	
125	Styrene	C100425	C_8H_8	1240	719.162	1.05951	
126	Myrcene	C123353	$C_{10}H_{16}$	1178.9	623.482	1.20043	
127	o-Xylene	C95476	C_8H_{10}	1195.8	657.167	1.09339	
128	(Z)-4-Heptenal	C6728310	$C_7H_{12}O$	1260.5	749.747	1.14745	
129	Hexyl acetate	C142927	$C_8H_{16}O_2$	1290.9	797.599	1.37921	
130	3-Furanmethanol	C4412913	$C_5H_6O_2$	1802.7	2409.312	1.1025	
131	Salicylic aldehyde	C90028	$C_7H_6O_2$	1737	2090.266	1.13654	
132	(E,E)-alpha-Farnesene	C502614	$C_{15}H_{24}$	1618.9	1619.575	1.43366	
133	Linalyl acetate	C115957	$C_{12}H_{20}O_2$	1589.4	1519.491	1.21298	
134	Benzaldehyde	C100527	C_7H_6O	1556.6	1415.508	1.1583	
135	Citronellal	C106230	$C_{10}H_{18}O$	1520.3	1308.926	1.35554	
136	2-Ethyl-1-hexanol	C104767	$C_8H_{18}O$	1542.3	1372.615	1.4278	
137	1-Octen-3-ol	C3391864	$C_8H_{16}O$	1487.5	1219.24	1.16025	
138	Ethyl isobutyrate	C97621	$C_6H_{12}O_2$	974.9	336.085	1.18508	
139	Methional	C3268493	C_4H_8OS	1462.4	1154.981	1.10453	

Table 2. The average area of VOCs in different varieties of hawthorn.

No	Compounds	CAS	Molecular Formula	SZ-01	SZ-02	GSZ	NSZ	Comment
1	Carvone	C99490	$C_{10}H_{14}O$	2193.56	541.96	823.61	642.46	
2	Geranyl acetate	C105873	$C_{12}H_{20}O_2$	5019.10	1463.21	5307.01	2248.35	
3	Neryl acetate	C141128	$C_{12}H_{20}O_2$	917.78	389.55	428.69	378.18	
4	2-Furanmethanol	C98000	$C_5H_6O_2$	5669.53	2512.74	972.12	632.27	Monomers
5	2-Furanmethanol	C98000	$C_5H_6O_2$	475.41	258.86	327.52	218.86	Dimers
6	Acetophenone	C98862	C_8H_8O	1023.00	669.31	1408.41	707.06	
7	alpha-Terpineol	C98555	$C_{10}H_{18}O$	240.43	84.07	97.26	108.22	
8	Diethyl succinate	C123251	$C_8H_{14}O_4$	1107.30	366.69	505.19	618.82	
9	gamma-Butyrolactone	C96480	$C_4H_6O_2$	1109.29	993.19	1921.69	975.16	
10	4-Terpinenol	C562743	$C_{10}H_{18}O$	2903.52	525.24	1004.35	445.81	
11	5-Methyl-2-furfural	C620020	$C_6H_6O_2$	2629.86	739.76	2082.71	597.91	
12	Propionic acid	C79094	$C_3H_6O_2$	1765.67	1282.93	2963.09	1424.15	
13	Linalool	C78706	$C_{10}H_{18}O$	4410.20	487.86	981.68	510.19	
14	(E)-2-Nonenal	C18829566	$C_9H_{16}O$	355.43	531.43	358.10	902.93	
15	2-Acetylfuran	C1192627	$C_6H_6O_2$	2575.65	922.08	1010.97	320.80	Monomers
16	2-Acetylfuran	C1192627	$C_6H_6O_2$	249.79	60.10	329.26	220.72	Dimers
17	1-Octanol	C111875	$C_8H_{18}O$	364.64	342.25	144.64	118.95	
18	Acetic acid	C64197	$C_2H_4O_2$	13428.62	13566.71	16484.00	16149.05	Monomers
19	Acetic acid	C64197	$C_2H_4O_2$	17099.25	15889.10	26134.83	20763.35	Dimers
20	Furfural	C98011	$C_5H_4O_2$	3003.64	3394.42	3123.56	2878.95	Monomers
21	Furfural	C98011	$C_5H_4O_2$	19115.34	8236.68	6371.44	5202.22	Dimers
22	alpha-Thujone	C546805	$C_{10}H_{16}O$	1330.87	140.79	82.44	2013.87	Monomers
23	alpha-Thujone	C546805	$C_{10}H_{16}O$	274.58	74.63	79.60	278.58	Dimers
24	Linalool oxide	C60047178	$C_{10}H_{18}O_2$	984.97	195.45	114.28	254.53	
25	5-Methyl-2(3H)-furanone	C591128	$C_5H_6O_2$	4697.91	5499.15	1945.43	247.30	Monomers

Table 2. *Cont.*

No	Compounds	CAS	Molecular Formula	SZ-01	SZ-02	GSZ	NSZ	Comment
26	5-Methyl-2(3H)-furanone	C591128	$C_5H_6O_2$	3054.88	2789.99	300.81	122.59	Dimers
27	(E)-2-Octenal	C2548870	$C_8H_{14}O$	620.58	399.09	618.27	481.19	
28	2-Ethyl-3-methylpyrazine	C15707230	$C_7H_{10}N_2$	355.92	216.98	107.11	49.75	
29	2-Cyclohexen-1-one	C930687	C_6H_8O	321.72	486.49	480.64	88.71	
30	Tetrahydrolinalool	C78693	$C_{10}H_{22}O$	328.98	104.38	140.67	159.87	
31	2,4-Hexadienal	C142836	C_6H_8O	685.98	101.89	769.25	89.31	
32	Methyl octanoate	C111115	$C_9H_{18}O_2$	739.69	205.33	149.65	134.87	
33	1-Hexanol	C111273	$C_6H_{14}O$	601.63	563.04	340.31	2029.10	Monomers
34	6-Methyl-5-hepten-2-one	C110930	$C_8H_{14}O$	1315.11	513.84	540.21	291.59	
35	2,5-Dimethylpyrazine	C123320	$C_6H_8N_2$	1888.63	162.56	612.51	133.67	
36	Ethyl heptanoate	C106309	$C_9H_{18}O_2$	948.68	69.21	135.69	71.20	
37	cis-2-Penten-1-ol	C1576950	$C_5H_{10}O$	222.05	232.78	109.55	73.88	
38	(E)-2-Heptenal	C18829555	$C_7H_{12}O$	445.84	487.73	883.89	518.61	
39	1-Hydroxy-2-propanone	C116096	$C_3H_6O_2$	1123.98	920.47	2165.55	492.28	Monomers
40	1-Hydroxy-2-propanone	C116096	$C_3H_6O_2$	351.30	142.87	972.10	38.89	Dimers
41	cis-3-Hexenyl acetate	C3681718	$C_8H_{14}O_2$	319.19	53.84	51.66	43.60	
42	Acetoin	C513860	$C_4H_8O_2$	520.34	365.75	1561.63	496.44	Monomers
43	2-Octanone	C111137	$C_8H_{16}O$	385.46	163.09	186.42	101.27	
44	2-Methyltetrahydrofuran-3-one	C3188009	$C_5H_8O_2$	1370.05	187.85	97.48	58.10	
45	2,6-Dimethylpyrazine	C108509	$C_6H_8N_2$	187.98	30.34	94.02	32.06	
46	1-Pentanol	C71410	$C_5H_{12}O$	1102.19	706.17	1512.61	1594.35	Monomers
47	1-Pentanol	C71410	$C_5H_{12}O$	284.89	109.55	502.58	583.88	Dimers
48	gamma-Terpinene	C99854	$C_{10}H_{16}$	880.50	83.20	133.79	123.04	
49	2,6-Dimethylpyridine	C108485	C_7H_9N	116.96	16.15	20.26	18.56	
50	2-Pentylfuran	C3777693	$C_9H_{14}O$	212.42	140.66	215.24	250.38	
51	(E)-2-Hexenal	C6728263	$C_6H_{10}O$	1503.23	1528.99	997.49	967.21	Monomers
52	(E)-2-Hexenal	C6728263	$C_6H_{10}O$	1610.16	1391.31	344.95	320.48	Dimers
53	3-Methyl-1-butanol	C123513	$C_5H_{12}O$	484.07	735.00	563.46	1280.72	Monomers
54	1,8-Cineole	C470826	$C_{10}H_{18}O$	4195.97	319.15	698.21	455.41	Monomers
55	1,8-Cineole	C470826	$C_{10}H_{18}O$	2361.24	75.23	49.00	78.11	Dimers
56	Diethyl disulfide	C110816	$C_4H_{10}S_2$	1966.57	5444.81	99.44	127.56	
57	Methyl hexanoate	C106707	$C_7H_{14}O_2$	1443.58	338.08	222.98	680.66	Monomers
58	2-Heptanone	C110430	$C_7H_{14}O$	356.57	215.85	180.50	454.44	Monomers
59	Methyl hexanoate	C106707	$C_7H_{14}O_2$	1080.43	38.42	241.15	679.24	Dimers
60	1,4-Cineole	C470677	$C_{10}H_{18}O$	663.29	78.52	129.59	298.24	Monomers
61	1,4-Cineole	C470677	$C_{10}H_{18}O$	221.18	21.13	22.03	44.06	Dimers
62	1-Penten-3-ol	C616251	$C_5H_{10}O$	1793.51	1585.35	1798.64	1259.45	Monomers
63	1-Penten-3-ol	C616251	$C_5H_{10}O$	2045.22	613.22	421.33	168.38	Dimers
64	3-Methylbutyl propanoate	C105680	$C_8H_{16}O_2$	1705.70	6833.11	103.03	70.18	
65	1-Butanol	C71363	$C_4H_{10}O$	1472.10	1392.35	2185.12	2529.75	Monomers
66	1-Butanol	C71363	$C_4H_{10}O$	273.52	337.82	838.25	1687.61	Dimers
67	Diallyl sulfide	C592881	$C_6H_{10}S$	367.91	553.83	204.87	334.69	
68	(E)-2-Pentenal	C1576870	C_5H_8O	655.72	757.33	988.04	515.80	Monomers
69	p-Xylene	C106423	C_8H_{10}	1189.37	1521.00	649.73	1639.21	
70	Cyclopentanone	C120923	C_5H_8O	3230.16	1918.17	411.52	2212.65	
71	(E)-2-Pentenal	C1576870	C_5H_8O	1004.82	443.33	467.68	274.07	Dimers
72	4-Methyl-3-penten-2-one	C141797	$C_6H_{10}O$	230.97	321.60	312.89	45.95	
73	beta-Pinene	C127913	$C_{10}H_{16}$	402.86	1641.86	85.24	97.13	
74	Isoamyl acetate	C123922	$C_7H_{14}O2$	593.09	164.07	214.31	497.19	
75	(Z)-2-Pentenal	C1576869	C_5H_8O	433.34	1556.07	357.24	603.99	Monomers
76	(Z)-2-Pentenal	C1576869	C_5H_8O	309.11	5188.38	93.61	349.68	Dimers
77	2-Methyl-1-propanol	C78831	$C_4H_{10}O$	583.87	852.56	628.62	1177.18	Monomers
78	Hexanal	C66251	$C_6H_{12}O$	1944.26	2803.51	3203.67	3114.45	Monomers
79	Hexanal	C66251	$C_6H_{12}O$	2331.59	3735.09	3204.66	5285.60	Dimers
80	(E)-2-Butenal	C123739	C_4H_6O	1630.63	910.14	567.29	438.94	
81	1-Propanol	C71238	C_3H_8O	328.31	322.34	1224.25	1262.25	
82	2-Butanol	C78922	$C_4H_{10}O$	1298.13	1530.66	105.56	203.92	

Table 2. *Cont.*

No	Compounds	CAS	Molecular Formula	SZ-01	SZ-02	GSZ	NSZ	Comment
83	1-Penten-3-one	C1629589	C_5H_8O	1272.76	578.92	725.65	220.10	
84	Isobutyl isobutyrate	C97858	$C_8H_{16}O_2$	1220.46	568.65	206.85	224.66	
85	3-Methyl-2-pentanone	C565617	$C_6H_{12}O$	932.60	826.90	24.55	29.98	
86	Isobutyl acetate	C110190	$C_6H_{12}O_2$	2312.15	4457.47	34.51	42.00	
87	Methyl 3-methylbutanoate	C556241	$C_6H_{12}O_2$	1182.99	127.62	38.62	82.05	
88	Pentanal	C110623	$C_5H_{10}O$	1427.20	1488.45	2802.92	2309.67	
89	2-Pentanone	C107879	$C_5H_{10}O$	1316.21	1179.53	570.79	1223.85	
90	Acetal	C105577	$C_6H_{14}O_2$	6995.16	1799.58	2950.66	2889.02	
91	2,5-Dimethylfuran	C625865	C_6H_8O	310.92	960.26	45.48	296.73	
92	Ethyl acetate	C141786	$C_4H_8O_2$	1302.33	2567.23	1565.33	3776.09	
93	Methyl acetate	C79209	$C_3H_6O_2$	15754.19	4078.68	5572.34	10602.05	
94	Acetone	C67641	C_3H_6O	5488.83	6531.17	4739.95	4322.90	
95	Propanal	C123386	C_3H_6O	1740.43	2326.29	3105.40	3006.52	
96	Butanal	C123728	C_4H_8O	228.42	73.54	76.38	60.76	
97	1-Propanethiol	C107039	C_3H_8S	1862.28	2854.62	30.31	18.37	
98	Dimethyl sulfide	C75183	C_2H_6S	1394.34	3434.90	791.13	1232.80	
99	2-Propenal	C107028	C_3H_4O	35.38	438.45	1562.47	364.22	
100	2-Methylbutanal	C96173	$C_5H_{10}O$	525.16	1865.45	1875.52	740.19	
101	2-Butanone	C78933	C_4H_8O	476.24	1406.98	1107.86	2272.08	
102	Ethanol	C64175	C_2H_6O	7729.16	6583.94	7764.54	7470.68	Monomers
103	Ethanol	C64175	C_2H_6O	2829.49	4731.35	5563.51	5461.13	Dimers
104	Methyl isobutyrate	C547637	$C_5H_{10}O_2$	145.10	618.81	19.14	21.81	
105	Thiophene	C110021	C_4H_4S	1051.26	688.43	2039.39	2418.48	
106	2-Methyl-2-propanol	C75650	$C_4H_{10}O$	209.37	524.83	655.14	925.26	
107	Heptanal	C111717	$C_7H_{14}O$	318.10	203.81	729.21	638.26	
108	2-Methyl-1-propanol	C78831	$C_4H_{10}O$	150.32	243.20	75.85	608.16	Dimers
109	3-Carene	C13466789	$C_{10}H_{16}$	191.59	253.67	43.51	646.50	
110	2-Methylpropanal	C78842	C_4H_8O	273.40	831.43	203.10	442.13	
111	3-Methylbutanal	C590863	$C_5H_{10}O$	188.66	766.00	365.90	1686.55	
112	Butanoic acid	C107926	$C_4H_8O_2$	1409.48	1213.95	5082.70	1589.55	
113	2-Methylpropanoic acid	C79312	$C_4H_8O_2$	1035.89	664.33	3651.45	1295.05	
114	Nonanal	C124196	$C_9H_{18}O$	358.80	871.42	779.24	553.25	
115	3-Ethylpyridine	C536787	C_7H_9N	166.39	102.28	412.40	62.95	
116	Acetoin	C513860	$C_4H_8O_2$	168.77	93.75	697.27	115.47	Dimers
117	1-Hexanol	C111273	$C_6H_{14}O$	66.88	66.76	63.03	314.99	Dimers
118	2-Methyl-3-furanthiol	C28588741	C_5H_6OS	100.71	54.57	77.06	91.73	
119	Octanal	C124130	$C_8H_{16}O$	164.88	328.96	289.08	251.28	
120	3-Methyl-1-butanol	C123513	$C_5H_{12}O$	196.78	210.35	104.43	983.60	Dimers
121	3-Methyl-2-butenal	C107868	C_5H_8O	100.87	789.99	224.29	885.33	Monomers
122	3-Methyl-2-butenal	C107868	C_5H_8O	128.27	332.57	34.78	299.32	Dimers
123	2-Heptanone	C110430	$C_7H_{14}O$	254.58	50.93	50.04	303.64	Dimers
124	Pyridine	C110861	C_5H_5N	71.09	95.75	86.10	258.29	
125	Styrene	C100425	C_8H_8	171.75	144.71	289.07	225.05	
126	Myrcene	C123353	$C_{10}H_{16}$	343.73	330.98	619.30	231.94	
127	o-Xylene	C95476	C_8H_{10}	100.68	170.75	119.71	96.15	
128	(Z)-4-Heptenal	C6728310	$C_7H_{12}O$	120.71	107.95	66.20	39.55	
129	Hexyl acetate	C142927	$C_8H_{16}O_2$	79.82	212.47	22.47	58.66	
130	3-Furanmethanol	C4412913	$C_5H_6O_2$	544.54	437.53	526.63	507.57	
131	Salicylic aldehyde	C90028	$C_7H_6O_2$	454.43	371.03	557.35	686.03	
132	(E,E)-alpha-Farnesene	C502614	$C_{15}H_{24}$	236.64	211.38	247.35	895.47	
133	Linalyl acetate	C115957	$C_{12}H_{20}O_2$	192.30	223.63	194.29	672.84	
134	Benzaldehyde	C100527	C_7H_6O	415.99	282.18	1085.63	761.03	
135	Citronellal	C106230	$C_{10}H_{18}O$	966.16	443.24	391.84	1637.24	
136	2-Ethyl-1-hexanol	C104767	$C_8H_{18}O$	379.36	212.65	2173.18	1290.30	
137	1-Octen-3-ol	C3391864	$C_8H_{16}O$	190.36	107.23	179.83	153.22	
138	Ethyl isobutyrate	C97621	$C_6H_{12}O_2$	1233.05	298.40	265.62	396.27	
139	Methional	C3268493	C_4H_8OS	80.68	185.66	73.75	155.20	

3.4. PCA of the Four Hawthorn Samples

The PCA method is a multivariate statistical analysis method that reduces the dimensions of a dataset by linearly combining variables and retaining the variability of the original data [27], in which the relationship between samples is more clearly and intuitively compared [28]. An analysis of PCA plots shows that the greater the similarity between samples, the greater the clustering degree, while the greater the difference between samples, the greater the distance [29]. As shown in Figure 5, the PCA of the VOCs in the four varieties of hawthorn was performed. Each color represents a different hawthorn sample, and the distance between each point indicates how similar they were.

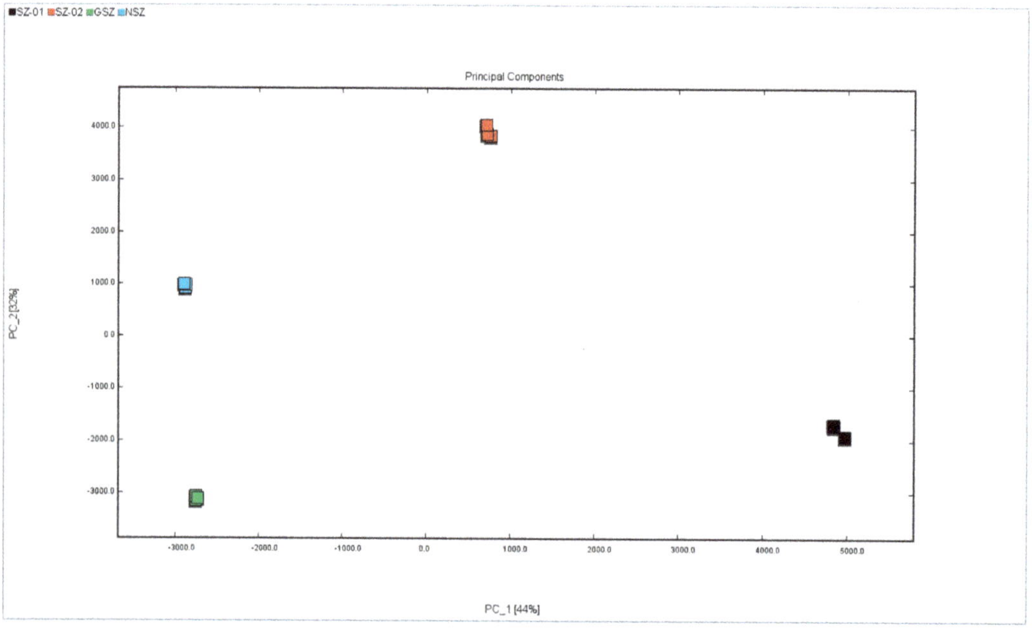

Figure 5. PCA analysis of different hawthorn varieties.

We can obviously see that the differentiation of the four samples and the distance between the Nanshanzha and Guangshanzha varieties was relatively close, which was because there were similarities in the VOC parts of the two samples, which was basically consistent with the signal peak information of the samples in the fingerprint spectrum. These results indicate that PCA combined with HS-GC-IMS could clearly distinguish differences in the VOCs among the four varieties of hawthorn.

3.5. Fingerprint Similarity Analysis Using Euclidean Distance

The distance coefficient is used to determine similarity in terms of the Euclidean distance, as is carried out in a PCA. This means that the difference is small if the distance between the samples is close, and this means that the difference is obvious if the distance is far [30]. Based on the Euclidean distance analysis, we created a fingerprint map with more intuitive characteristics than the PCA analysis. Based on the Euclidean distances between the four hawthorn varieties, Figure 6 depicts the fingerprint similarity. Thus, the Shanlihong and Nanshanzha varieties were close to each other, as were the Nanshanzha and Guangshanzha varieties, but the difference in VOCs between the Shanlihong and Nanshanzha varieties was of most significance. The results were in good agreement with the PCA analysis results. This may be because the Shanlihong and Shanzha varieties (the two are commonly known as "Northern Hawthorn") mostly grow in north China, while the Nanshanzha and Guangshanzha varieties mostly grow in south China, such as in Guangdong and Guangxi. The north and south of China have very different climates, soils, altitudes, and so on, which causes the VOC levels to differ between varieties of hawthorn.

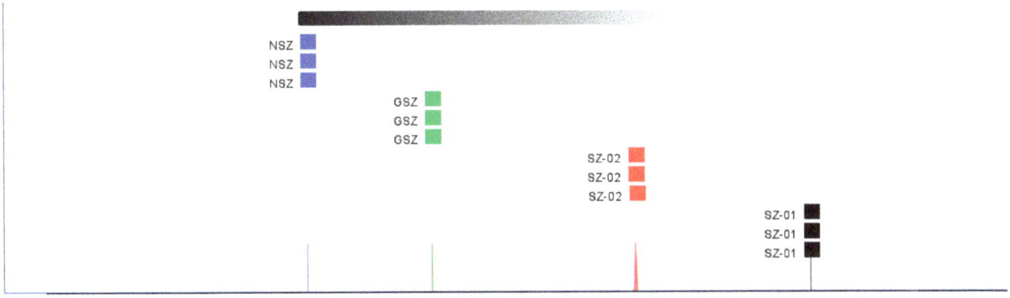

Figure 6. Fingerprint similarity based on the Euclidean distance of different varieties of hawthorn.

3.6. PLS-DA Analysis of the Four Hawthorn Samples

PLS-DA is a discriminant-analysis-supervised model that identifies hidden feature variables that damage model robustness and highlight the differences between groups [31]. Similarities and differences between samples are directly reflected in the PLS-DA score plot. In the score plot, a farther distance between two locations indicates a greater difference between two samples. The PLS-DA score plots of the four hawthorn samples were used to explore the differences in the VOC levels between the Shanlihong, Shanzha, Guangshanzha, and Nanshanzha varieties. The results are shown in Figure 7. It is evident that the samples of the different varieties of hawthorn were clearly distinguished based on the results of the principal component analysis. The SIMCA software (version 14.0) revealed that $R^2X = 0.993$ and $Q^2 = 0.999$, indicating that the model had a relatively accurate generalization and prediction ability and could both distinguish between the different varieties and identify their differences. Figure 8 shows the validation of the model carried out by the PLS-DA analysis after the permutation test ($n = 200$ times), with $R^2 = 0.0309$ and $Q^2 = -0.605$. It is evident from the slopes of the two regression lines that the model has a good prediction ability and does not show overfitting.

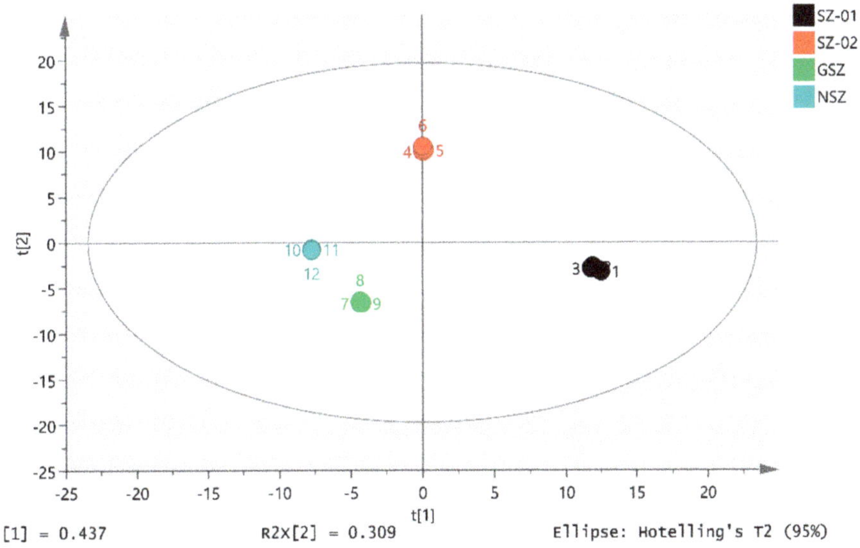

Figure 7. PLS-DA score chart.

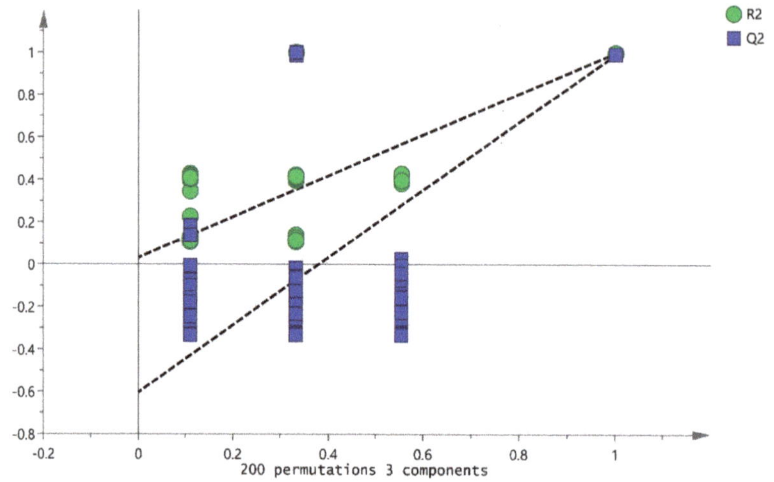

Figure 8. PLS-DA fitting curve.

It can be seen from the PLS-DA score chart that the supervised analysis method could better distinguish the four varieties of hawthorn and could obtain the variable weight value (variable important in projection: VIP). The results are shown in Figure 9. The VIP value is usually used to reflect the importance of PLS-DA model variables. The higher the column height and contribution value to the model in the figure, the more significant the difference [32]. The results show that there were 70 types of VOC with VIP >1 (marked in red boxes). Among them, the VOCs with the highest VIP values were 2-heptanone-M, alpha-thujone-M, (E)-2-pentenal-M, citronellal, 2-heptanone-D, and 2-cyclohexen-1-one. The abovementioned VOCs can be used as important indicators for the classification and identification of hawthorn, and they can provide a reference for the rapid identification of the four different types of hawthorn.

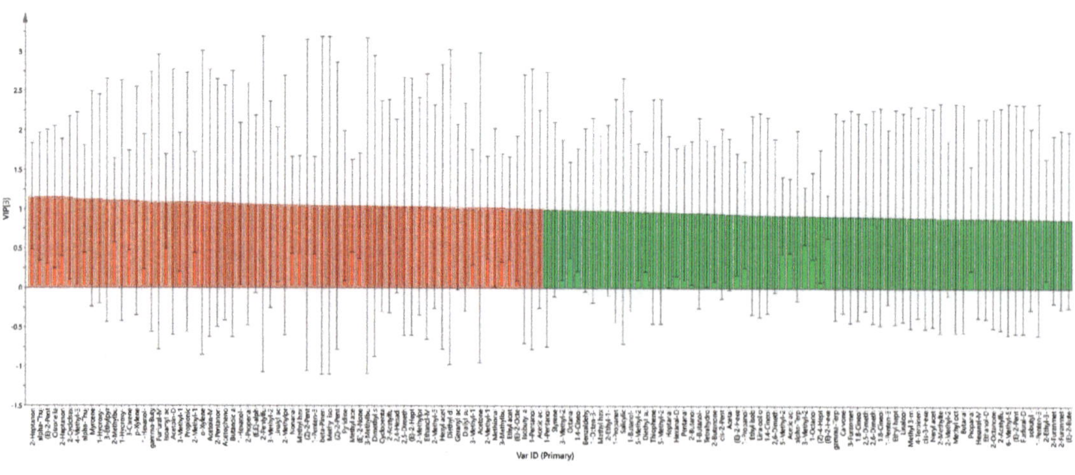

Figure 9. VIP diagram of the PLS-DA model.

4. Conclusions

This study used HS-GC-IMS for detecting and analyzing VOCs in hawthorn varieties. A total of 153 peaks were detected, and 139 VOCs were identified, including aldehydes, esters, alcohols, ketones, terpenes, acids, furans, pyridines, pyrazines, and thiophenes. By using the fingerprints, PCA, Euclidean distance, and PLS-DA, we were able to quickly and effectively distinguish the four hawthorns. According to the results, the four hawthorns could be distinguished accurately and objectively without using appearance features. This study used HS-GC-IMS to analyze the VOCs in four varieties of hawthorn, which is a simple, rapid, and accurate method. In addition, this approach requires fewer sample preparation steps and less analysis time than other analytical methods. As a result, the detection and comparison methods applied in this study provide a valuable reference tool for identifying hawthorns from food VOCs. Simultaneously, this research enriched the study of flavor compounds in hawthorn.

However, there were still some shortcomings in this study, such as the limited sample size of the test, the fact that established model may not be the optimal model, the fact that the ion migration spectrum database is not perfect, and the fact that the database of various commercial instruments is not universal, resulting in the identification of some ions. Next, we will carry out a study with a larger sample size and establish a comprehensive and detailed database of GC-IMS, which can also be combined with mass spectrometry to further improve the qualitative ability.

Author Contributions: Conceptualization, L.Z. (Lijun Zhu); methodology, L.Z. (Lijun Zhu) and F.O.; software, Y.M. and L.Z. (Lingfeng Zhu); validation, Y.X. and B.W.; formal analysis, L.Z. (Lijun Zhu) and F.O.; investigation, Y.X.; resources, B.W.; data curation, C.L.; writing—original draft preparation, L.Z. (Lijun Zhu); writing—review and editing, C.L.; visualization, Y.M. and Q.Z.; supervision, Q.Z.; project administration, Y.M. and C.L.; funding acquisition, C.L. All authors have read and agreed to the published version of the manuscript.

Funding: This research was funded by the Yunnan Key Laboratory of Screening and Research on Anti-pathogenic Plant Resources from Western Yunnan (No. APR202302), the Hunan Agricultural Science and Technology Innovation Fund Project (No. 2023CX30), the Hunan Provincial Natural Science Foundation of China (No. 2021JJ40406), the Undergraduate Research and Innovation Fund Project of Hunan University of Chinese Medicine (No. 2023), and the domestic First-Class Construction Discipline of Chinese Medicine in Hunan University of Chinese Medicine (No. 2024).

Data Availability Statement: Data are contained within the article.

Conflicts of Interest: The authors declare no conflicts of interest.

References

1. National Pharmacopoeia Committee. *Pharmacopoeia of the People's Republic of China*; China Medical Science and Technology Press: Beijing, China, 2020.
2. Lou, X.; Yuan, B.; Wang, L.; Xu, H.; Hanna, M.; Yuan, L. Evaluation of Physicochemical Characteristics, Nutritional Composition and Antioxidant Capacity of Chinese Organic Hawthorn Berry (*Crataegus pinnatifida*). *Int. J. Food Sci. Technol.* **2020**, *55*, 1679–1688. [CrossRef]
3. Duan, L.; Xiong, H.; Du, Y.; Wang, Z.; Li, Y.; Zhao, S.; Chen, J.; Si, D.; Pan, H. High-throughput LC-MS method for the rapid characterisation and comparative analysis of multiple ingredients of four hawthorn leaf extracts. *Phytochem. Anal.* **2022**, *33*, 635–643. [CrossRef] [PubMed]
4. Zhang, Q.; Zhao, P.; Sun, J.; Zhao, C.; Song, Y.; Wu, J. Research progress on chemical constituents and pharmacological action of hawthorn. *J. Northwest Pharm.* **2021**, *3*, 521–523.
5. Wang, S.Z.; Wu, M.; Chen, K.J.; Liu, Y.; Sun, J.; Sun, Z.; Ma, H.; Liu, L.T. Hawthorn Extract Alleviates Atherosclerosis through Regulating Inflammation and Apoptosis Related Factors: An Experimental Study. *Chin. J. Integr. Med.* **2019**, *25*, 108–115. [CrossRef] [PubMed]
6. Feng, Y.; Gao, S.; Zhu, T.; Sun, G.; Zhang, P.; Huang, Y.; Qu, S.; Du, X.; Mou, D. Hawthorn fruit acid consumption attenuates hyperlipidemia-associated oxidative damage in rats. *Front. Nutr.* **2022**, *9*, 936229. [CrossRef] [PubMed]
7. Kim, E.; Jang, E.; Lee, J.H. Potential Roles and Key Mechanisms of Hawthorn Extract against Various Liver Diseases. *Nutrients* **2022**, *14*, 867. [CrossRef] [PubMed]
8. Li, T.; Fu, S.; Huang, X.; Zhang, X.; Cui, Y.; Zhang, Z.; Ma, Y.; Zhang, X.; Yu, Q.; Yang, S. Biological Properties and Potential Application of Hawthorn and Its Major Functional Components: A Review. *J. Funct. Foods* **2022**, *90*, 104988. [CrossRef]
9. Nazhand, A.; Lucarini, M.; Durazzo, A.; Zaccardelli, M.; Cristarella, S.; Souto, S.B.; Silva, A.M.; Severino, P.; Souto, E.B.; Santini, A. Hawthorn (*Crataegus* spp.): An Updated Overview on Its Beneficial Properties. *Forests* **2020**, *11*, 564. [CrossRef]
10. Zhang, J.; Chai, X.; Zhao, F.; Hou, G.; Meng, Q. Food Applications and Potential Health Benefits of Hawthorn. *Foods* **2022**, *11*, 2861. [CrossRef]

11. Li, L.; Gao, X.; Liu, J.; Chitrakar, B.; Wang, B.; Wang, Y. Hawthorn pectin: Extraction, function and utilization. *Curr. Res. Food Sci.* **2021**, *4*, 429–435. [CrossRef]
12. Kwek, E.; Yan, C.; Ding, H.; Hao, W.; He, Z.; Liu, J.; Ma, K.Y.; Zhu, H.; Chen, Z.Y. Effects of hawthorn seed oil on plasma cholesterol and gut microbiota. *Nutr. Metab. (Lond.)* **2022**, *19*, 55. [CrossRef] [PubMed]
13. Jiang, H. Research Progress on Character Identification and Organic Acid Compositions of Shanzha (*Crataegus pinnatifida*), Nanshanzha (South *Crataegus*) and Guangshanzha (Cantonese *Crataegus*). *J. Liaoning Univ. Chin. Med.* **2023**, *25*, 132–137. [CrossRef]
14. Fei, C.; Dai, H.; Wu, X.; Li, L.; Lu, T.; Li, W.; Cai, B.; Yin, W.; Yin, F. Quality evaluation of raw and processed Crataegi Fructus by color measurement and fingerprint analysis. *J. Sep. Sci.* **2018**, *2*, 582–589. [CrossRef] [PubMed]
15. Gao, L.; Zhong, L.; Huang, R.; Yue, J.; Li, L.; Nie, L.; Wu, A.; Huang, S.; Yang, C.; Cao, G.; et al. Identification and determination of different processed products and their extracts of Crataegi Fructus by infrared spectroscopy combined with two-dimensional correlation analysis. *Spectrochim. Acta A Mol. Biomol. Spectrosc.* **2024**, *310*, 123922. [CrossRef]
16. Yin, J.; Wu, M.; Lin, R.; Li, X.; Ding, H.; Han, L.; Yang, W.; Song, X.; Li, W.; Qu, H. Application and development trends of gas chromatography-ion mobility spectrometry for traditional Chinese medicine, clinical, food and environmental analysis. *Microchem. J.* **2021**, *168*, 106527. [CrossRef]
17. Guo, Y.; Chen, D.; Dong, Y.; Ju, H.; Wu, C.; Lin, S. Characteristic volatiles fingerprints and changes of volatile compounds in fresh and dried Tricholoma matsutake Singer by HS-GC-IMS and HS-SPME-GC-MS. *J. Chromatogr. B Anal. Technol. Biomed. Life Sci.* **2018**, *1099*, 46–55. [CrossRef]
18. Song, H.; Liu, J. GC-O-MS technique and its applications in food flavor analysis. *Food Res. Int.* **2018**, *114*, 187–198. [CrossRef] [PubMed]
19. Huang, X.H.; Zheng, X.; Chen, Z.H.; Zhang, Y.Y.; Du, M.; Dong, X.P.; Qin, L.; Zhu, B.W. Fresh and grilled eel volatile fingerprinting by e-Nose, GC-O, GC-MS and GC×GC-QTOF combined with purge and trap and solvent-assisted flavor evaporation. *Food Res. Int.* **2019**, *115*, 32–43. [CrossRef]
20. Zhou, S.; Feng, D.; Zhou, Y.; Duan, H.; He, Y.; Jiang, Y.; Yan, W. Characteristic Volatile Organic Compound Analysis of Different Cistanches Based on HS-GC-IMS. *Molecules* **2022**, *27*, 6789. [CrossRef]
21. Dai, Y.; Liu, S.; Yang, L.; He, Y.; Guo, X.; Ma, Y.; Li, S.; Huang, D. Explorative study for the rapid detection of Fritillaria using gas chromatography-ion mobility spectrometry. *Front. Nutr.* **2024**, *11*, 1361668. [CrossRef]
22. Feng, X.; Wang, H.; Wang, Z.; Huang, P.; Kan, J. Discrimination and characterization of the volatile organic compounds in eight kinds of huajiao with geographical indication of China using electronic nose, HS-GC-IMS and HS-SPME-GC-MS. *Food Chem.* **2022**, *375*, 131671. [CrossRef]
23. Feng, D.; Wang, J.; He, Y.; Ji, X.J.; Tang, H.; Dong, Y.M.; Yan, W.J. HS-GC-IMS detection of volatile organic compounds in Acacia honey powders under vacuum belt drying at different temperatures. *Food Sci. Nutr.* **2021**, *9*, 4085–4093. [CrossRef]
24. Yang, B.; Huang, J.; Jin, W.; Sun, S.; Hu, K.; Li, J. Effects of Drying Methods on the Physicochemical Aspects and Volatile Compounds of *Lyophyllum decastes*. *Foods* **2022**, *11*, 3249. [CrossRef]
25. Duan, H.; Zhou, S.; Guo, J.; Yan, W. HS-GC-IMS Analysis of Volatile Organic Compounds in Different Varieties and Harvesting Times of *Rhizoma gastrodiae* (Tian Ma) in Yunnan Province. *Molecules* **2023**, *28*, 6705. [CrossRef]
26. Zhang, Y.; Tong, X.; Chen, B.; Wu, S.; Wang, X.; Zheng, Q.; Jiang, F.; Qiao, Y. Novel application of HS-GC-IMS for characteristic fingerprints and flavor compound variations in citrus reticulatae pericarpium during storage with different *Aspergillus niger* fermentation. *Food Chem. X* **2023**, *18*, 100653. [CrossRef] [PubMed]
27. Dong, W.; Tan, L.; Zhao, J.; Hu, R.; Lu, M. Characterization of Fatty Acid, Amino Acid and Volatile Compound Compositions and Bioactive Components of Seven Coffee (*Coffea robusta*) Cultivars Grown in Hainan Province, China. *Molecules* **2015**, *20*, 16687–16708. [CrossRef] [PubMed]
28. Christmann, J.; Rohn, S.; Weller, P. gc-ims-tools—A new Python package for chemometric analysis of GC–IMS data. *Food Chem.* **2022**, *394*, 133476. [CrossRef]
29. Zhou, Y.; Wang, D.; Duan, H.; Zhou, S.; Guo, J.H.; Yan, W.J. Detection and analysis of volatile flavor compounds in different varieties and origins of goji berries using HS-GC-IMS. *LWT* **2023**, *187*, 115322. [CrossRef]
30. Zhou, S.Q.; Feng, D.; Zhou, Y.X.; Zhao, J.; Zhao, J.Y.; Guo, Y.; Yan, W.J. HS-GC-IMS detection of volatile organic compounds in cistanche powders under different treatment methods. *LWT—Food Sci. Technol.* **2022**, *165*, 113730. [CrossRef]
31. Jiménez-Carvelo, A.M.; Martín-Torres, S.; Ortega-Gavilán, F.; Camacho, J. PLS-DA vs sparse PLS-DA in food traceability. A case study: Authentication of avocado samples. *Talanta* **2021**, *224*, 121904. [CrossRef]
32. Li, H.; Mao, Y.; Liu, Y.; Li, D.; Wang, M.; Ren, X.; Dou, Z. Comparative investigation of raw and processed products of Gardeniae Fructus and *Gardenia jasminoides* var. *radicans* using HPLC coupled with chemometric methods. *Biomed. Chromatogr.* **2021**, *35*, e5051. [CrossRef]

foods

Article

Increasing the Amounts of Bioactive Components in American Ginseng (*Panax quinquefolium* L.) Leaves Using Far-Infrared Irradiation

Xuan Wang [1,†], Myungjin Kim [2,†], Ruoqi Han [1], Jiarui Liu [1], Xuemei Sun [1], Shuyang Sun [1], Chengwu Jin [1,*,‡] and Dongha Cho [2,*,‡]

1 School of Food Engineering, Yantai Engineering Research Center of Green Food Processing and Quality Control, Ludong University, Yantai 264025, China; 2022121014@m.ldu.edu.cn (X.W.); 2023120979@m.ldu.edu.cn (R.H.); 2021120863@m.ldu.edu.cn (J.L.); 2875@ldu.edu.cn (X.S.); sysun81@ldu.edu.cn (S.S.)

2 College of Biomedical Science, Kangwon National University, Chuncheon 24341, Republic of Korea; gmrtka@kangwon.ac.kr

* Correspondence: jinchwu@ldu.edu.cn (C.J.); chodh@kangwon.ac.kr (D.C.)

† These authors contributed equally to this work.

‡ These authors contributed equally to this work.

Abstract: Both the roots and leaves of American ginseng contain ginsenosides and polyphenols. The impact of thermal processing on enhancing the biological activities of the root by altering its component composition has been widely reported. However, the effects of far-infrared irradiation (FIR), an efficient heat treatment method, on the bioactive components of the leaves remain to be elucidated. In the present study, we investigated the effects of FIR heat treatment between 160 and 200 °C on the deglycosylation and dehydration rates of the bioactive components in American ginseng leaves. As the temperature was increased, the amounts of common ginsenosides decreased while those of rare ginsenosides increased. After FIR heat treatment of American ginseng leaves at an optimal 190 °C, the highest total polyphenolic content and kaempferol content were detected, the antioxidant activity was significantly enhanced, and the amounts of the rare ginsenosides F4, Rg6, Rh4, Rk3, Rk1, Rg3, and Rg5 were 41, 5, 37, 64, 222, 17, and 266 times higher than those in untreated leaves, respectively. Moreover, the radical scavenging rates for 2,2-diphenyl-1-picrylhydrazyl and 2,2′-azino-*bis* (3-ethylbenzothiazoline-6-sulfonic acid) and the reducing power of the treated leaf extracts were 2.17, 1.86, and 1.77 times higher, respectively. Hence, FIR heat treatment at 190 °C is an efficient method for producing beneficial bioactive components from American ginseng leaves.

Keywords: American ginseng leaves; far-infrared irradiation; polyphenols; ginsenosides; antioxidant activity

1. Introduction

Panax quinquefolium L., also known as American ginseng, is a perennial herb from the family Araliaceae [1]. The beneficial effects of American ginseng products, such as antitumor and anti-inflammatory activities, immunity-improving effects, and blood-sugar-reducing activities, are mainly attributed to bioactive ingredients such as ginsenosides, polyphenols, and polysaccharides [2–5]. The cultivation period for American ginseng is long (harvesting is generally conducted after 4–5 years [6]), with the ginsenoside content in roots dependent on the cultivation period [7]. However, the leaves can be harvested every year since the ginsenoside content therein is less dependent on the age of the plant [8]. The activity of ginsenoside is closely related to the type, quantity, and positions of the sugar groups it carries; the fewer sugar groups in ginsenoside, the higher its physiological activity [9,10]. It has been proven that the ginsenoside content in American ginseng leaves is significantly higher than that in the roots [11]. In our previous research, we showed that heat treatment transforms the constant ginsenosides in ginseng leaves into rare ginsenosides with a higher physiological activity and significantly increases the total polyphenolic content (TPC) [12]. Moreover, American ginseng leaves are a high-quality source of ginsenosides, polyphenols, and other bioactive ingredients that are not only underused but also often discarded.

The most commonly used industrial methods for American ginseng processing involve physical, chemical, or biological treatment [13], with the physical method of heat treatment being the most commonly used. The outcomes from several studies have proven that heat treatment increases the amount of highly bioactive components, such as ginsenosides, polyphenols [14], and acidic polysaccharides [15], in American ginseng. Specifically, Zhang et al. [16] reported that heat treatment decreased the amounts of constant ginsenosides and increased those of rare ginsenosides.

Far-infrared irradiation (FIR) is a heat treatment method frequently used in the food industry that can penetrate plant materials, generate heat on the surface and inside the plant, and induce the decomposition of multi-chain molecular groups [17,18]. Compared to the traditional drying method, FIR drying offers advantages such as a shorter drying time, lower costs, and a more uniform temperature distribution, ensuring the quality and safety of food. However, it is important to acknowledge that FIR drying also has its drawbacks, including the generation of high heat, which can potentially lead to burns upon exposure [18]. In our previous research, we showed that FIR can significantly increase the TPC in ginseng leaves and promote ginsenoside transformations [12]. Geng et al. [19] found in their research that after infrared drying, the content of TPC in carrot slices increased, and the antioxidant activity also increased significantly. Ren et al. [20] confirmed that the antioxidant activity of ginger was improved after infrared drying and found that there was a high correlation between antioxidant activity and TPC. Rajoriya et al. [21] found that the antioxidant activity of FIR-dried apple slices increased significantly, and there was a main correlation between them and TPC. Moreover, FIR treatment of *Lycium barbarum* [22], rice [23], black rice [24], and mango [25] can significantly increase the amounts of bioactive components such as polyphenols, flavonoids, and anthocyanins. At present, research on American ginseng mainly focuses on the roots, with less focus on the leaves. Few studies have been conducted on the effect of FIR on the bioactive components and antioxidant activities in American ginseng leaves.

Since American ginseng leaves provide a high-quality source of polyphenols, ginsenosides, and other bioactive components, exploiting this resource is of utmost importance. The purpose of the present study is to determine the effects of FIR treatment on bioactive components (polyphenols, flavonoids, and ginsenosides) and the antioxidant activity of American ginseng leaves. Our approach could pave the way for the utilization of American ginseng leaves for biomedicine production and for other health-related benefits.

2. Materials and Methods

2.1. Chemicals

Organic solvents (methanol and acetonitrile) (high-performance liquid chromatography (HPLC)-grade) were purchased from Merck KGaA (Darmstadt, Germany). Folin phenol reagent was purchased from Wako Pure Chemicals (Osaka, Japan). Kaempferol and panasenoside standards were purchased from Sigma Chemical Co. (St. Louis, MO, USA). Ginsenosides were purchased from ChromaDex (Santa Ana, CA, USA) and Ambo Institute (Daejeon, Republic of Korea).

2.2. Sample Collection and FIR Treatment

The pretreatment process for raw materials is shown in Figure 1. Four-year-old American ginseng leaves were collected from Wendeng, Weihai, Shandong Province. American ginseng leaf samples with the following features were selected: top of the stem broadly ovoid or obovate, about 10–15 cm long and 5–6 cm wide, dark green on the surface, light green on the back, smooth, mature, and undamaged on the surface. Intact and undamaged leaves were washed with distilled water and then wiped with gauze. Fresh American ginseng leaves were fully dried in an oven (DHG-9143BS-III; CIMO, Shanghai, China) at 50 °C for 24 h and then ground with a grinder (FW100, Taisite instrument, Tianjin, China). The powder was sieved through a 200 micron sieve to obtain a uniform particle size. The powder was divided into two groups: (1) the non-FIR-treated control (Con; 0) and (2) the FIR-treated group (treatment in an FIR dryer (HKD-10; Korea Energy Technology, Seoul, Republic of Korea) for 30 min at 160 °C (FIR-160), 170 °C (FIR-170), 180 °C (FIR-180), 190 °C (FIR-190), or 200 °C (FIR-200)). The drying chamber consisted of a stainless-steel drying chamber, sample tray, centrifugal fan, and FIR heater. Two sets of FIR heaters were used, namely, one placed at the bottom of the drying chamber and the other at the top of the drying chamber. The sample tray was arranged between and parallel to the top and bottom heaters. Hot air was circulated in the drying chamber with a fan. The inlet air temperature flowing through the hot air heater was controlled with a PID controller (the accuracy was ±1 °C).

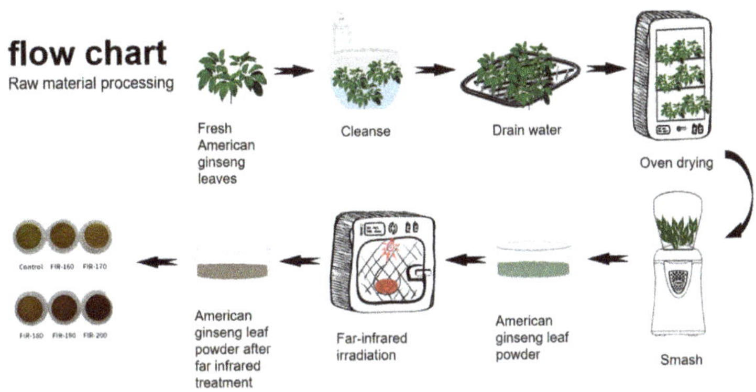

Figure 1. Flow chart of raw material pretreatment.

2.3. *Bioactive Compound Analysis*

2.3.1. Sample Extraction

The samples were extracted according to the method of Kim et al. [26] with some modifications. First, 2 g of sample was added to 100 mL of 80% (v/v) methanol. Sample extraction was carried out in an oscillating incubator (HNYC-202T; Ounuo, Tianjin, China) at 30 °C for 24 h. The mixture was centrifuged in a centrifugator (05pr-22 centrifuge, Hitachi, Tokyo, Japan) at 500× g for 10 min at room temperature. Afterward, the supernatant was collected and then filtered using Whatman No. 42 filter paper (Whatman Inc., Clifton, NJ, USA). The filtrate was concentrated using a vacuum rotary evaporator at 40 °C (Eyela Co., Tokyo, Japan). The sample was then freeze-dried using a vacuum freeze dryer (Marin Christ, Landkreis Osterode, Germany) at −60 °C and 0.071 mbar vacuum pressure [27] (the concentrated filtrate was placed in the refrigerator at −80 °C for one day before freeze-drying). The sample was stored in a refrigerator at 20 °C before being used in subsequent experiments.

2.3.2. TPC Determination

The TPC was determined by using the Folin-Ciocalteu method as follows. First, 1.9 mL of distilled water and 1.0 mL of Folin-Ciocalteu reagent were mixed in a tube, after which 0.1 mL of the sample solution (2 mg/mL dissolved in 80% (v/v) methanol) was added. Subsequently, 1.0 mL of 20% Na_2CO_3 was added, followed by incubation for 2 h at 25 °C. Afterward, the absorbance of the sample was recorded at 765 nm with a spectrophotometer (UV-2600i, Shimadzu, Kyoto, Japan). The results are expressed as milligrams of gallic acid equivalent (mg_{GAE}) per gram of dry weight.

2.3.3. HPLC Analysis of Panasenoside, Kaempferol, and Ginsenoside Amounts

The amounts of panasenoside and kaempferol in American ginseng leaves were determined by using HPLC. The equipment was an HPLC system (CBM-20A; Shimadzu Co, Ltd., Kyoto, Japan) with two gradient pump systems (LC-20AT; Shimadzu, Japan), an auto-sample injector (SIL20A; Shimadzu), a UV detector (SPD-20A; Shimadzu), and a column oven (CTO-20A; Shimadzu). Ground leaves (0.1 mg) were suspended in 1 mL of 80% (*v*/*v*) methanol and filtered through a 0.22 μm membrane filter before being injected onto the HPLC system. HPLC separation was performed on an Inertsil ODS-SP C18 column (250 mm × 4.6 mm, 5 μm; GL Sciences, Tokyo, Japan). The injection volume was 10 μL. The gradient running phase was programmed as a combination of solvent A (water with 0.1% trifluoroacetic acid) and solvent B (acetonitrile), during which solvent B was sequentially increased from 14% to 18% (0 to 10 min), 18% to 30% (10 to 20 min), 30% to 60% (20 to 30 min), 60% to 65% (30 to 33 min), 65% to 100% (33 to 40 min), 100% to 100% (40 to 50 min), and then finally adjusted from 100% to 14% (50 to 65 min). The operating temperature was set to 35 °C. The flow rate of the mobile phase was kept at 1.0 mL per minute. The detector was set at 355 nm to monitor panasenoside and kaempferol.

The prepared sample solution was filtered through a 0.22 μm membrane filter. The volume injected onto a Kinetex C18 column (100 mm × 4.6 mm, 2.6 μm; Torrance, CA, USA) was 10 μL. The gradient running phase was programmed as a combination of solvent A (water) and solvent B (acetonitrile), during which solvent B was sequentially increased from 17% to 23% (0 to 30 min), 23% to 24% (30 to 35 min), 24% to 32% (35 to 45 min), 32% to 44% (45 to 48 min), 44% to 44% (48 to 52 min), 44% to 55% (52 to 65 min), 55% to 100% (65 to 85 min), 100% to 100% (85 to 95 min), and finally adjusted from 100% to 17% (95 to 105 min).

2.4. Determination of the Free Radical Scavenging and Antioxidant Activities

2.4.1. Determination of the Free Radical Scavenging Activity

American ginseng leaf powder (90 mg) was added to 30 mL of 80% (*v*/*v*) methanol in a 50 mL tube and processed at 30 °C for 30 min via sonication (PS-40A; JeKen, Dongguan, China). Subsequently, the sample was centrifuged (J-E; Beckman Coulter Inc., Shanghai, China) at 3500 revolutions/min for 15 min. The supernatant was collected and filtered through a 0.22 μm membrane filter.

The DPPH free radical scavenging activity was measured using the procedure described by Kossah et al. [28] with some modifications. Briefly, 1 mL of sample solution at varying concentrations was mixed with 3 mL of DPPH solution. After reacting in the dark at room temperature for 30 min, the absorbance was measured at 517 nm.

The radical scavenging activity of ABTS was measured using the procedure described by Lim et al. [29] with some modifications. Briefly, 0.5 mL of the sample solution at varying concentrations was mixed with 5 mL of ABTS solution. After reacting in the dark at room temperature for 10 min, the absorbance was measured at 734 nm.

2.4.2. Determination of the Reducing Power

Sample processing followed the same procedure as that mentioned in Section 2.4.1. Measurement of the reducing power was conducted using the procedure described by Oyaizu et al. [30] with some modifications. Briefly, 0.4 mL of sample solution at varying concentrations was mixed with 1 mL of 0.2 mol/L phosphate buffer (pH 6.6) and 1 mL of 1% potassium ferrocyanide solution. The mixture was then incubated in a 50 °C water bath for 30 min, after which 1 mL of 10% trichloroacetic acid was added, followed by thorough mixing and centrifugation at 3000 revolutions/min for 10 min. After centrifugation, 2 mL of the supernatant was removed and 2 mL of distilled water and 0.4 mL of 0.1% ferric chloride were added. Next, the mixture was vortexed, and the absorbance of the mixture was measured at 700 nm.

2.5. Statistical Analysis

The data were analyzed statistically using SAS software (Enterprise Guide version 7.1; SAS Institute Inc., Cary, NC, USA). The data from the non-FIR-treated control and FIR treatment groups were compared by using a one-way analysis of variance. Differences between the experimental groups were evaluated by using Tukey's honestly significant difference (HSD) test at the $p < 0.05$ significance level.

3. Results and Discussion

3.1. The Effect of FIR Heat Treatment on the TPC in American Ginseng Leaves

The change in TPC in American ginseng leaves after FIR heat treatment is shown in Figure 2. The TPC gradually increased with the treatment temperature up to 190 °C and then declined. At 190 °C, the highest TPC was 15.89 mg/g, which was 1.56 times higher than that without treatment. According to Hwang et al. [31], phenolic compounds can scavenge free radicals by providing hydrogen; thus, promoting polyphenol generation in American ginseng leaves via FIR heat treatment improves the antioxidant activity of the extract. Our results are similar to those reported before. For example, Jeong et al. [32] found that the content of TPC increased significantly after baking dry ginseng powder, probably because Maillard reaction products with a phenol structure were produced during baking, which increased the content of TPC. Jin et al. [33] found that the content of TPC increased significantly after baking ginseng roots at 140–200 °C and speculated that baking would transform the combined polyphenol compounds into free polyphenols to increase the content of TPC. In the present study, the TPC in American ginseng leaves increased after FIR heat treatment. We speculate that there are two reasons for this. First, most phenolic compounds in plants are covalently bonded to insoluble polymers and are released by heat treatment [34,35]. Second, heating can promote the hydrolysis of macromolecular polyphenols, thereby producing smaller molecular polyphenols [36]. From the above experimental results, it can be concluded that the highest TPC content in American ginseng leaves was produced via FIR heat treatment at 190 °C.

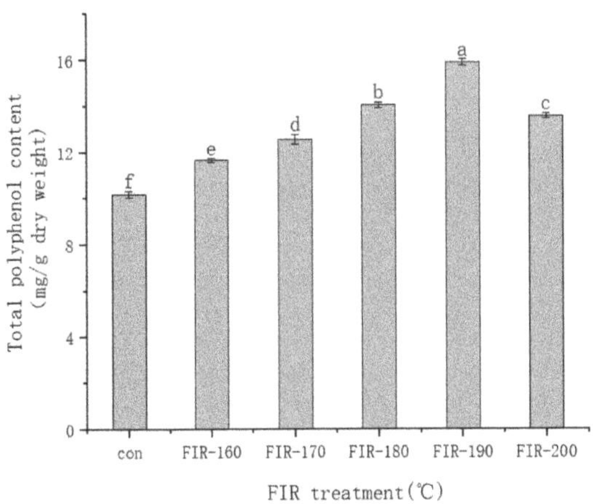

Figure 2. Changes in the total polyphenol content (TPC) of American ginseng leaves treated with FIR at various temperatures. The values are expressed as the mean and the standard error ($n = 3$). Letters a–f above the bar graphs indicate significant differences at $p < 0.05$ using Tukey's honestly significant difference (HSD) test.

3.2. The Effect of FIR Heat Treatment on the Amounts of Panasenoside and Kaempferol in American Ginseng Leaves

Figure 3 shows the kaempferol and panasenoside content in American ginseng leaves before and after FIR treatment at various temperatures. The panasenoside content in the leaves before FIR treatment was 18.52 mg/g. Increasing the FIR treatment temperature decreased the panasenoside content, with the lowest value observed after treatment at 200 °C (4.49 mg/g), which was 76% lower than that without treatment. Contrary to the change in panasenoside content, the kaempferol content increased from 0.026 mg/g without treatment up to 0.841 mg/g (FIR-190) at 190 °C, after which it decreased. The kaempferol content after FIR treatment at 190 °C was 32 times higher than that without treatment.

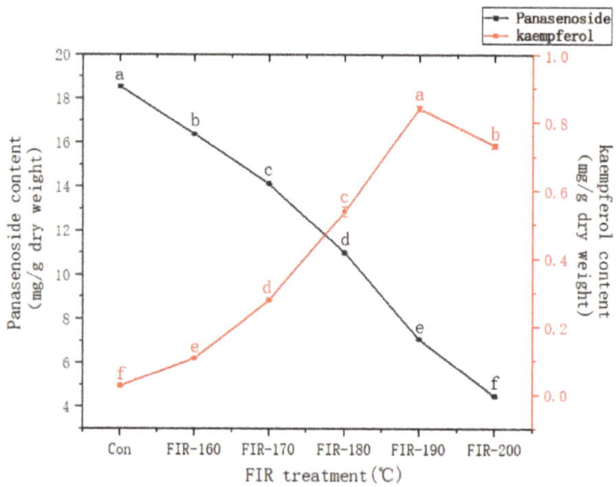

Figure 3. Changes in panasenoside and kaempferol contents in American ginseng leaves treated via FIR at various temperatures. The values are expressed as the mean and the standard error (n = 3). Letters a–f above the line charts indicate significant differences at $p < 0.05$ using Tukey's HSD test.

Panasenoside (kaempferol 3-o-glucosyl-(1→2)-galactoside) is a kaempferol glycoside [37]. According to the findings of our previous research on ginseng leaves, we speculate that during FIR treatment, energy can cause panasenoside deglycosylate to break the bond between kaempferol aglycone and the first glycosyl, resulting in a significant increase in kaempferol content. However, astragalin (kaempferol 3-β-D-glucopyranoside) was not detected. The deglycosylation of panasenoside is concurrent with kaempferol degradation; increasing the FIR temperature gradually caused the panasenoside deglycosylation rate to become lower than the kaempferol degradation rate. This explains why the panasenoside content in American ginseng leaves decreased rapidly while the kaempferol content increased relatively slowly, reaching the highest value after FIR treatment at 190 °C [12]. Although the kaempferol content in American ginseng leaves reached the highest value after FIR treatment at 190 °C, its content in ginseng leaves was different, and it stopped increasing after FIR treatment at 180 °C. This could have been caused by the different compositions and amounts of panasenoside in the raw material used. Kaempferol (a deglycosylation product) has high anti-inflammatory [2], antitumor [3], and antioxidant activities [38]. According to the above experimental results, it can be concluded that the kaempferol content in American ginseng leaves was markedly increased by FIR treatment at 190 °C.

3.3. *The Effect of FIR Treatment on the Ginsenoside Content in American Ginseng Leaves*

3.3.1. Protopanaxadiol (PPD)-Type Ginsenosides

The change in PPD-type ginsenoside content in American ginseng leaves after FIR heat treatment at various temperatures is shown in Figure 4A and Table 1. The amounts of the main ginsenosides (Rb1, Rb2, Rb3, Rc) were negatively correlated with the FIR treatment temperature. When the FIR temperature was 200 °C, their amounts (2.12, 4.10, 12.70, 1.70, and 8.66 mg/g, respectively) were even lower than in the untreated leaves (equating to reductions of 73%, 68%, 74%, 64%, and 73%, respectively). On the contrary, the amounts of rare PPD-type ginsenosides (Rg5, Rk1, and Rg3) showed an upward trend with increasing FIR temperature; their amounts were the highest after FIR treatment at 190 °C (13.37, 4.46, and 1.94 mg/g compared to 0.05, 0.02, and 0.11 mg/g in untreated leaves, equating to 266-, 222-, and 17-fold increases, respectively).

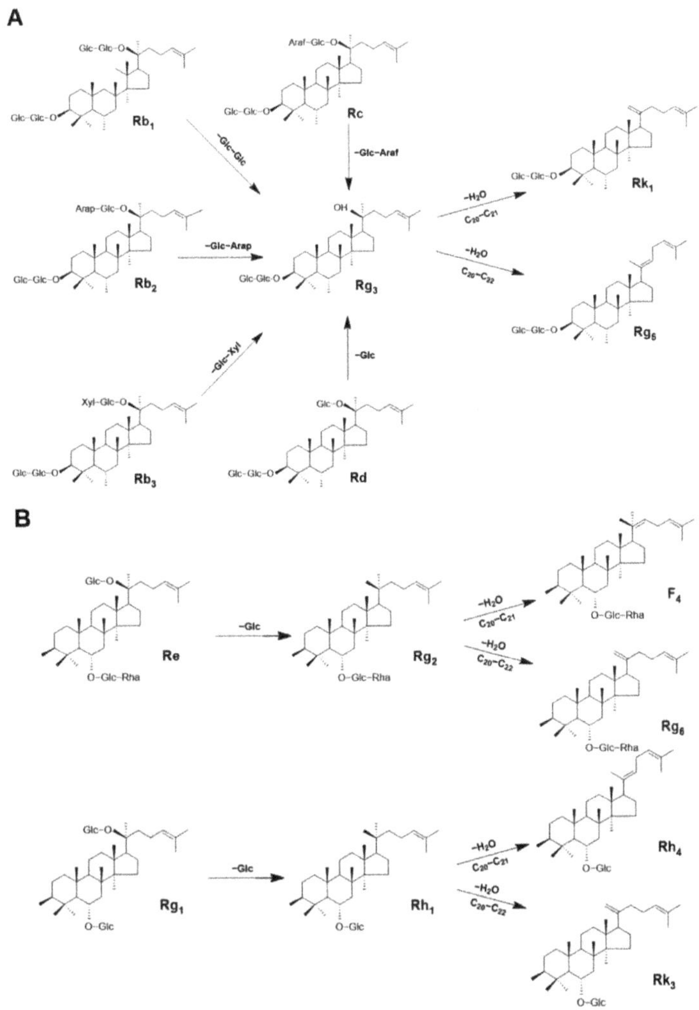

Figure 4. Schematic diagram of the transformation of (**A**) protopanaxadiol (PPD) and (**B**) protopanaxatriol (PPT) ginsenosides in American ginseng leaves after FIR treatment.

Table 1. Changes in the content of protopanaxadiol (PPD) ginsenoside in leaves of American ginseng treated with FIR at different temperatures.

Ginsenoside (mg/g)	FIR Treatment (°C)					
	Con	FIR-160	FIR-170	FIR-180	FIR-190	FIR-200
Rb1	7.88 ± 0.03 a	6.54 ± 0.05 b	5.91 ± 0.08 c	4.86 ± 0.04 d	2.82 ± 0.02 e	2.12 ± 0.02 f
Rb2	12.75 ± 0.03 a	11.69 ± 0.03 b	11.11 ± 0.05 c	10.13 ± 0.03 d	5.43 ± 0.02 e	4.10 ± 0.02 f
Rb3	48.25 ± 0.05 a	43.53 ± 0.06 b	40.65 ± 0.05 c	30.33 ± 0.04 d	17.46 ± 0.02 e	12.70 ± 0.02 f
Rc	4.71 ± 0.03 a	4.34 ± 0.04 b	4.11 ± 0.04 c	3.60 ± 0.05 d	2.18 ± 0.02 e	1.70 ± 0.02 f
Rd	31.73 ± 0.03 a	27.48 ± 0.04 b	25.48 ± 0.05 c	20.23 ± 0.04 d	11.75 ± 0.02 e	8.66 ± 0.02 f
Rk1	0.02 ± 0.00 e	0.56 ± 0.01 d	1.20 ± 0.01 c	2.62 ± 0.02 b	4.46 ± 0.01 a	4.46 ± 0.00 a
Rg3	0.11 ± 0.00 e	0.65 ± 0.03 d	0.93 ± 0.01 c	1.45 ± 0.02 b	1.94 ± 0.02 a	1.97 ± 0.01 a
Rg5	0.05 ± 0.01 e	1.81 ± 0.03 d	3.90 ± 0.03 c	8.21 ± 0.03 b	13.37 ± 0.02 a	13.38 ± 0.01 a

The values are expressed as the mean and the standard error (n = 3). Letters a–f above the line charts indicate significant differences at $p < 0.05$ using Tukey's HSD test.

The conditions applied can be dry heat, wet heat, etc. The main pathways for PPD-type ginsenoside conversion during dry heat treatment are as follows: Rb1, Rb2, Rb3, Rc, Rd→Rg3→Rk1, and Rg5 [39]. Increasing the FIR temperature significantly increased the amounts of Rk1, Rg5, and the other ginsenosides but not Rg3. This could be because Rg3, which is an intermediate produced by the deglycosylation of Rb1, Rb2, Rb3, Rc, Rd, and other ginsenosides, can also undergo dehydration and deglycosylation at similar reaction rates. Although ginsenosides have great potential, the bioavailability of the main ginsenosides is low because their absorption rates in the circulatory system are very low. On the contrary, rare ginsenosides exhibit good bioavailability because they can more easily permeate through cell membranes due to their small molecular weight [40]. In the present study, increasing the FIR treatment temperature decreased the main PPD-type ginsenoside content by 72% and increased the rare ginsenoside content hundreds of times compared to its value without treatment. Although many researchers have reported that heat treatment can promote ginsenoside transformations, different heat treatment methods and conditions affect their transformation rate [40–42]. For example, Park et al. [43] reported that after heat treatment at 120 °C, the main ginsenosides were converted into rare ginsenosides (Rg3, Rg5, and Rk1) with a lower polarity; they were difficult to detect in untreated samples and increased to 15.2, 3.6, and 2.9 μg/mg, respectively, after treatment. Pu et al. [44] reported that the rare ginsenosides Rg3, Rk1, and Rg5 could be produced effectively after the ginseng was expanded. Kim et al. [45] reported that the content of ginsenoside changed significantly with steaming temperature. With the increase in temperature, the contents of the ginsenosides Rg3, Rk1, and Rg5 increased from 0.69, 1.11, and 1.79 mg/g to 4.08, 5.16, and 8.98 mg/g, respectively. We found that the amounts of the rare ginsenosides Rg5 and Rk1 were higher, while the Rg3 content was lower than they reported. This could be because the reaction rate of Rd→Rg3 deglycosylation was different to that of Rg3→Rg5 and Rk1 dehydration due to the differences in the heat treatment method and temperature. Steam and drying treatments can increase the Rg3, Rg5, and Rk1 amounts and decrease the Rb1 amount [46]. The findings from the above research are similar to ours for FIR treatment of American ginseng leaves. When increasing the FIR treatment temperature from 160 to 190 °C, the amounts of PPD-type rare ginsenosides increased significantly. In particular, the amount of the rare ginsenoside Rg5 increased by 1.76 mg/g (con→FIR-160), 2.10 mg/g (FIR-160→FIR-170), 4.31 mg/g (FIR-170→FIR-180), and 5.16 mg/g (FIR-180→FIR-190). Heat treatment can transform the main PPD-type ginsenosides into rare ones with higher physiological activities and a higher clinical value.

3.3.2. Protopanaxatriol (PPT)-Type Ginsenosides

Changes in the amounts of PPT-type ginsenosides in American ginseng leaves according to the FIR treatment temperature shown in Figure 4B and Table 2 are similar to those observed for PPD-type ginsenosides; i.e., the main ginsenoside content decreased and the corresponding rare ginsenoside content increased with increasing temperature. When increasing the FIR treatment temperature, the amounts of the main ginsenosides Re and Rg1 decreased while those of Rg2 and Rh1 increased first and then decreased.

Moreover, after FIR treatment up to 190 °C, the amounts of the rare ginsenosides F4, Rg6, Rh4, and Rk3 did not increase (1.89, 1.82, 1.07, and 0.61 mg/g, respectively). Compared with untreated leaves, the amounts of the main ginsenosides Re and Rg1 decreased by 79.27% and 77.96%, respectively, while those of the rare ginsenosides Rk3, F4, Rh4, and Rg6 were 64, 41, 37, and 5 times higher, respectively.

Table 2. Changes in the content of protopanaxatriol (PPT) ginsenoside in leaves of American ginseng treated with FIR at different temperatures.

Ginsenoside (mg/g)	FIR Treatment (°C)					
	Con	FIR-160	FIR-170	FIR-180	FIR-190	FIR-200
Rg1	8.12 ± 0.03 a	5.47 ± 0.07 b	5.20 ± 0.02 c	3.99 ± 0.03 d	2.40 ± 0.02 e	1.79 ± 0.02 f
Re	29.18 ± 0.03 a	21.47 ± 0.04 b	19.93 ± 0.04 c	15.15 ± 0.04 d	8.35 ± 0.02 e	6.05 ± 0.02 f
Rg2	0.77 ± 0.01 e	1.74 ± 0.01 b	1.75 ± 0.03 b	1.83 ± 0.02 a	1.65 ± 0.01 c	1.35 ± 0.02 d
Rh1	1.36 ± 0.01 c	1.50 ± 0.02 b	1.76 ± 0.03 a	0.75 ± 0.01 d	0.61 ± 0.00 e	0.57 ± 0.01 f
F4	0.05 ± 0.00 e	0.56 ± 0.01 d	0.85 ± 0.01 c	1.41 ± 0.02 b	1.89 ± 0.00 a	1.88 ± 0.00 a
Rh4	0.03 ± 0.00 e	0.35 ± 0.00 d	0.54 ± 0.00 c	0.75 ± 0.00 b	1.07 ± 0.00 a	1.07 ± 0.00 a
Rg6	0.36 ± 0.00 f	0.83 ± 0.01 e	0.99 ± 0.01 d	1.42 ± 0.01 c	1.82 ± 0.01 a	1.79 ± 0.01 b
Rk3	0.01 ± 0.00 e	0.12 ± 0.00 d	0.20 ± 0.00 c	0.38 ± 0.00 b	0.61 ± 0.01 a	0.60 ± 0.00 a

The values are expressed as the mean and the standard error (n = 3). Letters a–f above the line charts indicate significant differences at $p < 0.05$ using Tukey's HSD test.

Under FIR heat treatment, the main transformation paths for PPT-type ginsenosides are as follows: Re→Rg2→Rg6 and F4 and Rg1→Rh1→Rh4 and Rk3 [39]. Our experimental results show that FIR treatment increases the amounts of the rare ginsenosides Rg6, F4, Rh4, and Rk3 by 4-, 40-, 36-, and 63-fold, respectively. Xue et al. [8] used an autoclave to heat American ginseng to a high temperature and obtained similar results. With increasing temperature and time, the PPT-type rare ginsenoside content in American ginseng also showed an increasing trend. Ji et al. [47] reported that processing a ginseng extract at 90 °C greatly increased the amounts of the rare ginsenosides F4, Rk3, and Rh4 compared to the untreated ginseng extract. Their results are similar to ours in that PPT-type main ginsenosides were transformed into rare ginsenosides via heat treatment. Rg2 and Rh1 are intermediate products of the transformation of Re and Rg1, respectively. Increasing the FIR treatment temperature caused the amounts of the Rg2 and Rh1 ginsenosides to first increase and then decrease, which could be because the deglycosylation rate of transformation of Re and Rg1 into Rg2 and Rh1, respectively, is lower than the dehydration rate of the latter two products [48]. As shown in Figure 4A,B, the highest amounts of rare PPD- and PPT-type ginsenosides were obtained at an FIR treatment temperature of 190 °C and decreased when the temperature was increased to 200 °C; this could have been caused by ginsenoside degradation due to the higher FIR energy [12]. However, the decreases in rare ginsenoside contents at 200 °C are slightly different from that after FIR treatment of ginseng leaves [12], which may have been caused by the different compositions and amounts of ginsenosides in the raw materials [49]. Although increasing the FIR treatment temperature from 160 to 190 °C increased rare PPT-type and PPD-type ginsenosides to different extents, the change in content increases was not significant. This could be due to the different structures of PPD-type and PPT-type ginsenoside aglycones, resulting in different dehydration rates.

3.4. The Effects of FIR Treatment on the Free Radical Scavenging and Antioxidant Activities of American Ginseng Leaves

2,2-Diphenyl-1-picrylhydrazyl (DPPH) and 2,2′-azino-bis (3-ethylbenzothiazoline-6-sulfonic acid) (ABTS) scavenging assays and the ferric reducing antioxidant power (FRAP) assay were used to determine the effect of FIR treatment on the antioxidant capacity of American ginseng leaf extracts. The results shown in Figure 5 indicate that FIR treatment significantly improved the antioxidant capacity of American ginseng leaf extracts. The DPPH and ABTS scavenging activities of American ginseng leaves treated with FIR at 190 °C were 2.17 times (32.08% vs. 69.62%) and 1.86 times (35.88% vs. 66.61%) higher than those in untreated leaves. The absorbance measured using the FRAP method increased 1.77-fold (0.212 vs. 0.374) after FIR treatment at 190 °C. When the FIR treatment temperature was raised to 200 °C, the DPPH and ABTS scavenging rates and the reducing power all exhibited a downward trend.

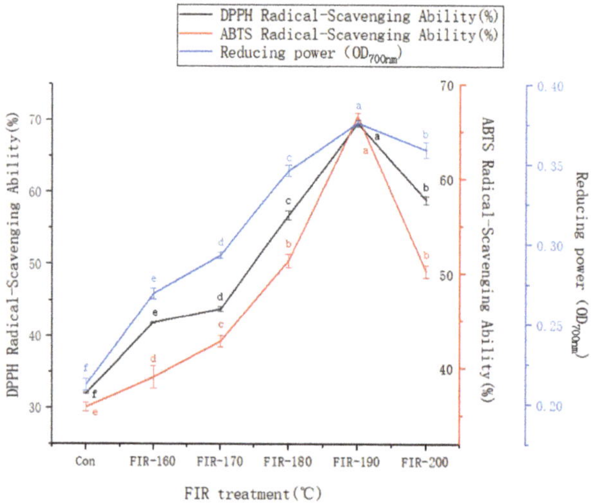

Figure 5. Changes in the 2,2-diphenyl-1-picrylhydrazyl (DPPH) and 2,2′-azino-bis (3-ethylbenzothiazoline-6-sulfonic acid) (ABTS) scavenging rates and reducing power in American ginseng leaves treated with FIR at various temperatures. The values are expressed as the mean and the standard error ($n = 3$). Letters a–f above the bar graphs indicate significant differences at $p < 0.05$ using Tukey's HSD test.

Many antioxidants naturally covalently bind to insoluble polymers. If the bonds are weak, FIR can release and activate natural low-molecular-weight antioxidants such as phenolic acids, flavonoids, and carotenes [50]. The results in Figures 2 and 5 indicate that FIR treatment up to 190 °C increases the TPC and antioxidant activity in American ginseng leaves, after which they both decrease. These findings are similar to our previous research results in ginseng leaves. We speculate that FIR treatment can improve the antioxidant activity of ginseng leaves by releasing bound phenols. When the FIR treatment temperature is too high, phenolic degradation is accelerated, resulting in decreases in the TPC and antioxidant activity [12]. Similar results have also been reported after FIR treatment of rice, peanut shells, and *Angelica sinensis* [23,50,51]. Although many studies have been conducted on American ginseng leaves, this is the first time that the antioxidant activity of American ginseng leaf extracts has been assessed after FIR treatment. The rational use of FIR treatment could improve the antioxidant activity of medicinal plants and increase their medicinal value. From the above experimental results, it can be concluded that 190 °C is the optimal FIR treatment temperature for processing American ginseng leaves to obtain a higher antioxidant activity.

4. Conclusions

In this study, the effects of FIR on bioactive components and the antioxidant activity of American ginseng leaves were investigated in detail. Our research results show that FIR treatment is an efficient processing method for producing beneficial bioactive components in American ginseng leaves. After FIR treatment at 190 °C, the highest TPC was 1.56 times higher compared to the untreated samples. The kaempferol content after FIR treatment at 190 °C was 32 times higher than that of the untreated samples. The contents of the ginsenosides Rk3, F4, Rh4, Rg6, Rg5, Rk1, and Rg3 were 64, 41, 37, 5, 266, 222, and 17 times higher, respectively, compared to the untreated group. Moreover, the antioxidant activity was significantly improved with an increase in the TPC. The results of this study could provide a reference for the rationalization and utilization of high-value biomedicines from American ginseng leaves. (In the following research, we will explore whether there is a direct correlation between the changes in ginsenoside content and antioxidant activity and try to analyze its mechanism.)

Author Contributions: Collected documents, drafted manuscripts, and organized data into tables, X.W.; Raw material pretreatment, M.K.; software and validation, R.H. and J.L.; investigation, X.S.; data curation, S.S.; writing—original draft preparation, J.L.; and conceptualization, supervision, funding acquisition, writing—review and editing, C.J. and D.C. All authors have read and agreed to the published version of the manuscript.

Funding: This study was supported by the Technological Innovation and Development Planning of Yantai (2023YD080), the National Natural Science Foundation of China (General program, No. 32272281) and the Youth Innovation Technology Project of Higher School in Shandong Province (Food Nanotechnology Innovation Team).

Institutional Review Board Statement: Not applicable.

Informed Consent Statement: Not applicable.

Data Availability Statement: The data presented in this study are available in the article.

Conflicts of Interest: The authors declare no conflicts of interest.

References

1. Cai, J.; Zhang, Q.; Huang, H. Ginseng extract and its constituents alleviate cisplatin toxicity and reverse cisplatin resistance. *Integr. Cancer Sci. Therap.* **2019**, *6*, 1–7. [CrossRef]
2. Wang, L.; Li, S.; Liu, H.; Bao, L. Advances in research on the effects of natural drugs with immune-promoting effects on immune function. *Eur. J. Inflamm.* **2020**, *18*, 2058739220926878. [CrossRef]
3. He, S.; Lyu, F.; Lou, L.; Liu, L.; Li, S.; Jakowitsch, J.; Ma, Y. Anti-tumor activities of *Panax quinquefolius* saponins and potential biomarkers in prostate cancer. *J. Ginseng Res.* **2021**, *45*, 273–286. [CrossRef]
4. Im, D.S. Pro-Resolving Effect of Ginsenosides as an Anti-Inflammatory Mechanism of *Panax ginseng*. *Biomolecules* **2020**, *10*, 444. [CrossRef]
5. Liu, Z.; Qu, C.Y.; Li, J.X.; Wang, Y.F.; Li, W.; Wang, C.Z.; Wang, D.S.; Song, J.; Sun, G.Z.; Yuan, C.S. Hypoglycemic and Hypolipidemic Effects of Malonyl Ginsenosides from American Ginseng (*Panax quinquefolius* L.) on Type 2 Diabetic Mice. *ACS Omega* **2021**, *6*, 33652–33664. [CrossRef]
6. Walsh, J.P.; Garnham, C.P.; Yeung, K.K.C.; Sumarah, M.W. Ilyonectria Root Rot of Ginseng Is Attenuated via Enzymatic Degradation of the Extracellular Fe^{3+}-Bound Siderophore N,N′,N″-Triacetylfusarinine C. *ACS Agric. Sci. Technol.* **2022**, *2*, 402–408. [CrossRef]
7. Fang, X.; Wang, M.; Zhou, X.; Wang, H.; Wang, H.; Xiao, H. Effects of growth years on ginsenoside biosynthesis of wild ginseng and cultivated ginseng. *BMC Genom.* **2022**, *23*, 325. [CrossRef]
8. Xue, P.; Yao, Y.; Yang, X.S.; Feng, J.; Ren, G.X. Improved antimicrobial effect of ginseng extract by heat transformation. *J. Ginseng Res.* **2017**, *41*, 180–187. [CrossRef]
9. Leung, K.W.; Wong, A.S.T. Pharmacology of ginsenosides: A literature review. *Chin. Med.* **2010**, *5*, 20. [CrossRef]
10. Park, C.S.; Yoo, M.H.; Noh, K.H.; Oh, D.K. Biotransformation of ginsenosides by hydrolyzing the sugar moieties of ginsenosides using microbial glycosidases. *Appl. Microbiol. Biot.* **2010**, *87*, 9–19. [CrossRef]
11. Kang, O.J.; Kim, J.S. Comparison of Ginsenoside Contents in Different Parts of Korean Ginseng (*Panax ginseng* C.A. Meyer). *Prev. Nutr. Food Sci.* **2016**, *21*, 389–392. [CrossRef]
12. Duan, S.; Liu, J.R.; Wang, X.; Sun, X.M.; Gong, H.S.; Jin, C.W.; Eom, S.H. Thermal Control Using Far-Infrared Irradiation for Producing Deglycosylated Bioactive Compounds from Korean Ginseng Leaves. *Molecules* **2022**, *27*, 4782. [CrossRef]
13. Jang, G.Y.; Kim, M.Y.; Lee, Y.J.; Li, M.; Shin, Y.S.; Lee, J.; Jeong, H.S. Influence of organic acids and heat treatment on ginsenoside conversion. *J. Ginseng Res.* **2018**, *42*, 532–539. [CrossRef]
14. Kim, K.T.; Yoo, K.M.; Lee, J.W.; Eom, S.H.; Hwang, I.K.; Lee, C.Y. Protective effect of steamed American ginseng (*Panax quinquefolius* L.) on V79-4 cells induced by oxidative stress. *J. Ethnopharmacol.* **2007**, *111*, 443–450. [CrossRef]
15. Kim, K.T.; Yoo, K.M. Effect of hot water boiling and autoclaving on physicochemical properties of American ginseng (*Panax quinquefolium* L.). *J. Ginseng Res.* **2009**, *33*, 40–47.
16. Zhang, F.; Tang, S.; Zhao, L.; Yang, X.; Yao, Y.; Hou, Z.; Xue, P. Stem-leaves of Panax as a rich and sustainable source of less-polar ginsenosides: Comparison of ginsenosides from Panax ginseng, American ginseng and Panax notoginseng prepared by heating and acid treatment. *J. Ginseng Res.* **2021**, *45*, 163–175. [CrossRef]
17. Zeng, Y.; Liu, Y.; Zhang, J.; Xi, H.; Duan, X. Effects of far-infrared radiation temperature on drying characteristics, water status, microstructure and quality of kiwifruit slices. *J. Food Meas. Charact.* **2019**, *13*, 3086–3096. [CrossRef]

18. Aboud, S.A.; Altemimi, A.B.; Al-Hilphy, A.R.S.; Yi-Chen, L.; Cacciola, F. A Comprehensive Review on Infrared Heating Applications in Food Processing. *Molecules* **2019**, *24*, 4125. [CrossRef]
19. Geng, Z.; Torki, M.; Kaveh, M.; Beigi, M.; Yang, X. Characteristics and multi-objective optimization of carrot dehydration in a hybrid infrared/hot air dryer. *LWT* **2022**, *172*, 114229. [CrossRef]
20. Ren, Z.; Yu, X.; Yagoub, A.; Fakayode, O.; Ma, H.; Sun, Y.; Zhou, C. Combinative effect of cutting orientation and drying techniques (hot air, vacuum, freeze and catalytic infrared drying) on the physicochemical properties of ginger (*Zingiber officinale* Roscoe). *LWT* **2021**, *144*, 111238. [CrossRef]
21. Rajoriya, D.; Shewale, S.; Bhavya, M.; Hebbar, H. Far infrared assisted refractance window drying of apple slices: Comparative study on flavour, nutrient retention and drying characteristics. *Innov. Food Sci. Emerg.* **2020**, *66*, 102530. [CrossRef]
22. Zhang, Q.; Wan, F.; Yue, Y.; Zang, Z.; Xu, Y.; Jiang, C.; Shang, J.; Wang, T.; Huang, X. Study on Ultrasonic Far-Infrared Radiation Drying and Quality Characteristics of Wolfberry (*Lycium barbarum* L.) under Different Pretreatments. *Molecules* **2023**, *28*, 1732. [CrossRef]
23. Ratseewo, J.; Meeso, N.; Siriamornpun, S. Changes in amino acids and bioactive compounds of pigmented rice as affected by far-infrared radiation and hot air drying. *Food Chem.* **2020**, *306*, 125644. [CrossRef]
24. Zhong, Y.; Zhang, Y.; Zhu, Z.; Wang, Y.; Zeng, Z.; Liu, C. Comparative study on physicochemical and nutritional properties of black rice influenced by superheated steam, far infrared radiation, and microwave treatment. *Innov. Food Sci. Emerg.* **2023**, *84*, 103282. [CrossRef]
25. Yao, L.; Fan, L.; Duan, Z. Effect of different pretreatments followed by hot-air and far-infrared drying on the bioactive compounds, physicochemical property and microstructure of mango slices. *Food Chem.* **2020**, *305*, 125477. [CrossRef]
26. Kim, C.; Kim, B.; Kim, C.; Baek, J.; Jung, I. Variations in Ginsenosides of Raw Ginseng According to Heating Temperature and Time. *J. Pharmacopunct.* **2020**, *23*, 79–87. [CrossRef]
27. Ghasemzadeh, A.; Jaafar, H.; Rahmat, A. Variation of the Phytochemical Constituents and Antioxidant Activities of Zingiber officinale var. rubrum Theilade Associated with Different Drying Methods and Polyphenol Oxidase Activity. *Molecules* **2016**, *21*, 780. [CrossRef]
28. Kossah, R.; Zhang, H.; Chen, W. Antimicrobial and antioxidant activities of Chinese sumac (*Rhus typhina* L.) fruit extract. *Food Control* **2011**, *22*, 128–132. [CrossRef]
29. Lim, Y.J.; Kwon, S.J.; Qu, S.; Kim, D.G.; Eom, S.H. Antioxidant contributors in seed, seed coat, and cotyledon of γ-ray-induced soybean mutant lines with different seed coat colors. *Antioxidants* **2021**, *10*, 353. [CrossRef]
30. Oyaizu, M. Studies on products of browning reaction antioxidative activities of products of browning reaction prepared from glucosamine. *Nutr. Diet.* **1986**, *44*, 307–315.
31. Hwang, I.G.; Kim, H.Y.; Joung, E.M.; Woo, K.S.; Jeong, J.H.; Yu, K.W.; Lee, J.; Jeong, H.S. Changes in ginsenosides and antioxidant activity of Korean ginseng (*Panax ginseng* C.A. Meyer) with Heating Temperature and Pressure. *Food Sci. Biotechnol.* **2010**, *19*, 941–949. [CrossRef]
32. Jeong, H.; Park, D.; Seo, H.; Choi, M.; Cho, Y. Effect of Roasting Time and Cryogenic Milling on the Physicochemical Characteristics of Dried Ginseng Powder. *Foods* **2020**, *9*, 223. [CrossRef]
33. Yu, J.; Jang, I.; Moon, J.; Jang, I.; Kim, Y.; Kim, D.; Suh, S. Change of Proximate Composition, Antioxidant Activity, and Ginsenoside Content of White Ginseng with Different Roasting Conditions. *Korean J. Med. Crop Sci.* **2022**, *30*, 1–11. [CrossRef]
34. Yoo, K.M.; Lee, C.; Lo, Y.M.; Moon, B. The Hypoglycemic Effects of American Red Ginseng (*Panax quinquefolius* L.) on a Diabetic Mouse Model. *J. Food Sci.* **2012**, *77*, 147–152. [CrossRef] [PubMed]
35. Musilova, J.; Lidikova, J.; Vollmannova, A.; Frankova, H.; Urminska, D.; Bojnanska, T.; Toth, T. Influence of Heat Treatments on the Content of Bioactive Substances and Antioxidant Properties of Sweet Potato (*Ipomoea batatas* L.) Tubers. *J. Food Quality* **2020**, *2020*, 8856260. [CrossRef]
36. Escobedo, R.; Miranda, R.; Martínez, J. Infrared irradiation: Toward green chemistry, a review. *Int. J. Mol. Sci.* **2016**, *17*, 453. [CrossRef]
37. Qian, Z.; Lu, J.; Gao, Q.; Li, S. Rapid method for simultaneous determination of flavonoid, saponins and polyacetylenes in Folium Ginseng and Radix Ginseng by pressurized liquid extraction and high-performance liquid chromatography coupled with diode array detection and mass spectrometry. *J. Chromatogr. A* **2009**, *1216*, 3825–3830. [CrossRef]
38. Nasanbat, B.; Uchiyama, A.; Amalia, S.N.; Inoue, Y.; Yokoyama, Y.; Ogino, S.; Torii, R.; Hosoi, M.; Motegi, S.I. Kaempferol therapy improved MC903 induced-atopic dermatitis in a mouse by suppressing TSLP, oxidative stress, and type 2 inflammation. *J. Dermatol. Sci.* **2023**, *111*, 93–100. [CrossRef]
39. Wang, L.; Yang, X.; Yu, X.; Yao, Y.; Ren, G. Evaluation of antibacterial and anti-inflammatory activities of less polar ginsenosides produced from polar ginsenosides by heat-transformation. *J. Agric. Food Chem.* **2013**, *61*, 12274–12282. [CrossRef]
40. Murugesan, M.; Mathiyalagan, R.; Boopathi, V.; Kong, B.M.; Choi, S.K.; Lee, C.S.; Yang, D.C.; Kang, S.C.; Thambi, T. Production of minor ginsenoside CK from major ginsenosides by biotransformation and its advances in targeted delivery to tumor tissues using nanoformulations. *Nanomaterials* **2022**, *12*, 3427. [CrossRef]
41. Li, X.; Li, F.; Pei, W.; Yang, J.; Gu, Y.; Piao, X. The Content and Principle of the Rare Ginsenosides Produced from Gynostemma pentaphyllum after Heat Treatment. *Molecules* **2023**, *28*, 6415. [CrossRef]
42. Ye, X.; Li, C.; Zhang, H.; Li, Q.; Cheng, S.; Wen, J.; Wang, X.; Ren, H.; Xia, L.; Wang, X.; et al. Saponins of ginseng products: A review of their transformation in processing. *Front. Pharmacol.* **2023**, *14*, 1177819. [CrossRef]

43. Park, E.H.; Kim, Y.J.; Yamabe, N.; Park, S.H.; Kim, H.K.; Jang, H.J.; Kim, J.H.; Cheon, G.J.; Ham, J.; Kang, K.S. Stereospecific anticancer effects of ginsenoside Rg3 epimers isolated from heat-processed American ginseng on human gastric cancer cell. *J. Ginseng Res.* **2014**, *38*, 22–27. [CrossRef]
44. Pu, J.; Ramadhania, Z.; Mathiyalagan, R.; Huo, Y.; Han, Y.; Li, J.; Ahn, J.; Xu, F.; Lee, D.; Zeng, X.; et al. Ginsenosides Conversion and Anti-Oxidant Activities in Puffed Cultured Roots of Mountain Ginseng. *Processes* **2021**, *9*, 2271. [CrossRef]
45. Kim, G. Changes of Ginsenoside Compositions in Cultivated Wild Ginseng with Different Steaming Temperatures and Time. *J. East Asian Soc. Diet. Life* **2023**, *33*, 415–421. [CrossRef]
46. Yoo, S.; Park, B.I.; Kim, D.H.; Lee, S.; Lee, S.H.; Shim, W.S.; Seo, Y.K.; Kang, K.; Lee, K.T.; Yim, S.V. Ginsenoside Absorption rate and extent enhancement of black ginseng (CJ EnerG) over red ginseng in healthy adults. *Pharmaceutics* **2021**, *13*, 487. [CrossRef] [PubMed]
47. Ji, Y.J.; Kim, H.D.; Lee, E.S.; Jang, G.Y.; Seong, H.A. Heat Treatment Enhances the Neuroprotective Effects of Crude Ginseng Saponin by Increasing Minor Ginsenosides. *Int. J. Mol. Sci.* **2023**, *24*, 7223. [CrossRef] [PubMed]
48. Park, S.E.; Na, C.S.; Yoo, S.A.; Seo, S.H.; Son, H.S. Biotransformation of major ginsenosides in ginsenoside model culture by lactic acid bacteria. *J. Ginseng Res.* **2017**, *41*, 36–42. [CrossRef] [PubMed]
49. Kim, M.S.; Jeon, S.J.; Youn, S.J.; Lee, H.; Park, Y.J.; Kim, D.O.; Kim, B.Y.; Kim, W.; Baik, M.Y. Enhancement of minor ginsenosides contents and antioxidant capacity of american and canadian ginsengs (*Panax quinquefolius*) by puffing. *Antioxidants* **2019**, *8*, 527. [CrossRef] [PubMed]
50. Lee, S.C.; Jeong, S.M.; Kim, S.Y.; Park, H.R.; Nam, K.; Ahn, D. Effect of far-infrared radiation and heat treatment on the antioxidant activity of water extracts from peanut hulls. *Food Chem.* **2006**, *94*, 489–493. [CrossRef]
51. Azad, M.O.K.; Piao, J.P.; Park, C.H.; Cho, D.H. Far infrared irradiation enhances nutraceutical compounds and antioxidant properties in *Angelica gigas* Nakai powder. *Antioxidants* **2018**, *7*, 189. [CrossRef] [PubMed]

foods MDPI

Article

Physical and Chemical Properties, Flavor and Organoleptic Characteristics of a Walnut and Purple Rice Fermented Plant Drink

Hongyu Mu [1,†], Tianyi Dai [1,†], Si Huang [1], Kuan Wu [1], Mingming Wang [1], Chunlei Tan [1], Feng Zhang [1], Jun Sheng [1,2,3,*] and Cunchao Zhao [1,2,3,4,*]

1 College of Food Science and Technology, Yunnan Agricultural University, Kunming 650201, China; muhongyu0710@163.com (H.M.); dtynongda@163.com (T.D.); 18870918872@163.com (S.H.); wink577@163.com (K.W.); wmm295813@163.com (M.W.); tcl98316@163.com (C.T.); 15762196909@163.com (F.Z.)

2 Yunnan Plateau Characteristic Agricultural Industry Research Institute, Kunming 650201, China

3 Engineering Research Center of Development and Utilization of Food and Drug Homologous Resources, Ministry of Education, Yunnan Agricultural University, Kunming 650201, China

4 Key Laboratory of Precision Nutrition and Personalized Food Manufacturing, Ministry of Education, Yunnan Agricultural University, Kunming 650201, China

* Correspondence: shengj@ynau.edu.cn (J.S.); g2021002@ynau.edu.cn (C.Z.); Tel.: +86-13354649242 (C.Z.)

† These authors contributed equally to this work.

Abstract: In recent years, green and healthy foods have attracted much attention. Plant-based foods have become an alternative to animal-derived foods. In this study, we used walnut and purple rice as the primary raw materials to produce a fermented plant drink. The process included boiling, mixing, grinding, inoculation, fermentation, and sterilization. We then analyzed the similarities and differences between the resulting walnut and purple rice fermented plant drink and an unfermented walnut and purple rice plant drink, as well as dairy-based yoghurt, in terms of physical chemistry, flavor, and sensory characteristics. We also examined the similarities and differences between the walnut and purple rice fermented plant drink and room-temperature yoghurt. The study results revealed that the walnut and purple rice fermented plant drink exhibited greater viscosity than the walnut and purple rice unfermented plant drink and room-temperature yoghurt. Additionally, the former displayed enhanced stability and recovery ability. Notably, distinguishable differences were observed between the three samples in terms of the presence of unknown volatiles and the umami signal, as indicated by electronic nose/tongue and GC-IMS analyses. The umami flavor of the walnut and purple rice fermented plant drink surpasses that of room-temperature yoghurt, while its taste is less salty than that of the walnut and purple rice plant drink. Despite possessing a weaker aroma than dairy-based yogurt, it is more potent than the walnut and purple rice plant drink. Additionally, its relative abundance of olefins, ketones, and alcohols enhances its unique flavor profile, surpassing both other options. Based on sensory analysis, it can be deduced that walnut and purple rice fermented plant drink has the highest overall acceptance rate.

Keywords: walnut and purple rice fermented plant drink; room-temperature yoghurt; rheology; flavor; sensory

1. Introduction

As the global population expands, there will be a significant increase in the demand for food, specifically protein-rich foods [1]. Milk is a nutritious source of protein, calcium, and vitamin D. However, the high levels of saturated fat in milk can adversely affect human health [2]. The escalation in global warming, the increase in chronic diseases, and the mass emergence of vegetarianism and light diets have all played a role in the rise of milk substitutes [3]. Plant-based proteins offer numerous benefits over animal-based sources and are increasingly becoming a groundbreaking and rapidly expanding aspect of various food sectors [4]. Traditional yoghurt is commonly produced using cow's milk [5], whereas plant-based yoghurt is typically derived from protein-rich plants. Plant-based yoghurt is frequently utilized as a milk alternative owing to its biologically active constituents and the advantages it has for human health [6]. Overall, plants and animals have complementary and symbiotic roles in healthy and sustainable food systems [7].

The walnut, belonging to the Juglandaceae family, has been acknowledged as a nutrient-dense food in various parts of the globe [8]. Walnut kernels contain a high concentration of protein, fat, vitamins, minerals, fiber, and polyphenolic compounds [9]. Containing unsaturated fats and polyphenols, they offer a myriad of health benefits, including reducing the risk of cardiovascular and coronary heart disease, treating type II diabetes, preventing and treating specific cancers, and alleviating symptoms associated with aging and neurological diseases [10]. Purple rice is a glutinous rice variety that is highly valued as a colored rice germplasm resource in China due to its long history of cultivation [11]. Research has demonstrated that the nutritional elements of colored rice, including protein, amino acids, and minerals, surpass those of regular white rice [12]. In addition, purple rice is also rich in polyphenols, flavonoids, anthocyanins, dietary fiber, vitamins, and essential minerals such as Fe, Zn, Ca, and P. These nutrients have been proven to be beneficial for human health [13]. Anthocyanins, the major components of pigmented rice, exhibit a range of biological activities, including free radical scavenging, antioxidation, anti-tumor, anti-atherosclerosis, hypoglycemic, and anti-allergic effects [14]. These activities are vital for preventing diseases and promoting health, making pigmented rice an attractive candidate for food and pharmaceutical applications [15]. Walnuts and purple rice are high-quality plant-based foods with significant nutritional and medicinal benefits. They contain numerous bioactive substances, including phenolic compounds, unsaturated fatty acids, and anthocyanins, which are beneficial to human health. The combination of walnuts and purple rice not only enriches nutrition but also enhances their unique flavors, making it an excellent choice for dairy-free milk alternatives.

Compared to traditional animal-based milk yoghurt, plant-based fermented drinks are from safe sources and have the benefits of no cholesterol, no trans fats, no lactose, and no or low levels of allergens and fiber, which are beneficial for cardiovascular health, weight management, and managing lactose intolerance and cow's milk allergy [16]. Plant-based fermented beverages offer unique nutritional benefits and excellent product safety, giving them great potential for market development. A growing body of research also shows that gentle fermentation can improve the taste and flavor of plant-based products [17,18]. The aim of this study was to investigate the effect of mild fermentation with *Lactobacillus plantarum* on the flavor and sensory properties of a walnut and purple rice plant fermented beverage and to further analyze the similarities and differences between the walnut and purple rice fermented plant drink and ordinary room-temperature yoghurt, in order to provide a theoretical basis for the development of alternatives to animal-sourced yoghurt.

2. Materials and Methods

2.1. Raw Materials

Hulled walnut kernels, purple rice, fresh milk, sucrose, carboxymethylcellulose CMC, modified starch, and rapeseed oil were purchased from a local market (Kunming, China). *Lactobacillus plantarum* (GDMCC accession number GDMCC1.2685) was purchased from Guangdong Provincial Microbial Strain Preservation Centre (Guangzhou, China). Direct-pitch commercial fermenters (Mild1.0 comprising *Lactobacillus bulgaricus* and *Streptococcus thermophilus*) were purchased from Chr. Hansen, Denmark (Chr. Hansen, Hersholm, Denmark). All other chemicals used were of analytical grade and are commercially available.

2.2. Strains and Growth Conditions

First, activated *Lactobacillus plantarum* was inoculated in 100 mL MRS broth medium, inoculated in 2% (*v*/*v*) inoculum, and incubated at 37 °C for 16 h to obtain starter cultures for fermentation. Afterwards, the concentration of *Lactobacillus plantarum* was adjusted to 1×10^8 CFU/mL and then centrifuged at 5000 rpm for 5 min at 4 °C, washed three times with sterile saline, and recovered as a seed solution for fermentation.

2.3. Sample Preparation

The method was slightly adjusted according to the existing formula and process in the laboratory.

Walnut and purple rice fermented plant drink: After washing, purple rice was soaked in deionized water for 2 h. After soaking, the purple rice was heated at 95 °C for 30 min, and an equal amount of peeled walnut kernels were added and then ground with deionized water (80 °C) at a ratio of 1:3. The walnut–purple rice milk was finely ground in a colloid mill (JM-F80, Shanghai Toppay Machinery Equipment Co., Ltd., Shanghai, China) for 15 min until there were no obvious large particles and the texture was fine and uniform. The 8.5% sucrose, 0.3% CMC, and 0.5% rapeseed oil were weighed and added to the walnut and purple rice milk. The mixture was stirred with a blender, heated at 95 °C for 1.5 h, cooled to 34 °C, and inoculated with 1% (*v*/*v*) culture. The walnut and purple rice fermented plant drink (H1) was fermented at 34 °C for 14 h. The fermentation was stopped when the pH reached about 4.5, and the milk was sterilized at 85 °C for 30 min and stored at room temperature.

Walnut and purple rice plant drink: Using the same process and formula, the unfermented walnut and purple rice plant drink (H2) was used as the control group and stored at room temperature.

Room-temperature yogurt: The milk was preheated to 45 °C, and 8.5% sucrose, 2% modified starch and 0.6% complex thickening emulsifier HBT-A8570 were added and heated to 65 °C for 15 min. The milk was then homogenized using a high-pressure homogenizer (GYB60-6S, High-Pressure Homogenizer, Shanghai Donghua High-Pressure Homogenizer Factory, Shanghai, China) at 25 MPa for 5 min. The milk was then pasteurized in a six-coupled electric thermostatic mixer (JJ-6A-6H, Changzhou Jibori Instrument Manufacturing Co., Ltd., Changzhou, China) for 2 min. The sterilized mixed milk was cooled to 37 °C and inoculated with 0.5‰ of the commercial direct pitch fermenter. Fermentation was stopped at 42 °C for 6 h to reach a pH of approximately 4.5, and then the milk was sterilized at 85 °C for 30 min to obtain room-temperature yoghurt (C1). It was stored at room temperature, and each treatment was repeated in triplicate.

2.4. Physicochemical Analysis

Titration acidity (TAA): Samples were prepared by diluting (10 g) with distilled water (20 mL). Phenolphthalein solution (2 mL, 5% *w*/*v*) was added to the diluent as an indicator. The mixture was titrated with 0.1 M NaOH to express a pink color that did not fade within 30 s. TAA (°T) was calculated as the volume of NaOH consumed (mL) multiplied by ten.

pH: After the samples were brought back to room temperature, they were stirred uniformly with a glass rod. The pH of the fermented milk was measured using a pH meter (pHS-3C, Shanghai Yitian Scientific Instrument Co., Ltd., Shanghai, China), and the samples were repeated three times.

The samples were analyzed for their chemical composition (protein, fat, total solids, and ash) using standard AOAC 2005 procedures [19,20].

2.5. Rheological Analyses

The samples were characterized for their rheological behavior according to the previous study conditions [21–24]. The rheological properties of the samples were analyzed using an MCR 302 rheometer, Anton Paar Austria, Graz, Austria) equipped with a temperature control system. A 50 mm diameter parallel plate configuration with an interplate gap of 1 mm was used.

Relationship between strain and time: The temperature of the rheometer was set at 25 °C, 10 Pa shear stress was applied to the sample, and the change in strain with time was recorded for 5 min. A total of 50 points were taken for data detection. Time was taken as the horizontal axis and shear stress as the vertical axis.

Relationship between apparent viscosity and shear rate: The samples were poured between the rheometer plates and placed at room temperature for 15 min. The test was carried out with a CC25 DIN fixture, and the test distance was adjusted to 500 μm. The shear rate was increased from 0.01 S^{-1} to 1000 S^{-1}, and 50 points were taken for data detection. The shear rate was used as the abscissa, and the apparent viscosity was used as the ordinate to draw the curve.

Determination of dynamic rheological properties: The detection temperature was set at 25 °C, the shear strain was 0.5%, the frequency was increased from 0.1 to 10 Hz, and the duration was 10 min. The changes in the elastic modulus (G′) and viscous modulus (G″) of the two starch pastes with the increase in frequency were measured.

Relationship between shear rate and shear stress: The samples were poured between the rheometer plates and allowed to stand at room temperature for 15 min. The test was carried out with a CC25 DIN fixture, and the test distance was adjusted to 500 μm. The shear rate increased from 0.01 S^{-1} to 1000 S^{-1}, from which 50 points were taken for data detection, and then decreased from 1000 S^{-1} to 0.01 S^{-1} for data detection. Taking the shear rate as the abscissa and the shear stress as the ordinate, the curve was drawn.

2.6. Electronic Nose and Electronic Tongue Determination

Electronic nose: 7 g of sample was weighed into a 30 mL headspace vial and detected using an electronic nose (cNose, Shanghai Baosheng Industrial Development Co., Ltd., Shanghai, China) with the following parameters: incubation at 50 °C for 20 min, an oscillation speed of 500 r/min, an injection volume of 5000 μL, an injection speed of 125/s, an inlet temperature of 200 °C, a pressure of 10 kPa, a flow rate of 30 mL/min, and an injection time of 45 s. Each sample was replicated three times.

Electronic tongue: 100 g of sample was weighed, and 100 mL of purified water was added. The water was stirred to mix the water and yoghurt samples thoroughly and evenly, and centrifugation was carried out at 3000 rpm for 10 min before removing the supernatant and testing with an electronic tongue instrument (Ctongue, Shanghai Bao Sheng Industrial Development Co., Ltd., Shanghai, China). Each group of samples was tested three times in parallel at room temperature.

2.7. HS-GC-IMS Volatile Flavor Compound Analysis

A gas chromatography–ion mobility spectrometer (FlavourSpec 1 H1-100053), GAS, Dortmund, Germany), with an Rtx-Wax column (30 m × 0.53 mm, 1 μm) from Restek, Bellefonte, PA, USA and an automated headspace injection unit (Combi PAL) from CTC, Zwingen, Switzerland, was used in this study.

Headspace sampling was used, and the sampling conditions were optimized. The optimized headspace sampling conditions were as follows: 0.5 mL of the sample was sealed in a 20 mL headspace vial and heated and shaken at 70 °C for 15 min, and then 200 μL of the sample was taken from the upper space of the headspace vial for detection. The temperature of the headspace injection needle was 85 °C. The temperature of the headspace injection needle was 85 °C. The GC-IMS test conditions were set as follows: the temperature of the GC column was 60 °C, and the temperature of the drift tube was 45 °C. The flow rate of the carrier gas for GC-IMS was 2 mL/min for 0~2 min, 2~10 min, 2~10 mL/min for 2~10 min, 10~90 mL/min for 10~30 min, and 90 mL/min for 30~40 min. The flow rate of the drift gas for ion mobility spectrometry (IMS) was kept constant at 150 mL/min. The carrier gas and the drift gas were both nitrogen (purity: 99.999%). Each sample was replicated 3 times.

2.8. Sensory Analysis

Sensory analysis of the samples was performed by a group of experienced sensory testers (n = 12, age range 20–59) recruited from the School of Food Science and Technology,

Yunnan Agricultural University. The team members were asked to rate the intensity at a level of 0 to 10 for taste and flavor descriptors. Team members were also asked to rate the overall acceptability of the samples at a level of 0 to 10. The questionnaire includes the definition of descriptors for sensory testing, which can be found in Table 1. Sensory analysis was carried out in the sensory analysis room and in duplicate within two days.

Table 1. Sensory word definitions and reference samples.

QDA	Description	Definition
Flavor	Green grass flavor	Odors associated with newly cut grass
	Nuts flavor	Odors associated with nuts
	Cereal flavor	Odors associated with cereals
	Fermented flavor	Acid substances produced by fermentation of dairy products, with an occasionally sour smell
Taste	Sweetness	Taste related to sucrose
	Sour taste	Taste associated with citric acid
	Bitter taste	Taste associated with caffeine
	Salty taste	Taste related to sodium chloride
	Astringent taste	The sample is placed on the tongue to make the tongue feel convergent
Taste	Dispersion	The extent of the sample in the mouth without chewing
	Thickness	The force, fluidity, and thinness of the tongue to suck the sample in the spoon into the mouth or expand the sample
	Adhesion	Mechanical texture characteristics related to the force required to move the adhesive object, stickiness
	Granular feeling	The number and size of particles that can be perceived in the sample
Overall	Acceptance	Acceptance of the sample

2.9. Data Analysis

SPSS version 23 (IBM SPSS Inc., Chicago, IL, USA) was used to conduct a one-way analysis of variance (ANOVA) with Tukey's test at a 95% confidence level. Values are expressed as the mean ± standard deviation. PCA and PLS-DA were performed using OriginPro 8.6 software (OriginLab Corporation, Northampton, MA, USA) and SIMCA 14.1 software (Biometric Software Developer Umetrics, Umeå, Sweden), respectively.

3. Results and Discussion

3.1. Physicochemical Analysis

Table 2 shows the comparison of the physical and chemical components of the three samples. The results showed that the contents of protein, fat, and solids in the walnut and purple rice fermented plant drink were higher than those in ordinary milk-based yoghurt ($p < 0.05$). Acidity is a very important indicator of yoghurt fermentation. An acidity of 4.5 is usually used as the end point of fermentation in yoghurt production [25]. The pH of unfermented H2 was 6.38 and was reduced to 4.45 (H1) after incubation with a mixture of *Lactobacillus plantarum* fermentation broth at 34 °C for 14 h, confirming the effectiveness of the fermentation process. Acidity and pH were mainly influenced by lactic acid produced by the growth of lactic acid bacteria [26]. The pH of dairy yoghurt is usually around 4.5, and the pH range in the dairy and non-dairy samples is 3.4 and 4.4. The total titratable acidity showed significant differences between dairy and plant samples. Compared to the walnut and purple rice fermented plant drink, the TTA of the dairy yoghurt was significantly higher, and the difference between the two was about 30. Because walnut and purple rice contain a large amount of fiber and other substances, the total solids content of walnut–purple rice yoghurt is much higher than that of ordinary dairy yoghurt. Walnut and purple rice yoghurt with a high fiber content enhances the effects of intestinal peristalsis and improves digestive function [27].

Table 2. Physicochemical analysis of the different samples.

	H1	H2	C1
Protein (%)	3.02 ± 0.04 [a]	2.99 ± 0.04 [b]	2.86 ± 0.02 [c]
Fat (%)	6.67 ± 0.08 [a]	6.02 ± 0.07 [b]	2.93 ± 0.04 [c]
Titration acidity	48.93 ± 1.70 [b]	4.07 ± 0.68 [c]	78.72 ± 1.41 [a]
pH	4.45 ± 0.01 [c]	6.38 ± 0.06 [a]	4.54 ± 0.03 [b]
Total solids (%)	29.09 ± 0.03 [a]	28.74 ± 0.04 [b]	20.55 ± 0.03 [c]

H1: walnut and purple rice fermented plant drink, H2: walnut purple plant drink, C1: room-temperature yoghurt. Values are expressed as the mean ± standard deviation. Different lowercase letters indicate significant differences between groups for the same indicator ($p < 0.05$), the same as below.

3.2. Rheological Measurements

Rheological properties can indicate changes in viscosity as the shear rate varies. Viscosity is a measure of a liquid's physical characteristics that resist flow under shearing conditions, which can help to elucidate the attributes of dairy products [28]. Newtonian fluid is an ideal fluid whose viscosity remains unaffected by the shear rate. On the contrary, non-Newtonian fluids do not exhibit this ideal flow property [29]. Figure 1a,b display the relationship between viscosity, shear rate, and shear time for all sample groups. The results indicate that increasing shear rate and time led to a rapid decrease in apparent viscosity for each group before stabilizing gradually. This observation revealed the characteristics of high pseudoplasticity and the shear-thinning phenomenon of non-Newtonian fluid [30]. When the rate of shear is low, the damage level to the gel network structure inside the sample is minimal. Subsequently, as time and the rate of shear increase, the level of damage to the gel network structure inside the sample increases until it is completely destroyed. Consequently, electrostatic interaction and hydrophobic interaction weaken, ultimately resulting in the phenomenon of shear thinning [31]. The viscosity of normal-temperature yoghurt is lower than that of the walnut and purple rice fermented plant drink, suggesting that despite the addition of stabilizer and modified starch to normal-temperature yoghurt, the higher total solids content of the walnut and purple rice fermented plant drink results in more binding force between the particles, leading to higher viscosity. Additionally, fermentation promotes the precipitation of polysaccharides and other substances. Compacted polysaccharides are capable of modifying the structural network via non-covalent interactions and of incentivizing thickening behavior. This leads to an increase in the viscosity of the sample system and a decrease in its fluidity [32]. The shear stress and shear stress values of the walnut and purple rice fermented plant drinks were higher than those of the room-temperature yoghurt samples among the three groups of samples, indicating that the overall viscosity of the former was higher. The shear stress and shear rate relationship curve of various samples can be seen in Figure 1c. The walnut and purple rice fermented plant drink yields somewhat higher shear stress than the regular dairy yoghurt and walnut and purple rice plant drink. When an external force is exerted, there is a significant alteration in the structural composition of the walnut and purple rice fermented plant drink. Meanwhile, the thixotropic ring area of this yoghurt is smaller as compared to the other two groups, which suggests that its structural composition remains largely unaffected upon the application of external force. After the changes in the organizational state, the energy needed to revert to the initial state is notably lower in the walnut and purple rice fermented plant drink group compared to the other two groups. This suggests that this group is comparatively more stable and exhibits a more robust recovery ability. Prior research has demonstrated that yogurt with fish gelatin produces higher shear stress than control yogurt lacking gelatin, resulting in a stronger gel structure, making the system more stable [33].

The viscoelastic features of yoghurt specimens can be analyzed further by conducting dynamic oscillation trials. Figure 1d demonstrates that upon increasing the frequency, the G' and G'' values of yoghurt samples increased. Furthermore, for the same type of samples, G' was found to be greater than G''. This indicates that the specimens exhibit a certain stiffness (elastic behavior), presenting a semi-solid weak gel structure in line with the characteristics of yoghurt [34]. The G′ and G″ measurements of walnut and purple rice yoghurt were higher than those of ordinary room-temperature yoghurt, implying that walnut and purple rice yoghurt exhibit more solid behavior than ordinary room-temperature yoghurt. The starch particle structure in the walnut and purple rice fermented plant drink was chain-like and dispersed, with a strong interaction between particles (gel formation). Consequently, the gel of the walnut and purple rice fermented plant drink was softer than that of ordinary room-temperature yoghurt [35]. Additionally, the storage modulus (G′) and loss modulus (G″) of the walnut purple plant drink were lower compared to those of the stirred walnut and purple rice yoghurt, demonstrating that the structure of the latter was denser. This further supports the notion that fermentation enhances the gelling properties and robustness of the walnut and purple rice fermented plant drink [36].

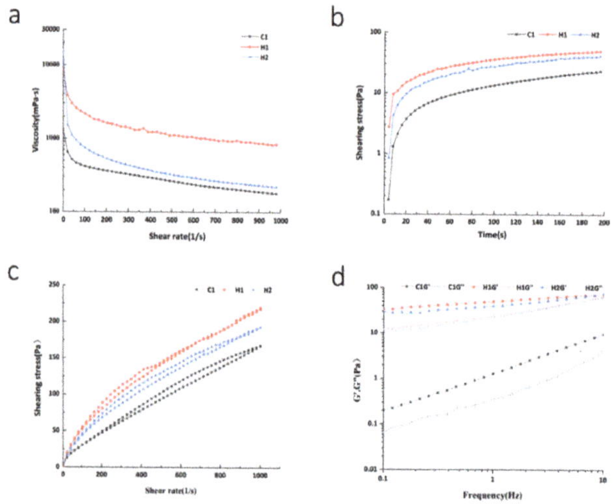

Figure 1. (**a**) Relationship between viscosity and shear rate; (**b**) relationship curve between time and shear stress; (**c**) the relationship curve between shear rate and shear stress; (**d**) viscoelastic characteristic curve.

3.3. Analysis of Electronic Tongue and Electronic Nose

3.3.1. Electronic Tongue Analysis

An electronic tongue is a bionic taste analysis device capable of objectively measuring food taste changes through alterations in the lipid membrane potential of an artificial taste sensor array [37]. The flavor radar images for various samples were acquired by extracting the response values from each sensor, as illustrated in Figure 2. Figure 2a shows that the acidity values for the three samples are not significant. Their acidity is lower than or proximate to the taste threshold. Additionally, other taste indicators are present above the taste threshold, substantiating their effectiveness as taste markers. H2 has a lower degree of astringency, bitterness aftertaste, and astringency aftertaste than the taste threshold. Acidity and sweetness are significant indicators of flavor across the samples. Differences between the samples' acidity and sweetness are also readily apparent. The study clearly demonstrates the distinct sweetness attributes of all three samples. The data indicate that ordinary room-temperature yoghurt possesses the strongest sour taste, approaching the threshold of tastelessness, while the walnut and purple rice yoghurt exhibits considerably less sourness compared to the plain yoghurt (a numerical difference of several units on a standard scale) The sour taste of the walnut and purple rice plant drink is significantly lesser than that of room-temperature yoghurt. Additionally, the sweetness of the walnut and purple rice plant drink is greater than that of normal dairy-based yoghurt, with the walnut and purple rice fermented plant drink exhibiting the least amount of sweetness.

The study results revealed higher bitterness, bitter aftertaste, astringency, and astringency aftertaste scores in yoghurts as compared to odorless points, except for the walnut and purple rice plant drink Additionally, technical term abbreviations should be explained upon their first usage. It is important to note that the terms 'bitterness' and 'astringency' do not imply that the dairy products tasted extremely bitter. Instead, these terms reflect the richness, milk flavor, and strong sense of taste in these products. For the walnut and purple rice fermented plant drink, it is essential to consider the bitter taste in the results, as walnut contains polyphenols and other compounds. It is crucial to account for the sensory situation of the raw material itself. The bitterness and astringency of the other two samples are lower than those of ordinary yogurt at room temperature. In particular, there is a notable difference in bitterness between the three samples, while the difference in astringency and aftertaste is minor. Richness refers to the aftertaste of umami, reflecting the persistence of umami, which can also be known as umami persistence. The study's findings reveal that the sample's richness is greater than its umami. This is because in calcium-rich samples such as yoghurt and cheese, the output of umami is inhibited by calcium, while its effect on aftertaste is unaffected The instrument test output indicates a higher aftertaste value than umami value, which aligns with the senses being unaware of the umami taste while perceiving the richness. The umami test revealed significant differences in umami, richness, and saltiness between the three yoghurts, with the dairy yoghurt having the lowest umami and richness and the highest saltiness. The umami of the walnut and purple rice fermented plant drink sample H1 is superior to that of C1, albeit less rich and with the lowest saltiness score. Sample H2 possesses the highest umami and richness, with a substantial saltiness score as well.

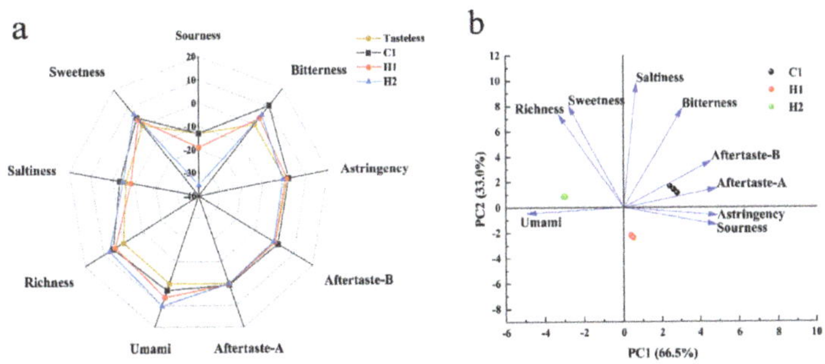

Figure 2. Radar (**a**) and PCA (**b**) of the electronic tongue's assessment of different samples. Note: The "tasteless point" is the output of the reference solution, which comprises KCL and tartaric acid. The sour taste tasteless point is −13, the salty taste tasteless point is −6, and the tasteless points of other indicators are 0. Consequently, when the sample's taste value is lower than the "tasteless point", the sample is considered to have no taste, and vice versa. The table's richness represents the aftertaste of the umami, indicating the persistence of the umami in the sample, called the umami persistence. The bitterness aftertaste reveals the degree of bitterness residue, and the astringent aftertaste indicates the degree of astringency residue.

Taste is a complex sensation perceived by the taste buds, making it narrow-minded to assess a specific taste in isolation. As such, we conducted a PCA analysis on the taste data from the three samples (refer to Figure 2). PCA, a widely used statistical analysis method, was employed [38]. The first and second principal components were used as the horizontal and vertical axes, respectively, with variance contribution rates of 70.2% and 29.8%, respectively. Following analysis of all taste indices in this test, cluster analysis was performed on the three yoghurt samples. The graph clearly indicates that the three samples display distinct taste variations and unique characteristics. There are significant variances in the first and second principal components of the three samples. Figure 2b depicts that the first principal component is mainly impacted by sourness, bitterness, and umami, while the second principal component is influenced by saltiness, sweetness, richness, and bitterness.

The variations in flavor profiles of the three yoghurt samples are largely discernible through these aspects.

3.3.2. Electronic Nose Analysis

The electronic nose and electronic tongue work in a similar way [39]. The aforementioned bionic olfactory analysis technology is capable of performing a comprehensive evaluation of volatile odor information within samples in a rapid, sensitive, and non-destructive manner [40]. The radar chart, which displays the average of the results from three tests conducted on the 10 sensors of the electronic nose at 70 s, is illustrated in Figure 1a. Figure 3 reveals that the electronic nose has a substantial response to the three sample groups. The scent of standard yoghurt at room temperature is powerful, with the electronic nose exhibiting a significant response. Additionally, standard yoghurt, when sealed and left stationary, accumulates a relatively high concentration of headspace. In contrast, the aroma of the walnut and purple rice plant drink is faint, and the concentration of enrichment during standing is minimal. The radar map indicates that sensors No. 2, No. 6, No. 7, No. 8, and No. 9 respond to the three samples. Sensor No. 2 is sensitive to small molecule nitrogen oxide gases, No. 6 is sensitive to short-chain alkanes such as methane, No. 7 is sensitive to inorganic sulfur gases, No. 8 is sensitive to alcohol ether aldehyde ketones, and No. 9 is sensitive to organic sulfur gases. Nonetheless, considerable variations are present in the response values of these sensors. Throughout the testing process, the C1 sensors for normal-temperature yoghurt showed the highest response values, with particular sensitivity found in sensors 2 (to small molecule nitrogen oxide gas), 7 (to inorganic sulfur gas), and 9 (to organic sulfur gas). Additionally, the volatile odor of the walnut and purple rice fermented plant drink was lower than that produced by C1; the response value from sensor H2 was especially lower than that of C1. The order of the three smells is C1 > H1 > H2. H1's response value was significantly greater than that of H2, even though most raw materials were identical. The key factor at play was lactobacillus fermentation, which stimulated the release of related compounds and led to the improvement in the response value [41].

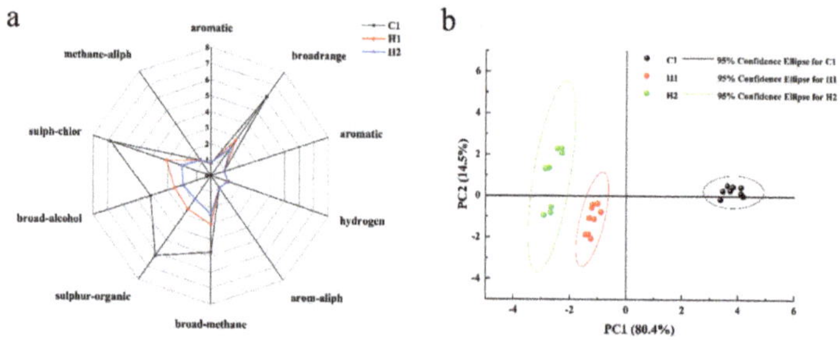

Figure 3. Radar (**a**) and PCA (**b**) of electronic noses' assessment of different samples.

According to the principal component analysis (PCA), the combined contribution rate of the first and second principal components amounts to almost 95%, effectively encompassing the majority of the sample's original information. Notably, the contribution rate of the first principal component is 80.4%, while that of the second is 14.5%. The scatter plots of the samples from the three groups clustered together, demonstrating good repeatability within each group and high similarity of the sample data. Furthermore, there was a clear distinction between the groups, as revealed by the mutual clustering. The electronic nose effectively discerned differences between the three samples. The normal room-temperature yoghurt exhibited the largest discrepancy in odor compared to the other two samples, as indicated by the sensor readings and depicted in the figure. Furthermore, the yoghurt was significantly distinct from the other two. On the X and Y axes, the walnut and purple rice plant drink was located at a lower position, which suggests that its volatile odor is comparatively weaker than that of the other two samples.

3.4. GC-IMS Analysis of Volatile Flavor Compounds

3.4.1. Analysis of Spectrum of Volatile Flavor Components

The fingerprint spectrum was reconstructed from all the peaks in the GC-IMS spectrum using the Gallery Plot plug-in, and the characteristic peak area was identified [42]. Figure 4 displays the VOC fingerprints of various samples. The horizontal axis presents the selected characteristic identification peak, and each column corresponds to a VOC. The vertical axis denotes the sample identification numbers, and each row represents a sample. The background is colored black and blue, whereas the red color indicates higher content, and the lighter color shows lower content [43]. The table displays the variation in VOC content across different samples. Each row represents the type of VOC contained in the sample, while each column shows the content variation. To aid observation and comparison, substances with similar variation rules are grouped and compared. These groups are then divided into four regions—A, B, C, and D. The A and D areas of the map exhibit the distinctive volatile components of the walnut and purple rice fermented plant drink H1. The main characteristics include acrylonitrile, 2-butanol, 2,2,4,6,6-pentamethylheptane-M, 2-acetyl-3-methylpyrazine, and 3-heptanol. The B area (shown in the diagram) indicates the distinct volatile compounds of H2, such as ethyl trans-2-hexenoate, propionaldehyde, 1-propanol, ethyl pentanoate, (E)-2-octenal, and isopentyl formate. The substances depicted in diagram b represent the specific volatiles of H2, mostly esters, with a concentration significantly higher than in other samples. It is worth noting that technical abbreviations are explained on their first use to ensure comprehension. The C region in the diagram includes volatile substances such as (−)-perillaldehyde, 2-phenylethyl acetate, methylpyrazine, 1-propanol, and ethyl benzoate, which are characteristic of C1. The characteristic volatile substances of C1 are represented by the peak colors, which are significantly darker in the three samples. This indicates a higher content of these substances than in the other samples. Objective evaluation is prioritized, and abbreviations of technical terms are explained.

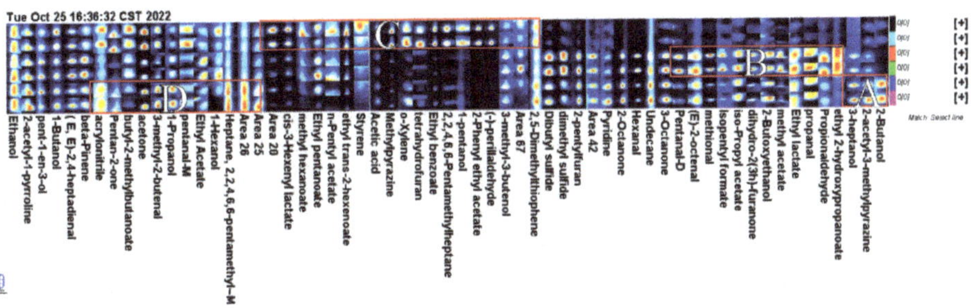

Figure 4. Gallery plot of samples. Note: The graph displays all the signal peaks chosen in a sample, with each column showing the signal peaks of a specific volatile organic compound in diverse samples. Certain substances have the abbreviations m and -D, referring to the monomer and dimer of the same element. The numerical value represents an unidentified peak. The parts of the different samples with higher content were classified and labelled with the letters A–D.

3.4.2. Analysis of Relative Contents of Volatile Substances

According to the examination of GC-IMS outcomes, there is a positive correlation between the characteristic peak intensity of an aroma substance and its content in the sample. The study identified a total of 55 chemical compounds, consisting of 5 ketones, 10 alcohols, 9 aldehydes, 13 esters, 2 acids, 2 ethers, 8 olefins, and 8 heterocyclic compounds. Among these compounds, ketones, alcohols, esters, and aldehydes were the primary volatile compounds of both samples. Relative content of different substances analysed through Table 3, we identified 21 significant flavor compounds (with a relative content greater than 1%) in the three samples. These compounds enhance the overall flavor of the product.

Table 3. Results of volatile flavor components of samples.

Compound	Relative Retention Index	Relative Amount			Odor Description
		H1	H2	C1	
		Aldehydes			
(−)-Perillaldehyde	1153.4	0.58 ± 0.01% [b]	0.49 ± 0.03% [b]	1.39 ± 0.04% [a]	Perilla incense
(E, E)-2,4-heptadienal	1009.5	0.56 ± 0.06% [a]	0.53 ± 0.031% [a]	0.33 ± 0.03% [a]	Melon and fruit fragrance
3-Methyl-2-butenal	762.6	2.94 ± 0.05% [a]	1.52 ± 0.06% [b]	1.83 ± 0.03% [b]	Almond, roasted
Hexanal	1071.2	1.24 ± 0.01% [b]	2.51 ± 0.08% [a]	0.40 ± 0.03% [c]	Apple, fat, fresh, green, oil
Pentanal-D	978.2	0.21 ± 0.00% [b]	0.41 ± 0.01% [a]	0.17 ± 0.00% [c]	Grass smell
(E)-2-octenal	1070.2	0.61 ± 0.01% [b]	0.70 ± 0.01% [a]	0.18 ± 0.01% [c]	Fat, fruit, grass, spice
Methional	906.1	0.82 ± 0.01% [b]	1.53 ± 0.06% [a]	0.38 ± 0.02% [c]	Savory
Propionaldehyde	819	0.44 ± 0.01% [b]	0.59 ± 0.02% [a]	0.11 ± 0.00% [c]	Floral, pungent, solvent
Pentanal-M	977	0.30 ± 0.01% [ab]	0.25 ± 0.01% [b]	0.39 ± 0.02% [a]	Grass smell
		Ketone			
Pentan-2-one	980	0.19 ± 0.03% [a]	0.15 ± 0.01% [c]	0.15 ± 0.06% [b]	Flowers, fruit fragrance
Dihydro-2(3h)-furanone	908.7	0.51 ± 0.02% [b]	0.82 ± 0.03% [a]	0.26 ± 0.01% [c]	-
Acetone	812.8	9.28 ± 0.02% [a]	6.06 ± 0.10% [b]	10.11 ± 0.12% [a]	Pungent
2-Octanone	1304.7	3.42 ± 0.03% [a]	2.97 ± 0.12% [b]	0.78 ± 0.04% [c]	Fat, fragrant, mold
3-Octanone	979.1	0.72 ± 0.02% [a]	0.72 ± 0.06% [a]	0.28 ± 0.02% [b]	Butter, herbs, mold
Pent-1-en-3-ol	1156.3	1.39 ± 0.06% [a]	1.40 ± 0.06% [a]	1.05 ± 0.07% [b]	Fruit fragrance, vegetable fragrance
1-Butanol	1116	0.36 ± 0.02% [a]	0.37 ± 0.02% [a]	0.35 ± 0.01% [a]	Fruit
1-Propanol	1023.9	0.99 ± 0.02% [a]	0.69 ± 0.02% [b]	0.41 ± 0.02% [c]	Alcohol, candy, pungent
1-Hexanol	873.6	0.23 ± 0.16% [a]	0.63 ± 0.12% [a]	0.33 ± 0.09% [a]	Banana, flower, grass, herb
Ethanol	948.2	7.29 ± 0.01% [a]	6.92 ± 0.03% [a]	4.98 ± 0.02% [b]	Alcohol, sweet
1-Pentanol	1242.4	0.35 ± 0.02% [b]	0.29 ± 0.02% [b]	0.89 ± 0.04% [a]	Balsamic, fruit, pungent, yeast
2-Butanol	1008.8	0.64 ± 0.01% [a]	0.10 ± 0.01% [b]	0.15 ± 0.01% [b]	Wine
3-Methyl-3-butenol	1245	0.72 ± 0.04% [a]	0.78 ± 0.02% [a]	0.62 ± 0.04% [a]	-
2-Butoxyethanol	906.2	0.95 ± 0.12% [a]	0.92 ± 0.04% [a]	0.45 ± 0.03% [b]	-
3-Heptanol	1304.1	0.40 ± 0.02% [a]	0.33 ± 0.02% [a]	0.15 ± 0.00% [b]	Herbs
		Esters			
2-Phenyl ethyl acetate	1240.9	1.27 ± 0.06% [a]	1.17 ± 0.05% [a]	4.90 ± 0.43% [b]	Fruit
Ethyl benzoate	1155.4	1.80 ± 0.03% [b]	1.29 ± 0.05% [c]	3.64 ± 0.00% [a]	Bitterness
Butyl-2-methylbutanoate	1041.3	0.40 ± 0.09% [b]	0.70 ± 0.01% [a]	0.83 ± 0.10% [a]	Fruit fragrance
Ethyl acetate	871.5	0.38 ± 0.05% [a]	0.43 ± 0.01% [a]	0.31 ± 0.02% [a]	Aromatic, brandy, grape
Methyl hexanoate	932.6	0.47 ± 0.09% [a]	0.82 ± 0.08% [a]	0.84 ± 0.22% [a]	Pineapple
Ethyl pentanoate	902	3.03 ± 1.85% [a]	6.61 ± 0.87% [a]	3.92 ± 0.80% [a]	Apple

Table 3. *Cont.*

Compound	Relative Retention Index	Relative Amount			Odor Description
		H1	H2	C1	
n-Pentyl acetate	913.3	0.14 ± 0.06% [a]	0.39 ± 0.06% [a]	0.33 ± 0.10% [a]	Pineapple, apple
Isopentyl formate	796.6	0.09 ± 0.00% [b]	0.13 ± 0.01% [a]	0.07 ± 0.00% [b]	Apple
Iso-Propyl acetate	857.2	0.32 ± 0.02% [a]	0.36 ± 0.02% [a]	0.12 ± 0.00% [b]	Banana
Methyl acetate	801.4	0.48 ± 0.01% [b]	0.79 ± 0.02% [a]	0.22 ± 0.02% [c]	Mint, cool
Ethyl lactate	794.1	0.29 ± 0.01% [b]	0.34 ± 0.00% [a]	0.09 ± 0.00% [c]	Cheese, floral, fruit, pungent
Ethyl 2-hydroxypropanoate	807.8	0.07 ± 0.00% [b]	0.21 ± 0.00% [a]	0.03 ± 0.00% [c]	Rum, fruit, and creamy aroma
Ethyl trans-2-hexenoate	1045	6.42 ± 1.80% [b]	12.46 ± 1.02% [ab]	18.81 ± 3.49% [a]	Fruit
Acids					
Acetic acid	1432	9.01 ± 0.55% [a]	7.97 ± 0.03% [ab]	6.71 ± 0.03% [b]	Acid, fruit, pungent, sour
Cis-3-hexenyl lactate	1234.7	1.74 ± 0.19% [c]	2.43 ± 0.03% [b]	4.60 ± 0.35% [a]	Green
Ethers					
Dibutyl sulfide	1069	2.61 ± 0.02% [b]	3.68 ± 0.01% [a]	0.45 ± 0.02% [c]	Green
Dimethyl sulfide	770.2	2.11 ± 0.03% [b]	3.78 ± 0.04% [a]	0.90 ± 0.01% [c]	Seafood flavor
Enolefins					
Undecane	1079.1	0.48 ± 0.01% [a]	0.47 ± 0.01% [b]	0.25 ± 0.02% [b]	-
2,2,4,6,6-Pentamethylheptane	977.7	0.65 ± 0.01% [b]	0.37 ± 0.03% [c]	1.41 ± 0.05% [a]	-
2,2,4,6,6-Pentamethylheptane-M	994	7.68 ± 0.06% [a]	5.04 ± 0.25% [b]	4.21 ± 0.02% [c]	-
Beta-pinene	977	1.02 ± 0.01% [ab]	1.15 ± 0.00% [a]	0.88 ± 0.02% [b]	Pine, polish, wood
o-Xylene	888.3	0.52 ± 0.16% [a]	0.21 ± 0.03% [a]	0.75 ± 0.12% [a]	Fragrance
Acrylonitrile	1001.6	2.21 ± 0.01% [a]	0.90 ± 0.01% [b]	1.87 ± 0.01% [b]	Peach kernel flavor
2-Acetyl-1-pyrroline	926.8	12.94 ± 0.31% [a]	9.01 ± 0.35% [b]	6.81 ± 0.33% [b]	Savory
Styrene	866	0.08 ± 0.01% [a]	0.09 ± 0.01% [a]	0.14 ± 0.03% [a]	Special aroma
Heterocyclic					
Tetrahydrofuran	887.2	0.61 ± 0.12% [a]	0.50 ± 0.04% [a]	1.06 ± 0.21% [a]	Ethyl ether flavor
Methylpyrazine	1244.7	3.37 ± 0.03% [b]	2.93 ± 0.05% [b]	8.02 ± 0.00% [a]	Nuts
2-Pentylfuran	1214.9	2.04 ± 0.07% [a]	2.09 ± 0.09% [a]	0.47 ± 0.04% [b]	Butter, floral, fruit
Pyridine	1174.7	1.21 ± 0.07% [a]	0.77 ± 0.01% [b]	0.47 ± 0.00% [c]	Odor
2-Acetyl-3-methylpyrazine	1074.1	0.30 ± 0.01% [a]	0.18 ± 0.00% [b]	0.11 ± 0.00% [c]	Wood fragrance
2,5-Dimethylthiophene	1168.6	0.61 ± 0.02% [a]	0.50 ± 0.01% [b]	0.43 ± 0.00% [b]	-

Values are expressed as the mean ± standard deviation. Different lowercase letters indicate significant differences between groups for the same indicator ($p < 0.05$). The odor descriptions are taken from the Odour Description Database (https://www.femaflavor.org/flavor-library and http://www.odour.org.uk (accessed on 1 October 2023)).

To enhance characterization of varying volatile compounds, we calculated the relative difference among the flavor compounds in the three samples based on signal intensity on the fingerprint, demonstrated in Figure 5. The similarity of the acquired VOC fingerprints was assessed using statistical approaches like the heat map and clustering method, as detailed in Figure 5a. The volatile organic compounds (VOCs) present in three samples were analyzed and vertically clustered. As can be seen in Figure 5b, there are some differences between the contents of the relative characteristic volatile classifications of H1, H2, and C1 in the three samples. From the figure, it is apparent that the primary components of the walnut and purple rice fermented plant drink are H1 (with 25.57% olefins, 15.15% esters, 14.12% ketones, and 13.31% alcohols), followed by H2 (with 25.69% esters, 17.26% olefins, 12.43% alcohols, and 10.71% ketones) and C1 (with 34%, 11% olefins, 16.32% alcohols, and 11.31%). The content of 2-acetyl-1-pyrroline was the highest in this product, and the threshold for this compound was lower, which gives it a savory taste and contributes to the overall flavor [44]. Compared to H2, the proportion of olefins in walnut and purple rice yoghurt H1 significantly increased, with a concurrent increase in ketone content. The ketones mainly resulted from microbial metabolism during fermentation [45]. They are produced mainly by the oxidation of esters and unsaturated fatty acids [46]. Many of these ingredients have a pleasant aroma, possibly accounting for the reduction in relative ester levels in H1 walnut and purple rice fermented plant drink, and the majority of ketones had a high threshold and did not significantly affect the flavor profile [47]. Fermentation can alter the composition and proportions of volatile flavor compounds, leading to a more intense flavor profile. A combination of flavor compounds give rise to the characteristic flavor of the walnut and purple rice fermented plant drink H1. In comparison to regular temperature yoghurt C1, esters exhibit the greatest disparity in their relative content between the two. The relative content of esters in plain normal-temperature yoghurt C1 is twice as high as in H1 yoghurt. Esters are key components of food flavor As well as giving foods a special "floral and fruity flavor", they can also mask the "unpleasant, irritating taste" caused by free fatty acids [48]. Aldehydes primarily arise from the oxidation of lipids [49]. The threshold is generally low, with a fragrant and fruity taste, which also gives H1 a different flavor from ordinary yoghurt C1 at room temperature.

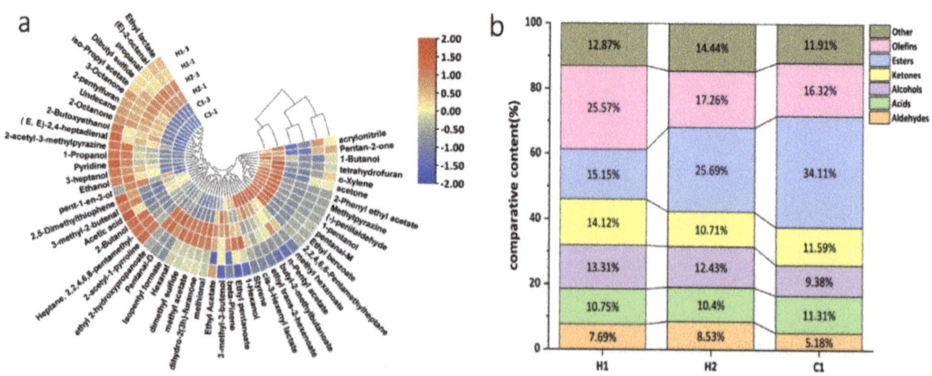

Figure 5. Cluster heat map (**a**) and difference map (**b**) of relative content of volatile flavor substances in samples. Note: in the clustered heatmap, red represents relatively high levels and blue relatively low levels, with darker colors representing greater differences.

3.4.3. Principal Component Analysis of Volatile Flavor Compounds

Principal component analysis (PCA) is a multivariate statistical approach that diminishes the dimensions of a dataset composed of numerous related variables, at the same time preserving as much variation within the data as possible This method has garnered broad usage in investigating variations in food aroma compounds [50]. Principal component analysis was conducted on the peak intensities of the volatile compounds in the three specimens, and Figure 6 displays the findings. The first principal component PC1 made a contribution rate of 59.5%, while the second principal component PC2 contributed 30.3%. The combined contribution rate of the first and second principal components was 89.8%, containing the majority of the information on the three samples and presenting the primary features of the volatile aroma. The principal component analysis chart illustrates that the three sample groups were significantly separated, denoting the utilization of GC-IMS technology in the identification of the volatile compounds in the samples. When combined with PCA, the samples are better distinguished. Simultaneously, the findings reveal significant contrasts between the three samples.

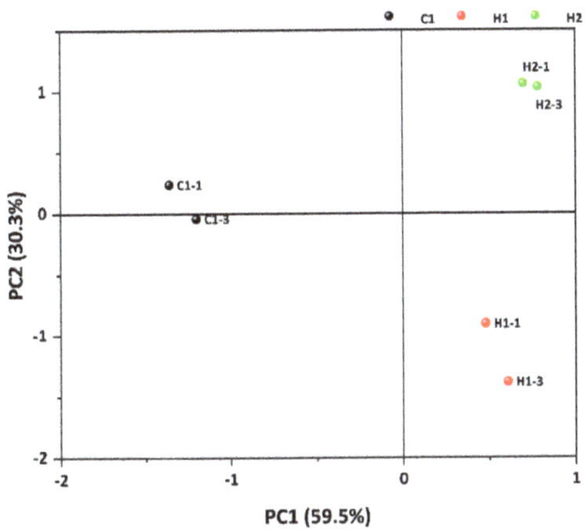

Figure 6. PCA analysis diagram of volatile substances in three samples.

3.5. *Organoleptic Investigation*

The sensory evaluations of three samples at varying levels of ripeness were analyzed using an OPLS-DA score plot. The results can be seen in Figure 7. The associated model was generated by randomly shuffling the order of categorical variables and yielded values of R2X = 0.699 and R2Y = 0.179. These R2 and Q2 values indicate a high level of general interpretation for the model when they fall within the range of 0.5 to 1 [51]. Q2 = 0.965 suggests the excellent predictive ability of the model. Clear stability and reliable predictability indicate that the model meets all the necessary requirements to identify the sensory differences of the three samples. To check for overfitting, we disrupted the categories of some samples and performed 200 instances of replacement fitting. Figure 5 demonstrates that R2 and Q2 intersect with the longitudinal axis at (0, −0.0068) and (0, −273), respectively. Additionally, the slopes of the two regression lines for R2 and Q2 are relatively high. The values for R2 and Q2 produced by the random arrangement on the left exhibit lower values compared to those on the right. Additionally, the fact that the Q2 regression line has a negative intercept suggests that the OPLS-DA model does not exhibit overfitting and demonstrates reliable predictive capabilities. This renders it suitable for discriminant analysis across diverse samples. Table 4 reveals that fermentation has significantly shifted the taste and flavor of walnut and purple rice yoghurt. The bitterness, sweetness, and astringency of H1 were significantly decreased ($p < 0.05$). In line with expectations, the fermented and sour tastes had the highest scores. Fermentation enhances food acidification, augmenting the sensory, nutritional, textural, and microbial safety of food. Furthermore, it negates the use of chemical preservatives and maintains the original natural state of the food. In our study, while fermentation significantly decreased the off-flavors of nuts and grains, it did not lower the unpleasant levels, which could be attributed to the new compounds generated by microbial fermentation [52]. Fermentation significantly enhanced the flavor, as well as the dispersibility and adhesion of the yoghurt, while reducing its graininess. The improved taste may be attributed to the fermentation process, which results in a smoother taste compared to the walnut and purple rice plant drink. The overall acceptance of the walnut and purple rice fermented plant drink is comparable to that of regular room-temperature yoghurt. Although the flavor of the walnut and purple rice fermented plant drink is slightly inferior to that of traditional yoghurt, its unique nutty and cereal notes add distinctive qualities. Moreover, team members rated it similarly to typical yoghurt in terms of degree of acceptance.

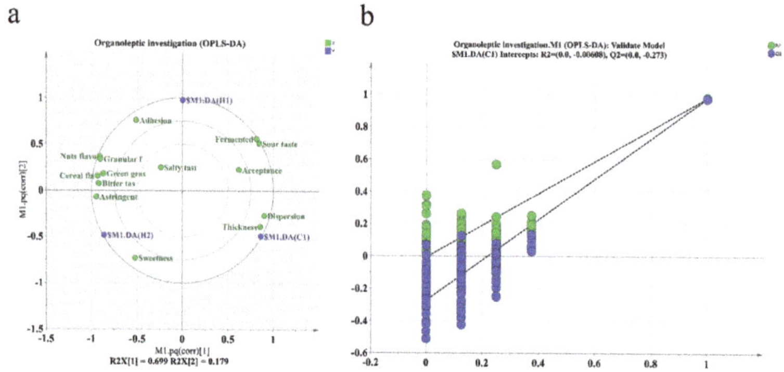

Figure 7. (**a**) OPLS-DA score plot and (**b**) OPLS-DA permutation test plot.

Table 4. Scores of 14 sensory attributes of three samples.

		C1	H1	H2
Flavor	Green grass flavor	1.25 ± 0.62 ^{c}	3.42 ± 0.67 ^{b}	4.50 ± 1.00 ^{a}
	Nuts flavor	0.00 ± 0.00 ^{c}	4.17 ± 0.72 ^{b}	5.00 ± 0.74 ^{a}
	Grainy flavor	0.00 ± 0.00 ^{c}	2.17 ± 0.58 ^{b}	3.33 ± 0.65 ^{a}
	Fermented flavor	6.50 ± 0.80 ^{b}	7.17 ± 0.58 ^{a}	0.00 ± 0.00 ^{c}
Taste	Sweetness	5.83 ± 0.58 ^{b}	4.83 ± 0.72 ^{c}	7.25 ± 0.45 ^{a}
	Sour taste	5.83 ± 0.94 ^{a}	5.92 ± 0.67 ^{a}	0.00 ± 0.00 ^{c}
	Bitter taste	0.33 ± 0.49 ^{c}	1.92 ± 0.51 ^{b}	3.17 ± 0.58 ^{a}
	Salty taste	0.42 ± 0.51 ^{b}	0.83 ± 0.39 ^{a}	0.67 ± 0.49 ^{ab}
	Astringent taste	0.25 ± 0.45 ^{c}	1.75 ± 0.45 ^{b}	3.75 ± 0.62 ^{a}
Textural attributes	Dispersion	7.25 ± 0.45 ^{a}	4.50 ± 0.52 ^{b}	3.67 ± 0.78 ^{c}
	Thickness	6.92 ± 0.51 ^{a}	3.67 ± 0.49 ^{b}	4.08 ± 0.67 ^{b}
	Adhesion	4.08 ± 0.51 ^{c}	6.50 ± 0.67 ^{a}	5.50 ± 0.52 ^{b}
	Granular feeling	0.00 ± 0.00 ^{c}	4.00 ± 0.60 ^{b}	4.83 ± 0.72 ^{a}
Overall	Acceptance	7.17 ± 0.58 ^{a}	6.92 ± 0.51 ^{a}	6.08 ± 0.67 ^{b}

$^{a–c}$ Means within a row with different superscripts are different at $p < 0.05$.

4. Conclusions

This study demonstrated the similarities and differences between the chemical, rheological, and organoleptic properties of a walnut–purple rice fermented plant beverage, walnut–purple rice fermented plant beverage, and room-temperature yoghurt. The results showed that the walnut–purple rice fermented plant beverage exhibited high pseudoplasticity and shear-thinning non-Newtonian fluid properties. Similar to the room-temperature yoghurt, the walnut–purple rice fermented plant beverage showed a semi-solid gelatinous structure with good viscosity and strong recovery properties. It is worth noting that the walnut–purple rice plant beverage before and after fermentation differed significantly from room-temperature yoghurt in terms of flavor and texture. Using an electronic nose, electronic tongue, GC-IMS, and sensory analysis, the walnut–purple rice fermented plant beverage was found to have a better taste. The results obtained in this study can show the potential of walnut and purple rice fermented plant beverages to be an alternative to animal-based yoghurt and can also provide a theoretical basis for the development of plant-based food products as well as the industrialization and further processing of walnut and purple rice.

Author Contributions: Conceptualization; methodology; writing—original draft, H.M.; Conceptualization; methodology; writing—original draft, T.D.; Investigation; visualization, S.H.; Writing—review and editing; formal analysis, K.W.; Investigation, M.W. Formal analysis, Software, C.T.; Methodology, Investigation, F.Z.; Supervision, Project administration, J.S. Funding acquisition, Supervision, Project administration, C.Z. All authors have read and agreed to the published version of the manuscript.

Funding: Yunnan Province-City Integration Project (202302 AN360002, Yunnan Provincial Department of Science and Technology), Yunnan Provincial Department of Education Research Fund Project (2023Y1011, Yunnan Provincial Department of Education), Yunnan Province Fengqing County Walnut Industry Science and Technology Mission (202204BI090012, Fengqing County Government, Yunnan Province).

Institutional Review Board Statement: All experimental procedures involving animals and humans were conducted in accordance with the recommendations of the Ethics Committee of Yunnan Agricultural University (Approval No. 202209012).

Informed Consent Statement: Informed consent was obtained from each subject prior to their participation in the study.

Data Availability Statement: Data is contained within the article.

Conflicts of Interest: The authors declare no competing financial interests.

References

1. Ingerslev, A.K.; Rasmussen, L.; Zhou, P.; Nørgaard, J.V.; Theil, P.K.; Jensen, S.K.; Lærke, H.N. Effects of dairy and plant protein on growth and growth biomarkers in a piglet model. *Food Funct.* **2021**, *12*, 11625–11640. [CrossRef] [PubMed]
2. Unger, A.; Torres-Gonzalez, M.; Kraft, J. Dairy Fat Consumption and the Risk of Metabolic Syndrome: An Examination of the Saturated Fatty Acids in Dairy. *Nutrients* **2019**, *11*, 2200. [CrossRef] [PubMed]
3. Montemurro, M.; Pontonio, E.; Coda, R.; Rizzello, C.J.F. Plant-Based Alternatives to Yogurt: State-of-the-Art and Perspectives of New Biotechnological Challenges. *Foods* **2021**, *10*, 316. [CrossRef] [PubMed]
4. Astolfi, M.L.; Marconi, E.; Protano, C.; Canepari, S. Comparative elemental analysis of dairy milk and plant-based milk alternatives. *Food Control* **2020**, *116*, 107327. [CrossRef]
5. *Codex Standard 243-2003*; Codex Satandard for Fermented Milk. FAO: Rome, Italy; WHO: Geneva, Switzerland, 2003.
6. Vanga, S.; Raghavan, V. How well do plant based alternatives fare nutritionally compared to cow's milk? *J. Food Sci. Technol.* **2018**, *55*, 10–20. [CrossRef]
7. Craig, W.J.; Fresán, U. International Analysis of the Nutritional Content and a Review of Health Benefits of Non-Dairy Plant-Based Beverages. *Nutrients* **2021**, *13*, 842. [CrossRef]
8. Zhang, Y.G.; Kan, H.; Chen, S.X.; Thakur, K.; Wang, S.; Zhang, J.G.; Shang, Y.F.; Wei, Z.J. Comparison of phenolic compounds extracted from Diaphragma juglandis fructus, walnut pellicle, and flowers of Juglans regia using methanol, ultrasonic wave, and enzyme assisted-extraction. *Food Chem.* **2020**, *321*, 126672. [CrossRef]
9. Sánchez-González, C.; Ciudad, C.J.; Noé, V.; Izquierdo-Pulido, M. Health benefits of walnut polyphenols: An exploration beyond their lipid profile. *Crit. Rev. Food Sci. Nutr.* **2017**, *57*, 3373–3383. [CrossRef]
10. Ni, Z.J.; Zhang, Y.G.; Chen, S.X.; Thakur, K.; Wang, S.; Zhang, J.G.; Shang, Y.F.; Wei, Z.J. Exploration of walnut components and their association with health effects. *Crit. Rev. Food Sci. Nutr.* **2022**, *62*, 5113–5129. [CrossRef]
11. Deng, G.F.; Xu, X.R.; Zhang, Y.; Li, D.; Gan, R.Y.; Li, H.B. Phenolic compounds and bioactivities of pigmented rice. *Crit. Rev. Food Sci. Nutr.* **2013**, *53*, 296–306. [CrossRef]
12. Tyagi, A.; Lim, M.J.; Kim, N.H.; Barathikannan, K.; Vijayalakshmi, S.; Elahi, F.; Ham, H.J.; Oh, D.H. Quantification of Amino Acids, Phenolic Compounds Profiling from Nine Rice Varieties and Their Antioxidant Potential. *Antioxidants* **2022**, *11*, 839. [CrossRef] [PubMed]
13. Danpreedanan, N.; Yamuangmorn, S.; Jamjod, S.; Prom, U.T.C.; Pusadee, T. Genotypic Variation of Purple Rice in Response to Shading in Yield, Anthocyanin Content, and Gene Expression. *Plants* **2023**, *12*, 2582. [CrossRef] [PubMed]
14. Gu, W.; Peng, Y.; Wang, R.; Wang, R.; Wu, H.; Zhu, J.; Ni, X.; Xiong, Q. Comparison of Metabolites and Main Nutritional Components between Uncooked and Cooked Purple Rice. *Metabolites* **2023**, *13*, 1018. [CrossRef] [PubMed]
15. Kasote, D.; Tiozon, R.N., Jr.; Sartagoda, K.J.D.; Itagi, H.; Roy, P.; Kohli, A.; Regina, A.; Sreenivasulu, N. Food Processing Technologies to Develop Functional Foods With Enriched Bioactive Phenolic Compounds in Cereals. *Front. Plant Sci.* **2021**, *12*, 771276. [CrossRef] [PubMed]

16. Liu, W.; Pu, X.; Sun, J.; Shi, X.; Cheng, W.; Wang, B. Effect of Lactobacillus plantarum on functional characteristics and flavor profile of fermented walnut milk. *LWT-Food Sci. Technol.* **2022**, *160*, 113254. [CrossRef]
17. Henchion, M.; Hayes, M.; Mullen, A.M.; Fenelon, M.; Tiwari, B. Future Protein Supply and Demand: Strategies and Factors Influencing a Sustainable Equilibrium. *Foods* **2017**, *6*, 53. [CrossRef]
18. Khezerlou, A.; Zolfaghari, H.; Forghani, S.; Abedi-Firoozjah, R.; Alizadeh Sani, M.; Negahdari, B.; Jalalvand, M.; Ehsani, A.; McClements, D. Combining Non-Thermal Processing Techniques with Edible Coating Materials: An Innovative Approach to Food Preservation. *Coatings* **2023**, *13*, 830. [CrossRef]
19. Berbert; Queiroz, D.M.; Melo, E.C. *Official Methods of Analysis of AOAC International*, 18th ed.; AOAC: Rockville, MD, USA, 2005.
20. Turek, K.; Wszołek, M. Effect of walnut oil on the fatty acid content of probiotic kefir produced either with kefir grains or kefir starter cultures. *Int. Dairy J.* **2022**, *127*, 105290. [CrossRef]
21. Ge, Z.; Yin, D.; Li, Z.; Chen, X.; Dong, M. Effects of Commercial Polysaccharides Stabilizers with Different Charges on Textural, Rheological, and Microstructural Characteristics of Set Yoghurts. *Foods* **2022**, *11*, 1764. [CrossRef]
22. Jia, S.; Zhao, H.; Tao, H.; Yu, B.; Liu, P.; Cui, B. Influence of corn resistant starches type III on the rheology, structure, and viable counts of set yogurt. *Int. J. Biol. Macromol.* **2022**, *203*, 10–18. [CrossRef]
23. Szołtysik, M.; Kucharska, A.Z.; Dąbrowska, A.; Zięba, T.; Bobak, Ł.; Chrzanowska, J. Effect of Two Combined Functional Additives on Yoghurt Properties. *Foods* **2021**, *10*, 1159. [CrossRef] [PubMed]
24. Ale, E.C.; Perezlindo, M.J.; Pavón, Y.; Peralta, G.H.; Costa, S.; Sabbag, N.; Bergamini, C.; Reinheimer, J.A.; Binetti, A.G. Technological, rheological and sensory characterizations of a yogurt containing an exopolysaccharide extract from Lactobacillus fermentum Lf2, a new food additive. *Food Res. Int.* **2016**, *90*, 259–267. [CrossRef] [PubMed]
25. Aryana, K.J.; Olson, D.W. A 100-Year Review: Yogurt and other cultured dairy products. *J. Dairy Sci.* **2017**, *100*, 9987–10013. [CrossRef] [PubMed]
26. Glibowski, P.; Karwowska, M.; Latoch, A.; Nosowska, K.; Udeh, K.O. Effect of different tea extracts on the physicochemical and sensory parameters of stirred probiotic yoghurts. *Acta Sci. Pol. Technol. Aliment.* **2019**, *18*, 185–193. [CrossRef] [PubMed]
27. Schlörmann, W.; Lamberty, J.; Ludwig, D.; Lorkowski, S.; Glei, M. In vitro-fermented raw and roasted walnuts induce expression of CAT and GSTT2 genes, growth inhibition, and apoptosis in LT97 colon adenoma cells. *Nutr. Res.* **2017**, *47*, 72–80. [CrossRef]
28. Lee, W.J.; Lucey, J.A. Formation and Physical Properties of Yogurt. *Asian-Australas. J. Anim. Sci.* **2010**, *23*, 1127–1136. [CrossRef]
29. Bao, H.; You, S.G.; Cao, L.; Zhou, R.; Wang, Q.; Cui, S.W. Chemical and rheological properties of polysaccharides from fruit body of Auricularia auricular-judae. *Food Hydrocoll.* **2016**, *57*, 30–37. [CrossRef]
30. Wang, X.; Kristo, E.; LaPointe, G. Adding apple pomace as a functional ingredient in stirred-type yogurt and yogurt drinks—ScienceDirect. *Food Hydrocoll.* **2019**, *100*, 105453. [CrossRef]
31. Chavhan, G.B.; Kanawjia, S.K.; Khetra, Y.; Puri, R. Effect of potassium-based emulsifying salts on sensory, textural, and functional attributes of low-sodium processed Mozzarella cheese. *Dairy Sci. Technol.* **2014**, *95*, 265–278. [CrossRef]
32. Chronakis, I.S. On the molecular characteristics, compositional properties, and structural-functional mechanisms of maltodextrins: A review. *Crit. Rev. Food Sci. Nutr.* **1998**, *38*, 599–637. [CrossRef]
33. Huang, T.; Tu, Z.; Shangguan, X.; Wang, H.; Zhang, L.; Bansal, N. Characteristics of fish gelatin-anionic polysaccharide complexes and their applications in yoghurt: Rheology and tribology. *Food Chem.* **2021**, *343*, 128413. [CrossRef] [PubMed]
34. Nguyen, P.T.M.; Kravchuk, O.; Bhandari, B.; Prakash, S. Effect of different hydrocolloids on texture, rheology, tribology and sensory perception of texture and mouthfeel of low-fat pot-set yoghurt. *Food Hydrocoll.* **2017**, *72*, 90–104. [CrossRef]
35. Gupta, M.K.; Torrico, D.D.; Ong, L.; Gras, S.L.; Dunshea, F.R.; Cottrell, J. Plant and Dairy-Based Yogurts: A Comparison of Consumer Sensory Acceptability Linked to Textural Analysis. *Foods* **2022**, *11*, 463. [CrossRef] [PubMed]
36. Brueckner-Guehmann, M.; Banovic, M.; Drusch, S. Towards an increased plant protein intake: Rheological properties, sensory perception and consumer acceptability of lactic acid fermented, oat-based gels. *Food Hydrocoll.* **2019**, *96*, 201–208. [CrossRef]
37. Dong, W.; Hu, R.; Long, Y.; Li, H.; Zhang, Y.; Zhu, K.; Chu, Z. Comparative evaluation of the volatile profiles and taste properties of roasted coffee beans as affected by drying method and detected by electronic nose, electronic tongue, and HS-SPME-GC-MS. *Food Chem.* **2019**, *272*, 723–731. [CrossRef] [PubMed]
38. Zhang, J.; Cao, J.; Pei, Z.; Wei, P.; Xiang, D.; Cao, X.; Shen, X.; Li, C. Volatile flavour components and the mechanisms underlying their production in golden pompano (*Trachinotus blochii*) fillets subjected to different drying methods: A comparative study using an electronic nose, an electronic tongue and SDE-GC-MS. *Food Res. Int.* **2019**, *123*, 217–225. [CrossRef]
39. Lu, L.; Hu, Z.; Hu, X.; Li, D.; Tian, S. Electronic tongue and electronic nose for food quality and safety. *Food Res. Int.* **2022**, *162*, 112214. [CrossRef] [PubMed]
40. Tiggemann, L.; Ballen, S.C.; Bocalon, C.M.; Graboski, A.M.; Manzoli, A.; Steffens, J.; Valduga, E.; Steffens, C. Electronic nose system based on polyaniline films sensor array with different dopants for discrimination of artificial aromas. *Innov. Food Sci. Emerg. Technol.* **2017**, *43*, 112–116. [CrossRef]
41. Yakubu, H.G.; Kovacs, Z.; Toth, T.; Bazar, G. Trends in artificial aroma sensing by means of electronic nose technologies to advance dairy production—A review. *Crit. Rev. Food Sci. Nutr.* **2023**, *63*, 234–248. [CrossRef]
42. Arroyo-Manzanares, N.; Martín-Gómez, A.; Jurado-Campos, N.; Garrido-Delgado, R.; Arce, C.; Arce, L. Target vs spectral fingerprint data analysis of Iberian ham samples for avoiding labelling fraud using headspace—Gas chromatography-ion mobility spectrometry. *Food Chem.* **2018**, *246*, 65–73. [CrossRef]

43. Ge, S.; Chen, Y.; Ding, S.; Zhou, H.; Jiang, L.; Yi, Y.; Deng, F.; Wang, R. Changes in volatile flavor compounds of peppers during hot air drying process based on headspace-gas chromatography-ion mobility spectrometry (HS-GC-IMS). *J. Sci. Food Agric.* **2020**, *100*, 3087–3098. [CrossRef] [PubMed]
44. Wei, X.; Handoko, D.D.; Pather, L.; Methven, L.; Elmore, J.S. Evaluation of 2-acetyl-1-pyrroline in foods, with an emphasis on rice flavour. *Food Chem.* **2017**, *232*, 531–544. [CrossRef] [PubMed]
45. Cao, M.; Fonseca, L.M.; Schoenfuss, T.C.; Rankin, S.A. Homogenization and lipase treatment of milk and resulting methyl ketone generation in blue cheese. *J. Agric. Food Chem.* **2014**, *62*, 5726–5733. [CrossRef] [PubMed]
46. Licón, C.C.; Mendoza, J.H.D.; Maggi, L.; Berruga, M.I.; Aranda, R.M.M.; Carmona, M. Optimization of headspace sorptive extraction for the analysis of volatiles in pressed ewes' milk cheese. *Int. Dairy J.* **2012**, *23*, 53–61. [CrossRef]
47. Li, M.; Yang, R.; Zhang, H.; Wang, S.; Chen, D.; Lin, S. Development of a flavor fingerprint by HS-GC-IMS with PCA for volatile compounds of Tricholoma matsutake Singer. *Food Chem.* **2019**, *290*, 32–39. [CrossRef] [PubMed]
48. El Hadi, M.A.; Zhang, F.J.; Wu, F.F.; Zhou, C.H.; Tao, J. Advances in fruit aroma volatile research. *Molecules* **2013**, *18*, 8200–8229. [CrossRef] [PubMed]
49. Lee, S.M.; Lim, H.J.; Chang, J.W.; Hurh, B.S.; Kim, Y.S. Investigation on the formations of volatile compounds, fatty acids, and γ-lactones in white and brown rice during fermentation. *Food Chem.* **2018**, *269*, 347–354. [CrossRef] [PubMed]
50. Liu, H.; Xu, Y.; Wu, J.; Wen, J.; Yu, Y.; An, K.; Zou, B. GC-IMS and olfactometry analysis on the tea aroma of Yingde black teas harvested in different seasons. *Food Res. Int.* **2021**, *150*, 110784. [CrossRef]
51. Powers, R.; Worley, B. Multivariate Analysis in Metabolomics. *Curr. Metabol.* **2013**, *1*, 92–107. [CrossRef]
52. Oh, Y.J.; Kim, T.S.; Moon, H.W.; Lee, S.Y.; Lee, S.Y.; Ji, G.E.; Hwang, K.T. Lactobacillus plantarum PMO 08 as a Probiotic Starter Culture for Plant-Based Fermented Beverages. *Molecules* **2020**, *25*, 5056. [CrossRef]

MDPI AG
Grosspeteranlage 5
4052 Basel
Switzerland
Tel.: +41 61 683 77 34

MDPI Books Editorial Office
E-mail: books@mdpi.com
www.mdpi.com/books

www.ingramcontent.com/pod-product-compliance
Lightning Source LLC
Chambersburg PA
CBHW061602220926
51850CB00005B/402
* 9 7 8 3 7 2 5 8 7 8 6 9 7 *